VOLUME 1

Surfactants in Solution

VOLUME 1

Surfactants in Solution

Edited by

K. L. Mittal

IBM Corporation, Hopewell Junction, New York

and

B. Lindman

University of Lund, Lund, Sweden

PLENUM PRESS • NEW YORK AND LONDON

7299-2803

CHEMISTRY

Library of Congress Cataloging in Publication Data

Main entry under title:

Surfactants in solution.

"Proceedings of an international symposium on surfactants in solution, held June
27–July 2, 1982, in Lund, Sweden"—T.p. verso.
Includes bibliographical references and indexes.
1. Surface active agents—Congresses. 2. Solution (Chemistry)—Congresses. 3.
Micelles—Congresses. I. Mittal, K. L., 1945–   . II. Lindman, Björn, 1943–
TP994.S88   1983   668'.1   83-19170
ISBN 0-306-41483-X (v. 1)
ISBN 0-306-41484-8 (v. 2)
ISBN 0-306-41485-6 (v. 3)

Proceedings of an international symposium on Surfactants in Solution,
held June 27–July 2, 1982, in Lund, Sweden

© 1984 Plenum Press, New York
A Division of Plenum Publishing Corporation
233 Spring Street, New York, N.Y. 10013

PREFACE

    This and its companion Volumes 2 and 3 document the proceedings of the 4th International Symposium on Surfactants in Solution held in Lund, Sweden, June 27–July 2, 1982. This biennial event was christened as the 4th Symposium as this was a continuation of earlier conferences dealing with surfactants held in 1976 (Albany) under the title "Micellization, Solubilization, and Microemulsions"; in 1978 (Knoxville) under the title "Solution Chemistry of Surfactants"; and in 1980 (Potsdam) where it was dubbed as "Solution Behavior of Surfactants: Theoretical and Applied Aspects." The proceedings of all these symposia have been properly chronicled.[1,2,3] The Lund Symposium was billed as "Surfactants in Solution" as both the aggregation and adsorption aspects of surfactants were covered, and furthermore we were interested in a general title which could be used for future conferences in this series. As these biennial events have become a well recognized forum for bringing together researchers with varied interests in the arena of surfactants, so it is amply vindicated to continue these, and the next meeting is planned for July 9–13, 1984 in Bordeaux, France under the cochairmanship of K. L. Mittal and P. Bothorel. The venue for 1986 is still open, although India, inter alia, is a good possibility. Apropos, we would be delighted to entertain suggestions regarding where and when these biennial symposia should be held in the future and you may direct your response to KLM.

    The response to these biennial events has been growing and as a matter of fact we had to limit the number of presentations in Lund. Even with this restriction, the Lund Symposium program had 140 papers from 31 countries by more than 300 authors. So it is quite patent that this meeting was a veritable international symposium both in spirit and contents. It should be added that the program contained a number of overviews by prominent researchers, as it is imperative to include some overviews to cover the state-of-knowledge of the topic under discussion.

    As for these proceedings [containing 126 papers (2156 pages) by 324 authors from 29 countries], these are arranged in nine parts. Parts I and II constitute Volume 1; Parts III–VI comprise Volume 2; and Parts VII–IX are the subject of Volume III. Apropos,

the papers in the proceedings have been rearranged (from the order
they were presented) in a more logical manner. Among the topics
covered include: Phase behavior and phase equilibria in surfactants
in solution; structure, dynamics and characterization of micelles;
thermodynamic and kinetic aspects of micellization; mixed micelles;
solubilization; micellar catalysis and reactions in micelles;
reverse micelles; microemulsions and reactions in microemulsions;
application of surfactants in analytical chemistry; adsorption and
binding of surfactants; HLB; polymerization of organized surfactant
assemblies; light scattering by liquid surfaces; and vesicles.

A few salient aspects of these proceedings should be recorded
for posterity. All papers were reviewed by qualified reviewers so
as to maintain the highest standard. As a result of this, most
papers were returned to respective authors for major/minor revi-
sions and some did not pass the review. In other words, these
proceedings are not simply a collection of unreviewed papers,
rather the peer review was an integral part of the total editing
process. It should be added that we had earnestly hoped to include
discussions at the end of each paper or group of allied papers, but
in spite of constant exhortation, the number of written questions
received did not warrant undertaking such endeavor. However, it
must be recorded that there were many spontaneous and brisk discussions both formally in the auditorium and informally in other
more suitable (more relaxed) places. Most often the discussions
were enlightening, but on occasions one could feel some enthalpy as
these tended to be exothermic.

Also a general concern was expressed about the possibility of
correlating research done in different laboratories. In particular
in the microemulsion field it was felt that a few selected, stable
and well-defined systems should be chosen for collaborative work
between a number of active groups using a variety of techniques.
The response to such discussion (initiated by Prof. M. Kahlweit,
Göttingen) was very heartening and culminated in the so-called Lund
Project (coordinator Prof. P. Stenius, Stockholm), which is a coor-
dinated collaboration between a number of research groups in different countries. A report meeting was hosted by M. Kahlweit in
Göttingen in the spring 1983 and further results will be presented
in Bordeaux in 1984. In addition, throughout the meeting, small
groups of people were seen to be leisurely discussing more specific
topics of mutual interest. In other words, there were ample and live-
ly discussions in various forms during the span of this symposium.

Coming back to the proceedings, even a cursory look at the
Table of Contents will convince even the most skeptic that the
field of surfactants in solutions has come a long way, and all
signals indicate that the accelerated tempo of interest and
research in this area is going to continue. Also it is quite clear
that as we learn more about the amphiphilic molecules, more excit-

ing research areas and pleasant applications will emerge. It should
be added that these proceedings cover a wide spectrum of topics by
a legion of prominent researchers and provide an up-to-date coverage of the field. The coverage is inter- and multidisciplinary and
both overviews and original unpublished research reports are included. Also it should be pointed out that both the aggregation and
adsorption of surfactants are accorded due coverage. These proceedings volumes along with the earlier ones in this vein (total $\sim$ 5000
pages) should serve as a repository of current thinking and research dealing with the exciting field of surfactants in solution.
Also these volumes should appeal to both veteran and neophyte researchers. The seasoned researchers should find these as the source
for latest research results, and these should be a fountainhead of
new research ideas to the tyro.

Acknowledgements: One of us (KLM) is thankful to the appropriate management of IBM Corporation for permitting him to parti-
cipate in this symposium and to edit these proceedings. His special
thanks are due to Steve Milkovich for his cooperation and understanding during the tenure of editing. Also KLM would like to
acknowledge the assistance and cooperation of his wife, Usha, in
more ways than one, and his darling children (Anita, Rajesh, Nisha
and Seema) for creating only low decibel noise so that Daddy could
concentrate without frequent shoutings. The time and effort of the
reviewers is sincerely appreciated, as the comments from the peers
are a desideratum to maintain standard of publications. We are
appreciative of Phil Alvarez, Plenum Publishing Corp., for his
continued interest in this project. Also we would like to express
our appreciation to Barbara Mutino for providing excellent and
prompt typing service. Our thanks are due to the members of the
local Organizing Committee (Thomas Ahlnäs, Thomas Andersson, Gunnar
Karlström, Ali Khan, Mary Molund, Gerd Olofsson, Nancy Simonsson
and Marianne Swärd) who unflinchingly carried out the various
chores demanded by a symposium of this magnitude. The financial
support of the Swedish Board for Technical Development, the Swedish
National Science Research Council, and the University of Lund is
gratefully acknowledged.

K. L. Mittal   B. Lindman
IBM Corporation   University of Lund
Hopewell Junction, NY 12533   Lund, Sweden

1. K. L. Mittal, Editor, Micellization, Solubilization and
    Microemulsions, Vols. 1 & 2, Plenum Press, New York, 1977.
2. K. L. Mittal, Editor, Solution Chemistry of Surfactants,
    Vols. 1 & 2, Plenum Press, New York, 1979.
3. K. L. Mittal and E. J. Fendler, Editors, Solution Behavior
    of Surfactants: Theoretical and Applied Aspects, Vols. 1 & 2,
    Plenum Press, New York 1982.

CONTENTS

CONTENTS OF VOLUME 2

PART III. THERMODYNAMIC AND KINETIC ASPECTS
OF MICELLIZATION

PART IV.   SOLUBILIZATION

PART V. MICELLAR CATALYSIS AND REACTIONS
IN MICELLES

xvi

CONTENTS OF VOLUME 2

CONTENTS OF VOLUME 3

PART VII. REVERSE MICELLES

PART VIII. MICROEMULSIONS AND REACTIONS
IN MICROEMULSIONS

PART IX.   GENERAL OVERVIEWS AND OTHER PAPERS

Part I
Phase Behavior and Phase Equilibria
in Surfactant Solutions

PRINCIPLES OF PHASE EQUILIBRIA IN SURFACTANT - WATER SYSTEMS

B. Jönsson, P.-G. Nilsson, B. Lindman,
L. Guldbrand and H. Wennerström

Division of Physical Chemistry 1, University of Lund
Chemical Center, S-220 07 Lund, Sweden, and
Division of Physical Chemistry, Arrhenius Laboratory
University of Stockholm, S-106 91 Stockholm, Sweden

The relation between the molecular properties of
a surfactant and the phase equilibria in the corresponding surfactant - water system is investigated. The
emphasis is on the qualitative features that emerge
from previous quantitative studies. The systems
discussed are: poly(ethyleneoxide)alkyl ether - water,
dialkylphosphatidylcholine - water, alkanoate - water,
and alkanoate - alcohol - water. While the interactions
involving the alkyl chains are essentially the same in
these systems, there are large differences in the
nature of the interaction between the polar groups. For
each system the dominating interactions involving the
polar groups are identified. For the ionic surfactant
this is the direct ion - ion interaction, while for the
zwitterionic phosphatidylcholines the headgroup - headgroup and headgroup - solvent interactions are both
important. It is argued that the hydrophilicity of the
ethyleneoxide chains is mainly due to the entropy of
mixing with the aqueous solvent, while there is an
effective repulsive interaction between an ethyleneoxide unit and the water above room temperature. The
repulsion increases strongly with increasing temperature which has important implications for the
temperature dependence of phase equilibria.

INTRODUCTION

In a aqueous medium, surfactant molecules self-associate to
to form large aggregates. These can be of a finite size and be
present in an isotropic (micellar) solution or they can form liquid crystalline structures of infinite extension. The rich polymorphism is probably best characterized via phase equilibrium
studies, which form a central topic in physical chemistry of association colloids. During the last thirty years the pioneering experimental studies by McBain and coworkers on surfactant phase
equilibria have been followed up by a number of groups[1-5]. Due to
these joint efforts, there exists today a large number of well documented phase diagrams of binary and ternary systems[1,6]. The experimental studies have inspired different theoretical approaches
to the phase equilibrium problem. These include rules about preferred curvatures[2], considerations about chain packing[7-9], analysis
of aggregate growth by monomer addition[10,11], and more quantitative descriptions of electrostatic effects[12,13]. There exists also
a sizeable literature dealing specifically with biological lipids
[9,14-18].

As one part of our studies of surfactant - water systems we
have recently presented models and model calculations of phase
equilibria for several different systems[19-25]. In the present paper we continue this work of trying to obtain the relation between
thermodynamic properties, as manifested in phase equilibria,
and intermolecular interactions. Due to the molecular complexity
of surfactant - water systems it is, at the present stage, necessary to apply rather simple models. The hope is that the essential features of the system are incorporated into the model, so
that it can account for the gross features of the phase equilibria.

In addition to the insight these studies provide into surfactant - water systems one should keep in mind that there is also
a wider applicability to models of interfaces between polar and
apolar regions. Such interfaces occur, for example, in biological
systems and the surfactant - water systems and can be considered as
useful model systems which are readily studied experimentally.

The paper is organized so that some general aspects are discussed first. Then it is shown, by way of example, how the phase
equilibria in four different types of surfactant - water systems
can be understood from a knowledge about the specific properties
of the polar headgroups combined with packing considerations.
These are the binary systems water - alkylpolyethylene oxides,
water - dialkylphosphatidylcholine, water - alkali alkanoate and
the ternary system water - alkali alkanoate - alkanol. The emphasis
is placed on the qualitative understanding, which has emerged
from quantitative treatments of the models[20-23].

GENERAL ASPECTS

In a broad sense of the term, a surfactant is a substance that accumulates at interfaces between water and an apolar fluid. In the more restricted sense used in this article a surfactant is a low molecular weight compound with a distinct hydrophilic head-group and a distinct apolar region. The apolar part usually consists of one (or two) hydrocarbon chains with 7 to 21 carbon atoms. The polar group can vary substantially from small strongly polar ionic to zwitterionic and dipolar to the weakly polar but large poly-ethyleneglycol groups[5]. The basic similarity between the behaviour of different surfactants in water arises from the tendency of the apolar chains to avoid contact with water through self-association, while the polar groups retain their contact with water. The association can lead to a number of aggregate structures depending on the interplay between the forces involving the polar groups and the organization of the apolar chains.

The role of the apolar chains is similar for different surfactants, and it is possible to arrive at rather general rules for preferred aggregate geometries on the basis of packing restrictions[8,10,11]. The size of the aggregate is, in at least one dimension, constrained by the length, l_0, of the hydrocarbon chain since polar groups should remain at the aggregate surface. The length l_0, the volume occupied by the apolar chains, v_0, and the cross sectional area, a_0, per surfactant are related so that large values of a_0/v_0 make spherical structures more favourable; whereas for smaller a_0/v_0 lamellar aggregates and phases of the inverse type are optimal[8,9,19]. In this description the effective cross-sectional area a_0 can be determined either by the alkyl chains or by the packing of the polar groups. Rules based on these geometrical packing considerations provide a very useful tool for the qualitative understanding of phase equilibria. Further insight is obtained by explicitly considering interactions involving the polar groups, but in this case it seems necessary to consider the different types of polar groups separately.

POLY(ETHYLENEOXIDE)ALKYL ETHER - WATER

In contrast to most other surfactants the aggregates formed by poly(ethyleneoxide) alkyl ethers $C_nH_{2n+1}(-OC_2H_4O)_mOH$ (C_nE_m) have a clear palisade layer, built by the polar groups, that separates the aqueous region from the alkyl chains (see Figure 1). Each ethylene oxide unit is only weakly polar so a number, m, of units is required to produce a functional surfactant. One possible model for the system is to consider three regions:

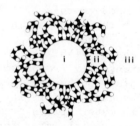

Figure 1. Model of non-ionic surfactant micelle showing the three regions mentioned in the text.

> i) the hydrocarbon core
> ii) the palisade layer with polar headgroups and water
> iii) the aqueous medium containing some monomers

Due to the presence of the palisade layer there is only a small water hydrocarbon contact when aggregates are formed. The region that is crucial in determining the properties of the system appears to be the palisade layer. The molecular organization of this region could be considered as similar to that of a concentrated polyethyleneoxide – water solution. For these polymer systems Kjellander and Florin[26] have shown that a modified Flory-Huggins theory gives a good description of the thermodynamic behaviour. In this model the contributions to the free energy of mixing can be assigned to the entropy of mixing of solvent (water) and the monomer (ethyleneoxide) units and to the effective pair interactions between the solvent and the monomers. The expression for the solvent (index 1) chemical potential is

$$\mu_1 - \mu_1^0 = RT(\ln X_1 + X_2) + wX_2^2 \tag{1}$$

not including the entropy of mixing of the whole aggregates. Here, X_i denotes the mole fraction of component i calculated on the basis of monomer units. The interaction parameter w is in the polyethylene oxide case strongly temperature dependent varying from 0.1 kT at 35°C to approximately 0.6 kT above 100°C[26]. The interaction is thus an effective repulsion in this temperature range and it is the configuration entropy term that promotes the mixing between the water and the ethyleneoxide units. The temperature dependence of the interaction term wX_2^2 is stronger than that of the $RT(\ln X_1 + X_2)$ term which means that in the range 0-100 °C the EO-water mixing is more favourable the lower the temperature.

Let us now apply this picture to the C_nE_m surfactant aggregates and combine it with the packing ideas presented in the preceding section. The poly(ethyleneoxide) chain will usually coil somewhat so that the thickness of the palisade layer is smaller than the length of an extended chain. At the θ-temperature and in the limit of large degrees of polymerization the average length of a polymer chain is[27]

$$\langle r^2 \rangle^{1/2} \sim m^{0.6} \tag{2}$$

For the short chains in the surfactant typically m = 4-10 and the limiting form of Equation (2) does not apply. However, the exponent is less than one which means that the effective area a_0 occupied by a chain increases with increasing m. Thus for a given alkyl chain length n, spherical aggregates will be favoured by higher m-values. Another factor is that the volume of the palisade layer increases when its water content increases. With increasing surfactant concentration the osmotic pressure increases somewhat and water will tend to leave the palisade layer, thus decreasing a_0. A more important effect is that the interaction parameter w of Equation (1) increases with increasing temperature and the palisade layer is dehydrated[28] again lowering a_0 and promoting a smaller curvature of the aggregate.

The qualitative features of the phase diagrams of the systems $H_2O-C_{12}E_8$ and $H_2O-C_{12}E_5$ shown in Figures 2 and 3 can be understood on the basis of these ideas. For $C_{12}E_8$ spherical aggregates are stable at low temperature and the liquid crystalline phases are of the cubic and hexagonal type. For $C_{12}E_5$ the lamellar phase region is much larger and the hexagonal phase has only a small stability region above 0° C. To account for the occurence of the critical point (cloud point) it is necessary to consider also inter-aggregate interactions. Such a theory has been developed by Kjellander[29] partly using the conceptual framework presented above.

DIALKYLPHOSPHATIDYLCHOLINE - WATER

In the biologically important phosphatidylcholines (lecithins) the headgroup is zwitterionic and intermediate in size between those of the alkylpolyethyleneoxides and the ionic surfactants. Due to the fact that the cross-sectional area of the two chains roughly matches the area occupied by the polar groups, lamellar

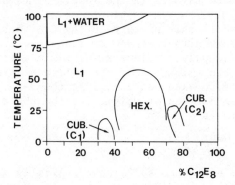

Figure 2. Schematic composition - temperature diagram for the system $C_{12}E_8$ - water. Hex. and cub. are hexagonal and cubic liquid crystalline phases respectively. L_1 is an isotropic phase (after ref. 45).

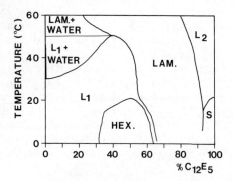

Figure 3. Schematic composition - temperature diagram for the system $C_{12}H_5$ - water. Hex. and lam. are hexagonal and lamellar liquid crystalline phases respectively. L_1 and L_2 are isotropic phases and S is solid surfactant (after ref. 46).

structures dominate. The chains can, however, exist both in a liquid-like and a crystalline state and this gives rise to a fairly
complex phase behaviour as illustrated in Figure 4 for dipalmitoyl
PC (DPPC). This diagram was determined using deuterium NMR[30] and
corresponding studies on the dimyristoyl PC system[31] have shown
that this system behaves similarily. One interesting feature of
Figure 4 is the occurence of two gel phases, Lβ' and Pβ, whose
structures are shown in Figure 5. For the crystalline chains the
cross sectional area is smaller than the area needed for the polar group and this creates a packing problem in a lamellar structure[32]. The structures of the Lβ' and Pβ phases represent two
different solutions to this problem as illustrated in Figure 5.

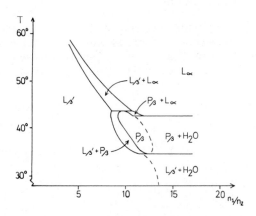

Figure 4. Experimentally determined phase diagram for the system
DPPC -D_2O. The abscissa gives the number of water molecules per
lipid molecule (after ref. 30). For notation see Figure 5.

        The DPPC - water system has most often been studied in excess
water, but as Figure 4 illustrates additional information about
the system is obtained by decreasing the water content. In a series of studies, Rand, Parsegian and coworkers[33,34] have studied
the relation between water activity and bilayer separation. Their
measurements show that there is an exponential repulsive "hydration" force between two opposing bilayers. As two bilayers approach,

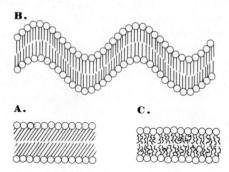

Figure 5. The molecular packing in the different lamellar structures
in the DPPC - H_2O system. A. Crystalline tilted chains, Lβ'; B.
Crystalline (straight) chains with a rippled superstructure , Pβ;
C. Liquid crystalline structure with melted chains, Lα.

this repulsion causes a substantial positive contribution to the
free energy. The force is weakest in the Lβ' and strongest in the
Lα phase. Since the hydration force is different in the three
phases, changes in water activity alters the relative stability
of the phases. Thus if one starts with the Lα phase in excess water and gradually decreases the water content at constant temperature, the system is brought to a point where it is favourable for
it to transform to one of the gel phases, usually the Lβ' phase. It
is possible to describe this effect quantitatively using measured
hydration forces, transition enthalpies and areas per polar group.
A calculated[24] phase diagram for the DPPC- water system based on
such a model is shown in Figure 6. The agreement with the experimental diagram of Figure 4 is quite satisfactory. One can even expect

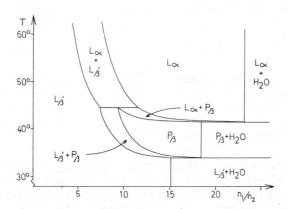

Figure 6. Calculated phase diagram for the DPPC - H_2O system (after ref. 24). Notation as in Figure 4.

that some features of Figure 6 are more accurate than in the experimental diagram. This applies particularly to the narrow P_β + Lα two phase region, which would be very difficult to detect experimentally.

ALKANOATE - WATER

The most studied ionic surfactants are probably the alkali alkanoates, $C_{n-1}H_{2n-1}COO^-M^+$. The headgroup has a small van der Waals' radius, but it is involved in strong ion-ion interactions. The electrostatic forces seem to dominate the interactions in the polar region except possibly at the lowest water contents and in a large excess of added salt. Observations supporting this conclusion are that the phase equilibria are insensitive to the counterion (except for the crystalline phases) and very sensitive to the presence of electrolyte[1].

On this basis it is reasonable to construct a thermodynamic model with two dominating contributions to the free energy. These are an electrostatic contribution, calculated on the basis of the Poisson-Boltzmann (PB) equation, and a contribution from the hydrocarbon - water contact, assumed to be proportional to the aggregate area, A, with a proportionality constant γ. The two contributions are counteracting each other so that the hydrophobic interaction tends to reduce the aggregate surface contact while the electrostatic forces are repulsive at the interface. For liquid crystalline systems the forces are balanced at equilibrium and this occurs when[21]

$$\gamma A = 2E_{el} \tag{3}$$

where E_{el} is the electrostatic energy. Since aggregate dimensions have been determined for the liquid crystalline phases[3,35] using X-ray diffraction, the value of γ can be determined by solving the PB equation. It was found that [21,36] $\gamma \simeq 18$ mJ/m^2 increasing slightly with increasing temperature.

For micellar systems the entropy of mixing of the micelles provides a third contribution to the free energy. The entropy term favours small aggregates and Equation (3) is not satisfied. This fact is sometimes expressed in terms of a surface tension of the micelles and it might be argued that an internal Laplace pressure is obtained.

The standard state of the molecules in the aggregates is assumed to be the same for all aggregate shapes. To complete the model it is necessary to specify the standard free energies of the surfactant as a monomer in water and in the crystal. The former determines the critical micelle concentration (cmc) but has minute effects on the phase equilibria at higher concentrations. The free energy of the crystal determines the Krafft boundary and below we have modelled it by fitting the Krafft point and the enthalpy difference, ~5 kJ/mole, for a surfactant in a crystal and in an aggregate.

Given a complete model expression for the free energy, the chemical potentials, μ_i, for the two components, water and surfactant, are determined for micellar solutions and normal hexagonal and lamellar liquid crystals. The calculations of the μ_i's are made as a function of composition at optimum aggregate size and over a range of temperatures. Two phases are in equilibrium at a given temperature when the chemical potentials of the two components are equal in the two phases. The phase diagram for the system $C_{11}H_{23}COO^-M^+$ shown in Figure 7 has been calculated using the method described above. A comparison with the experimentally determined diagram of Figure 8 reveals that the model accounts for the main features of the phase equilibria except at higher (>150°C) temperatures, where a thermal breakdown of the aggregate structures might occur.

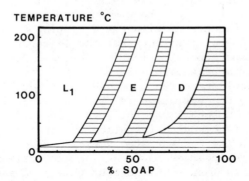

Figure 7. Calculated phase diagram for the Potassium myristate -
water system. The composition is given in weight percent.

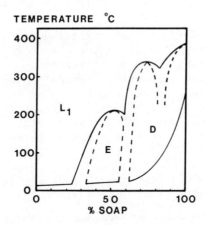

Figure 8. Experimentally determined phase diagram for the system
Potassium myristate - water. Data from ref. 47.

A merit of the model calculations is that one can extract the basic qualitative reasons behind the shift in stability of the different structures, with changes in concentration, temperature and surfactant molecule. At the lowest surfactant concentrations only monomers are present in the solution, while at the cmc the hydrophobic interaction is strong enough to compensate for the entropy decrease associated with the formation of an aggregate; this is so in spite of the fact that the aggregate is charged. When the surfactant concentration is increased further the contribution to the surfactant chemical potential from the entropy of mixing of the micelles decreases steadily. A more important effect of an increase in concentration is that the electrostatic free energy decreases in such a way that aggregates with high surface charge densities become electrostatically less unfavourable. The reason is that in the concentrated systems, the counterions cannot escape far from the charged surfaces for geometrical reasons and the ion distribution is less inhomogeneous than in a more dilute system. There is then a smaller entropy decrease associated with the creation of a charged surface in the more concentrated system. At some point, reproduced fairly accurately in the calculations, the addition of more surfactant leads to the appearance of a hexagonal liquid crystalline phase; this occurs when the higher electrostatic free energy associated with the higher surface charge density is compensated by a smaller aggregate surface. With further increase in surfactant concentration, the diameter of the cylindrical aggregates grows, decreasing the area per molecule. Thereafter the lamellar phase becomes more stable, since the electrostatic forces have been further weakened.

The Krafft boundary, representing the limit below which crystals appear, is fairly flat since the chemical potential of the soap is nearly constant when the concentration is increased in the micellar solution. The surfactant chemical potential becomes strongly concentration dependent only in the lamellar phase at the lowest water contents. In the model the Krafft point is interpreted as the temperature where the monomer solubility reaches the cmc[19,37]. The main factor that determines the solubility of an ionic surfactant is the stability of the crystal. The most efficient way to lower the Krafft point is consequently to chemically modify the surfactant in such a way that the packing in the crystal is perturbed (cf. the work of K. Shinoda).

In the model the micelles are assumed spherical, which is a simplification since micelles are known to grow to rods under certain circumstances. It has been shown[10] that such a growth is continuous in the absence of micelle-micelle interactions. In the real system the excluded volume of a rod-like aggregate is large, so that a transition to a hexagonal phase occurs[38].

In the calculations, packing constraints have not been explicitly included. For the micelles, the calculated optimal radius is close to the maximum value l_o. There is thus no great strain to

relieve by solubilizing an apolar molecule which would allow the
micelle to swell. Similarly at the high concentration limit of the
hexagonal phase the radius of the aggregates reaches l_0. It is quite
possible that the cylinders are actually strained and this opens
the possibility for less regular structures to be more stable[39-41].
In this concentration range, the area occupied by polar groups is
rather small, which might create a packing problem for surfactants
with large ionic groups such as SO_4^- and $-N^+(CH_3)_3$.

ALKANOATE - ALCOHOL - WATER

One of the most important properties of surfactant-water systems is their ability to solubilize a third component. The solubilizate can be a fully apolar substance such as a hydrocarbon or a
weakly polar one as a long-chain alcohol. Ternary systems alkanoate - alkyl alcohol - water have been extensively studied, particularly by Ekwall and coworkers[1,42,43]. A typical phase diagram is
presented in Figure 9.

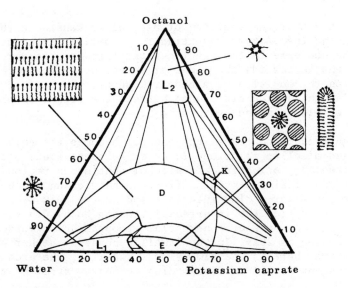

Figure 9. Experimentally determined phase diagram for the three
component system Potassium caprate - octanol - water. From ref. 42.

In constructing a model for this system we use the same
approach as for the binary water - ionic surfactant case with
additional features accounting for the role of the alcohol. When
an alcohol is added to a surfactant - water mixture several properties of the system change. With an alkyl chain of appreciable
length, $n \gtrsim 6$, the aqueous solubility of the alcohol is small and
it is incorporated into the surfactant aggregates. The volume of
the aggregates is increased and the area per charge is decreased
weakening the electrostatic interactions. The hydroxyl group goes
to the aggregate surface influencing the forces at the hydrocarbon -
water contact. To account for this effect we assume that the free
energy contribution, G_s, for this contact can be written

$$G_s = \gamma_{eff}A \tag{4}$$

where the proportionality constant is determined by the mole fraction of alcohol, X_{OH}, in the aggregate

$$\gamma_{eff} = (1-X_{OH})\gamma_{surf} + X_{OH}\gamma_{OH} \tag{5}$$

For γ_{surf} the value (18 mJ/m^2) determined for the binary system
is used. To obtain γ_{OH}, X-ray data for the lamellar phase have
been analyzed using Equation (3) and the value $\gamma_{OH} = 4$ mJ/m^2 was
obtained. This shows that the addition of alcohol decreases the
free energy contribution from the surface term. In addition there
is an entropy contribution due to the mixing of alcohol and surfactant in the aggregate.

It follows that the addition of alcohol weakens both the electrostatic term and the γA term resulting in a partial cancellation.
However, the decrease in the electrostatic term is stronger and
there is a change towards the lamellar phase, being the more
stable one on the addition of alcohol. However, the decrease of
γ_{eff} explains the marked difference between the effect on phase
equilibria of alcohols and hydrocarbons.

At high alcohol contents, the alcohol becomes the solvent as
in a reversed micellar solution. To describe these states, the
model is extended to incorporate alcohol continuous regions[44].
The standard chemical potential difference between an alcohol
molecule at the polar-apolar interface and in the bulk alcohol
may be calculated from the interfacial tension to be 0.5 kJ/mole
at 25°C. Furthermore the area per polar group in the reversed micelles was chosen as 20Å^2. In the lamellar phase at higher water
contents, where there is one-dimensional swelling, the area per
chain is constant at 25Å^2 corresponding to nearly straight chains
[1,42].

These features are included in the total thermodynamic model
of the system. Chemical potentials are derived for the three components in the micellar and reversed micellar solutions and the
hexagonal and lamellar liquid crystals. By an iterative procedure,
points are located where two phases are in equilibrium using the
criterion of equal chemical potential of all three components in the
two phases[23]. The calculated phase diagram, including tie-lines,
is shown in Figure 10. The water poor region has been excluded
since the model for the polar groups breaks down in this region.

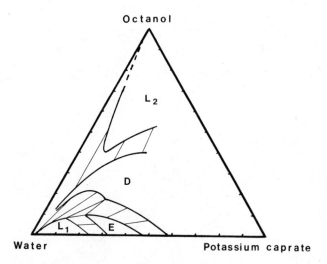

Figure 10. Calculated phase diagram for the three component system
Potassium caprate - octanol - water.

A comparison between Figures 9 and 10 shows that the model accounts for the basic features of the phase equilibria between different aggregate structures. In addition to the qualitative effects discussed in the preceding section, a couple of new features emerge on the addition of an alcohol. When alcohol is added to the hexagonal phase the cylinders can become strained and a lower symmetry might be preferred (Cf. Charvolin in this volume). In the lamellar phase, the thickness of the bilayers increases with increasing alcohol content until the limiting value of 25Å^2 cross sectional area per chain is reached. On addition of water to such a lamellar system there is a swelling of the water layers only. The lamellae repel each other electrostatically and the extent of swelling is controlled by a balance between the electrostatic repulsion and the van der Waals' attraction. The latter is able to compensate the electrostatic forces only if monomers go to the aqueous region and thus provide electrolyte screening.

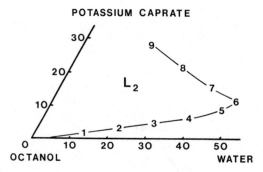

Figure 11. Calculated phase boundary for the reversed micellar (L$_2$) region.

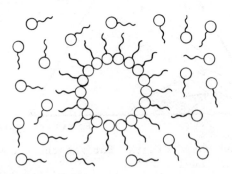

Figure 12. Schematic representation of a reversed micelle with an aqueous interior surrounded by a layer of surfactant and alcohol.

A significant feature of the calculated diagram is that the shape of the reversed micellar (L_2) region is similar to what is found experimentally. To understand the source of the "tongue" pointing towards the water corner, consider the details of Figure 10 shown in Figure 11. As surfactant is added to an octanol solution saturated with water, reversed micelles form with an aqueous interior and with a surrounding layer of surfactant and alcohol (see Figure 12). Thus large amounts of water can be incorporated into the phase. As more surfactant is added more inverted micelles form and they also become smaller in size. More alcohol is needed to form the interface at the water aggregates and ultimately a point is reached, where the larger part of the bulk alcohol reservoir has been used up and the chemical potential of the alcohol decreases rapidly. This occurs roughly at point 6 in Figure 11. Table I gives a quantitative prediction for the variation in chemical potential for points 1-9 of Figure 11 and the fraction of alcohol in the bulk phase is also presented.

Table I. The Deviation $\Delta\mu_{OH}$ of the Octanol Chemical Potential from the Value for Octanol Saturated with Water. The Fraction of Alcohol in the Continuous Phase n_{OH}^b/n_{OH}^{tot} is Also Shown. The Numbering of the Points Refers to Figure 11.

Point number	$\Delta\mu_{OH}$ (kT)	n_{OH}^b/n_{OH}^{tot}
1	~0	.94
2	~0	.85
3	$-3\cdot10^{-5}$	.78
4	$-2\cdot10^{-4}$	.61
5	$-1\cdot10^{-3}$	.41
6	$-8\cdot10^{-3}$	.13
7	$-4.8\cdot10^{-2}$	.06
8	$-1.2\cdot10^{-1}$	.07
9	$-2.1\cdot10^{-1}$	.08

CONCLUSIONS

We have tried to show, by way of example, how one can qualitatively understand the relation between molecular properties of surfactants and their phase behaviour. It appears that the gross features of the, at first sight rather complex, phase diagrams can be accounted for by using rather simple models. The crucial step is to recognize the essential interactions for each particular system. There is a large variation in the chemical nature of the polar groups and this has to be taken into account.

The qualitative ideas presented above have largely been sub-stantiated by calculations based on quantitative expressions, which are presented elsewhere[19-24].

There is a number of possible extensions of the applications of these ideas. They involve, for example, ternary system water – surfactant – hydrocarbon and quarternary systems water – surfactant – alcohol – hydrocarbon. Work along this line is in progess. Inte-resting problems of a more fundamental nature involve a detailed molecular interpretation of the γ-parameter. One significant con-tribution to this parameter concerns the packing of the alkyl chains in the aggregate (Cf. Gruen in this volume). Another funda-mental problem is to relate the continuum description of the water to a molecular description, which has to be applied at the lowest water contents.

REFERENCES

1. P. Ekwall in "Advances in Liquid Crystals", G. Brown, Editor, Vol. 1, p. 1, Academic Press, New York, 1975.
2. P. Winsor, Chem. Rev. $\underline{68}$, 1 (1968).
3. A. Skoulios, Adv. Colloid Interface Sci. $\underline{1}$, 79 (1967).
4. V. Luzzati and A. Tardieu, Ann. Rev. Phys. Chem. $\underline{25}$, 79 (1974).
5. R.G. Laughlin in "Advances in Liquid Crystals", G. Brown, Editor, Vol. 3, p. 41, Academic Press, New York, 1978.
6. G.J.T. Tiddy, Phys. Rep. $\underline{57}$, 1 (1980).
7. H.V. Tartar, J. Phys. Chem. $\underline{59}$, 1195 (1955).
8. C. Tanford, "The Hydrophobic Effect", John Wiley, New York, 1973.
9. J.N. Israelachvili, S. Marcelja and R.G. Horn, Quart. Rev. Biophys. $\underline{13}$, 121 (1980).
10. J.N. Israelachvili, D.J. Mitchell and B.W. Ninham, J. Chem. Soc. Faraday Trans. II $\underline{72}$, 1525 (1976).
11. H. Wennerström, J. Colloid Interface Sci. $\underline{68}$, 589 (1979).
12. V.A. Parsegian, Trans. Faraday Soc. $\underline{62}$, 848 (1967).
13. D.E. Mather, J. Colloid Interface Sci. $\underline{57}$, 240 (1976).
14. H. Traüble, M. Teubner, P. Wolley and H. Eibl, Biophys. Chem. $\underline{4}$, 319 (1976).
15. F. Jähnig, Biophys. Chem. $\underline{4}$, 309 (1976).
16. A.G. Lee, Biochem. Biophys. Acta $\underline{472}$, 237 (1977).
17. P.A. Forsyth, S. Marcelja, P.J. Mitchell and B.W. Ninham, Biochem. Biophys. Acta $\underline{469}$, 335 (1977).
18. B.R. Copeland and H.C. Andersen, J. Chem. Phys. $\underline{74}$, 2536 (1981).
19. H. Wennerström and B. Lindman, Phys. Rep. $\underline{52}$, 1 (1979).
20. G. Gunnarsson, B. Jönsson and H. Wennerström, J. Phys. Chem. $\underline{84}$, 3114 (1980).

21. B. Jönsson and H. Wennerström, J. Colloid Interface Sci. 80, 482 (1981).
22. B. Jönsson, G. Gunnarsson and H. Wennerström in "Solution Behaviour of Surfactants: Theoretical and Applied Aspects", K.L. Mittal and E.J. Fendler, Editors, Vol. 1, pp. 317-341, Plenum Press, New York, 1982.
23. B. Jönsson "The Thermodynamics of Ionic Amphiphile - Water Systems. A Theoretical Analysis". Thesis. Lund 1981.
24. L. Guldbrand, B. Jönsson and H. Wennerström, J. Colloid Interface Sci. 89, 532 (1982).
25. P.G. Nilsson, H. Wennerström and B. Lindman, J. Phys. Chem. In press.
26. R. Kjellander and E. Florin, J. Chem. Soc. Faraday Trans. I, 77, 2053 (1981).
27. P.J. Flory "Principles of Polymer Chemistry", Cornell University Press, Ithaca, 1953.
28. P.G. Nilsson, B. Jönsson and B. Lindman. To be published.
29. R. Kjellander, J. Chem. Soc. Faraday Trans. 2,78,2025 (1982).
30. J. Ulmius, H. Wennerström, G. Lindblom and G. Arvidson, Biochemistry 16, 5742 (1977).
31. J.M. Pope, L. Walker, B.A. Cornell and G.W. Francis, Biophys. J. 35, 509 (1981).
32. R.P. Rand, D. Chapman and K. Larsson, Biophys. J. 15, 1117 (1975).
33. D.M. LeNeveu, R.P. Rand and V.A. Parsegian, Nature 259, 601 (1976).
34. R.P. Rand, Ann. Rev. Biophys. Bioeng. 10, 277 (1981).
35. P.B. Gallot and A. Skoulios, Kolloid Z.Z. Polym. 208, 37 (1966).
36. P.J. Missel, N.A. Mazer, G.B. Benedek, C.Y. Young and M. Carey, J. Phys. Chem. 84, 1044 (1980).
37. B. Lindman and H. Wennerström in "Topics in Current Chemistry", Vol. 87, p. 1, Springer, Berlin, 1980.
38. L. Onsager, Ann. New York Acad. Sci 51, 627 (1949).
39. I.D. Leigh, M.P. McDonald, R.M. Wood, G.J.T. Tiddy and M.A. Trevethan, J. Chem. Soc. Faraday Trans. 1, 77, 2867 (1981).
40. K. Fontell, Prog. Chem. Fats other Lipids 16, 145 (1978).
41. C. Mandelmont and R. Perron, Bull. Soc. Chim. France 97, 425 (1974).
42. P. Ekwall, L. Mandell and K. Fontell, J. Colloid Interace Sci. 31, 508 (1969).
43. R. Friman, I. Danielsson and P. Stenius, J. Colloid Interface Sci. 86, 501 (1982).
44. J. Biais, P. Bothorel, B. Clin and P. Lalanne, J. Colloid Interface Sci. 80, 136 (1981).
45. K. Shinoda, J. Colloid Interface Sci 34, 278 (1970).
46. F. Harusawa, S. Nakamura, T. Mitsui, Colloid & Polymer Sci. 252, 613 (1974).
47. J.W. McBain, W.C. Sierichs, J. Amer. Oil Chem. Soc. 25, 221 (1948)

ON THE PHASE BEHAVIOR OF SYSTEMS OF THE TYPE

H_2O - OIL - NONIONIC SURFACTANT - ELECTROLYTE

M. Kahlweit and E. Lessner

Max-Planck-Institut fuer biophysikalische Chemie
Postfach 968, D-3400 Goettingen
Federal Republic of Germany

In a recent publication we had suggested study of
"simple" systems of the type H_2O - Oil - Nonionic Surfactant - Electrolyte as model for the pseudoquaternary
systems relevant in tertiary oil recovery, in order to
make qualitative predictions for the phase behavior of
such systems. In this paper the first results of such
studies are presented. It is shown in detail how the formation of the three-phase triangle arises from the
change of the phase diagram of such systems with temperature. It turns out that, within a limited temperature
range, the phase diagrams can show two critical points,
which are responsible for the ultra-low interfacial
tensions between neighbouring phases of the three-phase
triangle. Furthermore, a simple recipe is proposed
which permits to predict whether or not a system - with
given oil - will show a three-phase triangle with rising
temperature or addition of an appropriate electrolyte.
The predictions are compared with experiments using
short chain polyglycolethers (C_iE_j) as surfactants,
which have the advantage that these surfactants do not
form anisotropic phases.

23

INTRODUCTION

In a paper published in 1981[1] we had suggested study of
systems of the type
H_2O (A) - Oil (B) - "Simple" Nonionic Surfactant (C)
in order to understand the phase behavior of more complicated ternary and quaternary systems. In this paper we shall present some
results of such studies.

As we have claimed in our previous paper, the phase behavior
of such a system arises from the interplay of the miscibility gaps
of the three corresponding binary systems. Figure 1 shows the
Gibbs triangle A-B-C as basis with the three binary phase diagrams
A-B, B-C, and A-C versus temperature, which form the side planes
of the prism.

The miscibility gap between water and oil extends far beyond
0 °C. The UCT (T_α) of the gap between oil and surfactant is located
close to 0 °C. Its position depends - with given oil - on the
chemical nature of the surfactant: the higher its HLB, the higher the
T_α; the lower its HLB, the lower the T_α. The phase diagram of water

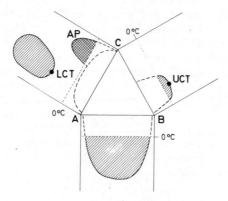

H_2O (A) - Oil(B) - $C_i E_j$ (C)

Figure 1.   Phase prism with Gibbs triangle A-B-C as basis and the
three binary phase diagrams A-B, B-C, and A-C versus temperature
(schematic).

and surfactant is the most complicated of the three. The UCT of
its lower miscibility gap appears to be located below 0 $^\circ$C, except
for the region of anisotropic phases (AP), which - for long chain
surfactants - extend beyond 0 $^\circ$C on the surfactant rich side of
the phase diagram. These anisotropic phases disappear between
about 30 and 50 $^\circ$C. At about the same temperature there appears
an upper miscibility gap on the water rich side, which seems to
be a closed loop.[2] Its LCT (T_β) again depends on the chemical
nature of the surfactant: the higher its HLB, the higher T_β, the
lower its HLB, the lower T_β. The LCT of the (upper) gap between
A and C, and the UCT of the (lower) gap between B and C, can thus
not be changed independently. The higher the hydrophilicity of the
surfactant -with given oil-the higher the critical temperatures(both);
the higher its oleophilicity,the lower the critical temperatures(both).

And now to the influence of the third component on these critical temperatures: the UCT of the gap between B and C increases,
in general, with the addition of water, moving on a critical line
which starts at T_α and ascends into the prism. The LCT of the
(upper) gap between A and C, on the other hand, may either increase
or decrease with the addition of oil, depending on the chemical
nature of the surfactant and the oil. The critical line, which
starts at T_β, may thus either ascend or descend into the prism.

THE THREE-PHASE TRIANGLE

How the interplay between these three miscibility gaps gives
rise to the formation of a three-phase triangle within the prism,
is shown in Figure 2. Below T_α, the gaps A-B and B-C form - probably - a connected miscibility gap (Figure 2.1). With rising
temperature, gap B-C narrows until its UCT T_α is reached (Figure
2.2). With further increase in temperature, the gap will disconnect
from the B-C side, its critical point moving on the ascending
critical line into the prism (Figure 2.3). In this temperature
range the phase diagram will thus show one critical point, which
faces the B-C side. One must, therefore, expect critical fluctuations in the oil rich region of the phase diagram, which should
not be mistaken for inverse micelles.

Now, let the critical line, which starts at the LCT of the
gap between A and C, descend into the prism. If it ascends, this
will not change the following discussion, as it will be shown
later. As it has been proved by Schreinemakers et al.[3] on the
basis of the Gibbs phase rule, gap A-C will then collide with gap
A-B at the endpoint of its critical line (Figure 2.4).

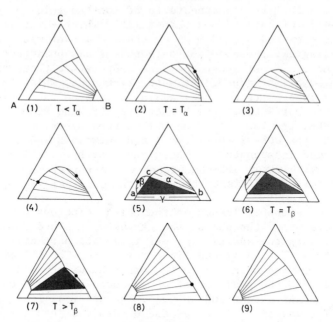

Figure 2.   Change of the phase diagram of the ternary system A-B-C
with temperature: formation of a three-phase triangle (schematic).

One then has to distinguish between two cases: if the critical
point of gap A-B has meanwhile moved so far counterclockwise that,
at this temperature, it now faces the A-C side, the two gaps will
merge at their critical points, thus forming a connected miscibility gap, which extends from the A-B to the A-C side without intermediate formation of a three-phase triangle.

If however, the critical point of gap A-B still faces the B-C
side at this temperature - as shown in Figure 2.4 - the collision
of the two gaps will give rise to the formation of a three-phase
triangle a-b-c (Figure 2.5).

The (lower) aqueous phase a, and the (middle) surfactant
phase c, thus emerge from a critical point. During the first stages
of this process, these two phases will, accordingly, be very similar in their properties, and there is nothing exotic about the
ultra-low interfacial tension between them: it rises from zero
according to the scaling law for the interfacial tension between
near-critical phases.

As one raises the temperature further, the two-phase region β
will grow towards the A-C side, its critical point moving on the
critical line, which   terminates on the A-C side at T$_\beta$ (Figure 2.6),
whereas the two-phase region α will retreat further from the B-C
side. Consequently, the a-c side of the three-phase triangle will
widen, whereas its b-c side will narrow. In this temperature range
the system will show two critical points, one facing the A-C side,
the other one the B-C side.

At T = T$_\beta$ the first of these two critical points will disappear. From then on the two-phase region β will extend from the
a-c side of the three-phase triangle to the A-C side of the prism
(Figure 2.7).

With further increase in temperature,the two phases b and c will
finally merge at the critical point of the two-phase region α
(Figure 2.8). The three-phase triangle disappears. During the last
stages of this process phases b and c will become very similar in
their properties, and the interfacial tension between them will
vanish according to the scaling law for the interfacial tension
between near-critical phases. From then on one will find a connected
miscibility gap, which extends from the A-B to the A-C side
(Figure 2.9).

The entire process can thus be looked at as the transition from
a connected miscibility gap A-B - B-C (Figure 2.1) to a connected gap
A-B - A-C (Figure 2.9). This transition will take place without
intermediate formation of a three-phase triangle, if at that temperature, at which the endpoint of the critical line of gap A-C
touches gap A-B (Figure 2.4), the tie-lines of gap A-B descend
towards the A corner. If, however, the tie-lines of gap A-B descend towards the B corner, the transition will take place via an
intermediate formation of a three-phase triangle as schematically
shown in Figure 2.

This result suggests a rather simple experiment which permits
to predict whether or not one can expect the formation of a three-
phase triangle (Figure 3).

For this purpose one fills a test tube with e.g. 1 cm^3 of
water and 1 cm^3 of oil - at room temperature. If one then adds
1 cm^3 of the surfactant, it will dissolve mainly either in the
aqueous phase or in the oil phase. If it dissolves mainly in the
oil, then the tie-lines of gap A-B descend towards the A corner,and
an increase in temperature or an addition of an appropriate electrolyte will - most probably - lead to the formation of a connected miscibility gap A-B - A-C without intermediate formation of
a three-phase triangle.

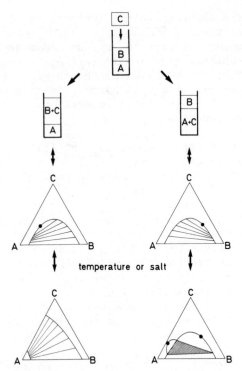

Figure 3.   How to predict whether or not one can expect the formation of a three-phase triangle.

If, however, the surfactant dissolves mainly in the aqueous
phase, then the tie-lines of gap A-B descend towards the B corner,
and an increase in temperature or an addition of salt will – most
probably – lead to the formation of a three-phase triangle. This
experiment takes only a minute and has worked for all systems,
which we have hitherto studied.

Figure 2 further predicts a two-phase region γ, which extends
from the a-b side of the three-phase triangle to the A-B side of
the prism, and which has been overlooked in previous discussions
of the phase behavior by other authors.[4]

Last but not least, it emphasizes the role of the critical
points in the phase diagram: the three-phase triangle arises from
the collision between gap A-B and A-C at the endpoint of the

critical line of the latter, and it disappears when the (middle)
surfactant phase and the (upper) oil phase merge at the critical
point of the two-phase region α. It also supports the statement
made in our previous paper,[1] namely that the phase inversion temperature is in fact a temperature range, the position of which on
the temperature scale - for a given system - depends on the mean
composition of the solution and thus has little physical significance. We also repeat that - for a given oil - the temperature
range, in which the three-phase region exists, can be shifted up-
wards or downwards by an appropriate choice of the surfactant:
the higher its HLB, the higher - in general - the temperature range;
the lower its HLB, the lower the temperature range.

TWO SIMPLE SYSTEMS

Figure 4 shows the phase diagram of the system
H_2O (A) - n-Decane (B) - C_4E_1 (C)
as it changes with temperature. The LCT T_β of the (upper) gap
between A and C is 48 $^{\circ}$C. The critical line descends into the
prism with the addition of n-Decane and touches the gap between
A and B at 24 $^{\circ}$C. The two points in the upper left figure denote
the positions of the two critical points at 25 $^{\circ}$C. At this tempe-

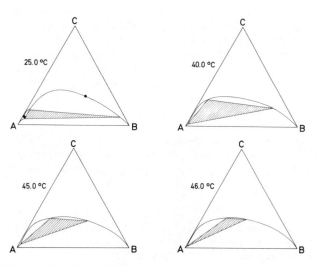

Figure 4.   Phase diagram of the system H_2O - n-Decane - C_4E_1 at
different temperatures (in mole fractions).

rature the phase diagram thus corresponds to the situation shown
schematically in Figure 2.5. It changes its shape with rising temperature in qualitative agreement with Figure 2, except, that the
three-phase triangle disappears at 46.3 $^{\circ}$C, i.e., at a temperature
slightly below T_β.

The fact that the three-phase triangle emerges from a critical
point and disappears at the second critical point can be readily
demonstrated by measuring the correlation length in the solution.
For this purpose we prepared a solution just outside the gap
between A and B with a composition at which we expected the critical line of gap A-C to touch gap A-B at 24 $^{\circ}$C. Below 24 $^{\circ}$C this
solution was homogeneous. As one raises the temperature, the solution approaches the endpoint of the critical line: the correlation
length increases to approach infinity (full points in Figure 5).

After the formation of the three-phase triangle at 24 $^{\circ}$C we
then took the middle phase c of the three-phase system, the composition of which is well defined at each temperature. Just above
24 $^{\circ}$C, the composition of this phase is still close to that of the
critical point of region β (Figure 2.5). Accordingly, the correlation length is still large, but decreases as one raises the temperature and thus increases the distance between phase c and the
critical point (empty points in Figure 5). With further increase in
temperature, phase c starts to approach the critical point of region
α (Figure 2.7): the correlation length passes through a minimum and
increases again to approach infinity at 46.3 $^{\circ}$C, at which temperature phases c and b merge at the critical point (Figure 2.8).

As it is well known, the addition of an appropriate electrolyte (D) has an effect equivalent to that of a rise in temperature. The phase diagram of such a quaternary system should be
represented in a tetrahedron with the Gibbs triangle A-B-C as basis.
For a representation in two dimensions one may then project the
coordinates x_i from the D corner onto the basis by $x_i' = x_i/(1-x_D)$,
i = A,B,C. For sufficiently low x_D (<< 1) one has
$x_i' \approx x_i$.[5]

Figure 6 shows the three-phase triangle in the above system
at 25 $^{\circ}$C, without electrolyte (shaded area) and with 30 g NaCl
per 10^3 cm^3 H$_2$O, setting $x_i' = x_i$. As one can see, the effect of
the added electrolyte is equivalent to a rise in temperature of
about 15 $^{\circ}$C.

We add that Cazabat et al [6] have measured the correlation
length in the five component system
H$_2$O - Toluene - Butanol - NaDS - NaCl
keeping the temperature constant and varying the NaCl concentration.

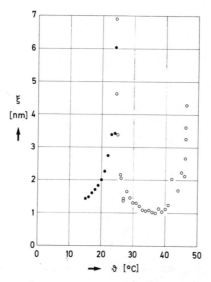

Figure 5.   Correlation length ξ versus temperature. Full points: homogeneous solution. Empty points: middle phase c (see text).

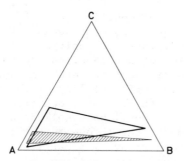

Figure 6.   Three-phase triangle in the system H_2O – n–Decane – C_4E_1 – NaCl (30 g per 10^3 cm^3 H_2O) at 25 °C.

The authors found a curve equivalent to the one shown in our Figure 5. Unfortunately, however, the phase diagram of a five component system is difficult to discuss in a two-dimensional representation.

As a second simple ternary system we have chosen H_2O (A) – n–Decane (B) – C_8E_3 (C) which has been studied by Kunieda and Friberg [7] for exactly the same reason for which we have performed our investigations, namely to prevent anisotropic phases. Figure 7 shows the phase diagram of this system (in weight fractions) as it changes with temperature.

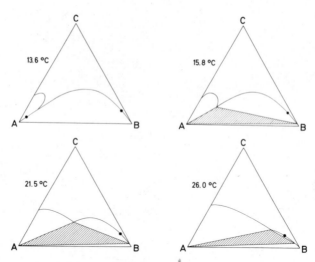

Figure 7.   Phase diagram of the system H_2O – n–Decane – C_8E_3 at different temperatures (in weight fractions).

Since C_8E_3 is more oleophilic than C_4E_1, the LCT of the (upper) gap between A and C is considerably lower than that for C_4E_1: $T_\beta = 8$ °C. Furthermore, the critical line ascends into the prism with the addition of n–Decane. As one can see, the phase diagram changes with rising temperature in qualitative agreement with Figure 2. Since the critical line of the two-phase region β ascends into the prism, the region grows from the A-C side towards the gap between A and B, until both collide at the endpoint of the critical line at 13.6 °C: the three-phase triangle appears. With further increase in temperature its a-c side increases in length, whereas its b-c side decreases until phase b and c merge with the critical point of region α at about 30 °C.

SOCALLED MICROEMULSIONS

In the literature the homogeneous solutions in such ternary and quaternary systems are called "microemulsions". When using this term, most authors refer to the paper by Hoar and Schulman published in 1943.[8] It should be noted, however, that quantitative determinations of (parts of) the phase diagrams of such quaternary systems were published by Bowcott and Schulman only in 1955.[9] These authors studied the system
H_2O – C_6H_6 – Hexanol – K-Oleate
but, unfortunately, did not use a Gibbs triangle to represent the results. From their Figure 2 one can, however, evaluate the mole fractions of the four components at the boundary of the miscibility gap. The result is shown in Figure 8 as curve (1).

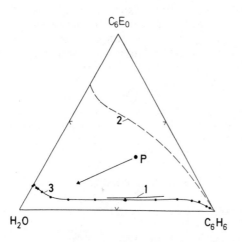

Figure 8. Phase diagram of the system H_2O - C_6H_6 - C_6E_0 (curve 2) and of the system H_2O - C_6H_6 - C_6E_0 - K-Oleate (curve 3) at 25 °C. Curve 1 from Ref. 9 (at 20°C). The small homogeneous regions in the H_2O corner are omitted for reasons of clarity. (in mole fractions)

Since the mole fraction of K-Oleate is rather small (x_D between .01 and .02), the curve runs close to the A-B-C basis, ascending slightly from the right to the left. As one can see, the authors determined only a rather small part of the phase diagram.

Also shown is the miscibility gap of the ternary system H_2O - C_6H_6 - C_6E_0 (curve 2). The comparison shows that the addition of the ionic detergent does increase the mutual solubility of water and benzene considerably,[10] but does not clarify whether or not the amount of added detergent was sufficient to make the miscibility gap along the A-C side disappear. For that reason we have repeated the experiment using the same amount of K-Oleate (curve 3), which showed that the gap becomes indeed much narrower but remains connected. This explains the fact that Bowcott and Schulman were not able to change from a "w/o emulsion" (e.g. point P in Figure 8) to an "o/w emulsion" in the A corner without passing through the miscibility gap.

Had the authors studied the phase diagram of the system H_2O - C_6H_6 - Ethanol which was determined as early as 1931,[11] they would have found that with C_2E_0 alone (i.e. without ionic detergent) one can obtain a comparable mutual solubility between water and benzene (in particular, if expressed in weight fractions instead of mole fractions) and, furthermore, avoid the miscibility gap along the A-C side.

They would have then possibly hesitated to propose their model of
submicroscopic droplets being stabilized by interfacial layers of
the nonionic cosurfactant and the ionic detergent, since the ternary systems with "simple" surfactants such as C_2E_0 or C_4E_1 contain
neither ionic groups nor micelles. The phases separate rather rapidly, and the middle phase c, in particular, shows no indications of
an anisotropic structure.

We, therefore, again suggest study of the microstructure of
"simple" ternary systems first, before increasing the complexity
of the systems.

ACKNOWLEDGMENTS

We are indebted to Mr. B. Faulhaber for his assistance with
the experiments.

REFERENCES

1. C.-U. Herrmann, G. Klar, and M. Kahlweit, J. Colloid Interface
    Sci., 82, 6 (1981).
2. J.C. Lang and R.D. Morgan, J. Chem. Phys., 73, 5849 (1980).
3. F.A.H. Schreinemakers, in "Die heterogenen Gleichgewichte",
    H.W. Roozeboom, Editor, Vol III/2, Vieweg, Braunschweig, 1913;
    R. Vogel, in "Handbuch der Metallphysik", G. Masing, Editor,
    Vol. II, Akademische Verlagsagesellschaft, Leipzig, 1937.
    For more recent literature in English we recommend:
    A. Prince, "Alloy Phase Equilibria", Elsevier, Amsterdam, 1966.
4. See e.g. M.L. Robbins, in "Micellization, Solubilization, and
    Microemulsions", K.L. Mittal, Editor, Vol. 2,   p. 713,
    Plenum Press, New York, 1977;
    M. Bourrell, Ch. Koukounis, R. Schechter, and W. Wade,
    J. Disperson Sci. Techn., 1, 13 (1980).
5. See e.g. J.C. Lang and B. Widom, Physica, 81 A, 190 (1975).
6. A.M. Cazabat, D. Langevin, J. Meunier, and A. Pouchelon,
    J. Physique Lett., 43, L-89 (1982).
7. H. Kunieda and S.E. Friberg, Bull. Chem. Soc. Japan,
    54, 1010 (1981).
8. T.P. Hoar and J.H. Schulman, Nature, 152,102 (1943).
9. J.E. Bowcott and J.H. Schulman, Z. Elektrochem., 59, 238 (1955).
10. See also: C.-U. Herrmann, U. Wuerz, and M. Kahlweit,
    Ber. Bunsenges. Phys. Chem., 82, 560 (1978).
11. E.R. Washburn, V. Hnizda, and R. Vold, J. Amer. Chem. Soc.,
    53, 3237 (1931).

PHASE EQUILIBRIA IN AND LATTICE MODELS FOR NONIONIC SURFACTANT-WATER MIXTURES

John C. Lang

The Procter & Gamble Company
Miami Valley Laboratories
Cincinnati, Ohio 45247

Three major points are made in this paper: (1) From the three-component phase equilibria we deduce there is important temperature dependence to the interactions between water and nonionic surfactant, and that these changes in interactions are unsymmetrical. (2) On comparison with model calculations and data for simpler chemical mixtures, the two-component phase equilibria show the source of the temperature dependence is directionality in hydrogen bonding. (3) Even without the effects of aggregation there may be something very special about lower consolute boundaries as a consequence of this directionality. There are numerous minor inferences, but these three points are the central conclusions.

I. INTRODUCTION

It has been shown, most clearly and exhaustively by Ekwall
and co-workers,[1] that phase equilibria in multi-component aqueous
mixtures of amphiphilic substances can be richly diverse and
intricate. These mixtures show a multiplicity of microscopic and
molecular structures, each a manifestation of nature's ingenuity
at finding new forms by which to reduce the free energy of the
total solution. The first figure, Figure 1, perhaps one of
Ekwall's most famous, illustrates the variety of phases existing
in a single three-component mixture: two isotropic liquid phases,
four or five distinct mesomorphic phases, and solid phases. The
precise characteristics and properties of these and similar phases
have themselves been of fundamental interest, from the dynamics of
molecules in different environments[2] to the collective properties
of phases.[3] But it is quite clear from Wennerström's thermodynamic
modeling that it is the molecular properties of packing and size,
charge, and aggregate geometry as well as bulk composition that
regulate stability of phases of amphiphilic substances.

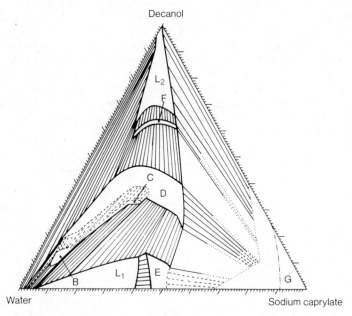

Figure 1. Three-component isothermal phase diagram of decanol-
water-sodium caprylate, after Ekwall.[1] The figure shows isotropic
phases L_1 and L_2, mesomorphic phases B, C, D, E, F, and solid
phases G.

Other groups[4,5] have shown the variety of isotropic phases produced by solutions of amphiphiles in the presence of alcohols, salts, and hydrocarbon. The Minnesota group[6] has demonstrated convincingly that the origin of such complexity is the variety of possible morphologies not the more common and expected diversity resulting from a large number of components, in particular surfactant homologues. It has been well-documented that these microemulsion phases, while distinct from one another, share common distinctions from the more ordered less labile mesophases.[7]

Our own work follows more closely the direction of McBain and co-workers[8,9] who were interested not so exclusively in the structural details of particular systems and phases but in those variables which control their evolution. In the upper left of Figure 2 is a schematic phase diagram of a mixture of salt-water-ionic soap, identifying the types of regions to be expected. The temperature dependence of mixtures of water-sodium chloride-sodium palmitate is shown in the other diagrams. With an increase in temperature the characteristic entropy-dominated decrease in immiscibility is manifested for all regions of coexistence: liquid crystalline phases shrink and finally disappear, isotropic regions expand. This type of behavior is characteristic of systems whose separate molecular interactions need not be changing or, the less likely possibility, whose interactions change symmetrically. We

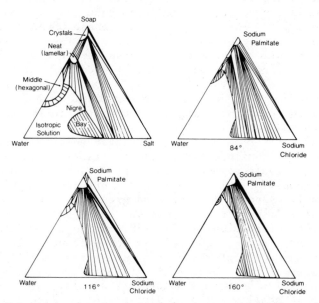

Figure 2. Schematic isothermal three-component phase diagram for soap-water-salt (upper left), and sequence of three-component sodium palmitate-water-sodium chloride phase diagrams at 84°C 116°C, and 160°C. This is the work of McBain and co-workers.[8,9]

shall observe this need not always be the case. However, one
might anticipate that members of different classes of amphiphiles[10] will follow common characteristic patterns of variation
as a response to changes in a thermodynamic field, for example
temperature or chemical potential of an extra component.

While a goal of our work is to understand the complexity of
chemical specificity, the emphasis of this study is to discern the
generality of phenomena dictated by the physics. Such generality
is deduced from common responses to parallel perturbations of the
same or analogous systems. The experimental techniques utilized
are contemporary methods for establishing phase boundaries and
compositions of conjugate phases.[11] Though the techniques are
modern, our approach at one level is classical and little different
from that of Ekwall et. al., or McBain and co-workers. The thermodynamic analysis relies heavily on the understandings of Gibbs,[13]
Schreinemakers[14] and the modern work of Griffiths and Wheeler.[15]
Inferences from the shapes of phase boundaries or the directions
of tie lines and their departure from or concurrence with solution
theory disclose the significance of subtle changes in free energy
on the state or phase of these mixtures, or conversely of the
significant changes in free energy following from apparently subtle
distinctions in phase equilibria. Studies of the structure of
molecular organization within phases are of fundamental importance
to the understanding of stability, or relative stability, of
phases. Changes can be anticipated from both the location and
shape of boundaries. While with accurate phase boundaries such
work has been able to continue on the material I shall discuss,[12,16]
I shall not emphasize the details of these structures or the
specific effects variations in them might produce.

In the remainder of this paper we shall highlight the features
of aqueous mixtures of the homogeneous nonionic surfactant $C_{10}E_4$,
3,6,9,12-tetraoxadocosanol. By comparing these systems with mixtures of other similar materials, e.g., near homologues or other
surfactants of the nonionic class, and contrasting them with distinct
materials, e.g., ionic components, we shall identify those properties and that behavior common to single short chain amphiphiles
with only weakly hydrogen-bonding head groups. In the following
section we shall discuss three-component phase equilibria, focusing
on the detailed information for $C_{10}E_4-H_2O-C_{16}H_{34}$. In Section III
we shall summarize some of the information presented elsewhere on
two-component aqueous mixtures of nonionic surfactants[11,17] and in
particular contrast the behavior of $C_4E_1-H_2O$ with $C_{10}E_4-H_2O$. There
we shall distinguish between absolute and specific hydrophilicity.[10]
In Section IV we shall use the predictions of lattice-gas models
to highlight the departure of mixtures of surfactant materials
from those of simpler molecularly dispersed substances and to
suggest distinctions between upper and lower critical points.

II. THREE-COMPONENT PHASE EQUILIBRIA

Phase equilibria in mixtures of $C_{10}E_4$-H_2O-$C_{16}H_{34}$ at three distinct temperatures 60°, 50°, and 19°C are shown in Figure 3-6. They cannot be compared directly, component by component, with the phase diagrams of McBain who examined the effects of salt and temperature on surfactant-water phase equilibria; they are more nearly the types of equilibria considered by Ekwall. However, on examining the effect of temperature on equilibria of water-surfactant-hydrocarbon we find radical departure from the observation of very gradual, sometimes nearly imperceptible, influence of temperature recorded by McBain for ionic substances.

Figure 3 shows there is but one type of immiscibility of surfactant and water at 60°C; all liquid crystalline phases have disappeared. This region of surfactant-water immiscibility grows and is continuous with the oil-water immiscibility; while not necessary, no islands of stable liquid crystalline phases were observed in the interior of the diagram. This simple diagram is the picture of a surfactant which can become richly hydrated, twelve water molecules for each surfactant at 60°C and fewer at higher temperature; yet even hydrated remains immiscible with water. This picture is of a microemulsion phase which varies

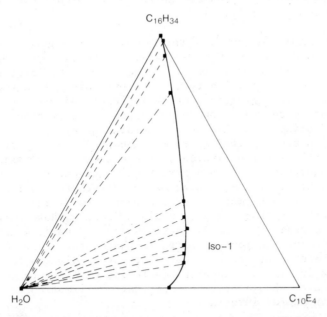

Figure 3.   Three-component phase diagram of $C_{10}E_4$-H_2O-$C_{16}H_{34}$ at 60°C showing single continuous two-phase region of water immiscibility with both oil and hydrated surfactant. Boundary is least squares fit through data points marking the ends of tie lines.

continuously from 0-40% water. Conversely, at any temperature the
less hydrated liquid surfactant is completely miscible with the
hydrocarbon. It should be clear that this would not be deduced if
the tie lines originated from the vicinity of the $C_{16}H_{34}$ vertex,
an impossibility at 60°C!

With a decrease in temperature of only 10°C, substantial
changes occur in the phase diagram; this is shown in Figure 4 and
in the interpretation of the lower left corner of Figure 4 shown
in Figure 5: (1) The number of one-phase regions changes from two
(water and microemulsion phase) to five (three are discontinuous),
with the appearance of a second microemulsion phase – the anomalous
phase,[11] and the lamellar liquid crystalline phase. (2) The number
of two-phase regions changes from one to seven. (3) And where
there were no three-phase regions there are now three. This complexity arises as a consequence of the liquid crystalline phase
intersecting the region of water-surfactant immiscibility, followed
by the second microemulsion phase intersecting the region of water-
lamellar phase coexistence, and with the first microemulsion phase
becoming immiscible with an inverted microemulsion phase. The
primary or dominant three-phase region--with equilibria between a
dilute aqueous phase, a concentrated microemulsion phase, and an
inverted microemulsion phase--dwarfs the minor three-phase regions
and is generated by the now two-phase region of oil-hydrated
surfactant immiscibility.

From a different perspective, Figure 4 and Figure 5 show that
mixtures of $C_{10}E_4$-H_2O-$C_{16}H_{34}$ can exist in one of four isotropic
phases. Three of these isotropic phases bound the large three-
phase region dominating Figure 4. The predominantly aqueous phase
has no detectable aggregates, which is in accord with concentration
of surfactant observed to be below the cmc. While both of the
phases coexisting with this phase contain aggregates, it is those
aggregates in the inverted microemulsion phase which appear larger.
Dynamic light scattering indicates these are both approximately an
order of magnitude larger and more highly anisotropic; however, a
portion of this increase in size may be attributed to proximity of
the plait point. The more isotropic particles of the microemulsion phase, a phase of intermediate density, have an equivalent
hydrodynamic radius of approximately 95Å.[12] Dynamic light scattering indicates the fourth isotropic phase, the second microemulsion phase, contains aggregates of intermediate size, approximately 260Å in radius. Neutron scattering experiments carried out
on these phases should provide important confirmation of this preliminary structural information.

Further very significant changes in the phase equilibria
occur on lowering the temperature another 30°C. The details will
be presented elsewhere.[12] While small three-phase regions are not

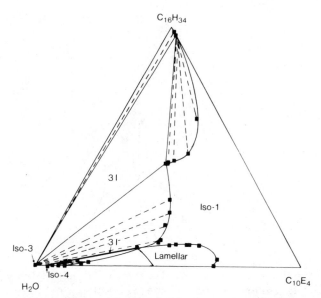

Figure 4.   Three-component phase diagram of $C_{10}E_4$–H_2O–$C_{16}H_{34}$ at 50°C showing the effects of the appearance of the oil-hydrated surfactant two-phase region: the creation of a large three-phase region and the shrinking of the two-phase region of immiscibility between water and microemulsion phases.   The lamellar liquid crystalline phase extends to very low surfactant concentration. Boundaries are least squares or least-square splines through points marking ends of tie lines.

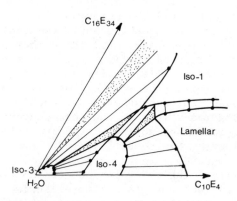

Figure 5.   Schematic diagram of lower left corner of Figure 4.   It shows (stippled) the three three-phase regions; only one was clearly discernible in Figure 4.   The anomalous phase, now identified as a second microemulsion phase, is shown here and is continuous with the front face.[11]

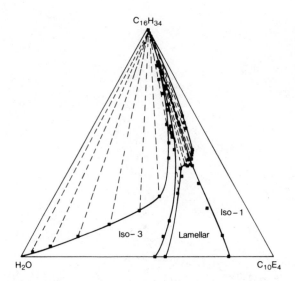

Figure 6.   Three-component phase diagram of $C_{10}E_4$-H_2O-$C_{16}H_{34}$ at
19°C showing the interruption of the oil-water, or oil-aqueous
surfactant, phase by the lamellar liquid crystalline phase. The
three-phase regions are indiscernible at this magnification. The
liquid crystalline phase here extends to high oil concentrations.
Boundaries are least squares or least squares splines through
points marking ends of tie lines.

included in Figure 6, the major consequences of the change in
temperature are evident: (1) The number of one-phase regions is
reduced to four (three are discontinuous), two isolated isotropic
regions and a lamellar mesophase.   (2) The number of two-phase
regions is presumably reduced to five, with the single large two-
phase region of oil-water immiscibility now more properly identified
as a region of immiscibility between oil and aqueous surfactant in
a microemulsion phase, which changes continuously from low to high
surfactant concentration.   This is the more conventional pattern
resembling that of ionic surfactants.   (3) Two three-phase regions,
not shown in the figure, can be anticipated where the liquid
crystalline phase intrudes into the region of oil-aqueous surfactant immiscibility.   The major consequence of reducing the
temperature is to replace a region of water-surfactant immiscibility with a region of oil-aqueous surfactant immiscibility.
However, in parallel with this the lamellar liquid crystalline
phase separates itself from the region of water-surfactant immiscibility at intermediate temperatures only to intrude into the
region of oil-aqueous surfactant immiscibility at lower temperatures.

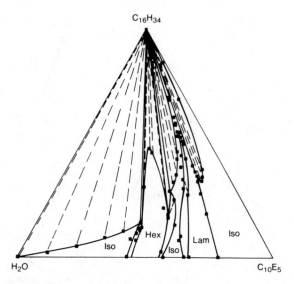

Figure 7. Three-component phase diagram of $C_{10}E_5$-H_2O-$C_{16}H_{34}$ at
20°C showing the added complexity created by the presence of two
mesomorphic phases. Only one of three three-phase regions can be
distinguished. Obvious parallels exist between this phase diagram
and that of the homologue, Figure 6; namely, the interruption of
oil-aqueous surfactant immiscibility by liquid crystalline phases.
Boundaries are least squares or least square splines through points
marking ends of tie lines.

In summary, the unconventional evolution of the $C_{10}E_4$-H_2O-
$C_{16}H_{34}$ phase equilibria contrasts profoundly with the gradual and
predictable evolution found for the salt-water-sodium palmitate
phase equilibria. The nonionic system not only is marked by the
rapid disappearance of the lamellar mesophase (40°C vs. 80°C for
the sodium palmitate mixtures), but also shows large distortions
in the shapes of corresponding boundaries, has entirely new phases
arise and then disappear, sees radical changes in the direction of
tie lines, and produces a liquid crystalline phase which at one
temperature disrupts the immiscibility of the surfactant with
water and at another, the immiscibility of the aqueous surfactant
and oil. The most profound change occurring in this 40° is under-
lined by the dislocation of the origin of the tie lines: surfactant-
water immiscibility is transformed to aqueous surfactant-hydrocarbon
immiscibility. Figure 7, the phase diagram of $C_{10}E_5$-H_2O-$C_{16}H_{34}$,
shows added complexity which can be expected for mixtures of nonionic surfactants which possess more than one liquid crystalline
phase. The clear parallels between this figure and Figure 6
emphasize the similarities and show how properties should be scaled
for these members of the same structural class of nonionics (viz.,
solubility, PIT,[18] etc.).

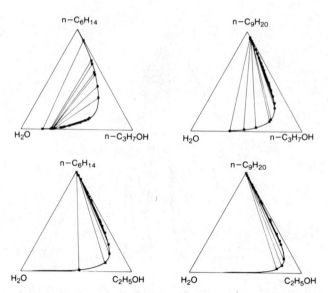

Figure 8. A series of three-component phase diagrams at 25°C showing the change in solubility and relative solubility of alcohol in the immiscible water and alkane phases as a function of structure of alcohol and hydrocarbon. [19]

   The origin of this contrast in behavior is suggested by the interesting Russian work on three-component aqueous mixtures of alcohols and alkanes. [19] The alcohol is either n-propanol or ethanol and the alkane is either n-hexane or n-nonane. All of these four three-component diagrams shown in Figure 8 are dominated by a single region of oil-water immiscibility. These systems show that as the alcohol and alkane become more distinct, the case of nonane and ethanol, the miscibility is diminished and the proto-surfactant, ethanol, interacts more strongly with water. The tie lines originate at the alkane vertex. Conversely, as the alcohol and alkane become more similar, the case of hexane and propanol, their miscibility is increased and the protosurfactant, propanol, interacts more strongly with alkane. The tie lines originate near the water vertex. Such contrast in direction of tie lines is caused by unsymmetrical or inequivalent changes in interactions between alcohol and water relative to alcohol and alkane. The inference from this work is that for nonionic surfactant systems temperature plays the role of structural variation and modifies the nonionic surfactant-water interactions relative to nonionic surfactant-alkane interactions in a profound and unsymmetrical way. It is these changes which are superposed on and far outweigh the effects of increased miscibility which accompany an increase in temperature.

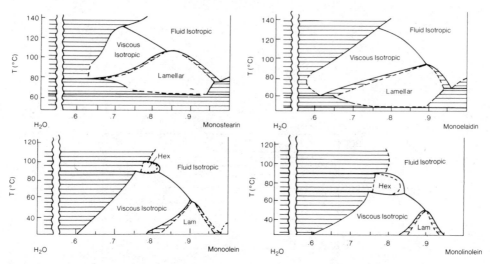

Figure 9.   Series of monoglyceride water phases diagrams showing
several examples of retrograde solubility.   (This is the work of
E. S. Lutton, J. Am. Oil. Chem. Soc., 42, 1068 (1965).)

III.   TWO-COMPONENT PHASE EQUILIBRIA

I have described a source of the temperature dependence of
hydrogen bonding elsewhere,[11] and will summarize our observations
only briefly here. Retrograde solubility, shown in Figure 9 for a
series of monoglyceride-water phase diagrams, might have several
causes: An internal equilibrium between several states would be
adequate. However for two types of nonionic surfactants, surfactants with polyoxyethylene head groups or phosphine oxide head
groups, we have observed not only retrograde solubility but whole
closed loops of coexistence that extend over from a few tens of
degrees to hundred of degrees.[11] Just as in simpler systems where
an increase in the water-hydrophobe interaction decreases the area
enclosed by the loop,[20] we observe that when the absolute hydrophilicity of the surfactant increases the area decreases. This is
shown in Figures 10 and 11 for aqueous mixtures of the homogenous
surfactants $C_{10}E_4$ (3,6,9,12-tetraoxadocosanol) and $C_{10}E_5$ (3,6,9,
12,15-pentaoxapentacosanol), respectively, where although the
specific hydrophilicity of the ether groups is the same the total
or absolute hydrophilicity of the latter is greater than the former
and results in a smaller loop of coexistence for $C_{10}E_5$.[11]

There are several contrasts between the phase diagrams of
these surfactant compounds and those for lower molecular weight
materials exhibiting closed loops of coexistence: (1) Most noticeable is the appearance of additional phases. In the case of

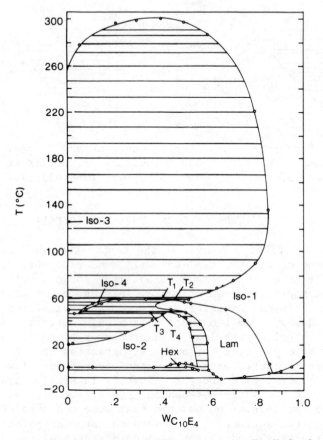

Figure 10. Two-component phase diagram of $C_{10}E_4$-H_2O showing
complete extent of water-surfactant immiscibility. Three isotropic,
two mesomorphic (hexagonal and lamellar), and one anomalous - or
isotropic second microemulsion - phases are identified.[11]

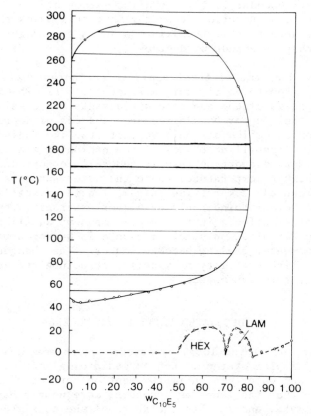

Figure 11. Two-component phase diagram of $C_{10}E_5$-H_2O showing complete
extent of water-surfactant immiscibility. It is smaller than the
loop of coexistence for $C_{10}E_4$-H_2O and neither mesophase interrupts
the coexistence curve.[11]

$C_{10}E_4$-H_2O there are two mesophases; the lamellar phase protrudes
into the loop of coexistence. As it does, there is created another
entirely new phase, a second micellar phase which on the addition
of hydrocarbon, in the three-component phase diagram, was identified
as a second or dilute microemulsion phase. The first micellar
phase appears at lower temperatures at the surfactant-rich side of
the lower consolute boundary. (2) The closed loops are not symmetric; the lower consolute boundary is displaced relative to the
upper consolute boundary. (3) The dilute sides of the closed
loops occur at concentrations near the CMC; this makes the surfactant loops appear as if the dilute side is a straight vertical
line, for perhaps hundreds of degrees.

     The phase diagrams of these nonionic surfactants also depart
significantly from those of ionic surfactants: (1) The liquid
crystalline phases of the nonionic surfactants are considerably
less stable; their upper boundaries are hundreds of degrees lower
than the upper eutectics of ionic surfactants like sodium palmitate.
(2) The liquid crystalline phases of the nonionics are capable of
considerably more distortion than is generally observed for the
ionics. At higher temperature, where the water-surfactant hydrogen
bonding is presumed weaker (vide infra), liquid crystalline
stability can be maintained only at higher water activity. This,
like the parallel effects in the three-component equilibria, is
the likely result of an unsymmetric change in hydrogen bonding;
the surfactant-water bonding becomes weaker. The changes in
liquid crystalline stability or micellar structure are consequences
of this temperature dependence.

IV.   DECORATED LATTICE-GAS MODELS

     One theory of closed-loop coexistence which adequately describes
the data for simpler systems, such as guaiacol-glycerine or shorter
chain ethers and water, is an improvement on earlier theoretical
treatments in that the loops are broader, more nearly matching
experimental data, and in that the shape of the coexistence curves
in the vicinity of upper and lower critical points are approximately
cubic, $\beta \sim 1/3$, in accord with theory of first order critical points.
This is the decorated lattice-gas model of Wheeler.[20] The model,
an extension of earlier work,[21,22] attributes the onset of immiscibility at lower temperatures to weakening of hydrogen bonding
between water and partially hydrogen bonding hydrophobe. This is
imagined to be a consequence of the angle averaging of the large
hydrogen bond interaction which has very strong angular dependence;
that is, the bond is strong when the hydrogen bond, or in the
model an arrow representing it, is pointed at the neighboring
molecule, and weak otherwise. The decorated lattice model
calculation first averages over all these angular contributions;
thereby it reduces the partition function to a sum over states of

different occupation. If there are only two types of occupation,
two types of molecules, this remaining sum is essentially the
simple spin 1/2 Ising model. In the model, weakening of the
average hydrogen bond produces immiscibility of the components at
lower temperature; and at higher temperature the increased importance of entropy of mixing reduces immiscibility.

The essence of the calculation is suggested by Figure 12.
This figure shows the values of a function ξ as a function of
reduced temperature. For the Ising model:

$$\xi = e^{-(2J/kT)} = e^{-(1/T_r)} \quad ,$$

which also defines the reduced temperature. Solution of the Ising
model results from evaluation of the partition function which will
be a function of ξ; from it one can identify the value of ξ, ξ_c,
of a phase transition. From the equation, this value of ξ_c
designates the transition temperature, T_c, which is scaled by the
interaction energy, J. The transition temperature is somewhat

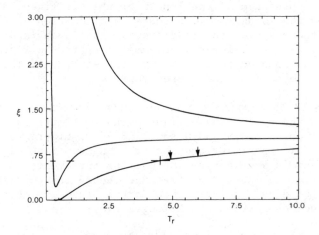

Figure 12. The reference temperature, ξ, plotted as a function of
the reduced temperature, kT/U_1. The lower curve is ξ for the
Ising model. The critical point occurs at values of ξ that differ
depending on the method of solution; ξ_c = .642 for the Ising model
discussed; ξ_c = .667 for a solution in the quasi-chemical approximation; ξ_c = .717 for a solution in the Bragg-Williams approximation.
The middle curve is ξ for a decorated lattice-gas model showing
both upper and lower critical points. A parameter ω designates
the number of distinguishable orientations available for bonding.
The well becomes steeper for larger ω, equivalent to increased
directionality in the model. The upper curve is an example of
$|U_2|>>U_1$; no phase transition occurs.[20] U_2 is the attractive
interaction; U_1, the repulsive.

different depending on the method of summing the partition
function or the nature of the approximations. While the 3-
dimensional Ising model has not been solved exactly, series
approximations provide reliable estimates of its properties. In
Figure 12, the lower curve shows the parameter ξ for the Ising
model as a function of the reduced temperature, and indicates the
value of ξ, ξ_c, at a transition. Below ξ_c, there are two phases;
and above ξ_c, there is one.

The upper and middle curves plot ξ for a decorated lattice
model. Here the two states of the lattice model correspond to the
lattice being occupied by one of two types of molecule. The sum
over the different hydrogen bonding states has been done and it is
for this reason the upper curves have a different shape from the
lower. Also, since there are two characteristic energies, for
example a hydrogen bonding energy and an energy when there is no
hydrogen bonding, these upper curves are accurately imagined to be
members of families of curves that result when the second energy
is assigned a particular value. Because the middle curve crosses
ξ_c twice, two critical points occur in the model. Also since the
temperature dependence of the length of the tie lines, again
deduced from the Ising model, appears as a function of $\xi(T)$, it is
clear from the shape of the middle curve that upper and lower
critical points are different, unsymmetrical in their temperature
dependence (as opposed to the observation earlier of the lack of
symmetry in composition and temperature).

Figures 13-15 compare experiment with predictions of a symmetric
form of this theory.[20] These figures show the fits of theory to
the closed loops for $C_{10}E_4$-H_2O and $C_{10}E_5$-H_2O; Figure 13 is clearly
the best fit of experiment and theory. On comparing Figures 13
and 15, large ω's are seen to improve the agreement; on comparing
with Figure 14, it is evident that the theory can not fit the very
sharp straight dilute solution boundary. Large ω's correspond to
greater directionality in the hydrogen bond.

Figure 16, presented earlier,[11] shows that even though the
lower consolute boundary is extraordinarily distorted, in the
vicinity of the critical point the boundary is curved and that
within the last degree of T_c the curve is nonclassical, characterized by the critical exponent $\beta \sim .36$. This also is in accord
with theory. So, while distortion of the lower consolute boundary
is the hallmark of surfactant systems, our experiments demonstrate
that in immediate proximity of the critical point the effects of
that distortion give way to the effects of critical fluctuations.

Table I is included to emphasize that the decorated lattice
model gives values of hydrogen bond energies which are of an
approximately correct magnitude, and which vary consistently with
the absolute hydrophilicity of the head group.[10,11,20]

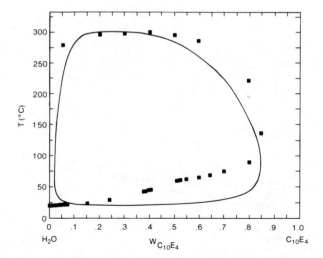

Figure 13. Closed-loop curve generated by the symmetric decorated
lattice-gas model[20] to fit the data for $C_{10}E_4-H_2O$ immiscibility.[11]
Asymmetry is generated by correction for inequivalent sizes of
surfactant and water molecules. The large angular dependence,
$\omega = 5000$, is required to provide broad loops.

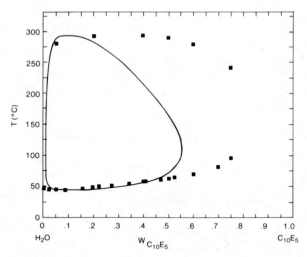

Figure 14. Closed-loop curve generated by the symmetric model[20]
to fit the data for $C_{10}E_5-H_2O$ immiscibility.[11] Large asymmetry is
used to attempt to fit the curve for dilute solutions. Curve fits
low concentrations somewhat better than curve in Figure 13 fits
that data, but is in considerable error at high concentrations.

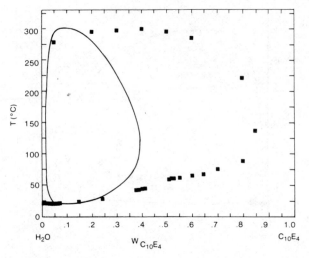

Figure 15. Closed-loop curve generated by the symmetric model[20]
to fit the data for $C_{10}E_4$-H_2O immiscibility. Compare with Figure
13. While there is some asymmetry it is comparable to that of
Figure 13. The primary reason for its poor fit is that the hydrogen
bonds are assumed less directional (ω=6); this produces a loop
which is too narrow to describe the experimental data.

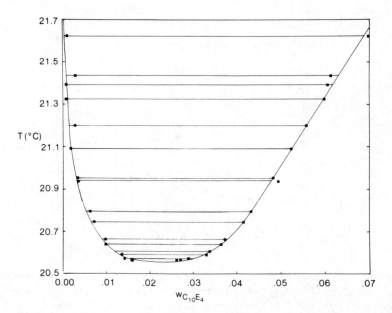

Figure 16. Phase diagram of $C_{10}E_4$-H_2O extending from 20.5 to
21.7°C emphasizing the shape of the bottom of the loop of coexistence. β is found to be .36 ± .02.[11]

Table I.   Interaction Energies from Decorated Lattice Models.[11]

	$C_{10}E_4-H_2O$	$C_{10}E_5-H_2O$	$C_{10}PO-H_2O$	$C_{12}PO-H_2O$
$U_1(6)$(kcal/mol)	+2.24	+2.27	+2.18	+2.15
$U_2(6)$(kcal/mol)	-0.358	-0.379	-0.411	-0.368
$U_1(5000)$(kcal/mol)	+1.18	+1.17	+1.02	+1.09
$U_2(5000)$(kcal/mol)	-4.07	-4.32	-4.84	-4.25

While the overall predictions of theory are in qualitative
accord with experiment, there are several points of divergence and
there remain some unanswered questions: (1) It is clear from
comparison of the fit of theory to data that more asymmetry is
required than simply a scaling of molecular volumes. This was
explicitly demonstrated in Figure 14, for $C_{10}E_5$, where the consequence of approaching the requisite asymmetry at low concentrations destroyed any possibility of good fits in the concentrated
regime. It will be interesting to learn if an asymmetric form of
the theory will be able to model the data more accurately. (2)
Purity of the materials can play a significant role in the location
of phase boundaries; we intend to investigate materials of even
higher purity than studied previously, and to investigate aqueous
mixtures of other nonionic surfactant compounds that have been
studied by other scientists. (3) In particular, the work of Corti
and Degiorgio[23] investigating mixtures of $C_{12}E_6-H_2O$ suggests that
for solutions with micelles, the exponents may not be nonclassical,
but classical. This is shown in Figure 17 in which both the
isothermal osmotic compressibility κ_T and the correlation length ξ
are plotted as functions of the reduced temperature, ϵ. The
exponents of both were approximately classical, $\gamma=.97$ and $\nu=.53$.

The question of the correct expected value of the exponents
or alternatively the range of validity for the value of an exponent
has been complicated further by our observation that even in systems
where there is no aggregation the exponent β is very unusual in
the vicinity of a lower critical point. This is illustrated in
Figure 18 which shows the exponent β plotted as a function of the
departure of temperature from the critical value. While all plots
extrapolate to the approximate Ising value of 5/16 sufficiently
near the critical temperature, the range of validity predicted by
the decorated lattice-gas model changes from circumstance to
circumstance. Figure 18 shows, firstly for contrast, β changes
almost imperceptibly within 15°C of the upper critical point.
However the range of validity of β for a lower critical point is
greatly variable depending on the directionality of the hydrogen
bonds, the discrepancy in volumes of the molecules, and presumably
the asymmetry of the bonding included in the model. Surprisingly,
within 15°C of the critical temperature β drops from the nonclassical value to about half that value when the directionality

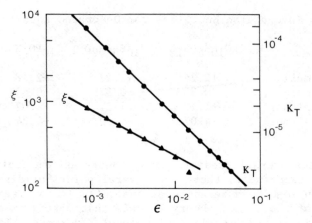

Figure 17. The isothermal osmotic compressibility κ_T and the
correlation length ξ as a function of the reduced temperature,
$\varepsilon = |T-T_c|/T_c$, for $C_{12}E_6$-H_2O mixtures in vicinity of the lower
critical point. This shows the exponent of κ_T, $\gamma = .97$; the
exponent of ξ, $\nu = .53$.[23]

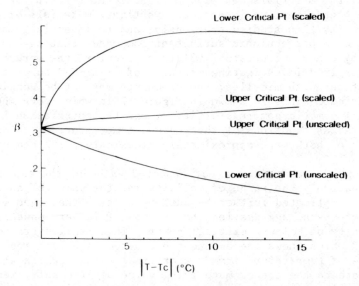

Figure 18. The critical exponent β calculated from the symmetric
decorated lattice-gas model for upper and lower critical points as
a function of deviation in temperature from the critical. The two
upper curves resulted when the coexistence curves were scaled to
account for differences in molar volumes. These emphasize the
large change in β over a small range in temperature in the
vicinity of the lower critical point. The results are approximate
for $C_{10}E_5$ and exact for $C_{10}E_4$. For all curves ω is 5000,
corresponding to highly directional hydrogen bonds.

of the bonds is chosen to give broad theoretical coexistence loops
more closely approximating experimental data. Nearly as startling,
any corrections to compensate for discrepancies in volumes of the
two types of molecule generate exponents which increase before
they decrease. This increase in β while generally not too large
could for some cases bring β close to 1/2, the classical value!
It also shows the value of the exponent can, further from T_c,
depend on the asymmetry of the loop, itself regulated by the size
of the aggregates. These recent results need further exploration
and confirmation with additional models, especially asymmetric
lattice-gas models;[20] however if they are borne out they suggest
that even in the absence of the complications of aggregation,
critical exponents in the vicinity of lower consolute points may
possess unique properties.

These observations indicate that critical exponents should be
established very close to the lower critical point, possibly within
1°C, and that the possibility of appreciable temperature dependence
of the exponents should be investigated outside this narrow regime.

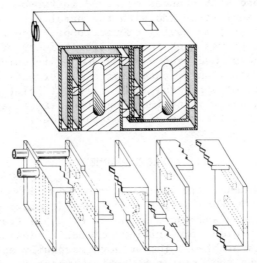

Figure 19. This figure shows two schematic cut-away drawings of
the central brass equilibrium chamber to be used in studies of
for the lower critical points in mixtures of nonionic surfactant
and water. Flow reaches this central chamber only until equilibrium is achieved approximately. The figures illustrate the path
of flow on four sides of the chamber; countercurrent heat exchange
occurs here. Flow in the outer brass enclosure (not shown) passes
through all six sides. Sealed pyrex cuvettes are inserted through
the holes in the top of the chamber and contain the gravimetrically
prepared samples.

Figure 19 shows a schematic diagram of the type of apparatus
required for investigating these questions experimentally; it is
the innermost portion of a constant temperature chamber which
should maintain long term stability of better than 1 millidegree.
Though flow to this chamber is indicated, it occurs only for a
short period of an hour or so until the temperature of this inner
chamber approximates the temperature desired. Subsequently flow
is maintained only through an outer brass enclosure and temperature
is controlled by an intermediate shield to which are mounted
heaters and sensors. With it we hope to investigate some of the
questions arising from the theory.

V. CONCLUSION

Investigation of three-component phase equilibria, and in
particular the directions of tie lines, have provided information
about changes in interactions between water and nonionic surfactant.
The complexity of phase equilibria these changes generated was
contrasted with phase equilibria for mixtures of ionic surfactants.
The source of the variation in these interactions was reviewed and
shown to correspond with changes observed in two-component water-
surfactant phase equilibria and with the parallels observed for
simpler molecular systems having closed loops of coexistence.
Theory of such loops was used to study the effects of variation in
angular dependence of hydrogen bonding and of inequivalence in
molecular size; theory anticipates large differences in the temperature dependence of upper and lower critical point exponents for
these systems. Finally, experiment and theoretical approaches
were outlined to resolve some of these questions.

VI. ACKNOWLEDGEMENTS

It is with pleasure the author acknowledges his gratitude to
both Professor J. C. Wheeler for useful and detailed discussions
of lattice-gas models and thermodynamics of phase transitions, and
Professor Benjamin Widom for critique of early ideas and suggesting
approaches to this problem. The author gratefully acknowledges
his indebtedness to the Procter & Gamble Company for its support
of this research.

VII. REFERENCES

1. P. Ekwall, in "Advances in Liquid Crystals," G. H. Brown,
   Editor, Vol. 1, p. 1, Academic Press, N.Y., 1975; L. Mandel
   and P. Ekwall, Acta Polytechnica Scandinavica, 74 (I), p. 1
   (1968); a more current phase diagram is that of R. Friman, I.
   Danielsson, and P. Stenius, J. Colloid Interface Science, 86,

    501 (1982) showing the corrected understandings of the "B" and "C" phases originally questioned by Tiddy, Lindman, and co-workers, cited in the reference.

2. H. Wennerström and B. Lindman, Physics Reports, $\underline{52}$, 1 (1979). B. Jönsson and H. Wennerström, J. Colloid Interface Sci., $\underline{80}$, 482 (1981).

3. J. D. Litster and R.J. Birgenau, Physics Today, $\underline{35}$ (5), 26 (1982); P. S. Pershan, ibid, 34; W. F. Brinkman and P. E. Cladis, ibid, 48; R. Pindak and D. Moncton, ibid, 57.

4. B. K. Das and R. B. Griffiths, J. Chem. Phys., $\underline{70}$, 5555 (1979); D. Furman, S. Dattagupta, and R. B. Griffiths, Phys. Rev. B, $\underline{15}$, 441 (1977).

5. K. E. Bennett, H. T. Davis, and L. E. Scriven, J. Phys. Chem. $\underline{86}$, 3917 (1982).

6. B. M. Knickerbocker, C. V. Pesheck, H. T. Davis and L. E. Scriven, J. Phys. Chem., $\underline{83}$, 1984 (1979); B. M. Knickerbocker, C. V. Pesheck, H. T. Davis and L. E. Scriven, J. Phys. Chem., $\underline{86}$, 393 (1982).

7. A. M. Bellocq, D. Bourbon and B. Lemanceau, J. Colloid Interface Sci., $\underline{79}$, 419 (1981). A. M. Bellocq, J. Biais, B. Clin, A. Gelot, P. Lalanne and B. Lemanceau, J. Colloid Interface, Sci., $\underline{74}$, 311 (1980); J. B. Rosenholm, T. Drakenberg and B. Lindman, J. Colloid Interface Sci., $\underline{63}$, 538 (1978).

8. J. W. McBain, L. H. Lazarus and A. V. Pitter, Z. Phys. Chem., $\underline{A147}$, 87 (1930). J. W. McBain and M. C. Field, J. Am. Chem. Soc., $\underline{55}$, 4776 (1933).

9. J. W. McBain, R. D. Vold and M. J. Vold, J. Am. Chem. Soc., $\underline{60}$, 1866 (1938).

10. R. G. Laughlin, in "Advances in Liquid Crystals," G. H. Brown, Editor, Vol. 3, p. 41, p. 99, Academic Press, N.Y., 1978.

11. J. C. Lang and R. D. Morgan, J. Chem. Phys., $\underline{73}$, 5849 (1980).

12. J. C. Lang, to be published.

13. J. W. Gibbs, "Thermodynamics, Collected Works," Vol. 1, Longmans Green and Company, N.Y., 1931.

14. F. A. H. Schreinemakers, in "Die Heterogenen Gleichgewichte vom Standpunkt der Phasenlehre," H. W. B. Roozeboom, Editor, Vol. I-III, Vieweg, Braunschweig, (1901-1911).

15. R. B. Griffiths and J. C. Wheeler, Phys. Rev. A, $\underline{2}$, 104 (1970).

16. P. G. Nilsson and B. Lindman, J. Phys. Chem., $\underline{86}$, 271 (1982).

17. J. C. Lang and R. D. Morgan, paper presented at the 181st National Meeting of the American Chemical Society, Atlanta, GA, March 29-April 3, 1981; PHYS 0069.

18. S. Friberg, I. Buraczewska and J. C. Ravey in "Micellization, Solubilization, and Microemulsions," K. L. Mittal, Editor, Vol. 2, p. 791, Plenum Press, N.Y., 1976.

19. A. I. Vorob'eva and M. Kh. Karapet'yants, Russ. J. Phys. Chem., $\underline{40}$, 1619 (1966); $\underline{41}$, 602 (1967); $\underline{41}$, 1061 (1967).

20. J. C. Wheeler, J. Chem. Phys., $\underline{62}$, 433 (1975); G. R. Andersen and J. C. Wheeler, J. Chem. Phys., $\underline{69}$, 2082 (1978); $\underline{69}$, 3403 (1978).

21. J. A. Barker and W. Fock, Disc. Faraday Soc., $\underline{15}$, 188 (1953).
22. J. Hirschfelder, D. Stevenson, and H. Eyring, J. Chem. Phys., $\underline{5}$, 896 (1937).
23. M. Corti and V. Degiorgio, Phys. Rev. Letters, $\underline{45}$, 1045 (1980).

AMPHIPHILIC AGGREGATES IN A LYOTROPIC NEMATIC PHASE

J.Charvolin, Y.Hendrikx and M.Rawiso[+]

Laboratoire de Physique des Solides, Université
Paris Sud, 91405 Orsay, France
[+]ILL, 38042 Grenoble Cedex, France

Lyotropic nematics are fluid and anisotropic phases
encountered in narrow domains of the phase diagrams
of some amphiphilic molecules-water systems.
Their nematic behavior, or the existence of an orientational order without translational order, was shown
to be related to the clustering of the amphiphilic molecules in finite anisotropic aggregates which are
either discoïdal or cylindrical. We shall study here
the aggregates in the discoidal phase of the system Na
decylsulfate-1-decanol-water.First we shall describe
the use of X-ray and neutron scattering methods for the
structural investigation of such concentrated systems
with intermediate state of disorder. Then we shall determine the characteristic dimensions of the aggregates.

INTRODUCTION

The polymorphism of lyotropic mesophases illustrates the wide range of states of organization for amphiphilic molecules in water. We shall discuss here some particular aspects related to the recently studied "nematic" phases formed by some surfactants.

Typical ordered liquid crystalline phases of amphiphiles-water systems are made of infinite aggregates of amphiphilic molecules which are arranged with some long range translational order (periodicity of the density along 1, 2 or 3 dimensions in the well known lamellar, hexagonal or cubic phases). On the other hand the totally disordered micellar phases are made of finite spherical or ellipsoïdal aggregates. In 1967 Lawson and Flautt [1] made some new phases by adding small amounts of a long chain alcohol (1-decanol) and salt (Na sulfate) to a classical water-amphiphile system (Na decyl sulfate, hereafter called SdS) close to the transition region from an ordered hexagonal phase to a disordered micellar phase. The new phases were fluid and anisotropic[1-2], some of their macroscopic properties (textures, orientation in a magnetic field) were quite similar to those of thermotropic nematics, they were therefore called "lyotropic nematics". Since then many other mixtures, forming lyotropic nematic phases, have been made and used as orientating media for NMR studies in structural chemistry[3-6]. The investigation of their macroscopic properties was marked by the discovery of biaxial phases [7], expected for a long time [8] but never encountered in thermotropic nematics.

We recently studied the structures of the nematic phases of the original system SdS-decanol-water by small angle X-ray scattering [9]. We first confirmed that they are aqueous solutions, without long range translational order but with long range orientational order, of anisotropic aggregates of amphiphilic molecules. This is a typical nematic ordering, an intermediate state of order between the ordered and disordered phases briefly described above. Then we showed that, according to the concentrations of the compounds, the aggregates may be cylindrical or discoidal and we tried to estimate their dimensions. We faced here a difficulty particular to concentrated systems of particles: as the intra and inter aggregate lengths are close, the information concerning the dimensions of the aggregates and that concerning their distances contribute to the scattered intensity in the same domain of angle and thus cannot be discriminated with only one observation[10]. To overcome this difficulty we complemented the X-ray data with neutron ones, taking advantage of the change of contrast associated with the change of radiation, as will be explained below. The comparison of the two spectra enables us to present a first-order quantitative model for the nematic phase of discoïdal aggregates.

SAMPLES

Sodium decyl sulfate was either prepared in the laboratory[11] or
was of commercial origin (Merck, p.p.a.99%), 1-decanol was commercial (Fluka, p.p.a.>99%), water was D_2O from Service des Molécules
Marquées (C.E.A. Saclay). The mixtures were prepared in sealed tubes by weighting of the appropriate quantities of the compounds and
homogenizing using ultra-sonication and centrifugation. To get
stable samples all glass material must be perfectly clean (without
fatty impurity) and furthermore the temperature during the homogeniza-
tion has to be maintained below 50°C in order to prevent the hydrolysis of SdS into decanol which displaces the sample from the nematic domain towards the lamellar one[12]. The phase diagram of the
SdS-decanol-water at room temperature was studied in the vicinity
of the nematic domain and presented in reference[12], we shall consider
here samples in the nematic domain with discoidal aggregates. Their
concentrations (percentages in weight) are: SdS 36.32, decanol 6.63,
D_2O 57.06. We also obtained similar results with a sample having the
same SdS/decanol ratio but containing a small quantity of salt:
SdS 36.4, decanol 6.56, D_2O 52.8, Na_2So_4 4.14. All our samples are
characterized by a negative anisotropy of magnetic susceptibility,
as shown by the NMR study of D_2O[13]. This means that the axes of
the aggregates, or the director $\vec{n}$ of the phase, orient   perpendicularly to the field. We have used this property to prepare samples
with uniform orientation.

EXPERIMENTAL SET-UP

X-ray diffraction patterns were obtained in the laboratory with
a photographic monochromatic Laue Camera with λ = 1.54 Å and a sample
to film distance of 80 mm. The results are analyzed as a func-
tion of the vector $\vec{s}$ of the Bragg relation, s =(2sinθ)/λ, where 2θ is
the angle between the incident and diffracted beams. With our equip-
ment the minimum value of s is $(120\ Å)^{-1}$. The microdensitometer mapping of the pattern which will be shown below was obtained from the
Service de Microdensitométrie du CNRS. The samples were in sealed
glass capillaries of diameter 1.5 mm held perpendicularly to the
X-ray beam, their temperature was 25°C, and they were oriented by the
12 kg. field of a permanent magnet mounted inside the camera.

Neutron diffraction patterns were obtained at Institut Laue
Langevin (ILL) on spectrometer D_{17} with λ = 8 Å and a sample to detector distance of 140 cm. Although neutron data are usually analyzed as function of the scattering vector $\vec{q}$ = 2π $\vec{s}$, we shall use
here the vector $\vec{s}$ to prevent confusion. The minimal value of s is
here $(400\ Å)^{-1}$. The samples were in quartz cells of thickness 2 or

1mm, their temperature was 23°C, they were oriented by a 12 kgauss magnetic field.

In both cases we shall discuss only the configuration where the magnetic field is along the equator, i.e. the director $\vec{n}$ of the phase, which is the mean direction of the axes of the aggregates, is perpendicular to it.

DIFFRACTION PATTERNS

As always in the studies of lyotropic liquid crystals the diffraction patterns are to be analyzed in two regions. In the large angle region($s \backsim (4.5 \text{ Å})^{-1}$), where the diffraction is due to the short range organization of the amphiphilic molecules, we observe a broad band characteristic of disordered paraffinic chains. This region was discussed in details in reference [9], we shall not consider it any more here. In the small angle region($s \backsim (40 \text{ Å})^{-1}$) the diffraction is due to the structure and organization of the aggregates, we shall focus our attention on this region.

X-Ray Diffraction

The densitometric mapping of the pattern obtained with an oriented sample is shown in Figure 1. The pattern is anisotropic and consists of two intense wide spots along $\vec{n}$ at $s \backsim 0.027 \text{Å}^{-1}$ joined by two weaker lateral streaks at $s \backsim 0.015 \text{ Å}^{-1}$. The position of the most intense band is close to that of the Bragg peak in ordered lyotropic liquid crystals. This shows definitely that nematic phases are assemblies of more or less defined anisotropic aggregates, here discoids, with orientational order but without long range translational order.

However, as already quoted in reference[9], the structural interpretation of this, in terms of shape and packing of the aggregates, is not straightforward. In such disordered concentrated systems, where the dimensions of the aggregates are of the same order as their distances, we cannot discriminate between the contributions from the intra and inter aggregate lengths to the scattered intensity with only one piece of information. This ambiguity is illustrated if we refer to the extreme situations of ordered and disordered materials: if the maximum at $s \backsim 0.027 \text{ Å}^{-1}$ has a position close to that of the first order Bragg peak in ordered systems, where it corresponds to the periodicity of distance between the aggregates, it is also close to that of the maximum of scattered intensity observed in disor-

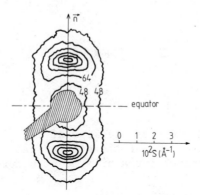

Figure 1. Small angle X-ray diffraction pattern for an oriented
nematic sample with discoïdal aggregates. $\vec{n}$ is the director of the
phase. The intensity levels are 48, 64, 96, 144, 192, 240 in arbitrary units. The black spot in the center of the pattern is the
shadow of the beam-stop.

dered diluted micellar solutions, where the shape of the band depends on the internal structure of the micelles[12]. Usually, in
structural studies of suspensions of rigid particles, the two contributions are discriminated by changing the concentration of the
particles so that, the inter particular contribution being affected
in a controlled way, models for the structure of the particle itself can be tested. This is not applicable here because the shapes
and size of the aggregates are concentration dependant. Therefore
we cannot dilute the system, we must keep the concentration constant and try to proceed the other way round varying the intra particular contribution, i.e.imagining some specific labelling of the

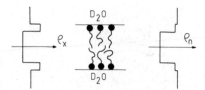

Figure 2.   Schematic representations of the electronic density ρ_x, which scatters X-rays, and of the scattering length density ρ_n, which scatters neutrons, for a bilayer structure in D_2O.

aggregate to change its electronic density, the property which scatters X-rays. We could have tried labelling the aggregate with heavy atoms introduced with a solute molecule. However, as solutes may induce structural modifications which we do not control at the moment, we preferred to work along another method. We took advantage of the change of scattering centers associated with the change of radiation. Moving from X-rays, scattered by the electronic density, to neutrons, scattered by the scattering lengths of the nuclei[14], the scattering density of the aggregate changes while its structure is kept constant: X-rays see the aggregate as an hollow shell of polar heads, because the electronic density is maximal here and at levels very close in the aqueous and paraffinic media [12], neutrons see the aggregate in D_2O as a bulky paraffinic core, because H and D scatter neutrons in very different ways. This is shown in Figure 2.

The comparison of X-ray patterns with neutron ones will therefore indicate which contribution dominates the scattering: if the patterns change drastically with the radiation the intra aggregate interferences are dominant; if not, the inter aggregate interferences are [15].

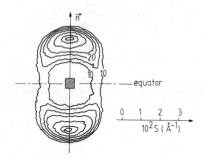

Figure 3.   Small angle neutron diffraction pattern for an oriented
nematic simple with discoidal aggregates. ñ is the director of the
phase. The intensity levels are 10, 15, 20, 22, 24, 50, 60, in
arbitrary units. The black spot in the center is the shadow of
the beam stop.

Neutron Diffraction

The intensity map provided by the X-Y detector of spectrometer
D_{17} is shown in Figure 3. The pattern looks similar to that obtained
with X-rays.   One can observe intense wide spots along $\vec{n}$ at $s{\sim}0.024\text{Å}^{-1}$
joined by two weaker lateral streaks at $s{\sim}0.016\text{Å}^{-1}$. Thus, although
the scattering density of the aggregate has changed considerably
moving from X-rays to neutrons, the pattern does not change appreciably. This indicates that the inter aggregate interferences are
dominant in determining the maxima of intensity; therefore, the latter give a rather direct access to the distance between the aggregates. (There are small changes in the positions and shapes of the
bands which suggest a weaker role of the intra term. We ignore such

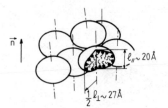

Figure 4.   Schematic representation of a nematic phase of discoïdal aggregates (SdS 36.32%,decanol 6.63%, D_2O 57.06% in weight).

corrections at the crude level of this first modelization, we shall
have to consider them when developing factor calculations.)

At this stage we must quote that similar X-ray studies are
being made by another team[16]. Their results differ very much from
ours. Particularly they see an additional band at $s \sim 0.01$ $Å^{-1}$
along the same direction than the band at $s \sim 0.025$ $Å^{-1}$, the relative
intensities of the two bands depend on the nature of the sample
cell and its thickness. We are unable to observe such effects what-
ever the configuration of the experiment is.

AGGREGATE SIZE

As the positions (s_M) of the intensity peaks are controlled
by the inter aggregate interferences we can estimate the mean
distance(d) between aggregates along directions parallel (//) to
$\vec{n}$ and perpendicular ($\perp$) to $\vec{n}$ as being given by $d = s_M^{-1}$. The volume given to an aggregate is therefore $V_{cell}=(\pi d_{//} d\perp^2)/4$ and the
volume of the aggregate is $V_a = (\pi d_{//} d\perp^2 \phi)/4$ where ϕ is the volume
fraction of amphiphilic molecules (SdS plus decanol). If we call
$\ell_{//}$ and $\ell_{\perp}$ the dimensions of the aggregate along $\vec{n}$ and perpendicular to $\vec{n}$, the   thickness and diameter, we can write:

$$4\ V_a = \pi d_{//}\ d_{\perp}^2 \phi = \pi \ell_{//}\ \ell_{\perp}^2$$

The dimensions $\ell_{//}$ and $\ell_{\perp}$ are not known independantly ;however ,a reasonable assumption consists in giving $\ell_{//}$ the value measured for the thickness of the"amphiphilic layer in an ordered lamellar phase", or $\ell_{//}$ = 20 Å, and the diameter of the discoid is $\ell_{\perp}$= 54 Å. Therefore we can propose the following representation of the nematic phase studied here (see Figure 4).

The aggregate in this phase is characterized by a rather low ratio of anisotropy $\ell_{\perp}/\ell_{//}$ = 2.7. When salt is added it increases slightly.

CONCLUDING REMARKS

At this stage the only points we can assert are that, for one point in the phase diagramm, the aggregates are discoids with small anisotropy and a rather low dispersity. Several discoidal shapes might have been thought of, oblate spheroids, ellipsoids with three axes, flat disks with rounded edge, biconcave disks, but the data we have are not sufficient to discriminate between them. Also it is not impossible that somewhere else in the phase diagramm the aggregates have a larger diameter. At last they might be fluctuating entities but the methods we have described here are not sensitive to the dynamics of the system.

REFERENCES

1.   K.D.Lawson and T.J.Flautt, J.Am.Chem.Soc.89,5490 (1967).
2.   F.B.Rosevear, J. Soc. Cosmet. Chem., 19, 581 (1968).
3.   A.Saupe, J.Colloïd Interface Sci.58, 549 (1977).
4.   K.Radley and A.Saupe, Mol.Cryst.Liq.Cryst.44, 227 (1978).
5.   N.Boden, P.M.Jackson, K.McMullen and M.C.Holmes, Chem.Phys. Lett.65, 476 (1979).
6.   B.J.Forrest and L.W.Reeves, Chemical Reviews 81, 1 (1981).
7.   L.J.Yu and A.Saupe, Phys.Rev.Lett.45, 1000 (1980).
8.   G.Toulouse, J.de Phys.Lett.38, L-67 (1977).
9.   J.Charvolin, E.T.Samulski and A.M.Levelut, J.de Phys.Lett. 40, L 587 (1979).
10. See note 12 of reference 9.
11. L.Strzelecki and C.Germain according to K.Radley, L.W.Reeves and S.J.Tracey, J. Phys. Chem., 80, 174 (1976).
12. Y.Hendrikx and J.Charvolin, J.de Phys.42, 1427 (1981).
13. J.Charvolin and Y.Hendrikx, in *Liquid Crystals of One and Two-Dimensional Order,* W. Helfrich and G. Hepke, Editors, Springer Series in Chemical Physics 11, 265 (1980).

14.   B.Jacrot, Rep.Prog.Phys.39, 911 (1976).
15.   The complete analysis will appear shortly, Y.Hendrikx, J.
      Charvolin and M.Rawiso, in preparation.
16.   L.Q.Amaral, C.Pimentel, M.R.Tavares and J.A.Vanin, J.Chem.
      Phys. 71, 2940 (1979).
      L.Q.Amaral and M.R.Tavares, Mol.Cryst.Liq.Cryst.Lett.56,203
      (1980).
      A.M.Figueiredo Neto and L.Q.Amaral, Mol.Cryst.Liq.Cryst.74,
      109 (1981).

LIQUID CRYSTALLINE STRUCTURES OCCURRING IN AQUEOUS SYSTEMS OF A TOTALLY FLUORINATED FATTY ACID AND SOME OF ITS SALTS

K. Fontell

Chemical Center, Food Technology
University of Lund
Box 740, S-220 07 Lund, Sweden

Liquid crystalline phases exist in binary fluoro-carbon surfactant/water systems above the T_c temperature of the surfactant. The internal structure is often lamellar but other structures may occur. The present report deals with X-ray diffraction studies at room temperature mainly of the lamellar phases existing in the binary system of heptadecafluorononanoic acid ($C_8F_{17}COOH$) and water, and in the corresponding systems of some of its salts, the counterion being sodium, lithium, cesium, ammonium, tetramethylammonium, dimethylammonium or diethylammonium. In the acid/water system and in the systems of the dimethyl and diethyl-ammonium salts, the behaviour of the lamellar phase is "ideal", that is, the thickness of the fluorocarbon (bi)layers and the areas per polar head group at the interfaces between polar and nonpolar regions are independent of concentration. The lamellar phases of systems where the counterion is ammonium or tetramethyl-ammonium show variations in these parameters. A few individual measurements have been performed in the liquid crystalline phases of systems where the counterion was sodium, lithium or cesium.

INTRODUCTION

Liquid crystalline structures are found in aqueous fluorocarbon surfactant systems. The phase diagrams of the binary systems of heptadecafluorononanoic acid ($C_8F_{17}COOH$)/water as well as
of some of its salts, the counterion being lithium, sodium, cesium
ammonium, tetramethylammonium, dimethylammonium or diethylammonium have been worked out at the Chemical Center, Lund[1,2]. This
report will present and discuss structural parameters estimated
from X-ray diffraction findings, especially in systems containing
extended regions of a lamellar liquid crystalline phase.

EXPERIMENTAL

The compounds and the methods of preparation of the samples
were the same as in the previous study[1,2]. The X-ray diffraction
results were obtained by film technique using either a pin-hole
camera after Kiessig[3] or a slit-collimated camera after Luzzati[4].
The X-ray radiation was copper K_α which was either nickel-fil-
tered or monochromatized with the aid of a bent quartz crystal.
Pycnometrically determined density values were used in the calculation of the structural parameters. The partial specific volume
of water was rather close to 1.00 cm^3/g, while the partial specific volumes of the different fluorocarbon surfactants were in the
range of 0.5 to 0.6 cm^3/g. (Figure 1). In the estimation of the
volume fraction of nonpolar material, additivity of volumes on
mixing was assumed and the whole fluorocarbon molecule including
the counterion was assumed to belong to the non-polar region
(the fluorocarbon (bi)layers). The experiments were performed at
room temperature, 293 K. In a few cases a higher temperature was
used.

RESULTS AND DISCUSSION

All samples with liquid crystalline structure exhibited in
the X-ray diffractograms a broad diffuse reflection having a
position corresponding to 1/(4.8-5.0) Å^{-1}, thus indicating a liquid nature of the fluorocarbon chains. (The occurrence of this
reflection parallels that obtained at 1/4.5 Å^{-1} in common liquid
crystalline amphiphilic systems. The difference in position is a
natural consequence of the replacement of the hydrocarbon chains
by fluorocarbon ones). In the low-angle region of the X-ray diffractograms one observes several sharp reflections indicating a
lamellar or hexagonal structure of the liquid crystalline phase(s).

The extent at room temperature of an homogeneous liquid
crystalline region was from 12 to 45% (w) of surfactant for the
acid/water system, from 5 to 90% (w) and from 8 to 88% (w) for the

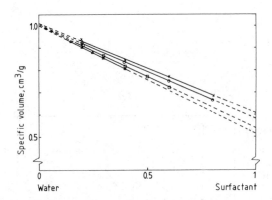

Figure 1. The dependence of the specific volume on concentration
in the lamellar liquid crystalline phases of the aqueous binary
systems of the heptadecafluorononanoic acid and its ammonium and
diethyl- and dimethylammonium salts. The partial specific volume
at 293 K of water is close to $1.00 \text{cm}^3/\text{g}$ and of $C_8F_{17}COOH$,   ,
$C_8F_{17}COONH_2(CH_3)_2$, $\bigcirc$, $C_8F_{17}COONH_2(C_2H_5)_2$, $\mathbf{x}$, and $C_8F_{17}COONH_4$,
  , (298 K) are 0.504, 0.580, 0.605 and 0.516 cm^3/g, respectively.

dimethylammonium and diethylammonium salt/water systems, respectively. The aqueous solubilities of these three compounds are low at
room temperature, < 0.2% (w), and increase very slightly with the
temperature. The thermal stability of the liquid crystalline regions is highest at high water contents[1]. The low surfactant
boundaries (± 1% (w)) of the liquid crystalline regions were determined by ocular inspection after centrifugation of the samples.
Samples from this phase region showed the mosaic texture 122.1 of
Rosevear[5,6] in the polarizing microscope thus indicating that the
structure is lamellar. Macroscopically the samples were greyish
turbid and possessed a rather loose consistency typical of liquid
crystalline phases having this structure. The conclusions were
confirmed by X-ray, as the sharp reflections in the low-angle
region were in the sequence 1:2:3:4. The values for the fundamental repeat were between 3 and 30 nm depending on the nature and
content of the surfactant (Figure 2).

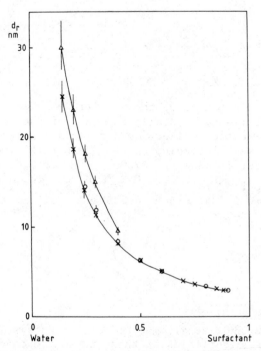

Figure 2. The dependence of the fundamental repeat, d_r, on the composition for samples from the lamellar phases in the aqueous binary systems of heptadecafluorononanoic acid and its dimethyl- and diethyl-ammonium salts, 293 K. Key: same as in Figure 1.

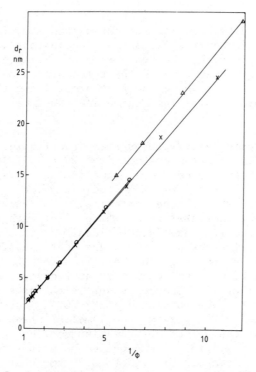

Figure 3. The fundamental repeat, d_r, plotted as function of the reciprocal volume of fluorocarbon surfactant for the lamellar phases in the aqueous binary systems of heptadecafluorononanoic acid and its dimethyl- and diethyl-ammonium salts, 293 K, Key: same as in Figure 1.

Plots of the repeat, d_r, <u>versus</u> the reciprocal volume fraction of
the surfactant showed a straight-line behaviour with slopes confirming the assumption of an "ideal" lamellar structure where the
fluorocarbon (bi)layer thickness as well as the area per polar
group, i.e. the fluorocarbon chain packing, does not vary with
concentration (Figure 3). The thickness values were about 2.6 nm
for the acid/water system and about 2.3 nm for the dimethylammonium and diethylammonium salt/water systems, respectively, while
the values for the area per polar group were 0.30, 0.41 and
0.46 nm^2, respectively (Figure 4).

In the liquid crystalline regions of the ammonium or tetramethylammonium salt/water systems, the conclusions were not as
clear cut. To begin with, the aqueous solubilities at room temperature are much higher than in the three previous systems, about
18 and 16% (w), respectively, and furthermore, the thermal stability of the liquid crystalline region increased with the surfactant content.

For the ammonium salt/water system, the liquid crystalline
region extended at room temperature from 20 to 60% (w). Microsco-
pically, the structure seems to be lamellar and this conclusion is
supported by the X-ray diffraction findings which show the existence of reflections in the sequence 1:2:3. The values for the
fundamental repeat increased from 4.5 to 10 nm with the water content. However, the structure is not "ideal" as the values for the
thickness of the fluorocarbon (bi)layers decreased from 2.4 to 1.1
nm with a concurrent increase in the values for the polar head
groups from 0.32 to 0.72 nm^2 when the water content of the phase
is increased. (Figure 5). The behaviour is the same in the system
containing the octanoate homologue ($C_7H_{15}COONH_4$) as has been shown
by Tiddy[7]. In the latter system, the lamellar phase extends from
55 to 82% (w) and the values for the fluorocarbon (bi)layer thickness decrease from 2.1 to 1.5 nm, while the values for the area
per polar group increase from 0.40 to 0.56 nm^2 when the water content is increased.

The behaviour of the tetramethylammonium salt/water system
is still more different. There seems to exist a homogeneous liquid
crystalline region at room temperature between 20 and 60% (w) surfactant. In the polarizing microscope a mosaic texture is observed.
However, the X-ray diffraction findings are ambiguous. They show
the existence of two different sets of reflections which have
different concentration dependencies. This could be taken as an
indication of the existence of two different phases but attempts
to separate them have so far been unsuccessful. The observed
spacings were 8.2 and 5.2; 6.9 and 5.0; 6.5 and 4.8; 5.9 and 4.7
nm when the lipid contents were 40, 45, 50 and 55% (w), respectively.

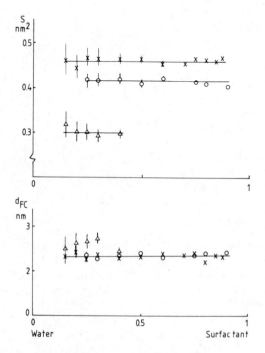

Figure 4.   The dependence of the thickness of the fluorocarbon
surfactant (bi)layers, d_{FC}, and of the area per polar head groups,
S, respectively, on the surfactant concentration in the lamellar
phases of the aqueous binary systems of heptadecafluorononanoic
acid and its dimethyl- and diethyl-ammonium salts, 293 K. Key:
same as in Figure 1.

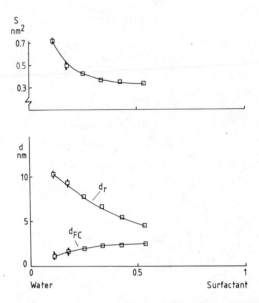

Figure 5. The dependence of the fundamental repeat, d_r, the thickness of the fluorocarbon surfactant (bi)layers, d_{FC}, and the area per polar head groups, S, on the surfactant concentration in the lamellar phase of the system ammonium heptafluoro-nonanoate/water (298 K).

Furthermore it should be noted that the behaviour of this system differs rather much from that reported for the $(C_7F_{15}COON(CH_3)_4$/water system. For this homologue system Hedge[8] et al. report the existence of a homogeneous liquid crystalline phase having a region of existence from 40 to 81% (w) of sur-factant . The structure is shown by polarizing microscopy and neutron diffraction to be lamellar with the thickness

of the fluorocarbon (bi)layers of 1.4-1.6 nm and the area per
polar head group of 0.5-0.6 nm^2.

Some isolated X-ray diffractograms were taken from systems
where the counterion was lithium, sodium or cesium. The T_c temperature curve for the sodium salt/water system lies above room
temperature, whereas for the two other systems it lies below the
room temperature. The aqueous solubilities are rather high and
the thermal stabilities of the liquid crystalline region increase
when the water contents are decreased. There seem to exist
several liquid crystalline structures. In the lithium salt/water
system a hexagonal structure occurs in addition to probably two
different lamellar ones. Within a certain temperature/concentra-
tion region the lamellar phase of the cesium salt/water system
transforms into a magnetically orientable phase of the same type
as those existing in a number of ternary aqueous amphiphile
systems which contain in addition an inorganic salt[9]. The octan-·
oate homologue has previously been shown to possess the same
property[10,11]. The sodium salt/water system becomes capable of
forming liquid crystalline structures first at temperatures above
about 325 K. No useful X-ray diffractograms have so far been obtained for this system.

CONCLUSIONS

The lamellar liquid crystalline phase occurring in the binary system of heptadecafluorononanoic acid and water, and in the
corresponding systems of the dimethylammonium and diethylammonium
salts and water, respectively, possess an "ideal" structure.
That is to say, the thickness of the fluorocarbon (bi)layers and
areas per polar head groups at the interfaces between polar and
non-polar regions are independent of concentration inside the extended regions of existence for homogeneous matter. These parameters show variations in the corresponding phase of the binary
system of the ammonium salt and water; the values for the thickness of the (bi)layers decrease and the values for the area per
polar head groups increase when the water content is increased.
The behaviour in the lamellar liquid crystalline phase of the
tetramethylammonium salt/water system at room temperature is unclear. It is possible that there exist two different
phases, which are difficult to separate.

The behaviour of the alkali salt/water systems where the
counterion is lithium, sodium or cesium is quite different.
Besides phases with lamellar structure there occur phases with
hexagonal or possibly other structures. The cesium salt/water
system possesses the capability of forming a magnetically
orientable structure.

ACKNOWLEDGEMENTS

The work has been supported by the Swedish Natural Science Council.

REFERENCES

1. K. Fontell and B. Lindman, J. Phys. Chem., In press 1983.
2. K. Fontell, in "Proc. Scandinavian Symposium on Surface
   Chemistry, 7th", Sept. 1981, Holte Denmark. K.S. Birdi,
   Editor, p. 205, Technical University, Lyngby, Denmark, 1982.
3. H. Kiessig, Kolloid.Z. Z.Polym. 98 213 (1942).
4. V. Luzzati, H. Mustacchi, A. Skoulios and F. Husson,
   Acta Cryst. 13 660 (1960).
5. F.B. Rosevear, J. Am. Oil. Chemists Soc. 31 628 (1954).
6. F.B. Rosevear, J. Soc. Cosmetic Chemists 19 581 (1968).
7. G.J.T. Tiddy, J. Chem. Soc. Faraday I, 68 608 (1972).
8. P.M. Hedge, R.K. Thomas, M. Mortimer and J.W. White,
   J. Chem. Soc. Faraday I, 76 236 (1980).
9. B.J. Forrest and L.W. Reeves, Chem. Rev. 81 1 (1981).
10. N. Boden, P.H. Jackson, K. McMullen and M.C. Holmes, Chem.
    Phys. Lett. 65 476 (1979).
11. N. Boden, K. Radley and M.C, Holmes, Mol. Phys. 42 493 (1981).

WATER ^{2}H AND ^{17}O NMR IN DODECYLAMMONIUM CHLORIDE/D_2O LYOTROPIC MESOPHASES

B. Robin-Lherbier, D. Canet, J.P. Marchal and J. Brondeau

Laboratoire de Chimie théorique
Université de Nancy I
B.P. 239, 54506 Vandoeuvre les Nancy cedex (France)

The system dodecylammonium chloride/D_2O has been investigated between 25°C and 70°C by ^{2}H and ^{17}O NMR of deuterium oxide. From the quadrupolar splittings of these nuclei, two different phases, which are macroscopically oriented by the magnetic field, have been detected : lamellar and lyotropic nematic. The latter appears only at low surfactant concentrations as an intermediate between the micellar medium and the lamellar phase. The evolution of quadrupolar splittings with concentration can be rationalized by means of a fast exchange two site model (free water and water bound to surfactant polar heads). In the lamellar phase, splittings decrease with temperature. This can be interpreted either by a tendency of water molecules to orient their symmetry axis parallel to the phase director, or by a decrease of the number of water molecules per polar head. Regardless of temperature and concentration, the ratio of ^{17}O and ^{2}H quadrupolar splittings is remarkably constant : $\Delta_O/\Delta_D = 7.13$. This can be tentatively explained by an orienting potential depending solely on the angle between the phase director and the molecular symmetry axis.

INTRODUCTION

^{2}H NMR of heavy water is one of the classical tools [1,2] for monitoring phase changes in aqueous solutions of surfactants. This is mainly because a characteristic doublet (quadrupolar splitting) appears as soon as the medium becomes anisotropic. This splitting as well as line shapes can yield information about water orientation with respect to the phase director [1-3], and also about the orientation of the directors themselves with respect to the magnetic field direction [4]. Complementary information should, in principle, be gained by studying the NMR spectrum of oxygen-17 which is the other quadrupolar nucleus of water. It is one of the purposes of this paper to explore the possibilities of these two spectroscopies on the simple dodecylammonium chloride/D_2O system : systematic measurements of both ^{2}H and [17]O quadrupolar splittings have been performed as functions of temperature and concentration. This should allow us to characterize the different mesophases of this system to which little attention has been paid [5]. In particular, the existence of a lyotropic nematic phase [6] will be demonstrated.

EXPERIMENTAL

Dodecylammonium chloride (referred as DdAC in the following) kindly supplied by CECA-SA (Paris) was washed in ether several times and subsequently dried. Solutions were prepared by weighing appropriate amounts of DdAC and D_2O (CEA Saclay, France) in a small glass container from which the solution can be transferred into a 10mm o.d. NMR tube after a convenient homogeneity has been attained by rotation at 70°C. Measurements were started at this temperature which, in all cases, corresponds to a micellar medium. Temperature is then lowered very slowly in order to accurately detect a phase change.

NMR measurements were performed at 12.2MHz and 13.8MHz for [17]O and ^{2}H respectively with the help of a home-made Fourier multinuclear spectrometer [7] built around a Bruker HX90 magnet and interfaced to a Nicolet 1180 computer. Switching from ^{2}H to [17]O (or inversely) does not necessitate any change of the probe configuration so that measurements on these two nuclei are performed under exactly the same conditions of temperature and orientation with respect to the magnetic field. Deuterium spectra were obtained with the following instrumental parameters : 8 transients accumulated in a 2K word array ; spectral window : 1000Hz ; 90° pulse duration : 18μs. Natural abundance oxygen-17 spectra free from baseline distortions were obtained according to a recently described procedure [8] with the following instrumental parameters : 200 000 transients accumulated in a 2K word array ; spectral window : 10 000Hz ; 90° pulse duration : 25μs. A typical spectrum exhibiting the quintuplet characteristic of an anisotropic medium is shown in Figure 1.

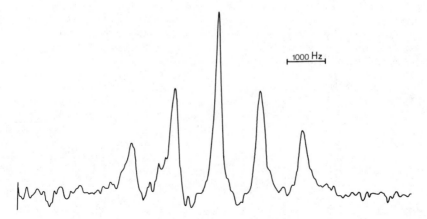

Figure 1. Typical ^{17}O NMR spectrum of D_2O [mixture of DdAC (33% in weight) and water at 305K]. The quadrupolar splitting is defined as the difference in Hz between the frequencies of two consecutive lines.

RESULTS AND DISCUSSION

Phase identification

Solutions up to 46% in weight of DdAC have been studied between 298 and 343K (beyond this concentration crystallization occurs). Lyotropic phases appear above 33% in a temperature range depending on the concentration. They are in all cases characterized by a well defined deuterium doublet whose linewidths however increase with concentration. This indicates that all directors are at the same angle with respect to the magnetic field (random orientation of domains would lead to a powder type spectrum). In other words macroscopic orientation has been induced by the magnetic field $\vec{B}_0$ due to a strong anisotropy of the magnetic susceptibility ($\Delta\chi$). It is well known that the sign of $\Delta\chi$ can be deduced from spectra obtained after rotation of the sample tube with respect to an axis perpendicular to $\vec{B}_0$. For instance if a rotation is such that all directors are set at the magic angle with respect to $\vec{B}_0$ (54.74°) the doublet transforms into a singlet. For Type I mesophase [9] ($\Delta\chi>0$) whose directors spontaneously align along $\vec{B}_0$, a singlet appears for a rotation angle of precisely 54.74°. In all cases encountered here, this does not occur but the evolution of deuterium spectra is characteristic of Type II mesophase [9] ($\Delta\chi<0$; directors perpendicular

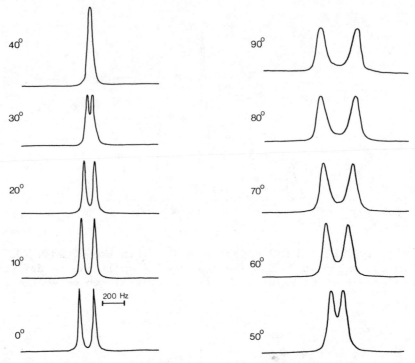

Figure 2. Evolution of the deuterium spectrum as a function of the rotation of the sample tube around an axis perpendicular to the magnetic field $\vec{B}_0$: lamellar phase whose orientation is frozen out (33% in weight of DdAC). Directors are initially perpendicular to $\vec{B}_0$ and to the rotation axis.

to $\vec{B}_0$). Indeed, in some favorable instances, a narrow singlet is observed after a rotation of ca.35° (Figure 2), indicating that directors were initially perpendicular not only to $\vec{B}_0$ but also to the sample tube axis (presumably because of wall effects [10]).

In fact two different phases, although they share the same diamagnetic susceptibility anisotropy, could be discerned. One (1), which is the first to be encountered when temperature is lowered, reorients when the sample tube is rotated i.e. a well defined doublet rapidly reappears. This phase is observed only at low concentrations. Another one (2) which does not reorient and which extends over a larger temperature range. Its orientation is therefore frozen out and remains, for very long periods (weeks), what it was when the phase formed. Such a behaviour is usual for ternary systems [9] and has been observed in very seldom instances for binary systems containing salt [10-13]. A further way of discrimination is provided by the evolution of the deuterium splitting with temperature. An example is given Figure 3: the break corresponds precisely to the formation of phase (2) and it is seen that the splittings present opposite variations for the two phases. As mentioned

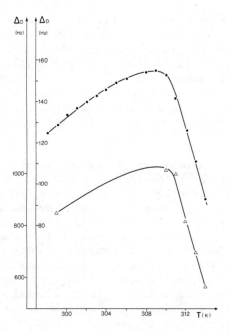

Figure 3. Quadrupolar splitting as a function of temperature
(● deuterium, Δ oxygen-17) at 33% weight of DcAC.

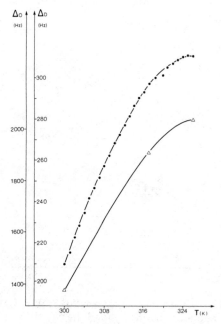

Figure 4. Quadrupolar splitting as a function of temperature
(● deuterium, Δ oxygen-17) at 46% weight of DdAC.

above phase (2) does not occur at higher concentration, this is illustrated in Figure 4. The final identification of these two phases
rests on observation by the polarizing microscope. Textures indicates[14] that phase (2) is <u>lamellar</u> i.e. possessing a long range
translational order whereas phase (1) is of the type <u>lyotropic</u>
<u>nematic</u> [9, 15] i.e. possessing only a long range orientational order.
Such a behaviour in a binary mixture has been previously reported is
only one instance by Boden et al. [16] for the system caesium per-
fluorooctanoate/water.

All these observations can be rationalized in the following
way : when temperature is lowered the system DdAC/D$_2$O goes from a
micellar medium to a lamellar lyotropic phase (bilayers of infinite
dimension). The viscosity change accompanying the formation of the
lamellar phase is responsible for its freezing out. At certain concentrations an intermediate phase occurs, it is constituted of finite size objects (presumably disc shaped) ; they prefigure the
lamellar phase which would build from their juxtaposition. This interpretation is consistent i) with the sign of the magnetic susceptibility anisotropy of each domain which reflects that of the molecule [17] and ii) with the viscosity change which is expected to be
much larger for infinite size aggregates.

A tentative phase diagram is presented in Figure 5. It is in
agreement with the one given by Broome et Al. [5] except for the
nematic lyotropic phase which they did not detect.

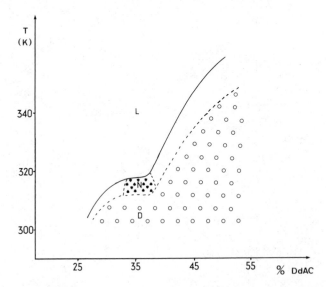

Figure 5. Phase diagram of the system DdAC/D$_2$O for the temperature
and concentration ranges investigated here. L : micellar medium ;
N : lyotropic nematic phase ; D : lamellar phase.

Evolution of quadrupolar splittings with concentration

A simple two state model involving free water, which is not oriented, and water bound to surfactant polar heads yields, by assuming that water exchanges rapidly between the two sites [1] :

$$\Delta_D = \frac{C_s}{C_w} \, n \, \Delta_{Do} \qquad (1)$$

where Δ_D is the observed deuterium quadrupolar splitting and Δ_{Do} corresponds to bound water molecules ; n is the number of molecules bound to one surfactant polar head : C_s and C_w are the surfactant and water concentrations respectively. Equation (1) can be expressed as a function of the commonly used surfactant proportion by weight wich will be denoted by r. M_w and M_s being the molar weights of water and surfactant respectively, one obtains :

$$\Delta_D = n \, \frac{M_w}{M_s} \, \Delta_{Do} \, \left(\frac{r}{1-r}\right) \qquad (2)$$

According to this model Δ_D should be a linear function of $\left(\frac{r}{1-r}\right)$ at a given temperature and for a given phase. As expected [18] this is actually observed, an example pertaining to the lamellar phase is given in Figure 6.

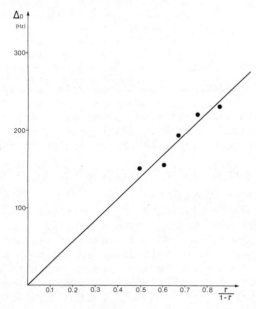

Figure 6. Representation of the deuterium quadrupolar splitting as a function of $\left(\frac{r}{1-r}\right)$, r is the surfactant proportion by weight- lamellar phase - 307K.

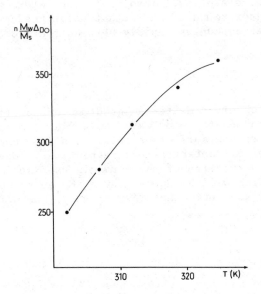

Figure 7. $n \dfrac{M_w}{M_s} \Delta_{Do}$ in the lamellar phase as a function of temperature.

Evolution of quadrupolar splittings with temperature

From Figures 3 and 4 the quadrupolar splittings are seen to increase with temperature in the lamellar phase. This temperature dependence is better represented by the evolution of the quantities $n \dfrac{M_w}{M_s} \Delta_{Do}$ (Figure 7) which are deduced from the application of (2). at different temperatures. If one assumes n to be temperature independent there is no doubt that Δ_{Do} increases with temperature indicating a priori that higher temperatures favor a "more structured medium".On the other hand the lyotropic nematic phase behaves as expected i.e. quadrupolar splittings increase when temperature is lowered. This paradox can be resolved by noticing [3] that "more structure" does not necessarily imply larger quadrupolar splittings. For this purpose let us consider three limiting cases :
1) the C_{2v} symmetry axis of the water molecule parallel to the phase director ;

2) this axis perpendicular to the phase director, the latter being in the molecular plane ; 3) the phase director perpendicular to the molecular plane. The corresponding quadrupolar splittings are in the ratio $\Delta_1 : \Delta_2 : \Delta_3 = 1 : 4.8 : 5.8$. A priori we cannot infer much from these three extreme cases since we ignore the form of the orienting potential. However it will be shown in the next section that this potential essentially depends on the angle Θ between the phase director and the symmetry axis of the water molecule. Referring to a potential of the Maier-Saupe type [19] and assuming a minimum for $\Theta = 0$ would explain our data. On the other hand, they could as well be interpreted by considering an increase of n which would result from a lateral expansion leading to an increased area per head group [20].

The opposite behaviour in the lyotropic nematic phase reflects the fact that orientation starts from zero and must necessarily build up, but indicates also that water probably orients differently with respect to the phase director than in the lamellar phase and/or n may be strongly temperature dependent.

Remarks about water orientation parameters in relation with oxygen-17 quadrupolar splittings

The conclusions reached so far rely essentially on deuterium splittings (see Figures 3 and 4). Nevertheless, the combination of both splittings, in principle, should lead to water orientation parameters or rather to four possible sets of these parameters [21, 22] (since the sign of quadrupolar splittings cannot be experimentally determined) and consequently shed some light on the water behaviour in lyotropic phases.

The analysis of water quadrupolar splittings necessitates two order parameters : $S_{xx} = \frac{1}{2} <3\cos^2\Theta_x-1>$ and $S_{zz} = \frac{1}{2} <3\cos^2\Theta_z-1>$ where Θ_x and Θ_z are the angles between the molecular frame x and z axes and the phase director (Figure 8). Analytical expressions for water oxygen and deuterium splittings in the bound state are given below [23, 24]

$$\Delta_{Oo} = -\frac{1}{40} \chi_o (3\cos^2\psi-1) \left[(3+\eta_o)S_{xx} + (3-\eta_o)S_{zz}\right] \quad (3)$$

$$\Delta_{Do} = \frac{1}{4} \chi_D (3\cos^2\psi-1) \{ \left[3\sin^2\alpha + \eta_D(1+\cos^2\alpha)\right]S_{xx}$$

$$+ \left[3\cos^2\alpha + \eta_D(1+\sin^2\alpha)\right]S_{zz} \} \quad (4)$$

ψ is the angle between the phase director and $\vec{B}_o$ and is here equal to $\pi/2$. The axis systems are defined in Figure 8. V_{ii} being a diagonal element of the field gradient tensor, one has for the quadrupolar coupling constants,

$$\chi_0 = e \ Q_0 \ (V_{yy})_0$$

$$\chi_D = e \ Q_D \ (V_{z'z'})_D$$

(Q_0 and Q_D are the oxygen and deuterium quadrupole moments), and for the asymmetry parameters,

$$\eta_0 = (V_{zz} - V_{xx})/V_{yy} \quad \text{and} \quad \eta_D = (V_{x'x'} - V_{y'y'})/V_{z'z'}$$

With appropriate numerical values [24] (χ_0 = 6.65MHz, η_0 = 0.935 ; χ_D = 222KHz, η_D = 0.11) formulae (3) and (4) yield :

$$\Delta_{0o} = 6.542 \ 10^5 S_{xx} + 3.433 \ 10^5 S_{zz} \qquad (5)$$

$$\Delta_{Do} = -1.123 \ 10^5 S_{xx} - 0.725 \ 10^5 S_{zz} \qquad (6)$$

$$\Delta_{Do} = -1.192 \ 10^5 S_{xx} - 0.657 \ 10^5 \ S_{zz} \qquad (7)$$

Equations (6) and (7) have been obtained with α = 52.2° (gas phase value) and α = 54.74° (half the tetrahedral angle), respectively.

Turning now to the experimental results, it seems better to consider the ratio $(\Delta_0/\Delta_D)_{exp.}$ in order to get rid of the factor $n \dfrac{M_w}{M_s} (\dfrac{r}{1-r})$ which is common to both splittings. It turns out that this quantity is remarkably constant for any phase, at any concentration and any temperature. The data are reported in Figure 9 ; and from these data $(\Delta_0/\Delta_D)_{exp.}$ = ± 7.13 (since the signs of the quadrupolar splittings cannot experimentally be determined). This could a priori suggest that oxygen-17 does not provide additional information as compared to deuterium in any case. In order to find out the origin of the Δ_0/Δ_D constancy, let us rewrite Equations (3) and (4) as :

$$\Delta_{0o} = a \ S_{xx} + b \ S_{zz}$$

$$\Delta_{Do} = c \ S_{xx} + d \ S_{zz}$$

This constancy can arise from

i) $\dfrac{a}{c} = \dfrac{b}{d} = \dfrac{\Delta_{0o}}{\Delta_{Do}} = \dfrac{\Delta_0}{\Delta_D}$

(linear dependence of the two equations in S_{xx} and S_{zz}) or

ii) S_{zz}/S_{xx} = constant = k ; in that case $\dfrac{\Delta_0}{\Delta_D} = \dfrac{\Delta_{0o}}{\Delta_{Do}} = \dfrac{a+bk}{c+dk}$

We first check case (i) by calculating the ratio of the S_{xx} and S_{zz} coefficients in Equations (3) and (4). They should be identical and equal to the value experimentally found for Δ_O/Δ_D. These ratio are given in table I for the selected α values.

Table I. Ratios of the Coefficients of S_{xx} and S_{zz} in Equations (5) -(7). These Ratios Correspond to Δ_O/Δ_D

α	S_{xx} coefficients	S_{zz} coefficients
52.2°	- 5.82	- 4.74
54.74°	- 5.49	- 5.23

Indeed the two ratios are almost identical for α = 54.74°. However these calculated ratios are very different from the experimental value of Δ_O/Δ_D. Therefore, unless the numerical values retained for the quadrupole coupling constants and the asymmetry parameters are very far from the ones used here, this type of explanation can be ruled out. We turn therefore to eventuality (ii). The possible relations between S_{xx} and S_{zz} are given in Table II. Two solutions are obtained for each of the α values.

Table II. Possible Relations Between S_{xx} and S_{zz} Arising from the Constancy of $(\Delta_O/\Delta_D)_{exp}$.

α	Δ_O/Δ_D = 7.13	Δ_O/Δ_D = - 7.13
52.2°	S_{xx} = - 0.60 S_{zz}	S_{xx} = - 1.18 S_{zz}
54.74°	S_{xx} = - 0.54 S_{zz}	S_{xx} = - 0.64 S_{zz}

Such two solutions would have as well been obtained by considering the curves given by Halle and Wennerström who plotted the ratio of splittings a function of the ratio of order parameters (Figure 4 cf reference 24). This was done for proton (dipolar splitting in that case), deuterium and oxygen-17. It turns out that the splitting ratio proton/deuterium is almost constant too (see experimental values quoted in Reference 24) and again two solutions for S_{xx}/S_{zz} can be obtained. One of the two possible values of the ratio S_{xx}/S_{zz} deduced from the ratio $^{17}O/^2H$ is close to one of the two possible values deduced from $^1H/^2H$ and this common solution is around S_{xx} / S_{zz} = - 0.5. Therefore, these considerations suggest that the constancy of the Δ_O/Δ_D ratio is attributable to a relation existing between the two order parameters at all temperatures and concentrations. We propose an explanation of this fact based on a peculiar property of the distribution function governing wa-

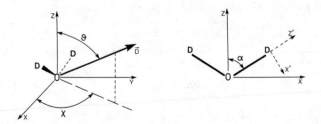

Figure 8. Definition of the axis systems and of the different
angles involved in the theoretical expressions of quadrupolar
splittings. $\vec{D}$ is the phase director.

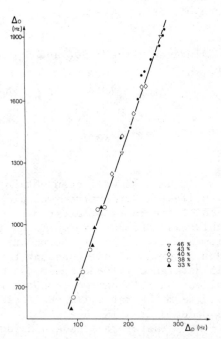

Figure 9. Oxygen-17 quadrupolar splitting vs deuterium quadrupolar
splitting for all investigated temperatures and concentrations.

ter orientation. Let us consider the expressions defining the orientation parameters (for the definition of the angles Θ and χ, see
Figure 8) :

$$S_{xx} = \frac{3}{2} \frac{\int_o^\pi \int_o^{2\pi} P(\Theta, \chi)\sin^3\Theta\cos^2\chi d\Theta d\chi}{\int_o^\pi \int_o^{2\pi} P(\Theta, \chi)\sin\Theta d\Theta d\chi} - \frac{1}{2} \qquad (8)$$

$$S_{zz} = \frac{3}{2} \frac{\int_o^\pi \int_o^{2\pi} P(\Theta, \chi)\sin\Theta\cos^2\Theta d\Theta d\chi}{\int_o^\pi \int_o^{2\pi} P(\Theta, \chi)\sin\Theta d\Theta d\chi} - \frac{1}{2} \qquad (9)$$

Let us suppose that the distribution function $P(\Theta, \chi)$ depends essentially on Θ, that is $P(\Theta, \chi) \simeq P(\Theta)$. It is a simple matter to
show that under this assumption :

$$S_{xx} = -S_{zz}/2 \qquad (10)$$

It can be seen that the relation deduced from experimental data,
for both splittings having the same sign and $\alpha = 54.74°$ is very
close to theoretical relation (10). Thus the constancy of the ratio
$(\Delta_o/\Delta_D)_{exp}$ can be interpreted on the basis of an orienting potential depending solely on the angle between the phase director and
the water molecule symmetry axis. This can be very reasonably interpreted by considering the pre-eminence of oxygen lone pairs.

ACKNOWLEDGEMENTS

     We are grateful to Dr. J. Charvolin for enlightning discus-
sions and to CNRS and DGRST for financial support.

REFERENCES

1. H. Wennerström, N.O. Persson and B. Lindman, in "Colloidal Dis-
    persions and Micellar Behavior", K.L. Mittal, Editor, p. 253,
    A.C.S. Symp. Ser. N°9, American Chemical Society, Washington
    D.C. 1975
2. G.J.T. Tiddy, in "Nuclear Magnetic Resonance", Specialist Pe-
    riodical Report, The Chemical Society, London, Vol. 4, p. 233,
    1975 ; Vol. 6, p. 207, 1977 ; Vol. 8, p. 174, 1979.
3. K. Abdolall, E.E. Burnell and M.I. Valic, Chem. Phys. Lipids,
    20, 115 (1977)
4. B. Mely, J. Charvolin and P. Keller, Chem. Phys. Lipids, 15,
    161 (1975)
5. F.K. Broome, C.W. Hoerr and H.J. Harwood, J. Am. Chem. Soc., 73,
    3350 (1951)

6. K.D. Lawson and T.J. Flautt, J. Am. Chem. Soc., $\underline{89}$, 5490 (1967)
7. J. Brondeau, B. Diter and D. Canet, Rev. Sci. Instrum., $\underline{52}$, 542 (1981)
8. D. Canet, J. Brondeau, J.P. Marchal and B. Robin-Lherbier, Org. Magn. Res., $\underline{20}$, 51 (1982)
9. F. Fujiwara, L.W. Reeves, M. Suzuki and J.A. Vanin, in "Solution Chemistry of surfactants", K.L. Mittal, Editor, Vol. 1, p. 63, Plenum Press, New York, 1977
10. F.Y. Fujiwara and L.W. Reeves, Can. J. Chem., $\underline{56}$, 2178 (1978)
11. A. Johansson and T. Drakenberg, Mol. Cryst. Liqu. Cryst., $\underline{14}$, 23 (1971)
12. F.Y. Fujiwara, L.W. Reeves, A.S. Tracey and L.A. Wilson, J. Am. Chem. Soc., $\underline{96}$, 5249 (1974)
13. W. Gauss, J. Kronenbitter, O. Lutz, A. Nolle and G.I. Siegloth, Zeit. Naturforsch., 33a, 105 (1978)
14. J. Charvolin, (1981), personal communication
15. Y. Hendrikx and J. Charvolin, J. Physique, $\underline{42}$, 1427 (1981)
16. N. Boden, P.H. Jackson, K. Mc Mullen and M.C. Holmes, Chem. Phys. Lett. $\underline{65}$, 476 (1979)
17. J. Charvolin and Y. Hendriks, in "Liquid Crystals of One-and Two-Dimensional Order", W. Helfrich and G. Heppke, Editors, p. 265, Springer-Verlag, Berlin, 1980
18. A. Khan, O. Söderman and G. Lindblom, J. Colloïd. Interface Sci., $\underline{78}$, 217 (1980)
19. W. Maier and A. Saupe, Z. Naturforsch., $\underline{14}$, 882 (1959)
20. N.O. Persson and B. Lindman, Mol. Cryst. Liq. Cryst., $\underline{38}$, 327 (1977)
21. W. Niederberger and Y. Tricot, J. Magn. Reson., $\underline{28}$, 313 (1977)
22. Y. Tricot and W. Niederberger, Biophys. Chem., $\underline{9}$, 195 (1979)
23. C.L. Khetrapal, A.C. Kunwar, A.S. Tracey and P. Diehl, in "NMR-Basic Principles and Progress", P. Diehl, E. Fluck, and R. Kosfeld, Editors, Vol. 9, Springer-Verlag, Berlin, 1975
24. B. Halle and H. Wennerström, J. Chem. Phys., $\underline{75}$, 1928 (1981)

THE INTERACTION BETWEEN WATER AND ETHYLENE OXIDE GROUPS IN OLIGO

(ETHYLENE GLYCOL) DODECYL ETHERS AS STUDIED BY ^{2}H NMR IN LIQUID

CRYSTALLINE PHASES

Tomas Klason and Ulf Henriksson

Department of Physical Chemistry
The Royal Institute of Technology
S-100 44 Stockholm 70 Sweden

Quadrupole splittings in ^{2}H NMR spectra from ^2H_2O have been determined in the binary systems ^2H_2O-$C_{12}H_{25}(O-CH_2CH_2)_xOH$ for x = 4 and 6. The average order parameter for the water molecules, as determined from the quadrupole splittings, is at a given number of water molecules per ethylene oxide group considerably higher for the surfactant with a short ethylene oxide chain. In the hexagonal phase, the order parameter decreases with increasing temperature. The temperature dependence is more complex in the lamellar phase. At low water contents the order parameter decreases with increasing temperature while at higher water contents the order parameter passes through a maximum. The results are discussed in terms of the temperature dependent repulsion between hydrated ethylene oxide groups and in relation to the stability regions for the different liquid crystalline phases.

Some lamellar phase samples containing solubilized benzene or cyclohexane have also been studied. The order parameter for the water molecules increases considerably with increasing concentration of solubilizate while the order parameter for the solubilized molecules decreases slightly.

INTRODUCTION

Knowledge about the short range forces between polar groups of nonionic surfactants is still limited. It is evident that the properties of aqueous solutions of poly (ethylene oxide) (PEO) are relevant for the understanding of aqueous systems containing surfactants with polar groups of ethylene oxide type. The interactions between water and PEO and between hydrated polymer segments in fairly dilute aqueous solutions have recently been treated theoretically by Kjellander and Florin[1] who were able to describe the thermodynamic properties of the system *e.g.* the phase separation observed at high temperature. The local concentration of ethylene oxide (EO) groups in the palisade layer of micelles and liquid crystals is, however, very high and the existing theoretical treatment can not therefore be applied directly to these systems.

NMR spectroscopy has contributed considerably to the understanding of aqueous amphiphilic systems[2]. Micellar solutions, liquid crystalline phases as well as microemulsions have been extensively studied. In this work we have measured ^2H quadrupole splittings for ^2H_2O in liquid crystalline phases in order to study the interaction between water and the nonionic polar groups in oligo (ethylene oxide) surfactants.

EXPERIMENTAL

Tetra and hexa (ethylene glycol) dodecyl ethers ($C_{12}E_4$ and $C_{12}E_6$) with uniform composition of the polar group (Nikko Co., Tokyo), ^2H_2O (Merck AG), benzene-d_6 (CIBA) and cyclohexane-d_{12} (CIBA) were used as received. The samples were prepared in sealed glass tubes and equilibrated several days before the NMR-measurements. The ^2H NMR spectra were recorded at 13.8 MHz on a Bruker CXP-100 spectrometer using the quadrupole echo method[3]. The sample temperature was controlled by a Bruker B-VT/1000 unit which was calibrated using a copper-constantan thermocouple.

RESULTS

The phase diagrams for the binary systems $C_{12}E_4-H_2O$[4] and $C_{12}E_6-H_2O$[5] are reproduced in Figure 1. Figure 2 shows the temperature dependence of the observed ^2H quadrupole splitting from samples prepared with ^2H_2O. Some samples containing solubilized benzene or cyclohexane were also studied. Figure 3 shows quadrupole splittings from ^2H_2O and order parameters for the deuterated solubilizates. In no case was it possible to detect separate NMR

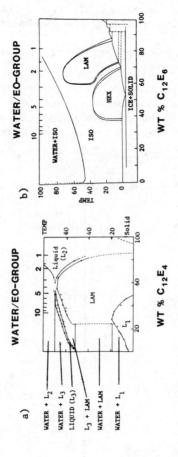

Figure 1. Phase diagrams for the binary systems $C_{12}E_4$-H_2O (a) and $C_{12}E_6$-H_2O (b) reproduced from references 4 and 5 respectively.

signals from the terminal −OD group of the surfactant indicating
that the deuterium exchange between this group and the water molecules is fast compared to the difference between the quadrupole
splittings.

THEORETICAL BACKGROUND

The ^{2}H nucleus has the spin quantum I = 1 and thus possesses
an electric quadrupole moment eQ. The quadrupole moment interacts
with the electric field gradient eq at the nucleus. The quadrupole interaction is characterized by the quadrupole coupling

constant $\dfrac{e^2Qq}{h}$, and the asymmetry parameter η. In the ^{2}H$_2$O mole-

cule the electric field gradient is mainly of intramolecular origin and the orientation of the principal axis system of the electric field gradient tensor is thus constant in a molecule fixed
coordinate system. The ^{2}H quadrupole coupling constant of super-
cooled ^{2}H$_2$O(l) is 214 kHz determined by NMR relaxation[6]. For

benzene and cyclohexane the values $\dfrac{e^2Qq}{h}$ = 187 kHz[7] and 174 kHz[8]

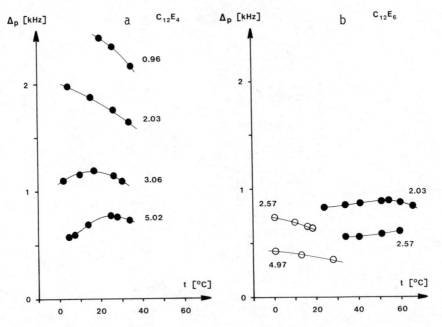

Figure 2. ^{2}H quadrupole splittings as function of temperature at
different concentrations. Lamellar phase is denoted by (•) and
hexagonal phase by (o). The numbers in the figure are the number
of water molecules per EO-group.
a) system C$_{12}$E$_4$−^{2}H$_2$O; b) system C$_{12}$E$_6$−^{2}H$_2$O.

respectively have been used. The asymmetry parameters are in all
cases small enough to be neglected in the following.

     The quadrupole interaction affects the nuclear spin energy
and for ^2H a doublet with the frequency separation

$$\Delta = \left| \frac{3}{4} \cdot \frac{e^2Qq}{h} \cdot S \cdot (3\cos^2\theta_{DL}-1) \right| = \left| \nu_Q \cdot S \cdot (3\cos^2\theta_{DL}-1) \right| \qquad (1)$$

is observed in the spectrum from liquid crystalline samples[9]. θ_{DL}
is the angle between the laboratory fixed magnetic field and the
director of the liquid crystal. S is an order parameter which for
the case of an axially symmetric electric field gradient is given
by the time average

$$S = \tfrac{1}{2} \cdot \langle 3\cos^2\theta_{MD}-1 \rangle \qquad (2)$$

where θ_{MD} is the time dependent angle between the molecule fixed
principal axis system of the electric field gradient tensor and
the director. In lamellar phases the director is perpendicular
to the lamellae and in hexagonal phases it is parallel to the
axes of the cylindrical aggregates. In polycrystalline samples, as
in this investigation, all orientations of the director with respect to the magnetic field are equally probable and contribute to

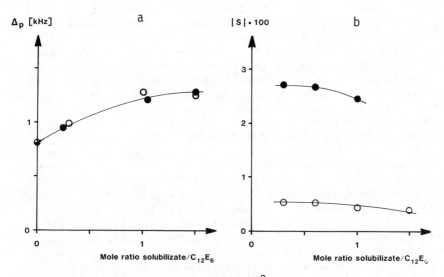

Figure 3. Quadrupole splittings from ^2H_2O (a) and order parameters
(b) for solubilized deuterated hydrocarbons in lamellar liquid crystals containing water, $C_{12}E_6$ and solubilized benzene (•) or
cyclohexane (o). The number of water molecules/EO-group = 2.03.
Temperature = 26.5°C.

the NMR spectrum which appears as a typical "powder spectrum" with two marked peaks with the frequency separation.

$$\Delta_p = |\nu_Q \cdot S| \tag{3}$$

DISCUSSION

The interaction between the polar groups of surfactant molecules plays an important role for the stability and properties of micelles and lyotropic liquid crystals[2,4,10,11]. For ionic surfactants the basic physical laws for the electrostatic interaction between the charged polar groups are well known and have recently been applied to calculate micellar size distributions[12,13] and stability regions for hexagonal and lamellar liquid crystalline phases[14]. The situation for nonionic surfactants is not that favourable. Firstly, the basic factors which determine the interaction between the hydrated polar groups are not well known. Secondly, the forces between the polar groups are rather weak which means that several different kinds of interactions might be important, while for ionic surfactants the interactions between the headgroups is dominated by the electrostatic repulsion.

The properties of aqueous nonionic surfactant systems differ in several respects from ionic systems. For example the critical micelle concentration (c.m.c.) is much lower than for ionic surfactants with the same alkyl chain length. This is a consequence of the comparatively weak repulsive forces between the polar groups. At higher temperatures nonionic surfactants in aqueous solutions exhibit phase separation ("clouding"). This phenomenon is similar to the phase separation that occurs in aqueous solutions of polyethylene oxide[15]. Kjellander and Florin[1] have shown that the phase separation in the PEO-H_2O system can be explained by a temperature dependent free energy of interaction between hydrated polymer segments. In aqueous solutions the PEO chains are surrounded by a shell of water more structured than the bulk water, i.e. with lower enthalpy and lower entropy. When two PEO-segments approach each other there is a change in free energy since some of the structured water is expelled from the overlap zone and the state of the water molecules in the overlap zone is different from that around isolated PEO-chains. The resulting interaction is repulsive at low temperatures. With increasing temperature the interaction becomes attractive due to the entropy dominance. The phase separation occurs when the attraction is strong enough to counterbalance the entropy of mixing. The clouding in micellar solutions of nonionic surfactants can be explained as a consequence of the interaction between aggregates using an interaction with the same type of temperature dependence as in the PEO-H_2O system[16-18].

The interaction between the water molecules and the EO-groups is reflected in the measured order parameters which contain information about the average molecular orientation. Several factors can in principle affect the order parameters and the detailed interpretation is therefore not straightforward. The water molecules can reside in different sites with different average orientation and since the exchange is rapid the observed quadrupole splittings is given by the average

$$\Delta_p = \left| \sum_i p_i (\nu_Q \cdot S)_i \right| \tag{4}$$

where p_i is the fraction of deuterium present at site "i" and $(\nu_Q \cdot S)_i$ is the average quadrupole interaction at site "i". It should be noted that $(\nu_Q \cdot S)_i$ can be both positive and negative. All quantities that appear in Equation (4) can change when the composition of the samples is varied. ν_Q can change if the degree of hydrogen bonding changes but this effect is probably small. The observed variation in Δ_p therefore mainly reflects changes in p_i and S_i which, however, can be difficult to separate. In the simplest model only two sites are considered: water molecules bound to the EO-groups and free water molecules tumbling without restrictions. Hence, the order parameter for the free molecules vanishes and Equation (4) is simplified to

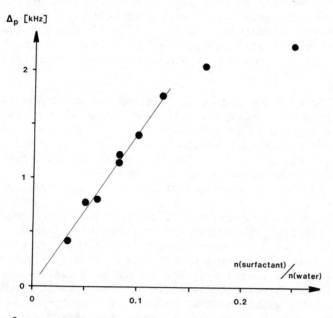

Figure 4. ^{2}H Quadrupole splittings in the system $C_{12}E_4$-^2H_2O at 26.5°C presented as function of the mole ratio surfactant/water.

$$\Delta_p = \frac{n_b}{n_w} \cdot \left| (\nu_Q S)_b \right| = n \cdot \frac{n_s}{n_w} \left| (\nu_Q S)_b \right| \tag{5}$$

where n is the number of water molecules bound to one surfactant
molecule. n_s and n_w are the number of surfactant and water mole-
cules present. If this simple model applies $n \cdot \left| (\nu_Q S)_b \right|$ can be cal-
culated from Equation (5) at each composition. Furthermore, if
$n \cdot \left| (\nu_Q S)_b \right|$ does not vary with composition a plot of the observed
quadrupole splitting *versus* the molar ratio surfactant/water will
give a straight line with zero intercept and the slope $n \cdot \left| (\nu_Q S)_b \right|$.
Figure 4 shows such a plot for the system $C_{12}E_4-{}^2H_2O$. It is seen
that a straight line through the origin can be fitted to the data
for the water rich liquid crystals. For the system $C_{12}E_4-2H_2O$ at
26.5°C the lamellar phase exists in a very wide composition range
and the linear part of the plot extends to around two water molecu-
les/EO-group. From available binary phase diagrams for nonionic
surfactants[4,5,17] it can be seen that the lowest water content for
which the lamellar phase can exist corresponds to around one
water molecule/EO-group. One interpretation of the data in Figure
4 is that most of the water will still be bound to the EO-groups
when the water content in the lamellar phase is increased to about
two water molecules/EO-group and it is only for higher water content that free water appears in the system. This is in agreement
with the observations of Kjellander and Florin that at least two
water molecules/EO-group are required to build a complete crosslinked water-PEO structure[1]. Also a study of both micellar solutions and liquid crystalline phases in the system Triton X-100-
${}^2H_2O$ indicates the presence of non-freezeable water corresponding
to about two ${}^2H_2O$/EO-group[19].

The effect of the number of EO-groups in the polar part on
the observed water quadrupole splitting can be seen by comparing
the results in Figures 2a and 2b. It is seen that at a given
number of water molecules/EO-group in the lamellar phase the observed splittings are greater for the surfactant with few EO-
groups. This indicates that the water molecules interacting with
the EO-groups closest to the alkyl chain are more ordered.

The temperature dependence of the observed water quadrupole
splittings for $C_{12}E_4-{}^2H_2O$ mixtures is shown in Figure 2a. In the
lamellar phases with low water content the splittings decrease
monotonically with increasing temperature. At higher water content the splitting increases, passes through a flat maximum and
decreases at the highest temperatures. These results indicate
that two effects with different temperature dependence might be
of importance. Higher temperature means increased amplitude of
the thermal motions of the water molecules and possibly also a
decrease in the number of bound water molecules, *i.e.* a partial

dehydration of the EO-chain. The order parameters are consequently reduced in agreement with the results at low water content.
On the other hand, the interaction between completely hydrated
EO-groups is, as mentioned above, repulsive at low temperature[1,16].
The polar groups are thus forced to be rather well separated and
the area/headgroup is large. When the temperature is increased
the interaction between the hydrated polar groups becomes less
repulsive. This is probably accompanied by a gradual change of
the conformation of the polar parts of the surfactants as the
distance between the head groups is decreased and the EO-chains
become more extended pointing out from the lamellar surface. This
will, of course, change the quantity $n \cdot |(\nu_Q S)_b|$. The number of
bound water molecules might, as pointed out above, decrease but
if $|(\nu_Q S)_b|$ increases more this can explain the temperature depen-
dence of the quadrupole splittings observed at high water content
and low temperatures.

It can be seen from Figure 2b for 2.57 ^2H_2O/EO-group that
the macroscopic structure has small effect on the observed quadrupole splitting. In hexagonal phases, the rapid diffusion around
the cylindrical aggregates causes a further averaging of the expression in Equation (2) and the absolute value of the order parameter is reduced by a factor of $\frac{1}{2}$ due to this effect upon transition
from a lamellar to a hexagonal phase if all other factors are unaffected[9]. Clearly, in the present case there is an increase in
the quantity $n \cdot |(\nu_Q S)_b|$ at the phase transition that more than
compensates for the factor $\frac{1}{2}$ due to the diffusion around the aggregates. It is not unreasonable that the dominating effect is
an increase in the number of bound water molecules/surfactant in
the hexagonal phase since the hexagonal structure is favoured when
the cross section area of the polar headgroup is large or when the
repulsion between the headgroups is strong[12,20] (vide infra). In
nonionic surfactant molecules of the type $C_n E_x$ these factors can vary
with the hydration of the EO-chains.

From Figure 2b it is also seen that in the hexagonal phase in
the system $C_{12}E_6$-^2H_2O the water quadrupole splitting decreases with
increasing temperature for the compositions studied although the water content is quite high. In cylindrical aggregates the radius of the
nonpolar core and thus also the area/headgroup is essentially determined by the hydrocarbon packing. At 22^oC the radius is approximately equal to the length of an extended hydrocarbon chain and independent of the water content[5]. This makes conformational changes in the
EO-chain of the type discussed above for the lamellar phase unlikely
to occur.

Figure 3a shows the water quadrupole splittings from lamellar
$C_{12}E_6$-^2H_2O samples with two ^2H_2O/EO-groups to which benzene or
cyclohexane has been added. It is seen that this causes a con-

siderable increase in the quadrupole splitting. For a sample with
this composition without solubilizate the quadrupole splitting increases with temperature as can be seen in Figure 2b. The analogy between these results should not, however, be pushed too far.
Several locations, possibly with different probabilities, are
available for the solubilized molecules in aqueous systems containing surfactants of this type. In addition to the hydrocarbon
core, various positions in the region containing the EO-chains,
can act as solubilization sites. The maximum solubilization capacity in micelle of $C_{12}E_x$ surfactants are different for different hydrocarbons[21]. For x = 27.6 and x = 40 the solubilization
capacity for n-hexane was ~ 0.6 moles per mole surfactant while
approximately four times as much benzene could be solubilized.
This indicates different solubilization mechanisms for the two
hydrocarbons. n-Hexane is probably solubilized mainly in the
hydrocarbon core while a considerable amount of benzene is present
in inner parts of the EO-chain layer which is in agreement with a
recent spectroscopic investigation[22]. However, the observed quadrupole splittings from ^2H_2O is found to increase both for benzene
and cyclohexane.

     Order parameters for solubilized benzene-d_6 and cyclohexane-d_{12}
are presented in Figure 3b. These results can be compared with
order parameters in the system hexadecyltrimethylammonium bromide
(CTAB)-water[23-25]. The order parameters for the solubilized
molecules relative to the normal to the surface of the aggregate are considerably smaller in the nonionic system; for benzene about one third of the value in the CTAB-system and about
one half for cyclohexane. The solubilization site for cyclohexane
is probably similar in both cases. The observed smaller order
parameter in the nonionic system therefore indicates a very
flexible microstructure of the hydrocarbon region, *i.e.* high
populations of *gauche* conformations in the alkyl chains. This is
in accordance with the large areas per headgroup reported by
Clunie *et al.*[5]. In the CTAB-system, benzene is at low concentrations preferentially localized close to the positively charged
polar groups due to its polarizable π-electron system[25]. This
imposes large restrictions on the rotational reorientation of
the molecule which are not present in the nonionic system neither
in the hydrocarbon core nor in the inner parts of the EO-chain
layer.

                          CONCLUDING REMARKS

     In the discussion above, the water quadrupole splittings have
been interpreted in terms of a temperature dependent free energy
of interaction between the hydrated EO-groups of the surfactant.
The phase diagrams of binary systems water-nonionic surfactants

show some interesting irregularities that can be seen in relation
to this discussion. Phase diagrams are available for bina-y sys-
tems containing C_8E_6[26], $C_{10}E_4$[17], $C_{10}E_5$[17], $C_{10}E_6$[26], $C_{12}E_3$[4], $C_{12}E_5$[4],
$C_{12}E_6$[5] and $C_{12}E_8$[4].

The extension of the liquid crystalline regions depends on the
interaction between EO-groups in the same aggregate and between EO-
groups in different aggregates. Some of the main features of the
phase diagrams can be qualitatively understood by assuming that
the intra-aggregate interactions dominate. The unfavourable contact
between water and the hydrocarbon chains is smaller in the lamellar
phase than in the hexagonal. Whether the lamellar or the hexagonal
phase will be the most stable depends on the repulsion between the
polar groups[12],[14]. At higher temperatures this repulsion is small
and the lamellar phase is formed since the effect of minimizing the
unfavourable contact between water and the hydrocarbon chains dominates. On the other hand. if the repulsion is sufficiently strong,
which is the case at low temperature, the hexagonal phase is formed.
It can be seen in the phase diagrams that at low temperature, the
hexagonal phase is stable in a wide composition range while the lamellar phase is stable only at low water content. This indicates
that the repulsion between not completely hydrated EO-chains is
weaker than between fully hydrated. At higher temperatures when
the repulsion between the polar groups is weaker, the lamellar
phase can be stable at higher water content. For example, it
can be seen in Figure 1b that in the composition range 1.8-2.8
water molecules/EO-group in the system $C_{12}E_6$-H_2O the lamellar
phase is stable at high temperature while the hexagonal phase is
stable at low temperature. The lamellar and the hexagonal phases
are separated by a narrow region with an isotropic cubic phase[27].

The lamellar phase existence region differs widely between
the different systems. For example, $C_{10}E_4$ has a large lamellar
region, for $C_{10}E_5$ it is considerably smaller and for $C_{10}E_6$
it is absent in the phase diagram. In the series C_8E_6, $C_{10}E_6$
and $C_{12}E_6$ it is only $C_{12}E_6$ that has a stable lamellar phase. It
seems that a necessary condition for the stability of a lamellar
structure is that the hydrocarbon chain is long enough to provide
sufficient attraction via the hydrophobic interaction to balance
the repulsion between the EO-groups.

                              SUMMARY

The water quadrupole splittings give information about the
average orientation of the water molecules and can be used to
study the interaction between water and ethylene oxide groups of
nonionic surfactants in liquid crystals. It has been possible to
rationalize qualitatively the experimental results by comparison

with aqueous polyethylene oxide solutions and by considering the temperature dependent interaction between the hydrated ethylene oxide groups.

ACKNOWLEDGEMENTS

This work has been supported by the Swedish Natural Science Research Council (NFR). The authors are indebted to J.C. Eriksson and R. Kjellander for valuable comments on the manuscript.

REFERENCES

1. R. Kjellander and E. Florin, J. Chem. Soc., Faraday Trans. I, 77, 2053 (1981).
2. H. Wennerström and B. Lindman, Phys. Reports, 52, 1 (1979).
3. J. H. Davis, K. R. Jeffrey, M. Bloom, M. I. Valic and T. P. Higgs, Chem. Phys. Letters, 42, 390 (1976).
4. G. J. T. Tiddy, Phys. Rep., 57, 1 (1980).
5. J. S. Clunie, J. F. Goodman and P. C. Symons, Trans. Faraday Soc., 65, 287 (1969).
6. E. Lang and H.-D. Lüdemann, Ber. Bunsenges. Phys. Chem., 84, 462 (1980).
7. P. Pyykkö, Ann. Univ. Turku, Ser. A 88, 93 (1966).
8. R. G. Barnes and J. W. Bloom, J. Chem. Phys. 57, 3082 (1972).
9. H. Wennerström, G. Lindblom and B. Lindman, Chem. Scripta, 6, 97 (1974).
10. B. Lindman and H. Wennerström. Topics in Current Chemistry, 87, 1 (1980).
11. R. G. Laughlin in "Advances in Liquid Crystals", G. H. Brown, Editor, Vol. 3, pp. 41-98, Academic Press, New York 1978.
12. J. N. Israelachvili, J. D. Mitchell and B. W. Ninham, J. Chem. Soc., Faraday Trans. II, 72, 1525 (1976).
13. G. Gunnarsson, Thesis, Lund 1981.
14. B. Jönsson, Thesis, Lund 1981.
15. S. Saeki, N. Kuwahara, M. Nakata and M. Kaneko, Polymer, 17, 685 (1976).
16. R. Kjellander, J. Chem. Soc. Faraday Trans. 2, 78, 2025 (1982).
17. J. C. Lang and R. D. Morgan, J. Chem. Phys., 73, 5849 (1980).
18. J. C. Lang, These proceedings.
19. K. Beyer, J. Colloid Interface Sci., 86, 73 (1982).
20. H. Wennerström, J. Colloid Interface Sci., 68, 589 (1979).
21. P. H. Elworthy, A. T. Florence and C. B. Macfarlane, "Solubilization by Surface Active Agents", p. 84, Chapman and Hall, London 1968.
22. H. Christenson and S. Friberg, J. Colloid Interface Sci., 75, 276 (1980).

23. U. Henriksson, T. Klason, L. Ödberg and J. C. Eriksson, Chem.
    Phys. Letters, 52, 554 (1977).
24. U. Henriksson and T. Klason, Finn. Chem. Letters, 139 (1982).
25. J. C. Eriksson, U. Henriksson. T. Klason and L. Ödberg, in
    "Solution Behavior of Surfactants - Theoretical and Applied
    Aspects". Vol. 2, pp. 907-920, K. L. Mittal and E. J. Fendler,
    Editors, Plenum Press, New York 1982.
26. J. S. Clunie, J. M. Corkill, J. F. Goodman, P.C. Symons and
    J. R. Tate, Trans. Faraday Soc., 63, 2839 (1967).
27. G. J. T. Tiddy, (1982), private communication.

SURFACTANT ALKYL CHAIN MOBILITY AND ORDER IN MICELLES AND MICROEMULSIONS

T. Ahlnäs, O. Söderman, H. Walderhaug and B. Lindman

Physical Chemistry 1, Chemical Center
University of Lund
S-220 07 Lund, Sweden

The NMR spin-lattice relaxation times (T_1) of ^{13}C were measured in two model surfactant systems, i.e., isotropic solutions of sodium octanoate in water (L_1) and of sodium octanoate and water in octanoic acid (L_2). From the experimental data at three magnetic fields (1.4, 2.3 and 8.5 T), information was deduced on the hydrocarbon chain mobility, order and on slow overall motions. This was achieved by applying the relaxation model of H. Wennerström et al.[1]. The measurements show clearly a frequency dependent relaxation over the entire concentration range in both systems. The fast local motion in the L_1 phase is comparable to that of the free monomer in water. In the L_2 phase, slightly slower local motions are obtained. As a general trend, the fast local motion is slowest at the headgroup and increases along the chain. The order parameter decreases along the chain towards the nonpolar end. The order parameter profile is quite invariable in the micellar solution but is found to change when water is added to the L_2 solution phase. The slow correlation time is fairly constant in L_1 but changes upon water addition in L_2. On the whole, the results emphasize the liquid character of the surfactant aggregates, i.e., characterized by low order and rapid molecular motions.

INTRODUCTION

Surfactant liquid crystalline phases are characterized by fast local motions and by long-range order. These characteristics are manifested in static effects in the NMR spectrum and in birefringence. The static effects show that the fast local motions do not average the interactions to zero, i.e. the fast local motions are anisotropic. The static effects contain useful information concerning both the packing of the amphiphile chains and the phase structure. In particular, the order parameter profiles, derived from the ^2H quadrupolar splittings in perdeuterated amphiphiles, have been used extensively to obtain information about chain packing and to test different models for amphiphilic aggregation[2]. In micellar solutions, these static effects are completely averaged out, giving rise to high resolution NMR spectra.

One nucleus often used to probe the properties of micellar solutions is ^{13}C. This is due to the large shift range which makes it possible to monitor almost every carbon along the hydrocarbon chain separately. However, the complex motional behaviour in amphiphile aggregates makes the interpretation of ^{13}C relaxation data difficult. Several theoretical approaches have been devised[3,4], but these models tend to be complicated in their application and, in particular, a comparison with results from lyotropic liquid crystals is not easy.

Wennerström et al.[1] have developed a relaxation model based on experimental information concerning the motions in micelles. This model can be used to extract from experimental data information about the dynamics and the packing of the amphiphile hydrocarbon chains in the aggregates. Furthermore, a comparison between order parameters of different phases provides structural information of particular interest for isotropic phases of unknown structure, like microemulsions.

In this work, we have applied this "two-step" relaxation model to two systems: ordinary micelles in the sodium octanoate-water system and the extensive solution phase (L_2) in the sodium octanoate-water-octanoic acid system.

EXPERIMENTAL SECTION

Samples were prepared by weighing the substances into NMR tubes. After shaking, the samples were kept at $25^{\circ}C$ for at least one week for equilibration. The chemicals used were sodium n-octanoate (99%) and octanoic acid (99%), both from BDH Chemicals Ltd.

The measurements were made at three magnetic field strengths: 1.40 T (15.1 MHz) with a Jeol FX-60 spectrometer, 2.34 T (25.2 MHz) with a Varian XL-100 spectrometer and 8.5 T (90.5 MHz) with a Nicolet spectrometer.

The relaxation times were measured using the Fast Inversion Recovery Technique (FIRT)[5]. The T_1 values were obtained by nonlinear least squares fit to Equation (1),

$$I = I_0(1 - Ae^{-\tau/T_1}) \tag{1}$$

where I_0 and I are the intensities of the NMR signal at equilibrium and at a delay time τ. A is a constant depending on the pulse repetition time and the pulse angle. For further details concerning the experimental technique, see Reference (6).

THEORY

Relaxation of carbons in alkyl chains can (in the absence of paramagnetic species) be referred entirely to the modulation of the ^1H-^{13}C dipolar coupling. The standard equation for analyzing ^{13}C relaxation is

$$\frac{1}{T_1} = \left[\frac{\mu_0 \hbar \gamma_C \gamma_H}{4\pi r_{C-H}^3}\right]^2 \cdot N \cdot \tau_c \tag{2}$$

Here, τ_c is the correlation time for the motion causing relaxation, N is the number of directly bonded protons and r_{C-H} is the carbon-proton distance (in this work, given the value of 1.09 Å). μ_0, $\hbar$ γ_C and γ_H have their usual meaning. Equation (2) assumes a single correlation time in the extreme narrowing range. If two or more fast processes modulate the same interaction or if the modulation of the interaction is nonexponential, it may be permissible to introduce an "effective" correlation time in the data analysis. However, it is immediately found that the application of equations like Equation (2) leads to inconsistencies. Applying Equation (2) to typical experimental micellar T_1: s gives correlation times well in the extreme narrowing range. Experimentally one finds T_1:s that depend on the field strength used[1]. In addition, one finds that the Nuclear Overhauser Enhancement (NOE) is lower than its maximal value[4,7]. It is evident, therefore, that a somewhat more complex relaxation model is needed.

We now proceed to examine the motions that an amphiphilic molecule is expected to undergo in a micelle. It is beyond any doubt that the hydrocarbon core of both liquid crystalline phases and micelles has a liquid-like character[8,9]. Furthermore, the local molecular motions appear to be the same in micelles and liquid crystals[10,11]. In the liquid crystalline phases, these motions are clearly anisotropic as evidenced by the quadrupolar and dipolar splittings. Therefore, the motions of the amphiphile in a micelle are also expected to be locally anisotropic. The frequency dependent T_1 relaxation times imply slow motions which average out the residual interactions. Prime candidates for these motions are micelle reorientation and/or diffusion of amphiphiles over the curved micellar surface.

A simple relaxation model should include fast local motions that are anisotropic and slower motions that are isotropic and due to aggregate tumbling and/or lateral diffusion. This is the physical basis of the model derived by Wennerström et al.[1]

For the detailed derivation of the relaxation expression, the reader is referred to Reference (1). The final expression is

$$\frac{1}{T_1} = \left[\frac{\mu_0 \hbar \gamma_C \gamma_H}{4\pi r_{C-H}^3}\right]^2 \left\{ 2(1-S^2)\tau_c^f + S^2 \frac{1}{5}\left[\frac{3\tau_c^s}{1+\omega_C^2\tau_c^{s2}} + \frac{\tau_c^s}{1+(\omega_H - \omega_C)^2\tau_c^{s2}} \right.\right. +$$

$$+ \left.\left. \frac{6\tau_c^s}{1 + (\omega_H + \omega_C)^2\tau_c^{s2}} \right]\right\} \qquad (3)$$

where τ_c^f and τ_c^s are the correlation times for the fast and the slow motions, respectively, ω_C and ω_H are the resonance frequencies in units of radians per second, and S is the local order parameter defined as the time average

$$S = \tfrac{1}{2}\langle(3 \cos^2\theta_{MD} - 1)\rangle \qquad (4)$$

Here, θ_{MD} is the angle between the local symmetry director and the C-H bond.

Equation (3) is plotted for the three different magnetic field strengths in Figure(1). Similar expressions can be derived for the NOE factor and T_2 from the theory in Reference (1).

For each carbon, there are three independent parameters, τ_c^f, τ_c^s and S, and hence three independent measurements are needed to solve Equation (3). Since T_2 is difficult to measure for ^{13}C, one is left with NOE and T_1. In this work, we have chosen to measure T_1 at three different fields.

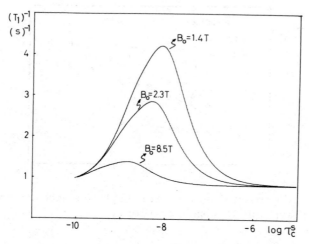

Figure 1. Equation (3) plotted for three different field strengths. The order parameter, S, is set at 0.2 and the fast correlation time, τ_c^f, at $2 \cdot 10^{-11}$s.

RESULTS AND DISCUSSION

The Micellar Solution Phase (L_1)

The relaxation times are found to be field dependent for samples above the critical micelle concentration (CMC). This manifests relaxation contributions from a slow motion and points to the presence of aggregates in the solution.

The field dependence is largest for the α-carbon i.e., the carbon next to the carboxylate group, and it decreases along the hydrocarbon chain. The relaxation rates (T_1^{-1}) for the α-carbon as a function of the amphiphile concentration and magnetic field strength are shown in Figure (2).

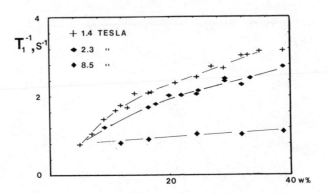

Figure 2.   Experimental values for $(T_1)^{-1}$ of the α-carbon as a function of amphiphile concentration.

Relaxation rates above the CMC contain contributions from the micellized amphiphile as well as from the free monomers.   In order to separate out the relaxation rates for the micellar mono- mers, the contribution from the free monomers has to be subtracted. An often used approximation is to assume a constant free monomer concentration (m_f) above the CMC - the pseuso-phase model.   How- ever, it is now well documented that m_f decreases above the CMC[12-14]. In a recent theoretical work, Gunnarson et al.[15] have calculated the free monomer concentration above the CMC.   If the cell model is used, a marked decrease in m_f above the CMC is found. The complicated thermodynamic equation giving the free monomer concentration can be replaced by a simple equation[16], that closely mimics the correct equation.

$$x = \frac{m_f}{m_t} = \left(\frac{m_{cmc}}{m_t}\right)^{3/2} \frac{V_w}{V_{tot}}$$
(5)

where m_f is the molal concentration of free monomer and m_t is the total concentration of amphiphile in the solution. V_w is the volume of the water molecules and V_{tot} is the total volume of both the water and amphiphile molecules.

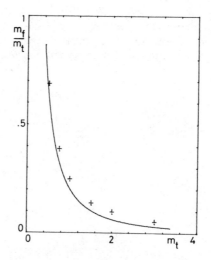

Figure 3.   Free amphiphile concentration divided by the total amphiphile concentration.   The points are experimental values from self-diffusion studies[13] on the system sodium octanoate - water.   The solid line shows Equation (5) plotted as a function of the amphiphile concentration.

    In Figure (3), results from self-diffusion measurements for the system sodium octanoate-water are compared with the theoretical values from Equation (5).

Relaxation rates for carbons in micelles are obtained from

$$R_{mic} = (R_{obs} - x R_{mon})/(1 - x) \qquad (6)$$

where $R = (T_1^{-1})$. Equation (6) assumes the well-demonstrated fast exchange situation.

Equation (3) is solved by iteration. Within the model, Equation (3) could be solved (i) for each carbon along the chain separately or (ii) by assuming that the slow correlation time is constant along the chain (which it should be); then, the total number of parameters is reduced.

In Figure (4), the fast correlation time (τ_c^f) is shown as a function of chain position and for various amphiphile concentrations. τ_c^f does not vary appreciably with concentration. The uncertainty is ±4 ps.

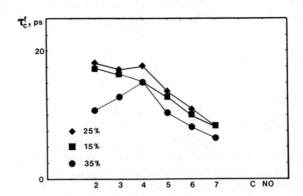

Figure 4. The deduced values for the fast correlation time (τ_c^f) as a function of position in the hydrocarbon chain.

The fast motion is slower at the polar head, indicating anchoring to the surface. Head group – head group and head group – water
interactions are mainly responsible for this reduction of the fast
motion. As the amphiphile concentration is increased, the electrostatic repulsion between the head groups will decrease which might
influence the fast correlation time. The hydration of the micelle
is expected to be roughly constant[17]. The deduced very rapid motions
are in agreement with the classical picture, the liquid-core Hartley
micelle[18].

In Figure (5), the order parameter profiles are shown for different amphiphile concentrations. The order parameter profiles
are s-shaped, with the largest values for S at the polar head.
Both the numerical values of S and the profiles are in agreement
with the predictions of Gruen's theoretical model[19]. In this model,
neither water nor head groups can exist in the micellar core. The

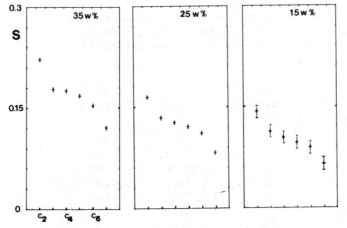

Figure 5. Order parameter profiles at three different amphiphile
concentrations.

hydrocarbon chains are assumed to be packed with the same density
as in a liquid hydrocarbon. Our findings are also in agreement
with experimental results of paramagnetically enhanced [13]C-
relaxation in this system[20,21] and other observations[22]. The order parameter value of the ω - methyl is very low (<0.15) and thus
seems not to be consistent with the Dill and Flory model.[23]
Aniansson[24] has pointed out that protrusion would cause an increase in disorder for the ω -methyl.

An important observation is that the order parameter profile
as well as the numerical values at the highest concentrations are
very closely the same as those obtained from deuterium splittings
in the hexagonal (E) phase[25]. Results are presented in Table I. In
order to compare the order parameters in the E-phase with those in
the L_1-phase, $S(L_1)$ has to be divided by a factor of 2. This follows since the local director in the L_1-phase makes an angle of
90^0 with the director of a hexagonal phase.

The molecular order is essentially equal in these two phases,
provided that the amphiphile concentration is high in the L_1 phase.
An hypothesis that might further be tested is that phases in equilibrium with each other over narrow two-phase regions have the same
order parameter values.

Table I. Order parameter values of sodium octanoate in water. A
comparison between values obtained from deuterium splittings in
the hexagonal phase and values obtained from ^{13}C relaxation in
the micellar phase.

Carbon atom number	E-phase	35 wt % L_1-phase
2	0.130	0.110
3	0.0918	0.089
4	0.0918	0.088
5	0.0861	0.084
6	0.0762	0.077
7	0.0570	0.060
8	0.0221	< 0.05

The slow correlation time (τ_c^s) is shown in Figure (6), as a function of the amphiphile concentration. The slow correlation time is essentially constant since the uncertainty is about ±30%.

If one assumes that only rotational and/or lateral diffusion are effective in averaging the residual interaction, theoretical values can easily be compared with the experimental ones. In Figure (7), values calculated from the Stokes-Einstein rotational

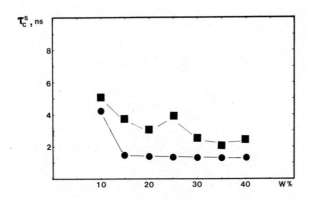

Figure 6. The deduced values for τ_c^s as a function of amphiphilic concentration. ■ indicates that T_1's for all methylene carbons have been used in the calculation. ● indicates that the T_1's for the α-carbon (No. 2) only have been used when evaluating τ_c^s.

diffusion equation and the lateral diffusion equation are plotted
against aggregate radius.

The Stokes-Einstein equation,

$$\tau_c^R = \frac{4\pi\eta R^3}{3kT} \tag{7}$$

where R is the aggregate radius and η is the viscosity of the
medium, assumes hard spheres in a continuous medium. The lateral
diffusion equation is given by

$$\tau_c^L = \frac{R^2}{6D_L} \tag{8}$$

where D_L is the lateral diffusion coefficient. For the present
system, a value of $2 \cdot 10^{-10} m^2 s^{-1}$ was used[26].

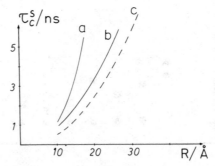

Figure 7. Correlation times (τ_c^S) calculated from rotational and/
or lateral diffusion as a function of aggregate radius. (a) denotes only rotational diffusion, (b) denotes only lateral diffusion, and (c) denotes a sum of both lateral and rotational diffusion: $\frac{1}{\tau_c^S} = \frac{1}{\tau_c^R} + \frac{1}{\tau_c^L}$.

Aggregate radii of 10-15 Å are obtained from the experimental
values if only rotational diffusion is assumed. However, for correlation times in the vicinity of 1 ns, lateral and rotational diffusion give about the same results. Therefore, it is difficult to
separate these motions from each other. Experimental results
from neutron scattering[27], classical light scattering and quasi
elastic light scattering[28] show that the aggregation number and the
radius increase when the amphiphile concentration is increased.
We could not detect this in the slow correlation time, perhaps
due to the high uncertainty, but both the fast correlation time and
the order parameter are concentration dependent.

Besides lateral and rotational diffusion, other mechanisms
might influence the slow correlation time. For short chain amphiphiles, monomer exchange[14,29] and protrusion[24] could affect τ_c^s,
especially near the CMC.

Large deviations from the assumed spherical aggregate to rod-
like form would change the relaxation behaviour completely and can
be excluded. Since S and τ_c^f are shown[6] to be quite insensitive to
moderate changes in τ_c^s, we believe that our approximation is reasonable.

The Water-Poor Solution Phase (L_2)

T_1 relaxation times were measured at three different magnetic
field strengths for the L_2 solution phase of the microemulsion model system, sodium octanoate - octanoic acid - heavy water. The
molar ratio between octanoic acid and sodium octanoate was kept
constant at 2.65 and the amount of water was varied. The phase diagram, determined by Ekwall et al.[30] is shown in Figure (8).

Experimental relaxation rates of the α-carbon are shown in
Figure (9). A clear frequency and composition dependence is
seen. This suggests that there are aggregates in the solution and
that their nature changes with the water content. The relaxation
rates were used in further calculations without corrections for contributions from free monomers, which can be assumed to be small.
Solving Equation (3) by iteration, we obtained values for the fast
and slow correlation times and the order parameter. The fast correlation times are shown as a function of water content in Figure
10.

The motion is, as in the micellar L_1 solution phase, slowest
at the polar head, demonstrating the anchoring of the polar head.
The fast motion is slower than for monomeric sodium octanoate in
water and slower than for micellar octanoate in the L_1 solution
phase. The fast correlation time profile is steeper than in the
L_1 solution phase. This could be interpreted as weaker chain-

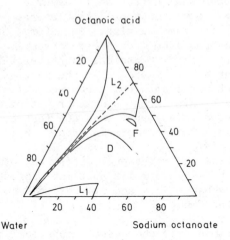

Figure 8. The three-component phase diagram sodium octanoate-octanoic acid-water (redrawn from Reference 30). The composition of the samples studied follows the dotted line.

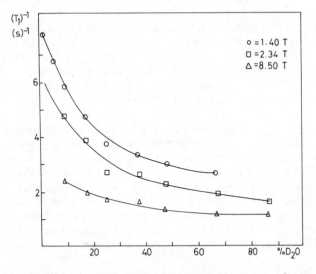

Figure 9. Relaxation rates for the α-carbon as a function of heavy water content (weight percent) at three different magnetic fields. The molar ratio between octanoic acid and sodium octanoate was kept constant at 2.65.

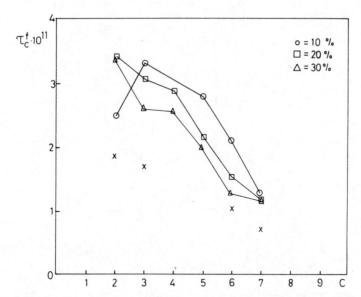

Figure 10. The fast correlation time at four different water contents. X indicates the correlation time for monomeric sodium octanoate in water.

chain interactions. The high values of τ_c^f close to the polar head
observed at low water contents, indicate a marked motional restriction for the head-group.

At low water contents, a nearly exponential decay of the order'
parameter is observed, see Figure 11. The absence of a plateau, observed in the lamellar phases[31] would, along with the reasoning of
Marcelja[32] and Levine[3], suggest that steric repulsions between
chains are relatively insignificant. Thus the chains should behave
roughly as isolated chains. No odd-even alternation in the order
parameter[33,34] is observed, so the molecular axis should be the
axis of symmetry. At higher water contents, a reduction of the order and a more plateau-like profile are obtained.

We now proceed to compare and discuss order parameter values
and profiles in different phases. At high water contents, the L_2
phase is in equilibrium with the lamellar D phase. The order

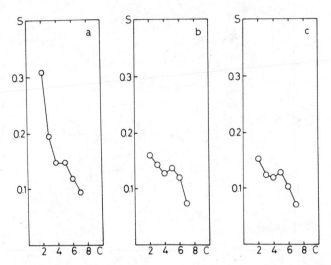

Figure 11. The order parameter profiles at three different
water contents: (a) 10 wt%, (b) 30 wt% and (c) 50 wt% in the L_2
solution phase.

parameter values in this part of the L_2 phase are much lower than
those found in the lamellar D phase of the system sodium octanoate - decanol - water[31]. On the other hand, unpublished results
for α-deuterated sodium octanoate in the lamellar phase of the
system sodium octanoate - octanoic acid - water[35] agree very well
with the values found in the L_2 phase. This is shown in Figure (12).
Decanol seems to have a different effect on the octanoic chain
order than octanoic acid. The alcohol enhances, while the
acid reduces or maintains, the chain order. This has important implications for understanding phase stability. As argued above, if
the tie-lines are short between two phases, one expects to find
similar order parameter values and profiles of adjacent samples.

The slow correlation time, τ_c^s, is shown as a function of
water content in Figure 13. As the water content is increased,
the motion is slowed down. The slow motion is assumed to consist
of an overall tumbling of the aggregate and/or the diffusion of the

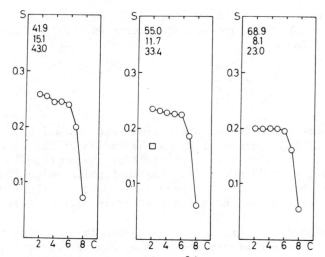

Figure 12.   Order parameter profiles[31] of octanoate at three different concentrations in the lamellar (D) phase of the system sodium
octanoate-decanol-water. □ indicates the corresponding value of the
α-carbon[35] in the system sodium octanoate-octanoic acid-water.   The
composition is given in wt% in the order water, sodium octanoate
and decanol or octanoic acid.

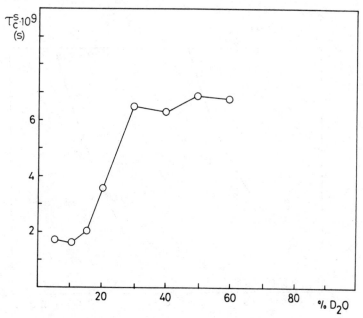

Figure 13.   The slow correlation time as a function of the heavy
water concentration.

amphiphile over the curved surface of the aggregates. If the aggregates are spherical and noninteracting and the motions are uncorrelated, one has

$$\frac{1}{\tau_c^s} = \frac{3kT}{4\pi\eta R_A^3} + \frac{6D_L}{R_D^2} \qquad (9)$$

where η is the viscosity of the medium taken as 4.7 cP[36]. R_A is the radius of the aggregate, R_D is the radius of the water core and D is the lateral diffusion coefficient of the amphiphile molecule taken to be $2 \cdot 10^{-10}$ m^2s^{-1} [26]. The amphiphile molecular length is taken to be 13.5 Å. The theoretical value of the slow correlation time as a function of R_D is plotted in Figure (14).

Comparing the theoretical values with the experimental ones, one finds that the contribution from aggregate reorientation is

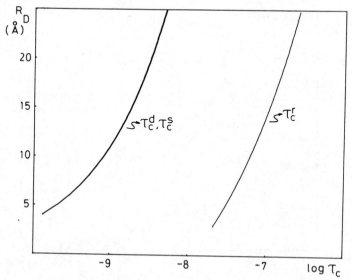

Figure 14. The theoretical value of the radius R_D as a function of the slow correlation time.

negligible. The τ_c^S values indicate a growth of the water core for
concentrations up to 40% water where the radius should be about 30Å.
At high water contents, no change in τ_c^S is seen. At very low water
contents, τ_c^S corresponds to an aggregate radius of an extended octanoate chain. This gives further evidence for the existence of small
complexes at the low water concentrations[37,38]. Hydrogen bonds are
argued to stabilize these complexes. This is also supported by the
high order parameter values of the α-carbon found at these concentrations.

It has been suggested[38] that reversed micelles exist in the concentration range 20-40% water and that a transition to normal micelles occurs above this concentration. Our results support a decrease of the aggregate curvature up to 40% water. Above this concentration, nothing dramatic happens when water is added.

Self-diffusion studies[39] show a continuous increase in the diffusion of all components upon the addition of water. This agrees
well with the decrease of the order parameter. The Cl^- coions were
found to diffuse slower than the Na^+ counterions at water concentrations below 40wt %, a further indication that reversed micelles are
present at these concentrations.

The following picture emerges: At low water contents (<10 wt %)
rather well-defined complexes exist; in an intermediate region
(10-40 wt %), there are aggregates that grow continuously with added water; and finally, there is a levelling off of τ_c^S due either
to the fact that the aggregates stop growing or that a change occurs in the aggregate geometry.

CONCLUSIONS

The present study shows that new insight into the structure
and dynamics of micellar solutions and microemulsions can be ob-
tined from multi-field ^{13}C NMR relaxation studies.

The analysis of the ^{13}C relaxation times was performed by means
of a two-step motional model, which seems to be adequate. This is
supported by the close agreement in order parameter values between
the L_1 micellar phase and the E hexagonal liquid crystalline phase.
Moreover, theoretical predictions of the Gruen model[19] are in agreement with our results in the micellar phase.

For the water-poor solution phase (L_2) of the system sodium
octanoate-octanoic acid-water, order parameter profiles are different from other phases that are not in equilibrium with the L_2
phase. It is demonstrated that a comparison of order parameter

values and profiles of different phases gives insights into structural features of microemulsions.

REFERENCES

1. H. Wennerström , B. Lindman, O. Söderman, T. Drakenberg and J.B. Rosenholm, J. Am. Chem. Soc., 101, 6860 (1979).
2. J. Seelig, Quart. Rev. Biophys., 10, 3 (1979).
3. Y.K. Levine, B. Partington and G.C.K. Roberts, Mol. Phys., 25, 497 (1973).
4. D. Canet, J. Brondeau, H. Nery and J.P. Marchal, Chem. Phys. Lett., 72, 184 (1980).
5. D. Canet, G.C. Levy and I.R. Peat, J. Magn. Res., 18, 199 (1975).
6. T. Ahlnäs, O. Söderman, C. Hjelm and B. Lindman, J. Phys. Chem., in press.
7. H. Walderhaug and O. Söderman, (to be published).
8. V. Luzzati, H. Mustacchi, A. Skoulios and F. Husson, Acta Cryst., 13, 660 (1969).
9. L. Benjamin, J. Phys., 70, 3790 (1966).
10. J. Ulmius and H. Wennerström, J. Magn. Res., 28, 309 (1977).
11. M. Schara, F. Pasnik and M. Sentjurc, Croat. Chem. Acta., 48, 147 (1976).
12. B. Lindman, C. Puyal, N. Kamenka, B. Brun and G. Gunnarson, J. Phys. Chem., 86, 1702 (1982).
13. B. Lindman and N. Kamenka, (to be published).
14. E.A.G. Aniansson, S.N. Wall, M. Almgren, H. Hoffmann,I. Kielmann, W. Ulbricht, R. Zana, J. Lang and C. Tondre, J. Phys. Chem., 80, 905 (1976).
15. G. Gunnarsson, B. Jönsson and H. Wennerström, J. Phys. Chem., 84, 3114 (1980).
16. B. Jönsson, personal communication, 1982 .
17. B. Halle and G. Carlström, J. Phys. Chem., 85, 3142 (1981).
18. G.S. Hartley, "The Aqueous Solution of Paraffin Chain Salts", Hermann, Paris, 1936.
19. D.W.R. Gruen, J. Colloid Interface Sci., 84, 281 (1981).
20. T. Zemb and C. Chachaty, Chem. Phys. Lett., 88, 68 (1982).
21. T. Zemb and C. Chachaty, (these proceedings).
22. B. Cabane, J. Physique, 42, 847 (1981).
23. K.A. Dill and P.J. Flory, Proc. Natl. Acad. Sci. U.S.A., 78, 676 (1981).
24. G.E.A. Aniansson, J. Phys. Chem., 82, 2805 (1978).
25. U. Henriksson, L. Ödberg and J.C. Eriksson, Mol. Cryst. Liq. Cryst., 30, 73 (1975).
26. G. Lindblom and H. Wennerström, Biophys. Chem., 6, 167 (1977).
27. J.B. Hayter and T. Zemb, personal communication.
28. M. Drifford, T. Zemb, M. Hayoun and A. Jehanno,(these proceedings).
29. G.D.J. Phillies, J. Phys. Chem., 85, 3540 (1981).
30. P. Ekwall and L. Mandell, Kolloid-Z.u.Z. Polymere, 233, 938 (1969).

31. T. Klason and U. Henriksson, "Solution behaviour of surfactants:
     Theoretical and Applied Aspects", K.L. Mittal and E.J. Fendler,
     Editors, Vol. 1, pp. 417-429, Plenum Press, New York, 1982.
32. S. Marcelja, Biochim. Biophys. Acta, 367, 165 (1974).
33. S. Marcelja, J. Chem. Phys., 60, 3599 (1977).
34. J.H. Davis and K.R. Jeffrey, Chem. Phys. Lipids, 20, 87 (1977).
35. L.B.-Å. Johansson, O. Söderman, K. Fontell and G. Lindblom,
     (to be published).
36. P. Ekwall and P. Solyom, Kolloid-Z.u.Z. Polymere, 233, 945
     (1969).
37. S. Friberg, L. Mandell and P. Ekwall, Kolloid Z.u.Z. Polymere,
     233, 945 (1969).
38. P. Ekwall, J. Colloid Interface Sci., 29, 16 (1968).
39. H. Fabre, N. Kamenka and B. Lindman, J. Phys. Chem., 85, 3493
     (1981).

THERMODYNAMICS OF PARTIALLY MISCIBLE MICELLES AND LIQUID CRYSTALS

R.F. Kamrath and E.I. Franses

School of Chemical Engineering
Purdue University
West Lafayette, Indiana 47907, U.S.A.

We have used the thermodynamic models we developed earlier to describe the mixing in aqueous solution of binary nonionic/nonionic and ionic/ionic surfactants with the same polar group and counterion. We have also derived general conditions for demixing of mixed micelles for the Redlich-Kister mixing model with one, two, and three parameters, mainly for the pseudo-phase separation model of mixed micellization. These models have been used to construct diagrams of total surfactant concentration versus the mole fraction of one surfactant in the mixture. These diagrams depict regions where one or two types of micellar pseudo-phases are present, as in mixtures of hydrocarbon surfactants with fluorocarbon surfactants. With the three-parameter model, there can be up to two two-phase regions, which were observed by Yedgar et al. (1974) for mixtures of Triton X-100 with a sphingolipid. The diagrams also depict lines of first mixed cmc's, second mixed cmc's, and of critical demicellization concentrations. We have used the same models when a mixed micellar pseudo-phase is at equilibrium with a mixed liquid crystalline phase, as in mixtures of nonionic surfactants with phospholipids and in sodium dodecylsulfate with sodium 4 - (1' -heptylnonyl) benzenesulfonate.

INTRODUCTION

Most surfactant systems in practice are mixed. Hence, understanding of how surfactants interact in mixed micelles and mixed liquid crystals is essential for the many industrial, biological, and
house-hold applications of surfactants. Inasmuch as surfactant
micelles in aqueous solution have a fluid hydrocarbon (or fluorocarbon) core, [1,2] they should be fully or partially miscible. The
same applies to lamellar lyotropic liquid crystals of hydrated surfactants above their "gel--liquid crystal" transition temperature.[2]

Nonideality of mixing in mixed micelles is quite common.
Negative deviations from ideality have been reported and modeled.[3-5]
Substantial positive deviations from ideality have been reported
mainly for mixtures of hydrocarbon surfactants with fluorocarbon
surfactants.[6-10] Based on comparisons between measured cmc's and
cmc's calculated via a micellization model, several authors have
suggested that two types of mixed micelles, hydrocarbon-rich and
fluorocarbon-rich, can coexist.[7-9] Evidence for two types of coexisting mixed micelles has also been reported for mixtures of
Triton X-100 (a nonionic surfactant) with sphingomyelin[11] and for
bile salt with lecithin.[12]

Mixed micelles can coexist at equilibrium with mixed liquid
crystals.[13-16] In these systems there must be substantial positive
deviations from ideality (see Theory section). We argue that the
thermodynamic models of partially miscible mixed micelles apply
also to a mixed liquid crystalline phase. These models also can describe component segregation in multicomponent phospholipid membranes.[17]

THEORY

Review of Modeling

Since in previous papers we have presented general models
for the thermodynamics of micellization of binary surfactants,
we will only review them here briefly. We have examined the following cases: (i) nonionic/nonionic with the pseudo-phase separation
model (PSM);[18] (ii) ionic/ionic PSM;[18] (iii) ionic/ionic and nonionic/
nonionic with the mass action model (MAM).[19] By PSM is meant that
the micelle is approximated as a separate phase. In reality, however, micelle sizes are distributed and finite and can depend on
salt concentration (for ionic surfactants). In the MAM, all micelles
are taken to have a fixed aggregation number N, which is finite,
concentration-independent, and for ionic surfactants independent of
the concentration of salts with common ion. Such approximations
and assumptions are often used to simplify the mathematical models
of micellization while retaining much of the physics. As shown

in Ref. 19, there is little difference between the predictions of the MAM with N > 50 and the PSM. Hence, here we will focus on the predictions of ~the PSM. The PSM is valid when the micelles are approximated as a 'phase', or two 'phases' in the case of demixing. Hence, the model is valid when one (or both) of the dispersed 'phases' in equilibrium with monomers in solution is a true thermodynamic phase such as a lamellar liquid crystal.

In the PSM model, we consider the case of two ionic surfactants.

$$x \ R_1^- + (1 - x) \ R_2^- + \beta M^+ \rightleftharpoons (R_1 R_2 M)_{x,\beta}^{-(1-\beta)} \qquad (1)$$

R_1^- and R_2^- are the surfactant monomers of components 1 and 2 and $(R_1 R_2 M)$ is the mixed micelle, with x being the mole fraction of component 1 in the mixed micelle and β being the fraction of counterion association ($0 < \beta < 1$). When two types of micelles coexist, we use superscripts A and B. For nonionic/nonionic surfactant mixtures, the monomers are denoted by R_1 and R_2 and of course the micelle $(R_1 R_2)_x$ is uncharged and $\beta=0$. In our models we use the usual component mass balances and the minimization of the Gibbs free energy of the surfactants/water system. In the PSM, the latter condition surely leads to the chemical potential of each component being equal in the bulk solution and in the micelles. This condition leads to the following equation for the monomer concentration c_i (i = 1 or 2) in terms of the pure cmc c_i^*, the counterion concentration c_M^+, and the activity coefficient γ_i in the micelle:

$$c_i = \gamma_i \ x_i c_i^* \left(\frac{c_i^*}{c_M^+} \right)^\beta \qquad (2)$$

We have taken the activity coefficients of monomers in the bulk solution to be 1.[18,19] By analogy to bulk solution thermodynamics, γ_i's are calculated from the excess free energy of mixing in the micelles or equivalently by the dimensionless function w(x).

$$g_E(x) \equiv g(x) - \{xg_1 + (1-x)g_2 + RT[x\ln x + (1-x)\ln(1-x)]\} \qquad (3)$$

$$RT \ x(1-x)w(x) \equiv g_E(x) \qquad (4)$$

We will use a Redlich-Kister expansion, which can formally describe all cases of nonideality.[20]

$$w(x) = A + B \ (2x-1) + C \ (2x-1)^2 + \ . \ . \ . \qquad (5)$$

If w(x) and the parameters A, B, C, etc. are given direct physical significance, then with this model there is an excess enthalpy of mixing and ideal entropy of mixing. If A=B=C...=0 then we have ideally mixed micelles, which by the entropy argument

are miscible at all proportions. If $B = C = \ldots = 0$ and $A \neq 0$, we
have the strictly regular solution model (RSM); A is the same quantity as w_o of our previous papers and the parameter β in Rubingh's
paper.[4] This solution model is symmetric in the components 1 and 2.
Asymmetric mixtures are handled by $B \neq 0$. We have made derivations
with up to three-parameter and calculations with up to two-parameter
models. Few mixtures can probably conform to the one-parameter model,
as we shall see.

Micelle Demixing

A binary mixed micelle of mole fraction x is unstable with respect to diffusional demixing if[21]

$$\left(\frac{\partial^2 g}{\partial x^2}\right) < 0 \tag{6}$$

By solving Equation (7) we can determine the limits of unstable
compositions w.r. to diffusion by this criterion.

$$\left(\frac{\partial^2 g}{\partial x^2}\right)_{T,p} = \frac{RT}{x(1-x)} + RT \frac{\partial^2}{\partial x^2}\left[x(1-x)w(x)\right] = 0 \tag{7}$$

These limits we call spinodals. The x-range(s) in which Condition
(6) is satisfied is termed unstable range. Outside this range,
the solution may still not be in a stable thermodynamic equilibrium
state. This state corresponds to the minimum free energy and hence
to the equality of chemical potentials of the components in the coexisting pseudo-phases. The compositions of the latter are called
binodals. They are found by solving the following equations for
x^A and x^B. (see Equation (2)):

$$\gamma_1^A\bigg|_{x^A} x^A = \gamma_1^B\bigg|_{x^B} x^B \tag{8}$$

$$\gamma_2^A\bigg|_{x^A} (1-x^A) = \gamma_2^B\bigg|_{x^B} (1-x^B) \tag{9}$$

We show here certain geometrical criteria for existence of
solutions $(0<x<1)$ to Equation (7).[22] For $w(x)=0$, $A=B=C=0$, this
equation has no solution. For $A \neq 0$, it has two solutions (the spinodals) if $A > 2$ and one at $A=2$ $(x=0.5)$, Figure 1a. Between these
two solutions the system is unstable. For $B \neq 0$, there can be two
solutions if the straight line c, Figure 1a, intersects the curve
$1/x(1-x)$. For given B, there is an intersection if A is above a
minimum value. For $C \neq 0$, there can be up to four solutions (spinodals), which are the intersections of a parabola with the previous

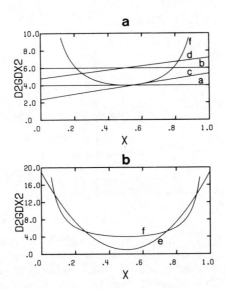

Figure 1. Graphical representation of the solution of Equation
(7). Curve f is $1/x(1-x)$. Curves a, b, c, d, and e are
$\partial^2[x(1-x)w(x)]/\partial x^2$ for the following respective sets of values of
A, B, C (Equation (5)): (2, 0, 0), (3, 0, 0), (1.932, 0.2496, 0),
(3, 0.2, 0), and (2, 0, 1.5).

$1/x(1-x)$ curve.[22] Certain calculated values of A, B, and C for
solutions to exist are given in Refs. 22 and 23. The case of four
solutions is relevant for the mixture of aqueous Triton X-100 with
a sphingolipid.[11]

In Figure 2 we show spinodal and binodal lines for symmetric
and asymmetric mixtures. Now, when demixing occurs, a second cmc
c** is defined as the total mixture concentration, at which a second
type of mixed micelles appears, at a fixed overall mixture composition
α. With our model we have constructed micellar pseudo-phase diagrams
which show   regions where there are no micelles or where one or two

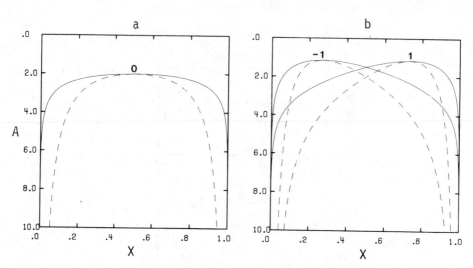

Figure 2. Binodals (full lines) and spinodals (broken lines) for various values of A with C=0; (a) B=0, symmetric mixtures; (b) B = -1 and 1, asymmetric mixtures. Mixed micellar pseudo-phase or mixed liquid crystalline phase is unstable between spinodals, stable outside binodals, and can be metastable between the spinodal and the binodal lines.

types of micelles are present. We distinguish two cases. In the first, two types of micelles coexist at all concentrations above the second cmc(Figures 3a, 3b, and 4b). In the second, there is a concentration below which there are two types of micelles and above which there is only one type of micelles (Figures 3c, 3d, and 5). Mysels first recognized the possibility of such a concentration which he called <u>critical</u> <u>demicellization</u> <u>concentration</u>, or <u>cdc</u>.[24] Mysels gave a detailed physical interpretation of a cdc.

In Figures 4 and 5 we show typical inventory and composition plots for various mole fractions α including the cases just discussed. For ionic surfactants with no added salt both monomer concentration curves have a maximum. These maxima are due to both counterion and mixture effects. For mixed nonionic surfactants, only one maximum is observed.[18,25] For a single ionic surfactant a maximum which is due to the counterion effect is observed, as has been well documented.[18,26]

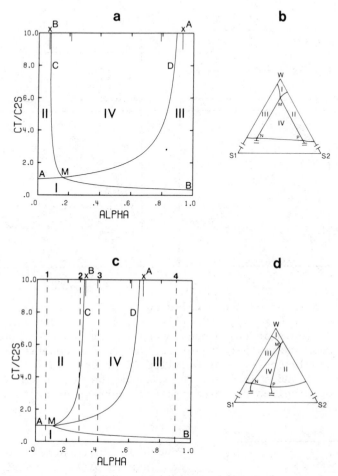

Figure 3. Micellar pseudo-phase diagrams for mixed ionic surfactants with B = C = 0 and β = 0.5: (a) c_2^*/c_1^* = 3.0 and A = 3.0; region I, no micelles; region II, micelles rich in component 1; region III, micelles rich in component 2; region IV, coexisting mixed micelles; curve AMB is the first cmc ; curve CMD is the second cmc; (b) equivalent representation of (a) as part of the overall ternary diagram; (c) and (d): same as (a) and (b), but for c_2^*/c_1^* = 4 and A = 2.1; curve MC is the cdc.

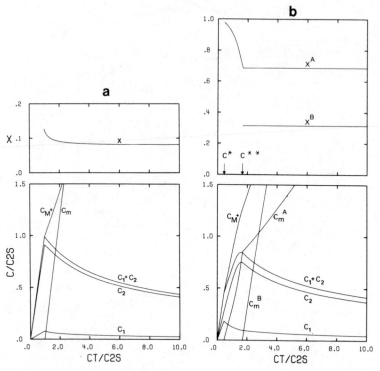

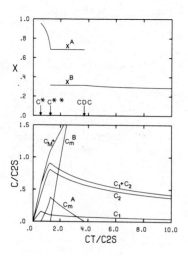

Figure 4.   Inventories and micelle compositions for mixed ionic surfactants with $\beta = 0.5$, $c_2^*/c_1^* = 3.0$, $A = 3.0$ (demixing possible), but $\alpha = 0.08$ (outside the binodal range); both monomer concentrations c_1 and c_2 show maxima because of the increasing counterion concentration c_M^+.   Only one type of micelles is present in (a), corresponding to lines 1 and 4 in Figure 3c. Above the second cmc c** in (b) two types of micelles are ever present, corresponding to line 3 in Figure 3c.

Figure 5.   Inventories and micelle composition plots as in Figure 4 but for the case of demicellization, corresponding to line 2 in Figure 3c.

APPLICATION TO SPECIFIC SYSTEMS

When there is no demixing, the solution model parameters are to be determined from the cmc's, since direct information on micelle compositions is difficult to obtain. Rubingh[4] and others[8,9] have used the pseudo-phase separation model for nonionic mixtures not only for nonionic/nonionic but also for ionic/nonionic binary mixtures, for which ionic effects are substantial. They determined the value of A for which the data collectively most closely agree with the calculated cmc-curves. We have calculated instead the value of A from each piece of data as a more sensitive test of the model. We have used the model for ionic mixtures and calculated solution model parameters for the ionic/ionic mixture sodium laurate (SL) - sodium perfluorooctanoate (SPFO).[7] We used the value $\beta = 0.53$, which is recommended by the authors. We observe that the calculated values of A (B=C=...=0) are very sensitive to small errors in the measured cmc, Table I. A not uncommon error of $\pm$ 1% in the measured cmc can lead to a substantial variation in the calculated value of A   (e.g. 1.50 to 2.19 for $\alpha = 0.165$) but to a rather small variation in the micelle mole fraction x* at the cmc. Moreover, an uncertainty of 0.1 in β can lead to an uncertainty of 15% in A.[18] Hence, if more precise cmc-data are unavailable, additional data on micelle composition and monomer concentrations are needed to evaluate the validity of the model and calculate A more precisely. The value of A was consistently and significantly smaller for SL-rich micelles than for SPFO-rich micelles. One can draw the limited conclusion that the one-parameter model is inadequate. Because of the uncertainties, however, A is not clearly larger than 2, which is a necessary criterion for demixing. Hence, although we can conclude from such data that there are substantial positive deviations from ideality, we cannot ascertain that demixing occurs. For that one would need additional and more precise data on cmc's and β and an established model probably with more than one parameter. We have tried to fit these data to the two-parameter model. To avoid the highly complex nonlinear least square method, we have deduced the parameters A and B from pairs of the cmc-data. As shown in Table II, most pairs yield consistent values for A and B, although some do not. Hence, no pair of two-parameters can adequately represent these data. Because of the sensitivity of the parameters to small errors in the cmc, we have not attempted to fit the data to the three-parameter model.

We have applied the two-parameter model to mixtures of sodium dodecylsulfate (SDS), which is an ionic surfactant forming micelles with a cmc of 8.3 mM, and sodium 4-(1' heptylnonyl) benzenesulfonate, which is an ionic surfactant forming liquid crystals with a solubility of 1.5 mM.[16] Although some evidence of aggregation has been reported,[28,29] no clear evidence of a cmc is available for SHBS in

Table I.   Calculated A-Values from Experimental Cmc's for Mixtures of Sodium Laurate (SL) with Sodium Perfluorooctanoate[a]

α_{SL}	c^* (mM)[b]	A	x^*	A(-1%)	x^*(1%)	A(+1%)	x^*(+1%)
0.000	30.6						
0.165	33.4	1.79	0.049	1.50	0.065	2.19	0.032
0.334	37.6	2.14	0.111	1.99	0.131	2.30	0.090
0.430	37.5	1.75	0.474	1.69	0.479	1.81	0.466
0.510	35.7	1.70	0.777	1.61	0.757	1.79	0.798
0.661	32.0	1.72	0.914	1.54	0.897	1.94	0.931
0.830	27.9	1.02	0.940	0.78	0.925	1.33	0.956
1.000	25.7						

(a): We use the ionic PSM with β = 0.53 for the regular solution model. For given c^* we calculate A and the micellar mole fraction at the cmc. We then vary c^* by - 1% and
+ 1% and calculate A and x^* again.

(b)   Cmc-values for various mole fractions of SL; they were
determined conductimetrically by Mukerjee and Yang.[7]

Table II.   A and B Solution Model Parameters Calculated From Pairs of Cmc's of Table I.

α_{SL}	0.334	0.430	0.510	0.661	0.830
0.165	*	(1.75, -0.05)	(1.73, -0.06)	(1.75, 0.04)	(1.40,-0.43)
0.334	——	(1.83, -0.40)	(1.89, -0.32)	(1.94,-0.26)	(1.63,-0.68)
0.430		——	(1.75, -0.10)	(1.75,-0.03)	*
0.510			——	(1.65, 0.09)	*
0.661				——	*

*: For B between -1 and 1, there are no values of A and B which are
consistent with the pair of cmc's.

water.[28] Because the matter is still unsettled, we will assume here
that only monomers exist below the SHBS solubility. For this
mixture, the binodals at concentrations 2 to 10 wt% have been
determined[16] as x^A=0.80+0.05, the mole fraction of SHBS in mixed
liquid crystal, and $x^B=\overline{0}.26$+0.06, which is the mole fraction of SHBS
in the mixed micelles at equilibrium with the liquid crystal. For
the most probable values of the binodals, we have found A = 2.23 and
B = 0.12. In Figure 6, we show the micellar pseudo-phase diagrams
for β=0.5 (ionic model). The actual mixture behavior should lie
between these two extremes, because the mixed micelles are
substantially charged (β=0.5) and the mixed liquid crystal particles

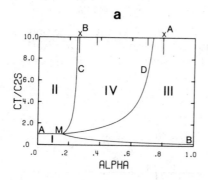

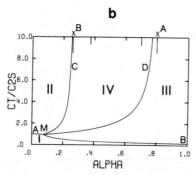

Figure 6. Diagram (see Figure 3) for aqueous SDS/SHBS;[16] in region IV mixed micelles coexist with mixed liquid crystals, see text; α is the mole fraction of SHBS. Note that for both $\beta = 0$ (a) and $\beta = 0.5$ (b) there is a cdc.

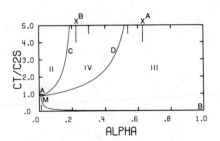

Figure 7.   Same as Figure 6a but for the system aqueous Triton X-100
mixed with phosphatidylcholine.   One binodal is 0.50;[14] the other was
taken as 0.33; on this scale, line MB appears to overlap with the
α - axis.

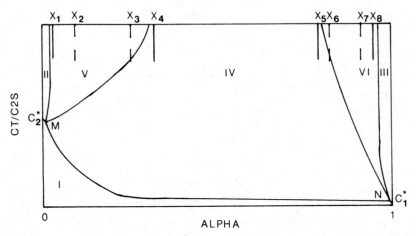

Figure 8.   Schematic representation (not to scale) of the diagram
of aqueous Triton X-100 with sphingomyelin.   There are two two-pseudo-
phase (or phase) regions, V and VI; regions II and IV are mixed
micellar and III is mixed liquid crystalline.   From Yedar el al.[11]
we have $x_4 = 0.32$ and $x_5 = 0.79$.   The other two binodals x_1 and x_2
are unknown but believed to be close to 0 and 1 respectively;
x_2, x_3, x_6, and x_7 are the spinodals.   Only a three-or-more parameter
model can represent such a diagram, see text.

are charged only at their surface, leading to a very small overall value of β. For both models, there is a cdc (line MC), above which there are no mixed liquid crystals. We are presently doing experiments to confirm this point. Furthermore, we feel that a hybrid ionic/nonionic model should represent the actual system more realistically.

In Figure 7 we present a diagram for Triton X-100 and dipalmitoylphosphatidylcholine. The first surfactant is a non-ionic micellizer and the second a zwitterionic liquid-crystal former. Hence, the nonionic/nonionic model (PSM) should best apply. We expect that $c_2{}^* \simeq 3\times10^{-4}M^{30}$ and $c_1{}^* = 0$ $(10^{-8}M)$.[2] Point M is very close to $\alpha = 0$. Hence, a cdc should surely exist in this system although we know of no specific data indicating so. For lack of precise cmc data, we have made a demonstration calculation taking $c_1{}^*/c_2{}^* = 100$.

Finally, in Figure 8 we show a schematic representation of a possible diagram for the mixture of Triton X-100 with sphingomyelin.[11] Since there appear to be two two-phase regions, the proper solution model $w(x)$ in our formalism must have at least three parameters; see Figure 1b and related discussion.

CONCLUSIONS AND REMARKS

1. For binary mixed ionic/ionic and mixed nonionic/nonionic surfactants, demixing occurs if Condition (6) is satisfied. We have examined specifically the Redlich-Kister model (5) with up to three parameters. There can be only one two-phase region if $C = 0$ and up to two two-phase regions if $C \neq 0$.

2. We have constructed pseudo-phase separation diagrams depicting regions of two-types of coexisting mixed micelles or mixed micelles coexisting with mixed liquid crystals.

3. Demicellization is possible if point M, Figure 3, lies outside the binodal range.

ACKNOWLEDGMENTS

We are grateful to the School of Chemical Engineering and the Computer Center of Purdue University for partially supporting this work.

REFERENCES

1. G.S. Hartley, in "Micellization, Solubilization, and Microemulsions," K.L. Mittal, Editor, Vol. 1, p. 23, Plenum Press, New York, 1977.

2.   C. Tanford, "The Hydrophobic Effect," 2nd Ed., Wiley, New York, 1980.

3.   H.Lang and K.H. Beck, Kolloid Z.-Z. Polym. 251, 424 (1973).

4.   D.N. Rubingh, in "Solution Chemistry of Surfactants," K.L. Mittal, Editor, Vol. 1, p. 337, Plenum Press, New York, 1979.

5.   B.T. Ingram, Colloid Polym. Sci., 258, 191 (1980).

6.   P. Mukerjee and K.J. Mysels, in "Colloidal Dispersions and Micellar Behavior," K.L. Mittal, Editor, p. 239, ACS Symposium Series No. 9, American Chemical Society, Washington, D.C., 1975.

7.   P. Mukerjee and A.Y.S. Yang, J. Phys. Chem., 80, 1388 (1976).

8.   M. Ueno, K. Shioya, T. Nakamura, and K. Meguro, in "Colloid and Interface Science," Vol. II, M. Kerker, Editor, p. 411, Academic Press, New York, 1976.

9.   N. Funasaki and S. Hada, J. Phys. Chem., 84, 736 (1980).

10. G. Perron, R. DeLisi, I. Davidson, S. Genereux, and J.E. Desnoyers, J. Colloid Interface Sci., 79, 432 (1981).

11. S. Yedgar, Y. Barenholz, and V.G. Cooper, Biochim. Biophys. Acta, 363, 98 (1974).

12. N.A. Mazer, G.B. Benedek, and M.C. Carey, Biochemistry, 19, 601 (1980).

13. A.A. Ribeiro and E.A. Dennis, Biochim. Biophys. Acta, 332, 26 (1973).

14. E.A. Dennis, Arch. Biochem. Biophys., 165, 764 (1974).

15. R. Hertz and Y. Barenholz, J. Colloid Interface Sci., 60, 188 (1977).

16. J.E. Puig, E.I. Franses, and W.G. Miller, J. Colloid Interface Sci., 89, 441 (1982).

17. S.H.-W. Wu and H.M. McConnell, Biochemistry, 14, 847 (1975).

18. R.F. Kamrath and E.I. Franses, I & EC Fundamentals, accepted (1982).

19. R.F. Kamrath and E.I. Franses, J. Colloid Interface Sci., submitted (1982).

20. J.M. Prausnitz, "Molecular Thermodynamics of Fluid-Phase Equilibria," Prentice-Hall, Englewood-Cliffs, New Jersey, 1969.

21. I. Prigogine and R. Defay, "Chemical Thermodynamics," Longmans, London, 1954.

22. H.S. Caram and L.E. Scriven, Chem. Eng. Sci., 31, 163 (1976).

23. J.M. Prausnitz and S.A. Shain, Chem. Eng. Sci., 18, 244 (1963).

24. K.J. Mysels, J. Colloid Interface Sci., 66, 331 (1978).

25. J.H. Clint, J. Chem. Soc. Farad. Trans. I, 71, 1327 (1975).

26. K.M. Kale, E.L. Cussler, and D.F. Evans, J. Phys. Chem., 84, 593 (1980).

27. E.I. Franses, J.E. Puig, Y. Talmon, W.G. Miller, L.E. Scriven, and H.T. Davis, J. Phys. Chem., 84, 1547 (1980).

28. W.G. Miller, F.D. Blum, H.T. Davis, E.I. Franses, E.W. Kaler, P.K. Kilpatrick, K.J. Nietering, J.E. Puig, and L.E. Scriven, see article in Proceedings of this Symposium.

29. L.J. Magid, R. Triolo, J.S. Johnson, Jr., and W.C. Koehler, J. Phys. Chem., 86, 164 (1982).

30. M.J. Rosen, "Surfactants and Interfacial Phenomena," p. 103, Wiley, New York, 1978.

^{31}P AND ^{2}H NMR STUDIES OF PHASE EQUILIBRIA IN THE THREE COMPONENT SYSTEM: MONOOLEIN-DIOLEOYLPHOSPHATIDYLCHOLINE

H. Gutman[1], G. Arvidson[2], K. Fontell[3] and G. Lindblom[4]

[1]Physical Chemistry 2, Chemical Center, Lund, Sweden
[2]Physiological Chemistry, Biomedical Center, Uppsala, Sweden
[3]Food Technology, Chemical Center, Lund, Sweden
[4]Physical Chemistry, University of Umeå, Umeå, Sweden

^{2}H and ^{31}P NMR have been used in the determination of the phase behaviour at 28°C of the ternary system monoolein/dioleoylphosphatidylcholine/heavy water. The isothermal phase diagram shows the existence of four one-phase areas containing homogeneous liquid crystalline matter and one area with isotropic solution. Two of the liquid crystalline phases have a lamellar structure, one has a cubic and the fourth a reversed hexagonal structure. A non-aqueous mixture of 3 to 5 moles of monoolein per mole of dioleoylphosphatidylcholine is an isotropic oily fluid.

INTRODUCTION

The phospholipids are basic components of most biological membranes. In many important physiological processes involving fusion
of membrane segments, such as endo- and exo-cytosis, transient
local rearrangements of the phospholipid bilayers into non-lamellar
structures are bound to occur. It is known that in the presence
of water many membrane lipids in isolation under physiological conditions in addition to lamellar bilayer structures may form other
types of structures[5,6,7]. However, the conditions for the formation of non-lamellar aggregates and their structures are not yet
completely elucidated. As a suitable system for studies of these
phenomena we have chosen the three-component system monoolein/
dioleoylphosphatidylcholine/water. The present paper will present
the phase equilibria of this system. The temperature was $28^{\circ}C$ unless otherwise stated.

EXPERIMENTAL

Materials

The lecithin, dioleoylphosphatidylcholine (DOPC), was synthetized by established procedures[8]. The purity was > 99% as judged
by thin layer chromatography. The monoolein (MO, 90% α-isomer)
was obtained from Sigma , and deuteriumoxide (minimum 99.7 atom-%
D) from Ciba Geigy.

Sample preparation

Appropriate amounts of the lipids were weighed into glass-
tubes (10 mm in diameter) and the lipids were mixed in a solution
of diethylether. The ether was evaporated under nitrogen atmosphere at about $30^{\circ}C$ and the last traces removed in high vacuum.
Heavy water was added and after weighing the glass-tubes were
flame-sealed. Alternatively, the samples were prepared without
mixing the lipids in ether. The sealed samples (lipids + heavy
water) were heated for an hour at $60^{\circ}C$ and thereafter centrifuged
in a desk-centrifuge at room temperature. The centrifugation was
repeated several times with 24 hours intervals, and the samples
were then stored at $25^{\circ}C$ for at least one week before the first
series of measurements, which were repeated after an interval of
one to two weeks. It was found that the samples had reached equilibrium. After the final measurements the decomposition of the
samples was routinely checked by thin layer chromatography. No
signs of oxidative or hydrolytic degradation were found.

Visual observations

The samples were placed between crossed polarizers and examined for homogenity and birefringence by ocular inspection. Visible
changes in the appearance of the samples could be followed day by
day until equilibrium was obtained. The microscopical textures of
the different anisotropic phases were observed in an Olympus
Vannox polarizing microscope. The texture of the lamellar liquid
crystalline phases was the mosaic one 122.1 in Rosevear´s classification[9,10], while the hexagonal phases showed the non-geometric
striated texture 231 or 232 of Rosevear. The isotropic phases,
both cubic and solution, showed no birefringence and had no texture.

The NMR measurements

All NMR measurements were performed on a modified Varian
XL-100-15 spectrometer operating in a pulsed Fourier transform
mode using an external lock. The ³¹P and ²H spectra were recorded
at resonance frequencies of 40.508 and 15.351 MHz, respectively.
The ³¹P measurements were performed with heteronoise proton de-
coupling.

The NMR spectra show different features for different phases
of a lipid/water system. When a sample contains two or more phases
the resulting spectrum will often be a mere superimposition of the
individual spectra. By recording the ²H and ³¹P spectra of a systematic series of samples one may obtain the features of the phase
diagram. This is a non-destructive method which does not need the
physical separation of the individual phases of a mixture. The
method has been successfully employed in several recent studies
of the phase behaviour of lipid/water systems[11-17].

The interaction of water deuteron quadrupole moments with
the electric field gradients of the nuclei determines the ²H
NMR spectra of a sample containing D_2O. An anisotropic liquid
crystalline sample shows a quadrupole splitting because these interactions have not been averaged to zero by the molecular motion.
The magnitude of the splitting depends on the internal structure
of the phase. For isotropic phases, the cubic liquid crystalline
and solution phases, the rapid molecular motion results in a
singlet. The conditions leading to the quadrupole splitting phenomenon in lyotropic liquid crystalline samples have been thoroughly discussed[18-20].

The ^{31}P NMR spectra of liquid crystalline samples containing phospholipids show also special features[21]. The spectral shape depends on the chemical shift anisotropy of the phosphorous nuclei, which is only partially averaged. The residual shift anisotropy $\Delta\sigma = \sigma_{||} - \sigma_{\perp}$ is a measure of the orientation and average fluctuation of the phosphate segments. The geometry of the liquid crystalline structure (hexagonal or lamellar) determines the sign of the phosphorous shielding anisotropy as well as its magnitude (it may be altered by a factor of two) under the assumption of a constant head group conformation. In isotropic samples "isotropic" motion will result in a complete averaging of the phosphorous chemical shift anisotropy and the spectra will accordingly show only a singlet.

Illustrative spectra showing these principles utilized in the determination of the regions and structures of the different phases are presented in Figures 1 and 2.

X-ray diffraction

The low-angle X-ray diffractograms were obtained with a Guinier-camera according to Luzzati as described elsewhere[5,22]. The temperature was 25°C.

RESULTS AND DISCUSSION

The phase equilibria in the three-component system monoolein/ dioleoylphosphatidylcholine/heavy water are presented in the isothermal phase diagram in Figure 3. Previous knowledge of phase boundaries in binary lecithin/water systems[23] as well as in the binary monoolein/water system is utilized[22,24]. The low-water DOPC corner of the phase diagram has not been studied.

The lamellar phase of the binary MO/D_2O system (84-96% (w) MO) has at most the capability to incorporate about 10% (w) of DOPC. Contrary to this the cubic phase (57-76% (w) MO) can incorporate large amounts of DOPC without loosing its structure. This is evidenced by the low-angle X-ray diffraction findings which show the same body-centered pattern in the whole region. The diffractograms may be indexed in the same way as those in the previous study of the binary MO/D_2O system[22]. The unit cell dimensions are furthermore of the same magnitude (a_o = 150-250 Å). These values are more dependent on the content of D_2O in the samples than of the ratio between the two lipids, being smallest for samples with low D_2O content. So far we have been able to incorporate about 75% (w) DOPC in the samples and still retain the same cubic structure. If it is possible to further increase

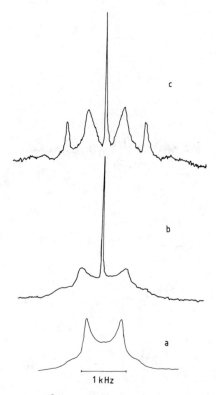

Figure 1.   Illustrative ^{2}H spectra demonstrating the principles
for the determination of the phase diagram for MO/DOPC/D$_2$0.
a. Lamellar phase, b. Two-phase mixture of lamellar and isotropic
phase (either pure D$_2$0 or cubic phase), c. Three-phase mixture
of lamellar, hexagonal and cubic phase.

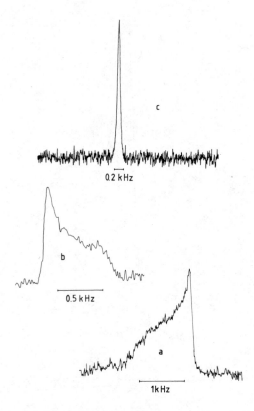

Figure 2.   Illustrative ^{31}P NMR spectra showing the principles for the determination of the phase diagram for $MO/DOPC/D_2O$. (The samples must contain DOPC). The chemical shift anisotropy is measured from edge to edge of the spectrum. a. Lamellar phase, b. Hexagonal phase, c. Cubic phase.

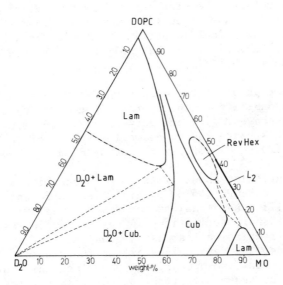

Figure 3.   Tentative isothermal phase diagram for the system mono-
olein/dioleoylphosphatidylcholine/heavy water at $28°C$. The borders
of the different phase regions are determined by inspection of ^2H
and ^{31}P NMR spectra and the findings complemented by X-ray diffraction and polarizing microscopy. Notations: Lam., anisotropic liquid crystalline phases with lamellar bilayer structure; Rev. Hex.,
anisotropic liquid crystalline phase with a reversed two-dimensional hexagonal structure; Cub., liquid crystalline isotropic
viscous phase with cubic structure; L_2, isotropic solution phase,
"oil". The locations of the two three-phase triangles, D_2O+Lam+
+Cub and Rev. Hex+Cub+Lam are very approximate.

the DOPC content the interesting question arises : is there any
connection between this cubic structure and that studied by
Luzzati et al. in binary lecithin/water systems[5,25]? Both structures are body-centered, belonging to the space groups Im3m and
Ia3d, respectively. The cubic phase of the monoolein/water system
is proposed to be composed of small bilayer entities forming a
three-dimensional network which separates two identical three-
dimensional water-channel systems[22,26]. The bilayer entities will
possess a smooth curvature and this will endow strain-free conditions to the structure. The structure of Luzzati et al. is proposed to be composed of short rod-like entities joined three and
three forming two identical intertwined networks[5,25]. However,
the structure of Luzzati et al. has only been observed at elevated temperatures in lecithin/water system.

The lamellar phase of the binary DOPC/D_2O system (55-96% (w)
DOPC) has the capability to incorporate about 37% (w) of MO. The
borders of this phase region are very approximate. The structural
parameters of this phase region as well as those of the other
lamellar phase (MO/D_2O + DOPC) were not studied.

Samples containing about 5% (w) of D_2O and with a mole ratio
around 3:1 of MO to DOPC (45/50/5) are anisotropic and possess
a liquid crystalline hexagonal structure. This is evidenced by
the existence of sharp reflections in the low-angle region of the
X-ray diffractograms as well as a hexagonal texture in the polarizing microscope. The reflections are in the sequence $1:\sqrt{3}:\sqrt{4}$
and the value for the two-dimensional hexagonal unit cell is
about 43 Å. If one assumes that the structure is of the reversed
variant (and unit density) and includes the glycerol and choline
groups in the polar core, the cross-section area per hydrocarbon
chain at the polar/non-polar interface will be around 37 $Å^2$.

Another feature of the presently studied phase diagram is
that a mixture of 3-5 moles of MO per mole of DOPC (26 to 42%
(w) DOPC) results in an isotropic easily flowing liquid. It is
possible that this liquid phase contains small amounts of water.
The internal structure of this solution phase, L_2, is unknown. The
low-angle X-ray diffractogram exhibits one diffuse reflection with
a position corresponding to about 34 Å.

CONCLUSION

The isothermal phase diagram of the system MO/DOPC/D_2O is
dominated by an extended cubic region which emanates from the
MO/D_2O axis and extends towards the DOPC corner. The two liquid
crystalline lamellar phase regions, emanating from the binary

MO/D_2O and $DOPC/D_2O$ axes, respectively will thus be separated from each other. In addition there exists a liquid crystalline anisotropic hexagonal phase at low water contents and furthermore, a mixture of the nonaqueous lipids over a particular concentration range is an oily isotropic liquid.

ACKNOWLEDGEMENTS

Financial support from the Swedish Natural Science Research Council, Magn. Bergvalls Stiftelse and from the Foundation of Bengt Lundqvists minne (Hélène Gutman) is gratefully acknowledged.

REFERENCES

1. Physical Chemistry 2, Chemical Center, Box 740, S-220 07 Lund, Sweden.
2. Physiological Chemistry, Biomedical Center, S-751 23 Uppsala, Sweden.
3. Food Technology, Chemical Center, Box 740, S-220 07 Lund, Sweden.
4. Physical Chemistry, University of Umeå, S-901 87 Umeå, Sweden.
5. V. Luzzati, in "Biological Membranes", D. Chapman, Editor, p. 1, Academic Press, New York 1968.
6. T. Gulik-Krzywicki, E. Rivas and V. Luzzati, J. Mol. Biol. 27, 303 (1967).
7. Å. Wieslander, A. Cristiansson, L. Rilfors, A. Khan, L.B.-Å. Johansson and G. Lindblom, FEBS Lett. 124, 273 (1981).
8. C.M. Gupta, R. Radhakrishnan and G. Khorana, Proc. Natl. Acad. Sci., USA 74, 4315 (1977).
9. F.B. Rosevear, J. Am. Oil Chemists Soc. 31, 628 (1954).
10. F.B. Rosevear, J. Soc. Cosmetic Chemists 19, 581 (1968).
11. J. Ulmius, H. Wennerström, G. Lindblom and G. Arvidson, Biochemistry 16, 5742 (1977).
12. N.O. Persson, B. Lindman and G.J.T. Tiddy, J. Colloid Interface Sci. 78, 217 (1980).
13. K. Beyer, J. Colloid Interface Sci. 86, 73 (1982).
14. A. Khan, L. Rilfors, Å. Wieslander and G. Lindblom, Eur. J. Biochem. 116, 215 (1981).
15. A. Khan, K. Fontell and G. Lindblom, J. Phys. Chem. 86, 383 (1982).
16. P.O. Eriksson, A. Khan and G. Lindblom, J. Phys. Chem. 86, 387 (1981).
17. A. Khan, K. Fontell, G. Lindblom and B. Lindman, J. Phys. Chem. 86, 4266 (1982).
18. H. Wennerström, G. Lindblom and B. Lindman, Chem. Scripta 6, 97 (1974).

19. H. Wennerström, N.O. Persson and B. Lindman, in "Colloidal Dispersions and Micellar Behavior", K.L. Mittal, Editor, p. 253, ACS Symposium Ser. No. 9, Washington D.C. 1975.
20. C.L. Khetrapal, A.C. Kunwar, A.S. Tracey and P. Diehl, "Lyotropic Crystals", Springer Verlag, Heidelberg 1975.
21. J. Seelig and A. Seelig, Quart. Rev. Biophys. 13, 19 (1980).
22. G. Lindblom, K. Larsson, L. Johansson, K. Fontell and S. Forsén, J. Am. Chem. Soc. 101, 5465 (1979).
23. V. Luzzati and A. Tardieu, Ann. Rev. Phys. Chem. 25, 79 (1974).
24. E.C. Lutton, J. Am. Oil Chemists Soc. 42, 1068 (1965).
25. V. Luzzati and A. Spegt, Nature 215, 710 (1967).
26. K. Larsson, K. Fontell and N. Krog, Chem. Phys. Lipids 27, 321 (1980).

MICELLE FORMATION AND PHASE EQUILIBRIA OF SURFACE ACTIVE

COMPONENTS OF WOOD

P. Stenius, H. Palonen, G. Ström and L. Ödberg

Institute for Surface Chemistry
Box 5607
114 86 Stockholm, Sweden

Studies of surfactant properties of sodium
abietate and sodium oleate at 60°C show that both
compounds form small micelles at low concentrations.
At high concentrations of oleate, rod-shaped micelles
are formed. Addition of salt causes the oleate micelles
to grow while the abietate micelles remain small.
Abietate and oleate when mixed in roughly equimolar
proportions form small mixed micelles that remain soluble
at much higher salt concentrations than either of the
pure compounds. All types of micelles show comparable
solubilization capacities. Oleate and abietate also form
lamellar liquid crystalline phase with water. With oleic
acid the oleate forms hexagonal and isotropic liquid
crystalline phases as well as reversed micellar solutions,
while sodium abietate/abietic acid forms only lamellar
phase. The results make it possible to draw some tentative conclusions concerning the behaviour of these
surface active components in black liquor, their removal
in the washing of pulp and their dispersability. These
conclusions are illustrated by comparison with real black
liquor systems.

INTRODUCTION

Wood contains a few percent of ether soluble compounds. These
are commonly, although somewhat loosely, called "wood extractives".
They are the source of tall oil, which is an important raw material
for the production of fatty and rosin acids. In the manufacture
of pulp and paper, they often cause problems because they tend to
form dispersions. The formation of such dispersions has been
correlated with the formation of sticky deposits on machinery at
various stages of pulp and paper manufacture and the occurrence
of lumps in the final product (so-called "pitch problems").[1,2]

Thus, a proper understanding of the surfactant properties of
amphiphilic components in wood extractives is of interest both in
view of the use of tall oil for commercial production of surfactants and in view of a better control of pitch problems.

The composition of wood extractives has been the subject of
numerous quite detailed investigations [3-12] but studies of their
surfactant properties are relatively few.[13-18] Table I shows a
typical composition of ether extractives from Scandinavian spruce,
pine and birch. All three extractives contain substantial amounts
of glycerides (which are subsequently partially or completely
saponified in the pulping process) as well as mixtures of fatty
acids and rosin acids.

Typical compositions after saponification are given in Table
II, which compares the ether extractives from a Kraft pulp and
from deposits on the screens of the pulp washing filters. As a
result of the pulping process, the glycerides have been saponified
and large amounts of fatty acid and soap occur in the extractives.
We also note that birch, as shown in Table I, contains virtually
no rosin acids. Softwood (pine and spruce) contains considerable
amounts of them. The remaining neutral substances are mainly
polycyclic compounds with betulinol and sitosterol as particularly
important components.

Thus, softwood as well as hardwood resin contains or can give
rise to considerable amounts of amphiphilic compounds. Although
this was realized many years ago[13-16] the implications for soap
recovery and pitch problems have not been considered in detail.[17-19]
In the following, we will describe some of the basic surfactant
properties of model compounds and also briefly touch upon some
possible applications of this knowledge in pulping and papermaking.

The model systems used in our investigations are given in
Table III. Sodium abietate, with the structure

Table I.   Typical Composition of Ether Extractives from Fresh Wood.

	Norwegian Spruce	Pine	Birch
Ether extract, % of wood	1.0	2.3	2.0
Fatty acids, % of ether extract	7.5	7.6	0
Oleic & lineolic acid, % of fatty acid	61	75	
Rosin acids, % of ether extract	27	29	0.3
Neutral components, % of ether extract	62	61	66
Glycerides, % of neutral components	47	67	57

Table II.   Typical Composition of Extracts and of Pitch Deposits from Screens in Hardwood Kraftmill (Birch).

	Ether Extract from Pulp	Ether Extract from Deposits
Total acids in extract	40	79
Fatty acids, % of acids	95	98
Unsaturated acids, % of fatty acids	46	60
Acids: soap, w:w	6:1	1:2
Neutral components, % of extract	60	21
Betulin, % of neutral comp.	14	79

Table III. Model Systems and Conditions Used in Investigations of
Surfactant Properties of Amphiphilic Components in Wood.

Oleic acid/sodium oleate
Abietic acid/sodium abietate
Abietic acid/sodium oleate
Sodium abietate/sodium oleate
Temperature 60°C
Electrolytes: 0 - 0.75 mol/dm^3 NaCl or Na$_2$SO$_4$;
0.001 - 0.1 mol/dm^3 CaCl$_2$
Solubilizates: Decalin, decane and decanol

and sodium oleate are model compounds of the rosin and fatty acid
soaps, respectively, in wood extractives, while the temperatures
and salt concentrations have been chosen so as to represent typical
conditions experienced in pulp washing. Solubilization of decalin,
decanol and decane was studied because these compounds may serve
as models of some of the neutral components in resins. The temperature is well above the Krafft temperature of sodium oleate
($\approx$ 20°C).[20]

EXPERIMENTAL

Materials

    Sodium oleate (NaOl) was prepared by neutralization of oleic
acid (HOl) (Fluka AG, puriss) with sodium ethylate in dry ethanolic
solution. The molecular weight of the soap was determined by
titration with perchloric acid in dioxane/acetic acid solvent.
The molecular weight differed less than 1% from the theoretical
value.

    Abietic acid (HAb) (Fluka AG. pract.) was further purified by
recrystallization.[19] The purified product melted at 170 ± 2°C.

    Sodium abietate (NaAb) was prepared from the purified abietic
acid by neutralization with sodium ethylate in dry ethanol. The
synthesis was performed in nitrogen atmosphere. Contact with
oxygen and exposure to light was avoided as much as possible. The
purity was checked by gas chromatography and mass spectrometry.[12]
The final product contained about 91.8% sodium abietate and about
6.6% other rosin acid soaps.[19] In our experiments, the use of
NaAb older than 5 days was avoided.

All other chemicals used were analytical reagent grade. The water was doubly distilled with a conductivity of ≈ 0.8 µS/cm.

Light Scattering

Dynamic light scattering (DLS) measurements were performed at 90° scattering angle with a Malvern light scattering photo-meter equipment with a 60-channel autocorrelator (Langley Ford) and a Spectra Physics 185 Ar-laser as light source.

The autocorrelation function was analyzed using the cumulant method.[21] The reproducibility of calculated hydrodynamic radii was ± 3%.

All solutions were ultracentrifuged at 35,000 x g for 0.5h before analysis.

Phase Equilibria and Solubilization

The phase equilibria and solubilization capacities were de-termined by mixing the components, whenever possible at tempera-tures high enough to obtain single-phase systems, and then cooled to 60°C. They were then equilibrated at this temperature, some-times for several weeks. In samples containing sodium abietate, however, care was taken to heat the samples as little as possible above 60°C. The samples were examined for optical birefringence between crossed polarizers. Multi-phase systems were separated by centrifugation at 60°C in a Beckman L-80 ultracentrifuge. The separated liquid crystals were identified by examination between crossed polarizers in the microscope. In some cases their struc-ture was verified by X-ray diffraction using the methods described earlier[22]. The occurrence of liquid crystals with lamellar or hexa-gonal structure was also in some cases verified by determining the deuterium nuclear quadrupole splitting in samples prepared with D_2O instead of H_2O[23] using a Bruker CXP-100 nmr spectrometer.

In an anisotropic phase the motion of the water molecules is such that the quadrupole interaction is not averaged to zero and the deuterium signal is split into a doublet which does not occur for isotropic phases. The relative magnitude of the splitting in different phases can also provide some help in identifying the phases.

The nuclear magnetic relaxation of a sodium counter ion in a micellar system is dependent both on the electrical field gradients the sodium ion is experiencing and on the time scale of the fluc-tuations of these fields. This offers some possibilities to obtain information on the process of micelle formation.[37]

Solubilities and Surface Tensions

The determination of solubilities was described in detail earlier[19]. Surface tensions were determined with a du Nouy ring tensiometer as described in ref. 19.

RESULTS AND DISCUSSION

Solubilities

Figure 1 shows the solubility of sodium abietate in sodium chloride solutions. The solubility decreases rapidly with increasing salt concentration. A lamellar liquid crystalline phase is formed in equilibrium with saturated aqueous solution. The formation of lamellar liquid crystals by compounds with this type of polycyclic, stiff hydrocarbon moiety has to our knowledge not been reported before.

Israelachvili et al.[24] have suggested that a bulky hydrophobic moiety (compared to the hydrophilic end group) should promote the formation of lamellar or reversed hexagonal structures rather than normal hexagonal structures (usually found) in equilibrium with saturated aqueous solutions of surfactants with normal or slightly branched aliphatic chains). The rather large hydrocarbon moiety of sodium abietate, accordingly, is expected to promote the formation of a lamellar structure.

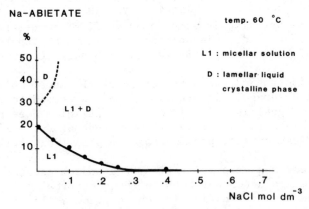

Figure 1. Solubility of sodium abietate in NaCl solutions at 60°C.

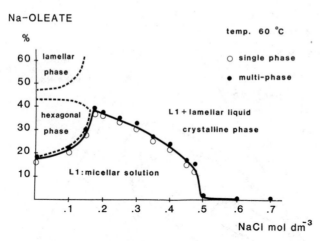

Figure 2.   Solubility of sodium oleate in NaCl solutions at 60°C.

The structure has been clearly identified by its microscopic behaviour and its X-ray diffraction pattern (diffraction lines corresponding to Bragg distances in the ratio 1:1/2:1/3...[25]

Figure 2 shows the solubility behaviour of sodium oleate as NaCl is added; this type of diagram has been well-known for many years.[26] Depending on the salt concentration, the aqueous solutions are in equilibrium with hexagonal or lamellar liquid crystalline phase.

The solubility behaviour of NaAb/NaOl mixtures is shown in Figure 3 for several NaCl concentrations. The addition of small amounts of oleate strongly increases the solubility of sodium abietate. Sodium abietate also increases oleate solubility. Thus, while neither soap is soluble in 0.75 mol/dm^3 NaCl a roughly equimolar mixture of the two compounds dissolves to about 20%.

Figure 4 shows the solubility of a 1:1 NaAb/NaOl mixture when Na$_2$SO$_4$ is added. The solubility is reduced to a very low value when the concentrations of NaCl or Na$_2$SO$_4$ are higher than 0.84 mol/dm^3 or 0.725 mol/dm^3, respectively. At these high salt concentrations, the aqueous solutions are in equilibrium with lamellar liquid crystalline phase. The water activity in 0.84 mol/dm^3 NaCl is 0.986; in 0.725 mol/dm^3 Na$_2$SO$_4$ it is 0.983[1]). Thus the transition

---
[1]) Calculated from osmotic coefficients given in ref. 36.

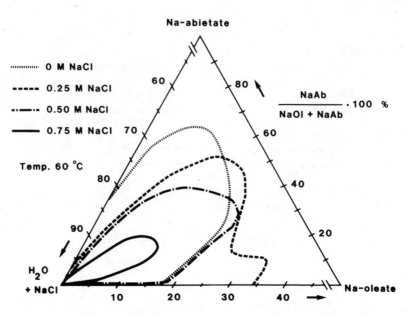

Figure 3.   Partial phase diagram of the system NaAb/NaOl/aqueous
NaCl solution at 60°C.   Each curve corresponds to the solubility
limit of NaOl/NaAb at constant concentration.

from equilibrium between micellar solution and lamellar liquid
crystalline phase to equilibrium with non-micellar solution is
determined by the chemical potential of the water rather than by
the specific type of electrolyte.

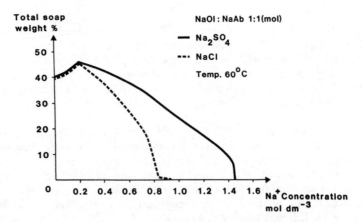

Figure 4.   Solubility of NaAb/NaOl mixtures in Na_2SO_4 and NaCl
solutions at 60°C.

Preliminary investigations of systems containing 0.75 mol/dm^3 NaCl indicated that the solution region at all NaAb/NaOl ratios is in equilibrium with lamellar liquid crystalline phase.

Micelle Formation

A straightforward explanation for the solubility at high salt concentrations is that sodium abietate and oleate form highly stable mixed micelles.

Figure 5 shows surface tension of aqueous NaAb solutions at pH 11.5; the pH was adjusted by addition of NaOH. The abietate shows a well-defined cmc (0.01 mol/dm^3 at 60°C) which is lowered by the addition of NaCl. The dependence of the cmc on the salt concentration indicates a counter ion binding of about 60%.

The formation of micelles was further investigated by dynamic light scattering. Figure 6 shows hydrodynamic radii for NaAb micelles. These were calculated assuming monodisperse, spherical micelles which follow the Stokes'-Einstein equation. The hydrodynamic radius does not (within experimental accuracy) depend on C_{Na}. The radius does not change very much with the salt concentration (1.7 nm in 0.0 mol/dm^3 NaCl, 1.8 nm in 0.1 mol/dm^3 NaCl) and is roughly two times the length of the hydrocarbon moiety of the NaAb.

Figure 7 shows the surface tensions of sodium oleate solutions and mixed NaOl/NaAb solutions.

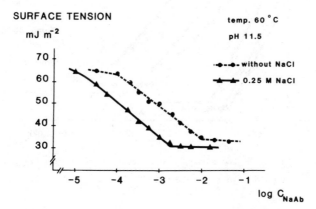

Figure 5.   Surface tension of sodium abietate solutions at 60°C.

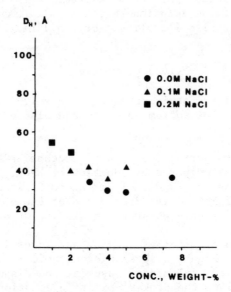

Figure 6.   Hydrodynamic radii of sodium abietate micelles at 60°C.

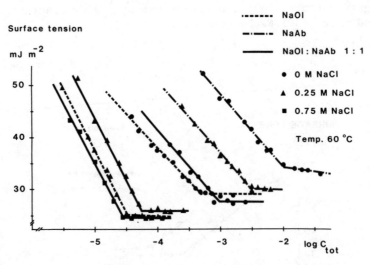

Figure 7.   Surface tensions of NaOl, NaAb and NaAb/NaOl solutions at 60°C at different NaCl concentrations.

NaOl forms micelles at much lower concentrations than NaAb.
The cmc in salt-free solutions, 0.48 mmol/dm^3 is higher than the
value 0.32 mmol/dm^3 reported by Tamamushi et al [27] who, however,
did not use a highly purified material. The lowering of the cmc
by addition of NaCl indicates a counter ion binding of about 60%.

Figure 8 shows hydrodynamic radii of sodium oleate micelles
at different NaCl concentrations. In salt-free solutions
the diameter is not strongly dependent on the oleate concentration
and is equal to about two times the oleate chain length (2.3 nm).

The slight decrease in the apparent hydrodynamic radius with
increasing C_{NaOl} is probably due to electrostatic intermicellar
interactions. There is no indication that large, rod-like micelles
are formed in salt-free solutions. We note that it was very
difficult to obtain completely clear solutions at the lowest NaOl
concentrations, probably because of the very low solubility of
oleic acid. The high value of the radius obtained at the lowest
concentration, therefore, is probably not significant.

Additional evidence that micelles of sodium oleate at this
temperature and low concentration are not rod-like is given by
studies of the sodium ion relaxation rates. Figure 9 shows the
peak half-widths of sodium ion nmr signals. At low NaOl concentrations there is no indication of a strongly enhanced relaxation
rate, as would be expected for large, rod-like micelles. Rather,

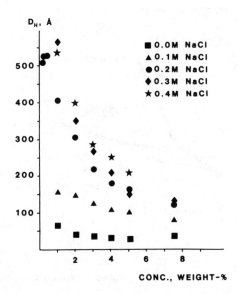

Figure 8. Equivalent hydrodynamic radii of sodium oleate micelles,
assuming spherical shape and Stokes-Einstein diffusion.

the sodium ions in the oleate solutions at low concentrations (the range studied by light scattering) behave in the same way as those in the solutions of small sodium abietate micelles.

This result differs from the results obtained by Reiss-Husson and Luzzati [28] whose interpretation of small angle X-ray scattering indicated that NaOl micelles are rod-like at all concentrations. However, their study was made at 27°C, which is much closer to the Krafft point of NaOl. The rapidly increasing line width at higher concentrations of NaOl at 60°C (Figure 9) could indicate that rod-like micelles are formed even at this temperature at high concentrations. It appears quite reasonable that a lowering of the temperature should promote the formation of rod-like micelles.

As shown in Figure 8 the apparent hydrodynamic radii increase very rapidly with increasing salt concentration, in particular at low NaOl concentrations. A quantitative interpretation of these results is difficult. However, it is obvious that the addition of NaCl causes the formation of very large (rod-like?) micelles which, as is shown by the rapidly decreasing apparent size of the micelles at high concentrations, are strongly interacting.

The solubilities in Figure 3 indicate that NaAb can be to some extent incorporated in the oleate micelles (curves for 0 and 0.25 M NaCl). However, at higher salt concentrations the formation of micelles by pure sodium oleate is completely suppressed

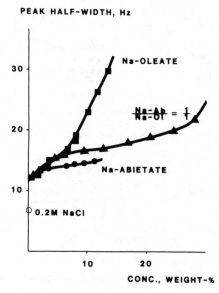

Figure 9. Relaxation rates of sodium ions in NaAb, NaOl and NaAb/Ol (1:1 by weight) solutions, given as the half-width of the nmr peak.

while NaAb/NaOl mixtures are still quite soluble. Figure 7 also
shows the surface tension of mixtures of NaAb/NaOl (1:1 by weight,
which is quite close to a molar ratio of 1:1) at different salt
concentrations. The curves show that the mixtures form micelles
at salt concentrations much higher than those that completely salt
out the pure compounds. The cmc's are close to those of NaOl at
corresponding salt concentrations.

    The formation of micelles is clearly confirmed by DLS measure-
ments (Figure 10). They indicate that rather small micelles (hydrodynamic radius between 2.5 and 3 nm) are formed. At high salt
concentrations the electrostatic interactions in the micelles are
decreased and this leads to a considerable increase in the micellar
size. It does not, however, depend very strongly on concentration.
The Na^+ relaxation behaviour is shown in Figure 9. The relaxation
rate does not rise rapidly at high concentrations, in sharp contrast
to the pure oleate micelles. We note also that the relaxation rates
are considerably higher than those of Na^+ in NaCl, which is clear
evidence that the binding of Na^+ to large (micellar) aggregates
plays an important role in the relaxation behaviour.

    Thus, the size of the mixed NaAb/NaOl micelles is independent
of concentration and, at low salt concentrations, is compatible
with a spherical or oblate micelle with a radius equal to about the
NaOl chain length. The micelles grow when salt is added but the

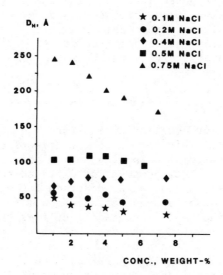

Figure 10.   Hydrodynamic radii of sodium oleate/sodium abietate
mixed micelles (1:1 by weight). The abscissa gives the total concentration of abietate + oleate.

solution is highly stable towards addition of salt. The cmc is the same as for pure oleate at corresponding salt concentrations. The stability towards salt is further emphasized by the fact that much higher concentrations of $CaCl_2$ can be added to the mixed micelles than to either pure abietate or oleate.[19]

Depending on the salt concentration, aqueous NaOl solutions are in equilibrium with either hexagonal or lamellar liquid crystalline phase. It was recently shown[29-31] that in systems forming highly charged micellar aggregates in aqueous solution, the conditions governing the transition between aggregates in aqueous solution and in liquid crystalline phases are predominantly electrostatic. At low concentrations of surfactants, because of strong electrostatic interactions, spherical micelles will always be more stable than the rod-shaped aggregates occurring in the hexagonal phase. Addition of salt decreases the repulsive forces between the micelles in oleate solutions. This makes it possible to accommodate more micelles in the randomly ordered micellar state existing in the solution, i.e. the solubility increases, and at the same time the formation of rod-like aggregates is promoted. At a sufficiently high salt concentration, repulsion between the charged end-groups decreases to such an extent that it becomes energetically favourable to form lamellar aggregates. These represent a geometry for which a larger portion of the hydrocarbon chains of the oleate molecules is shielded from the energetically unfavourable contact with water than in cylindrical or spherical aggregates. As the electrostatic forces decrease this factor becomes more important. Hence, at high salt concentrations, lamellar rather than hexagonal phase is in equilibrium with aqueous solution. Finally, the repulsive forces are decreased to such an extent that lamellar phase is always more stable than micellar aggregates. As predicted by the theory of Wennerström et al.[30,31] this transition takes place at a water activity which is the same for NaCl and Na_2SO_4.

Phase Equilibria

The attainment of equilibrium in systems containing fatty and rosin acids is slow. This implies an additional difficulty in investigations of systems containing sodium abietate because the chemical stability of this compound at elevated temperatures is limited. Figures 11-13 show phase diagrams for NaOl/HOl, NaAb/HAb and NaOl/HAb at constant concentration of NaCl (0.25 mol/dm^3) and $60^\circ C$. Because of the experimental difficulties the phase diagrams are not complete. However, the structure of the phases and the occurrence of more than one phase in the multiphase regions shown in the diagrams have been clearly established by optical microscopy, X-ray scattering and determination of nuclear quadrupole splitting of deuterium in samples prepared with D_2O.

When plotting these phase diagrams it was assumed that all of the NaCl is dissolved in the aqueous solution. The pH of the aqueous solutions in equilibrium with the different phases is indicated in the diagrams.

Sodium oleate/oleic acid. Figure 11 shows the NaOl/HOl system. Normal hexagonal phase is not formed in the system NaCl solution/ NaOl. The lamellar phase, upon addition of HOl, is able to swell considerably with water despite the high salt concentration. In addition, a reversed hexagonal phase and an isotropic liquid crystalline phase is formed. Both phases are in equilibrium with dilute NaOl solutions (below the cmc). The isotropic phase is extremely viscous; the structure of this phase has not been determined.

The isotropic solution region L2 (solution of water and NaOl in HOl) extends towards the water corner in a salient area which roughly corresponds to a molar ratio NaOl:HOl = 1:2. It has been shown for other fatty acid/soap systems (showing similar phase behaviour) that this is due to the formation of carboxylic acid/ sodium carboxylate complex consisting of two soap and four acid molecules in the solution.[32-34]

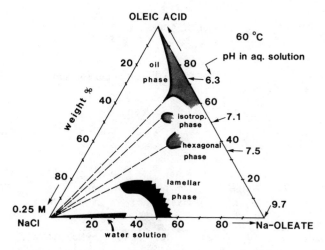

Figure 11. Phase diagram of the system sodium oleate/oleic acid/ 0.25 mol/dm^3 NaCl at 60°C.

Two-phase regions between the three different liquid crystalline phases and the two solutions have been clearly identified by
separation and by nmr studies. However, although the nmr spectra
give unambiguous information in two-phase regions (a typical spectrum
is given in Figure 14), three-phase areas involving two liquid crystalline phases have not been observed experimentally, indicating that
the areas must be very narrow. The dashed lines in Figure 11 indicate
that it has been experimentally confirmed that all three phases may
be in equilibrium with NaOl solutions below the cmc.

Sodium abietate/abietic acid. The phase diagram for this
system is shown in Figure 12. It has been determined only for
relatively low concentrations of the amphiphilic components. The
lamellar phase is able to incorporate considerable amounts of
abietic acid. Except for lamellar phase and aqueous solution, the
only phases occurring in this system are crystalline abietic acid
and sodium abietate.

Sodium oleate/abietic acid. As Figure 13 shows, the lamellar
phase of NaOl is able to incorporate considerable amounts of abietic
acid, which is also to some extent solubilized in the oleate
micelles while it is completely insoluble in the NaCl solution.
The phase diagram in Figure 13 is quite incomplete. In particular,
it is obvious that a further decrease of the pH will precipitate
the phases shown in Figure 11. The influence of abietic acid on
the regions of existence of these phases has not been investigated.

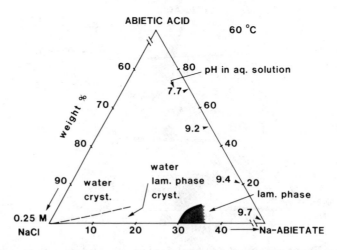

Figure 12. Phase diagrams of the system sodium abietate/abietic
acid/0.25 mol/dm^3 NaCl at 60°C.

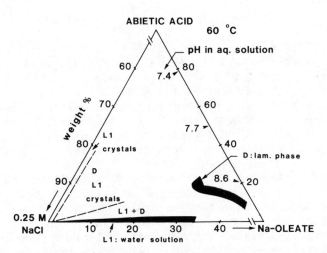

Figure 13.   Phase diagram of the system sodium oleate/abietic acid/0.25 mol/dm³ NaCl at 60°C.

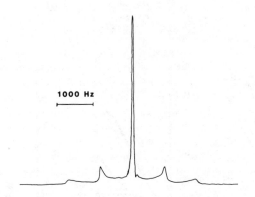

Figure 14.   Deuterium quadrupole splitting in two-phase region between lamellar and micellar phases in the NaAb/NaOl/0.25 mol/dm³ NaCl system. The single peak is from micellar solution and the double peak from lamellar phase.

Note, however, that abietic acid dissolves in the lamellar oleate phase to a molar ratio of about 1:2.

Solubilization in Micellar Solutions

The ability of NaAb, NaAb/NaOl and the NaOl micelles to solubilize water-insoluble compounds was investigated for three compounds that represent a saturated linear aliphatic hydrocarbon (decane), a cyclic hydrocarbon (decalin) and an amphiphilic compound (decanol) of about the same molecular weight. These compounds may also serve as models of the neutral components occurring in wood extractives. Figure 15 shows the maximum solubilization of these compounds in the three types of micelles.

We note first of all that decalin is solubilized to a larger extent than the other compounds; the solubilizing capacity is about 1 mol decalin/mol oleate in NaOl solution and about 0.7 mol/ mol surfactant for the other compounds. The solubilizing capacity for decane is also higher in the oleate solutions than in the other micellar solutions. Somewhat surprisingly, solubilization of decanol is considerably smaller (on a mol/mol basis) than solubilization of decalin. However, the molar volume of decalin is considerably smaller than the molar volume of either decane or decanol

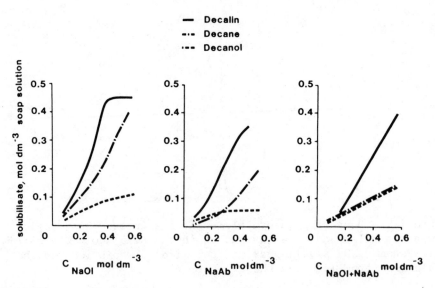

Figure 15.   Solubilization of decalin (——), decane (—·—) and decanol (———) in solutions of NaOl, NaAb and NaOl + NaAb (1:1 w/w) at 60°C.

(155, 194, and 191 cm^3/mol, respectively) which could explain the higher solubilizing capacity. The high solubilizing capacity of NaOl solutions is clearly due to the large micellar size. We also note that the solubilization in the NaAb or NaAb + NaOl solutions constitutes additional evidence that micelles are formed in these solutions.

Some Practical Applications

Some rather speculative suggestions can be made concerning practical problems in pulp manufacture which may be understood on the basis of the phase equilibria described. In technical systems, of course, the number of components is very much larger than can possibly be used in a model system. In particular, more extensive studies of the influence of the neutral components in wood extractives on the phase equilibria should be made to make the system more realistic. Nevertheless, some interesting conclusions are suggested by the results.

1. Dissolution of wood extractives in pulping. In the Kraft pulping process, lignin and wood extractives are dissolved at elevated temperatures in an alkaline sodium hydroxide/sodium sulphide solution. When birch is used as wood pulp it is often found that it is difficult to achieve complete dissolution of the extractives. They are later on removed from the solution by acidification; they separate from black liquor as a viscous oil ("tall oil"). As a matter of practical experience, addition of tall oil from pulping of softwood (pine) often improves the solubility of birch extractives. Table I and Figure 4 suggest an explanation for this behaviour: with the softwood tall oil large amounts of rosin acids are added and Figure 4 shows that this may considerably increase the solubility of fatty acid soaps at high salt concentrations. At the same time the soap to neutral component ratio is increased. This question is discussed in some detail in ref. 19.

2. Liquid crystal formation in black liquor. Roberts and Osterlund[35] have investigated tall oil from black liquor from a sulphate mill; the wood used was mainly birch and the proportion of fatty acids in the extractives, hence, is high. The acid number of the tall oil was 128. Figure 16 shows the results. The abscissa gives the weight-% of tall oil in a tall oil/water mixture. The ordinate shows the amount of NaOH added, which chemically corresponds to neutralization of the fatty (and rosin) acid soaps. The figure shows that aqueous micellar solutions of the tall oil components are formed at high soap concentrations, that solutions of water and soap in tall oil are formed at low soap concentrations and that liquid crystalline structures occur at intermediate concentrations. In addition, very stable emulsions are formed if all three phases are in equilibrium.

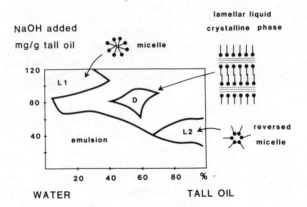

Figure 16.   Phase equilibria for tall oil (in black liquid) from
a sulphate mill.[35]

        To compare this diagram with the model system we have replotted
Figure 11 as shown in Figure 17. The abscissa shows the total percentage of acid + soap (which is taken as a model of the "tall oil"
in Figure 16).   The ordinate shows the fraction of the total acid
+ soap that has been neutralized.   The diagram is surprisingly
similar to the corresponding diagram for the technical tall oil.
The large area in the left-hand lower corner corresponds to two-
phase equilibrium between aqueous and oleic acid solutions or

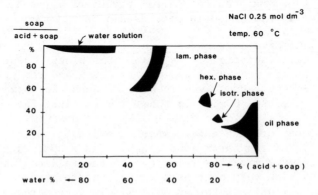

Figure 17.   The phase diagram in Figure 11 plotted analogously
to Figure 16.

three-phase equilibrium between the two solutions and liquid
crystals. It is well known from studies of analogous systems of
typical emulsifiers that extremely stable emulsions are easily
formed if oil, water and lamellar liquid crystal are in equilibrium.
This is, indeed, what is found in the corresponding region of the
technical system (Figure 16).

3. Pitch problems. The viscosities of the different phases
vary widely. Some of them are easily dispersible in aqueous
solution while others are virtually impossible to redisperse once
they have been separated from multi-phase mixtures. The pH values
of the aqueous solutions in equilibrium with the different phases
are shown in Figures 11-13. Clearly, small changes in pH, salt
concentration and the relative amounts of the different components
may have a strong influence on the tendency of phases which are
in equilibrium with dilute aqueous solutions to disperse and to
adhere to different surfaces. We think that this may be of importance for an understanding of the conditions that sometimes
lead to the formation of large amounts of dispersed pitch and
pitch deposits in paper mills. Further investigations to verify
this are in progress.

ACKNOWLEDGEMENTS

This work was supported by the Swedish Pulp Industry's Foundation for Forestry Research. Britt-Marie Müller and Sven Forsberg
are thanked for experimental assistance.

REFERENCES

1. L. A. Allen, Trans. Tech. Sect.Can.Pulp Paper Assoc., 3,32(1977).
2. L. A. Allen, Colloid Polymer Sci., 257, 533 (1979)
3. P. O. Bethge and B. O. Lindgren, Svensk Papperstidning, 65,
    640 (1962)
4. B. O. Lindgren, Acta Chem. Scand., 19, 1317 (1965)
5. J. Bergman, B. O. Lindgren, and C. H. Svahn, Acta Chem. Scand.
    19, 1661 (1965)
6. L. Westfelt, Acta Chem. Scand., 20, 2829 (1966)
7. S. K. Kahila, Paperi Puu, 48, 529 (1966)
8. B. Kimland and T. Norin, Acta Chem. Scand., 21, 825 (1967)
9. B. Kimland and T. Norin, Acta Chem. Scand. 22, 943 (1968)
10. B. Holmbom and E. Avela, Acta Acad. Aboensis Ser. B, 31, 13
    (1971)
11. B. Holmbom and V. Erä, J. Amer. Oil Chem. Soc., 55, 342 (1978)
12. B. Holmbom, Ph.D. Thesis, Åbo Akademi, Åbo, Finland, 1978
13. P. Ekwall and R. Lindström, Finnish Paper Timber J., 23, 64
    (1941)

14. O. Harva, Acta Acad. Aboensis Math. et Phys., XVII, 4 (1951)
15. P. Ekwall and H. Bruun, Acta Chem. Scand., 9, 412, 424 (1955)
16. H. Bruun, Acta Acad. Aboensis Math. et Phys., XIX, 3 (1953)
17. H. Bruun and S. Friberg, TAPPI, 51, 482 (1968)
18. K. Roberts and R. Österlund, TAPPI, 59, 156 (1976)
19. H. Palonen, P. Stenius, and G. Ström, Svensk Papperstidning, 85, R93 (1982)
20. J.W. McBain and W.C. Sierichs, J. Amer. Oil Chem.Soc., 25, 221 (1948)
21. D. E. Koppel, J. Chem. Phys., 57, 4814 (1972)
22. P. Ekwall, L. Mandell, and K. Fontell, Acta Polytech. Scand. Chem. Techn. Ser., 74I,1(1968)
23. J. Ulmius, H. Wennerström, G, Lindblom, and O. Arvidson, Biochemistry, 16, 5742 (1977)
24. J.N. Israelachvili, D. J. Mitchell, and B. W. Ninham, J. Chem. Soc. Faraday Trans. II, 72, 1525 (1976).
25. J. Husson, H. Mustacchi, and V. Luzzati, Acta Cryst., 13, 668 (1960)
26. J. W. McBain, R. Vold, and W. Jameson,J.Amer.Chem. Soc., 61, 30 (1938)
27. B. Tamamushi, M. Shirai, and K. Tamaki, Bull. Chem. Soc. Japan, 31, 467 (1958)
28. F. Reiss-Husson and V. Luzzati, J. Phys.Chem., 68, 3504 (1964)
29. G. Gunnarson, B. Jönsson, and H. Wennerström, J. Phys. Chem., 84, 3114 (1980)
30. B. Jönsson, H. Wennerström, and B. Halle, J. Phys. Chem., 84, 2179 (1980)
31. B. Jönsson, Ph.D. Thesis, Lund University, Lund, Sweden, 1981.
32. P. Ekwall and L. Mandell, Kolloid-Z. Z. Polymere, 233, 937 (1969)
33. G. Söderlund and S. Friberg, Z. Physik. Chemie, Neu Folge, 70, 139 (1970)
34. S. Friberg, L. Mandell, and P. Ekwall, Kolloid Z.u. Z. Polymere, 233, 955 (1969).
35. K. Roberts and R. Österlund, TAPPI, 59, 156 (1976)
36. R. H. Robinson and R. H. Stokes, "Electrolyte Solutions", 2nd rev. ed., Butterworths, London, 1965
37. B. Lindman, G. Lindblom, H. Wennerström, and H. Gustavsson, in "Micellization, Solubilization and Microemulsions", K. L. Mittal, Editor, Vol. 1, p. 195, Plenum Press, New York, 1977.

FLUID MICROSTRUCTURES OF

SODIUM 4-(1'-HEPTYLNONYL)BENZENESULFONATE MIXTURES

W.G. Miller, F.D. Blum,[*] H.T. Davis, E.I. Franses,[**]
E.W. Kaler,[†] P.K. Kilpatrick, K.E. Nietering,
J.E. Puig[††] and L.E. Scriven
Departments of Chemical Engineering and Materials
Science and Chemistry, University of Minnesota
Minneapolis, Minnesota 55455, USA

The phase behavior of sodium 4-(1'-heptylnonyl)-
benezenesulfonate, SHBS, in water or brine from 20 to
90°C is dominated by a broad biphasic region consisting
of a lamellar liquid crystalline phase in equilibrium
with a dilute isotropic SHBS solution. In the isotropic phase the SHBS is not micellar, though conduc-
timetry and nmr spectroscopy indicate weak, concentration dependent aggregation. At 25°C the lamellar phase
is approximately 100° above the gel-liquid crystal
phase transition temperature and the sulfonate head
group undergoes rapid though anisotropic motion. The
lamellar phase is very fluid and easily deformable,
whether dispersed in the saturated aqueous phase or as
the bulk lamellar phase. Mechanical agitation and
other sample preparation techniques give closed shell
liposome dispersions; the mean size and size distribution depend markedly on the sample history. Sonication of a dilute dispersion leads to long lived, non-
equilibrium vesicles. Addition of a suitable cosurfactant, sodium dodecylsulfate, leads to a mixed
micellar solution while the addition of a cosolvent,
e.g., n-pentanol leads to a non-micellar solution.

Many preparations of SHBS-H_2O(NaCl)-(cosolvent,
cosurfactant) show ultralow tension against hydrocarbon. In formulations in which the tension measurements depend on time and preparation technique, an
opaque, viscous, thin, surfactant-rich third phase
forms at the oil-water interface. The microstructures
present in this low volume fraction surfactant-rich

175

middle phase are unknown. The tension behavior, however, correlates with the nature of the microstructure
present in the precontacted aqueous surfactant dispersion. Formulations containing alcohols which produce
low, equilibrium tensions that do not depend on the
method of preparation yield a low-viscosity, translucent middle phase, a microemulsion. The microemulsion
middle phase is shown in this case to be continuous in
both water and oil; it is not a swollen micellar
microstructure.

INTRODUCTION

The literature on aqueous-surfactant and aqueous-hydrocarbon-surfactant systems is rich in proven and suggested microstructures:
single phase and multiphase; equilibrium and nonequilibrium. These
include, for example, molecular solutions; normal and inverted
micelles; microemulsions; lamellar, hexagonal, and cubic ordered
phases; dispersions of liposomes and vesicles. Our interest in
surfactant microstructures came from studies on ultralow ($< 10^{-3}$
mN/m) interfacial tensions between hydrocarbon and water induced
by the presence of low concentrations of surfactants. In some sulfonate systems we found that the apparent interfacial tensions (as
measured with a spinning drop tensiometer) depends on the order of
mixing components,[1] on the hydrocarbon drop size,[1] and on the
length of contact time of hydrocarbon with aqueous surfactant before tension measurement[2] (Figure 1). Through the use of multiple
experimental techniques we were able to correlate these unusual
observations with the surfactant microstructures present in the
aqueous phase prior to tension measurement.

The pure surfactant sodium 4-(1'-heptylnonyl)benzenesulfonate,
SHBS, together with cosurfactants and cosolvents, was used for
many of our studies. In this paper we review the microstructures
present, their size and their stability in surfactant systems containing SHBS.

NATURE OF MICROSTRUCTURES IN AQUEOUS SHBS
Phase Behavior

The progressive addition of surfactant to water at temperatures
above the Krafft-point typically results in an orderly sequence of
microstructures: molecular solution, micellar solution and, eventually, liquid crystalline phases. This occurs both with surfactants having a single hydrocarbon tail[3] and with ones having a
short double tail, such as Aerosol OT[4,5] and the di-alkanoyl lecithins.[6]

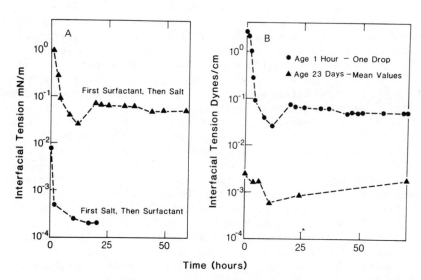

Figure 1. Spinning drop tension measurements against decane of 0.07 wt. % SHBS, 0.3 wt. % NaCl aqueous dispersions as dependent on order of mixing (A) and on contact time before tension measurement (B).

Our early efforts[1,7] yielded a simple temperature-composition phase diagram for SHBS in water. At low concentration at 25°C SHBS dissolves to form a molecular solution. Conductance data show deviation from Onsager limiting law behavior, but no evidence of micellization.[2,8] No Krafft boundary is observed to 90°C, yet at all temperatures investigated a lamellar liquid crystalline rather than a crystalline phase is formed.[7] The isotropic liquid crystalline biphasic region is very broad; at 25°C a 0.06 wt. % SHBS solution is in equilibrium with the hydrated lamellar liquid crystal (ca. 75 wt. % SHBS; 1 mole SHBS to 7.5 mole H_2O). In the presence of NaCl the equilibrium isotropic phase contains much less SHBS.[7]

It was recently proposed that SHBS forms normal micelles of 22Å radius at 45°C.[9,10] Inasmuch as the 45°C conductivity data,[9] on which the presence of a cmc is based, are similar to our 25°C data,[2] we sought to probe for the existence of a cmc using ^{23}Na nmr. With sodium salts of anionic surfactants the chemical shift of the sodium ion associated with micelles is well known to differ from the shift in molecular surfactant solutions.[11] Due to rapid exchange of the Na⁺ from the molecular solution to the micellar environment, the observed chemical shift is a weighted average. Thus the chemical shift undergoes a rapid change with concentration after the cmc is reached, and has proved to be a reliable method for cmc determination.[11] The chemical shift of the sodium ion in a lamellar phase also differs from that in the molecular surfactant solution. The sodium ion is not exchanged rapidly with the sodium in the equilibrium isotropic phase. However, the observed chemical shift changes to that of the lamellar phase shortly after the solubility limit is passed.[12] The observed ^{23}Na chemical shift at 47°C for SHBS in water, Figure 2, shows no break when crossing the suggested cmc but does reveal the appearance of the lamellar phase. The continuous smooth change in the chemical shift up to the solubility limit does suggest aggregation, but not classical micelle formation. This is consistent with both the conductivity and neutron scattering data. Thus over a wide range of surfactant concentration, the aqueous SHBS that is contacted with hydrocarbon for tension measurement consists of the liquid crystalline phase dispersed in an aqueous phase devoid of surfactant micelles. The nature of the dispersed liquid crystalline phase is, then, the microstructure of relevance with respect to the observed anomalous tension behavior.

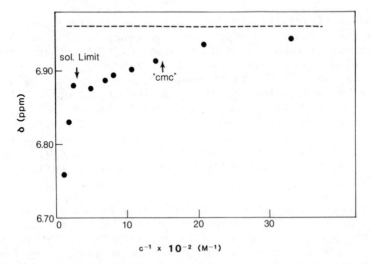

Figure 2. Sodium—23 nmr chemical shift as a function of SHBS concentration at 47°C, compared to the solubility limit and reported[9] cmc. The dashed line is the ^{23}Na chemical shift of NaCl at 47°C.

The Lamellar Phase

Optical microscopy of aqueous SHBS dispersions between crossed polars shows the birefringence expected of smectic liquid crystal.[7] Freeze fracture electron microscopy, Figure 3, reveals clearly its lamellar nature.[13] X-ray diffraction, Figure 4, yields a lamellae repeat distance of 33Å, which is independent of added electrolyte up to 5 wt. % NaCl.[14]

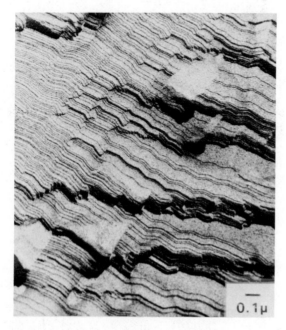

Figure 3. Freeze fracture electron micrograph of a 75 wt. % SHBS smectic phase revealing the lamellar organization.

Differential scanning calorimetry,[15] Figure 5, coupled with temperature dependent nmr[7,12,15] and x-ray diffraction[14] data reveals a number of features of the molecular dynamics and thermo-dynamics of the lamellar phase:

1) the hydrocarbon chain melting occurs at ca. -75°C, while the water between the surfactant bilayers melts over a broad range from -50 to -10°C.

2) the motion of the benzenesulfonate head group is highly an-isotropic, and becomes frozen-in at temperatures where the water between bilayers freezes.

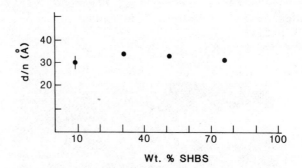

Figure 4. Concentration dependence of the x-ray spacings of SHBS dispersions in water.

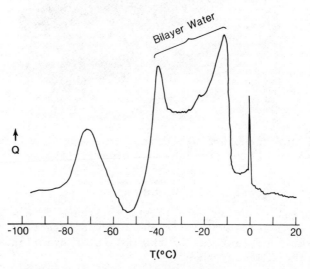

Figure 5. Differential scanning calorimetry of 75 wt % SHBS. The sharp peak at 0°C is due to a trace of extraneous water.

3) the x-ray diffraction pattern of the lamellar phase below
-50°C indicates that the frozen water is not normal hexagonal ice.
The SHBS concentration must be in the biphasic region (< 75 wt. %
SHBS) before the normal ice lattice is observed.

4) the ^2H, ^{13}C, and ^{23}Na nmr line widths at 25°C indicate the
water between bilayers, the sodium ion, and the benzenesulfonate
head group are all moving rapidly though anisotropically while the
hydrocarbon tails are moving rapidly and nearly isotropically.

The gross features of the lamellar phase are in common with
those of long double-tail surfactants. A number of the details
are, however, different:

1) the hydrocarbon chain melting, i.e. the gel-liquid crystalline phase transition, occurs at a temperature below the freezing
of the bilayer water, and is the lowest such transition temperature
reported. At room temperature the lamellar phase is about a hundred degrees above the gel-liquid crystalline transition temperature.

2) the equilibrium amount of water between surfactant bilayers
which contributes only ~ 8Å to the lamellar thickness, is insensitive to electrolyte (NaCl) concentration, and when frozen is not
ice-like. The distance between sulfonate sheets is so small that
simple electrostatic theory cannot be used to explain the spacing.

3) compared, for example,[4] to Aerosol OT, the isotropic-smectic
biphasic region is extremely broad. The x-ray data verify the
result obtained earlier[7] by isopiestic measurements.

Vesicles

Sonication of liquid crystalline dispersions of double-tailed
surfactants often produces spheroidal vesicles consisting of a
fluid core surrounded by at least one surfactant bilayer or lam-
ella.[16] The presence of vesicles in dilute (< 2 wt. %) sonicated
dispersions of SHBS was first revealed by fast-freeze cold-stage
transmission electron microscopy.[17] Further electron microscopy[18]
has shown that these vesicle populations have approximately a log-
normal distribution of sizes. Quasielastic light scattering measurements[18] indicate that SHBS vesicles have an average diameter of
ca. 450Å in water and that the average vesicle diameter decreases
when the vesicles are prepared in 0.3 wt. % NaCl brine or at higher
surfactant concentrations (5 or 10 wt. %). A combination of differential scanning calorimetry and nmr results, supported by small-
angle x-ray scattering evidence, indicates that most of the
vesicles contain a single bilayer (Figure 6).[18]

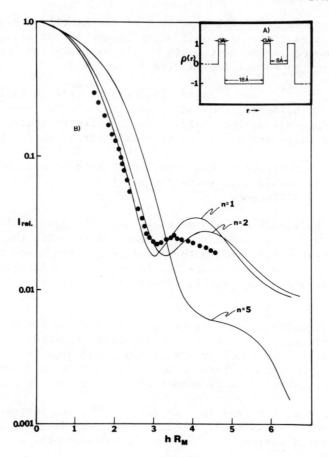

Figure 6. (A) Schematic electron density profile for a SHBS bilayer and water; (B) comparison of the scattered intensity calculated for vesicles with 1, 2, and 5 bilayers (solid lines) with the
measured intensity scattered from a 2.0 wt. % vesicle dispersion
in water (dotted line).

Vesicles are of scientific importance as, for example, simple
models of biological membranes. However, the thermodynamic and
kinetic stability of vesicular dispersions has not been fully determined. SHBS vesicles are ideally suited for long-term vesicle
stability studies because, unlike most biological vesicle-forming
surfactants, SHBS is chemically stable and biologically inert in
aqueous solutions at room temperature.

Optical microscopy,[17] conductance,[17] and light scattering[17,18]
measurements along with electron microscopy[18] were used to follow
the stability of SHBS vesicle dispersions upon aging. Although
vesicles were in some instances found to be stable for months when
left undisturbed after sonication, in all cases a reversion to the

smectic phase (present as dispersed liquid crystallites) was eventually observed by several techniques. These aging studies, as
well as those on phospholipid vesicle dispersions in the literature,[19] were carried out in the presence of some dispersed liquid
crystallites. Some liquid crystallites remain after sonication,
and our attempts to completely remove them failed. Both ultrafil-
tration and ultracentrifugation were unsuccessful, as was the traditional method of gel permeation chromatography (GPC). Indeed,
we showed[18] that flow through a glass bead GPC column actually
hastened the reversion of vesicles to liquid crystallites. Thus,
it has been impossible to investigate the kinetic stability of an
isolated vesicle dispersion in the absence of liquid crystallites.
Nonetheless, the reversion of SHBS vesicles to liquid crystallites
clearly indicates that these vesicles are not equilibrium structures in water or brine.

Vesicle stability was rapidly altered by osmotic shocks, and
by contact with hydrocarbon.[8,20] Tension behavior of vesicle dispersions against hydrocarbon differs from that observed with liposome dispersions.[20]

                              Liposomes

The state of the smectic phase depends markedly on sample
preparation and history. Sonication of dilute dispersions, as
discussed previously, produces vesicles. Mechanical agitation,
whether by hand or by a vortex mixer, alters the microstructure
even in the bulk smectic phase. Our first indication of this came
from ^{2}H nmr studies,[21,15] shown in Figure 7, where mechanical agitation changed the spectral line shape of a sample agitated at
ambient temperature. Upon standing, even for periods of up to two
years, the line shape did not revert to that observed before
shaking. However, annealing at 90°C for a few minutes resulted
in a return to the original line shape. Similar effects were
difficult to observe at lower concentration, where the system is
biphasic, as the spectrum was dominated by the water in the equilibrium isotropic phase. We conjectured, however, that even with
the single-phase smectic sample, agitation resulted in formation
of smaller domains of unknown morphology. Electron microscopy of
freeze-fractured, replicated samples shows clearly that this is
the case.[14] In the nonagitated sample, freeze-fracturing revealed
many examples of the fracture plane running through the bilayer
plane, as, for example, Figure 3. Upon mechanical agitation the
typical micrograph is as in Figure 8, where the fracture plane has
run around closed-shell smectic particles, liposomes, rather than
through them. Thus in the agitated samples the smectic phase consists of concentric shell liposomes, the mean size of which depends
on the agitation history. In the concentrated unshaken samples it
is difficult to determine if the much larger particles are closed
shell liposomes or planar.

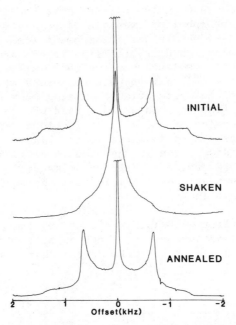

Figure 7. Deuterium nmr spectra at ambient temperature of the hydrated smectic phase (75 wt. % SHBS, 25 wt. % D_2O) before and after mechanical agitation at room temperature on a vortex mixer, and after heating to 90°C.

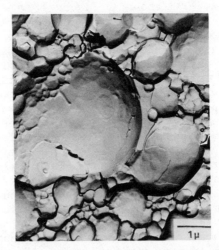

Figure 8. Electron micrograph of the hydrated lamellar phase after mechanical agitation.

The nmr lineshape is easily understood.[14] Annealing at 90°C
results in electron micrographs similar to the unshaken material,
indicating that the liposome size depends on thermal history as
well as mechanical treatment. Similar behavior was observed at
lower concentrations, where the system is biphasic. The mean size
distribution of the liposomes is highly dependent on sample preparation[14,18] and can be easily altered by even gentle handling
techniques. The size distribution can be quite broad with sizes
ranging from hundreds of angstroms to hundreds of microns in the
same sample.

EFFECTS OF COSOLVENTS AND COSURFACTANTS

The addition of short-chain alcohols to dispersions of SHBS
liposomes alters the microstructure. At low alcohol-to-SHBS
ratios, the alcohol is incorporated into the lamellar phase and
aligns itself with its polar hydroxyl group near the mean plane of
surfactant head groups and it short alkyl tail penetrating the
hydrocarbon lamellae. This was determined by a combination of
centrifugation and ^{13}C nmr.[22] The size and number of liposomes
are altered, as observed by spectroturbidimetry.[2,8] As the alcohol content is increased, the lamellar phase is solubilized into
an isotropic phase; the amount of alcohol necessary to completely
solubilize the liposomes decreases as the lipophilicity of the
alcohol increases in the series n-butanol, n-pentanol, n-hexanol.
Conductance data for the isotropic solution (Figure 9A) show marked
deviation from Onsager limiting law behavior but no sharp break
to indicate the onset of micelles. Low-angle x-ray scattering
measurements confirmed the absence of micelles.[22]

Addition of the cosurfactant sodium dodecylsulfate (SDS) to
a SHBS dispersion in water or brine results in formation of a
mixed lamellar phase, as deduced by conductivity, nmr and calorimetry.[2,23] These mixed liquid crystals form at weight ratios R of
SDS/SHBS below ca. 0.2. At R above ca. 2.0, a single isotropic
phase of mixed micelles of SDS and SHBS forms, while for 0.2 < R
< 2 both mixed liquid crystals and mixed micelles are present.
Figure 9B shows that the microstructures formed upon solubilization of SHBS lamellar liquid crystalline dispersions by a micelle-
forming surfactant are different from those formed upon solubilization by an alcohol.

Quasielastic light scattering measurements indicate that the
mixed micelles formed by mixtures of SDS and SHBS are non-spherical in both 0.3 M and 0.6 M NaCl and that the apparent molecular
weight of the micelle increases as the ratio R decreases.[24]

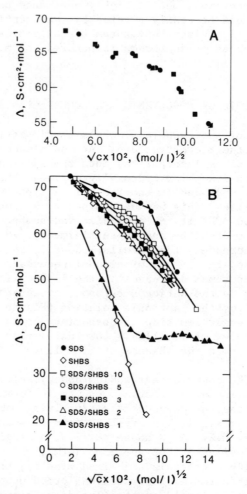

Figure 9. (A) Equivalent conductivity as a function of SHBS concentration for SHBS-NAA-water solutions; (B) Equivalent conductivity as a function of surfactant concentrations for aqueous preparations of SDS and SHBS in different weight ratios.

SHBS-BRINE-HYDROCARBON

The phase behavior in the presence of hydrocarbon depends markedly on the components present. With SHBS-H$_2$O(NaCl)-hydrocarbon,
and with SHBS-SDS-H$_2$O(NaCl)-hydrocarbon formulations which give
ultralow interfacial tensions, a viscous phase which is turbid, frequently opaque, low volume fraction, and of high surfactant content is formed between the water-rich lower phase and the oil-rich
upper phase.[1,8,23] The microstructures present in this third
phase are unknown. It is these systems which display the order-of-
mixing and contact time effects shown in Figure 1. Similar effects
have been observed with commercial, multicomponent sulfonates.[25]

The occurrence of apparently ill-behaved tension measurements,
the appearance of low volume fraction surfactant-rich phase after
contact with hydrocarbon, and the presence of liquid crystalline
dispersions in the aqueous surfactant phase before contact with
hydrocarbon are related. The history dependent tension measurements can be correlated with the size and type of SHBS dispersion
in the aqueous phase before contact with hydrocarbon. The dispersed surfactant must reach the surface of the hydrocarbon droplet
in the spinning drop measurement. The delivery time depends on
the size and density of the particle, and on the flow times in the
spinning drop tube. In addition, the dispersed surfactant must
spread on the hydrocarbon droplet and produce a thin coat. Thus
the order-of-mixing effect is a manifestation of the fact that the
SHBS liposomes are less finely dispersed when surfactant is added
to a salt solution than when salt is added to the dispersed surfactant. When the third phase was present we found ultralow tensions between oil-rich and water-rich phases;[1,2] otherwise, the
tension was not ultralow.

SHBS-BRINE-ALCOHOL-HYDROCARBON

When a low molecular weight alcohol is added as a cosolvent,
SHBS-alcohol-brine-hydrocarbon mixtures often yield microemulsion
phases in equilibrium with excess oil or brine or both.[26] These
microemulsions can incorporate substantial amounts of both oil and
water. They are typically transparent or translucent, isotropic,
and have low viscosities. Ultralow interfacial tensions are frequently found between these microemulsions and the excess phases
with which they are in equilibrium. Because these tension measurements are made on equilibrium systems, no order-of-mixing or aging
effects are observed. There is ample evidence that the ultralow
interfacial tension states and the patterns of phase behavior
found in microemulsion-forming mixtures such as these result from
the nearness of microemulsions to critical points.[27-29] However,
the microstructure of microemulsions and the way it changes with
relative proportions of oil and water remain open questions, par-

ticularly for microemulsions containing substantial amounts of both oil and water.

There is convincing evidence[30] that the addition of small amounts of oil to an aqueous surfactant solution often produces an oil-in-water microemulsion containing surfactant micelles swollen with oil. The analogous water-in-oil microemulsion likewise can contain inverted surfactant micelles swollen with water. Both types of micelles are presumed to be stabilized by surfactant that separates the oil(or water)-rich interiors from the continuous water(or oil)-rich exterior. As appreciable amounts of oil (or water) are incorporated into the microemulsion, the micelles swell; such micellar solutions have been described as equilibrium dispersions of "droplets".[31] This picture of microemulsion microstructure as disjoint "droplets" is viable at low volume fractions of dispersed component. However, at intermediate volume fractions of oil and water it is possible that microemulsions organize into equilibrium <u>bicontinuous</u> structures in which both oil-rich and water-rich regions span the sample.[32-40]

Bicontinuous water-oil-surfactant states have been described in detail elsewhere.[32,33] One convenient representation of microemulsion structure that can naturally incorporate water/oil, bicontinuous, and oil/water equilibrium microdispersions is the Voronoi model, a random tessellation of space into convex polyhedra filled by water and oil and separated by surfactant sheets. First used by Talmon and Prager[34,35] to build a free energy function for microemulsion systems that predicts phase behavior in qualitative agreement with experiment, it has been extended to model the dynamic and static scattering from a bicontinuous microemulsion microstructure[36] and to represent the electrical conductivity of microemulsions.[37] Experimental results show that the representation of a microemulsion as a bicontinuous microstructure is appropriate over a wide range of compositions in both SHBS and other surfactant mixtures.[36-41]

CONCLUSIONS

1) The temperature-composition phase diagram of SHBS with brine or water is dominated by a large biphasic region in which a lamellar liquid crystalline phase is in equilibrium with a dilute isotropic SHBS solution.

2) The isotropic SHBS solution is not micellar; however, conductance and nmr spectroscopy suggest that non-micellar aggregates are present.

3) The lamellar phase is very fluid, easily deformable, and can be readily dispersed in water to form a broad distribution of closed shell liposomes.

4) The size distribution of liposomes found in dispersions depends markedly on the method of preparation.

5) ^{2}H, ^{13}C, and ^{23}Na nmr indicate that at room temperature the sodium counterions and the sulfonate head group undergo rapid though anisotropic motion in the lamellar phase while the hydrocarbon tail groups move rapidly and nearly isotropically.

6) Sonication of dilute dispersion leads to long-lived, non-equilibrium vesicles which probably consist of a single bilayer.

7) Addition of a suitable alcohol to a liposome dispersion solubilizes the lamellar phase and yields an isotropic phase which is weakly aggregated but not micellar.

8) Addition of a micelle-forming cosurfactant, SDS, produces either mixed liquid crystals, mixed micelles, or both.

9) SHBS, brine, and hydrocarbon combine to produce an opaque, viscous, surfactant-rich third phase.

10) Interfacial tension behavior which is time and preparation dependent correlates with the presence of the viscous third phase and to the nature of the microstructure present in the unprecontacted aqueous surfactant phase.

11) With the addition of a suitable alcohol, SHBS, brine, and hydrocarbon can combine to form a microemulsion phase which solubilizes substantial amounts of both oil and water and which appears from conductance and small-angle x-ray scattering measurements to be bicontinuous.

ACKNOWLEDGMENTS

This work was supported by the Fossil Energy Division of the U.S. Department of Energy as part of the University of Minnesota program on fundamentals of enhancing petroleum recovery.

REFERENCES

1. E. I. Franses, J. E. Puig, Y. Talmon, W. G. Miller, L. E. Scriven, and H. T. Davis, J. Phys. Chem. 84, 1547 (1980).
2. J. E. Puig, L. E. Scriven, H. T. Davis, and W. G. Miller, in "Interfacial Phenomena in Enhanced Oil Recovery," D. Wasan and A. Payatakes, eds., AIChE Symposium Series S-212, 1 (1982).
3. K. Fontell, Mol. Cryst. Liq. Cryst. 63, 59 (1981).
4. J. Rogers and P. A. Winsor, J. Colloid Interface Sci. 30, 247 (1969).

5. E. F. Williams, N. T. Woodberry, and J. K. Dixon, J. Colloid
    Sci. 12, 452 (1957).
6. R. J. M. Tausk, I. Karmiggelt, C. Oudshorn, and J. Th. G.
    Overbeek, Biophys. Chem. 1, 175 (1974).
7. E. I. Franses, H. T. Davis, W. G. Miller, and L. E. Scriven,
    in "Chemistry of Oil Recovery," R. T. Johansen and R. L. Berg,
    eds., ACS Symposium Seires 91, Amer. Chem. Soc., Washington,
    D.C., 1979, p. 35.
8. J. E. Puig, Ph.D. Thesis, University of Minnesota, 1982.
9. L. J. Magid, R. J. Shaver, E. Gulari, B. Bedwell, and S.
    Alkhafaji, Prep. Div. Pet. Chem., Am. Chem. Soc. 26, 93
    (1981).
10. L. J. Magid, R. Triolo, J. S. Johnson, Jr., and W. C. Koehler,
    J. Phys. Chem. 86, 164 (1982).
11. H. Gustavsson and B. Lindman, J. Amer. Chem. Soc. 100, 4647
    (1978).
12. F. D. Blum, Ph.D. Thesis, University of Minnesota, 1981.
13. M. J. Costello and T. Gulik-Kryzwicki, Biochim. Biophys. Acta
    455, 412 (1976).
14. K. E. Nietering, P. S. Russo, F. D. Blum, and W. G. Miller,
    to be submitted to J. Colloid Interface Sci.
15. F. D. Blum and W. G. Miller, J. Phys. Chem. 86, 1729 (1982).
16. C. Huang, Biochemistry 8, 344 (1969).
17. E. I. Franses, Y. Talmon, L. E. Scriven, H. T. Davis, and
    W. G. Miller, J. Colloid Interface Sci. 86, 449 (1982).
18. E. W. Kaler, A. H. Falls, H. T. Davis, L. E. Scriven, and
    W. G. Miller, J. Colloid Interface Sci., in press.
19. A. L. Larabee, Biochemistry 18, 3321 (1979).
20. J. E. Puig, E. I. Franses, Y. Talmon, H. T. Davis, W. G.
    Miller, and L. E. Scriven, Soc. Pet. Eng. J. 22, 37 (1982).
21. E. I. Franses, K. Rose, F. D. Blum, R. G. Bryant, and W. G.
    Miller, in preparation.
22. P. K. Kilpatrick, F. D. Blum, H. T. Davis, A. H. Falls, E. W.
    Kaler, W. G. Miller, J. E. Puig, L. E. Scriven, and N. A.
    Woodbury, in "Microemulsions," I. D. Robb, ed., Plenum Publ.
    Corp., 1982, p. 143.
23. J. E. Puig, E. I. Franses, and W. G. Miller, J. Colloid
    Interface Sci., in press (1982).
24. E. W. Kaler, J. E. Puig, and W. G. Miller, to be submitted to
    J. Phys. Chem.
25. A. C. Hall, Colloids and Surfaces 1, 209 (1980).
26. K. E. Bennett, C. H. K. Phelps, H. T. Davis, and L. E. Scriven,
    Soc. Petr. Engr. J. 21, 747 (1981).
27. P. D. Fleming, III and J. E. Vinatieri, AIChE J. 25, 493
    (1979).
28. P. D. Fleming, III, J. E. Vinatieri, and G. R. Glinsmann, J.
    Phys. Chem. 84, 1526 (1980).
29. H. T. Davis and L. E. Scriven, Soc. Petr. Engr., paper #9728,
    Dallas, Texas, 21-24 September, 1980.

30. E. Gulari, B. Bedwell, and S. Alkhafaji, J. Colloid Interface
    Sci. 77, 202 (1980).
31. M. Zulauf and H. F. Eicke, J. Phys. Chem. 83, 480 (1979).
32. L. E. Scriven, Nature 263, 123 (1977).
33. L. E. Scriven, in "Micellization, Solubilization and Micro-
    emulsion," K. L. Mittal, ed. Plenum Press, New York, 1977,
    p. 877.
34. Y. Talmon and S. Prager, Nature 207, 333 (1977).
35. Y. Talmon and S. Prager, J. Chem. Phys. 69, 2984 (1978).
36. E. W. Kaler and S. Prager, J. Colloid Interface Sci. 86, 359
    (1982).
37. K. E. Bennett, J. C. Hatfield, H. T. Davis, C. W. Macosko,
    and L. E. Scriven, in "Microemulsions," I. D. Robb, ed.,
    Plenum, 1982, p. 65.
38. B. Lindman, N. Kamenka, T. M. Kathopoulis, B. Brun, and P-G.
    Nilsson, J. Phys. Chem. 84, 2485 (1980).
39. E. W. Kaler, K. E. Bennett, H. T. Davis, and L. E. Scriven,
    J. Chem. Phys., in press (1982).
40. E. W. Kaler, H. T. Davis, and L. E. Scriven, J. Chem. Phys.,
    in preparation.
41. P. Stilbs, M. E. Mosely, and B. Lindman, J. Magn. Resonance
    40, 401 (1980).

Current address:

*   Department of Chemistry, Drexel University, Philadephia, PA
     19104.

**   School of Chemical Engineering, Purdue University, West
     Lafayette, IN 47907.

†   Department of Chemical Engineering, University of Washington,
     Seattle, WA 98195.

††   Instituto Mexicano del Petroleo, 07730, Mexico, D.F.

PHASE STRUCTURES AND PHASE DIAGRAMS OF SOME SURFACTANT SYSTEMS WITH DIVALENT COUNTERIONS: EFFECT OF Ca^{2+} AND Mg^{2+} COUNTERIONS ON THE STABILITY OF LIQUID CRYSTALLINE PHASES

Ali Khan[1a], Krister Fontell[1b], and Björn Lindman[1c]

Chemical Center, University of Lund

P.O. Box 740, S-220 07 Lund, Sweden

The phase behaviour of some surfactant systems with divalent counterions is examined using ^2H nuclear magnetic resonance. Systems investigated are calcium octyl sulphate-decanol-water, calcium di-2-ethylhexyl sulpho-succinate-water, and magnesium di-2-ethylhexyl sulpho-succinate-water. The phase diagrams are compared with those of the corresponding systems with monovalent counterions. As regards the existence and stability ranges of isotropic solutions, normal and reversed hexagonal liquid crystalline phases, and cubic liquid crystalline phases there are only minor influences of counterion valency. On the other hand, the stability range of the lamellar liquid crystalline phase is dramatically reduced on replacing sodium with calcium or magnesium. The very much reduced capability of the lamellar phase to swell and take up water with divalent counterions is in agreement with recent theoretical Poisson-Boltzmann and Monte Carlo studies by Wennerström et al.

The observation of ^{25}Mg NMR quadrupole splittings for the lamellar liquid crystalline phase of the magnesium di-2-ethylhexyl sulphosuccinate-water system opens a new possibility to study counterion binding and counterion competition in surfactant systems.

193

INTRODUCTION

Surfactants usually undergo self-association in water to form micelles of different shapes and sizes, and liquid crystalline phases of different structures[2-6]. The structures of micelles in dilute solutions as well as liquid crystalline phases of normal hexagonal, reversed hexagonal and lamellar type have been well characterized. There are additionally a number of optically isotropic cubic liquid crystalline phases[6,7]. The transition between any two phases is usually of the first order and can arise as a result of concentration and/or temperature changes.

The locations and extensions of phases in the phase diagrams can be very different for different multicomponent surfactant systems. However, the phase diagram for a given ionic surfactant system is often found roughly unaltered, especially at high water contents, by substitution of one monovalent counterion for another[5,8-9]. This agrees with the theoretical work which demonstrates that the long-range electrostatic interactions play a dominant role in determining the aggregate shape[10-12]. Ionic surfactants with different monovalent counterions having similar chemical nature are expected to show quite small differencies in electrostatic interactions leading to similar phase stability in these systems. The valency of the counterion, on the other hand, influences the electrostatic effects dramatically[13], so replacing a monovalent counterion by a divalent one is expected to change the self-association behaviour and thus the phase diagram extensively.

In addition to the applied aspects, it is important to study surfactant systems with divalent counterions experimentally for a general understanding of the principles of phase diagrams. In this paper, we present a general account of phase diagrams for three surfactant systems with divalent counterions and compare them with the corresponding systems with monovalent counterions. The systems studied are the ternary system of calcium octylsulphate-decanol-water and the binary systems of calcium and magnesium di-2-ethylhexyl sulphosuccinate-water; all studies refer to 300K. As regards the first system we review here merely the results of a previous comprehensive report[14].

The phase diagrams of these systems have been studied by following the water deuteron (^{2}H) NMR spectra. The theoretical principles of ^{2}H NMR in amphiphile systems have been documented[15-17]; and the method has previously been applied successfully to study phase structures and phase diagrams of surfactant[14,18-22] and biological lipid systems[23-25].

The technique is non-perturbing, rapid and avoids the often cumbersome separation of different phases present in the sample and the results obtained are of relatively good precision.

EXPERIMENTAL SECTION

Calcium octylsulphate, calcium di-2-ethylhexyl sulphosuccinate, and magnesium di-2-ethylhexyl sulphosuccinate were prepared, respectively, from sodium octylsulphate and sodium di-2-ethylhexyl sulphosuccinate (Aerosol OT) by an ion-exchange method as described previously[14],[16]. Deuterium oxide (99.7 at %[2]H) was bought from Ciba-Geigy, Switzerland.

The preparation and mixing of samples to equilibrium were carried out in the way described elsewhere[14],[16].

^{2}H NMR studies were made et 15.351 MHz on a modified Varian XL-100-15 pulsed spectrometer working in the Fourier transform mode using an external lock. The samples were thermally equilibriated at least for one hour before the spectrum was recorded and the error in temperature was within $1^{\circ}C$. The ^{2}H NMR spectra were analyzed as reported earlier[14].

RESULTS AND DISCUSSION

Phase diagrams

The isotropic phases such as the normal micellar solution (L_1), the reversed micellar solution (L_2) and the cubic liquid crystalline phase (I) produce a singlet in the ^{2}H NMR spectrum (Figure 1a). Due to the phase anisotropy, the hexagonal (E) and the lamellar (D) liquid crystalline phases give a quadrupolar splitting in the spectrum (Figure 1b). The value of splitting in the D phase is, in agreement with theoretical results, much larger than that in the E phase region. An equilibrium mixture of an isotropic phase, L_1, L_2, or I, with an anisotropic liquid crystalline phase, E or D, produces a splitting and a central peak (Figure 1c). The values of the splittings were used to identify the E and D phases. When the two liquid crystalline phases, E and D, occur in the sample, two superimposed splittings are obtained in the ^{2}H NMR spectrum (Figure 1d). A spectrum consisting of two splittings and a central peak (only for three component systems) is obtained in a three phase triangle of one isotropic and two anisotropic phases (Figure 1e). On the basis of these simple principles it is possible to determine phase diagrams for simple and complex surfcatant systems.

For the systems of this study the identification of the various phases was straight-forward. It was then possible by a systematic variation of composition to establish with relatively good precision the entire phase diagram except for the surfactant-rich part of the ternary system. This difficulty arises because of the low water content which results in large splittings giving low peak amplitudes. In this part of the phase diagram, the crystals

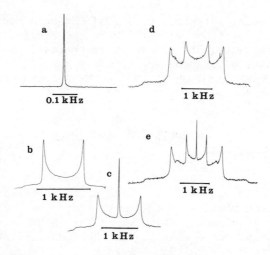

Figure 1. Series of typical ^{2}H NMR spectra for various samples
in the ternary system calcium octylsulphate-decanol-water at 300K
(actual sample compositions (w/w) are given in parenthesis):
(a) isotropic solution (L_1) (30.0/5.0/65.0); (b) hexagonal phase
(E) (50.0/8.0/42.0; (c) a mixture of one anisotropic liquid
crystalline phase and one isotropic phase (20.0/40.0/40.0);
(d) a mixture of two liquid crystalline phases (E and D) (50.0/
14.0/36.0); (e) a mixture of two anisotropic liquid crystalline
phases (E and D) and one isotropic solution phase (47.8/13.9/38.3)

of calcium octylsulphate are in equilibrium with the liquid crystalline phases. Polarizing microscopy studies, which rather easily
distinguish the presence of crystals of the surfactant from the
texture of the liquid crystalline phase, were used to establish
the phase diagram of the system.

The phase diagram for the ternary system calcium octylsulphate-
decanol-^2H_2O at 300K is shown in Figure 2a. In this ternary system,
there appear 5 homogeneous phases, 8 two-phase regions, and 4 three-
phase triangles. The liquid crystalline material in the E phase has
an angular microscopic texture and that in the D phase, a mosaic
texture. ^2H splitting values in the D phase are about twice as large as those in the E phase (Table I). X-ray diffraction studies
showed that the bilayer thickness in the D phase is about 24Å and
that the mean cross-section of the hydrocarbon chains is about
$26Å^2$. In the E phase, one obtains a rod diameter of about 26Å and
a mean cross-section for the hydrocarbon chains of about 52Å at
the polar end. These values agree well with those obtained for
analogous monovalent surfactant systems[5,27-28].

A comparison of this calcium surfactant system with the corresponding one (Figure 2b) with sodium as counterion[29] shows that
there are great similarities in the phase behaviour of the two systems. There is no problem with a comparison based on weight percentages since the equivalent weights of sodium and calcium are closely the same. It is clear that the loacation and extension of the
two micellar solution phases-aqueous L_1 and decanolic L_2-are little
affected by the counterion substitution. The same is also true for
the hexagonal liquid crystalline phase, region E. Minor differences
are that for the calcium system, the L_2 region extends towards higher water contents and that the E boundary against L_1 is more curved.
The main difference is, however, in the behaviour of the liquid
crystalline region D. The long salient of the D phase towards the
water corner for the sodium system is absent for the calcium system.
It appears that the lamellae do not have the same capability to
swell and incorporate large amounts of water in the presence of
calcium.

A very similar effect on the lamellar liquid crystalline phase
has been observed for binary systems of di-2-ethylhexyl sulphosuccinate and water when sodium is replaced by a divalent counterion.
The binary phase diagram of this system with sodium as a counterion
was studied by Rogers and Winsor[30], and by Ekwall et al.[31]. This surfactant has a very low solubility in water - ~1.3% at 300K. The
anhydrous compound is liquid crystalline and its structural building
blocks are rod-like units of reversed nature in hexagonal array[31,32].
The location and the extension of phases were recalculated to apply
for ^2H_2O at 300K and are shown in Figure 3. This system forms a
reversed hexagonal liquid crystalline phase, region F (~0-17% ^2H_2O),

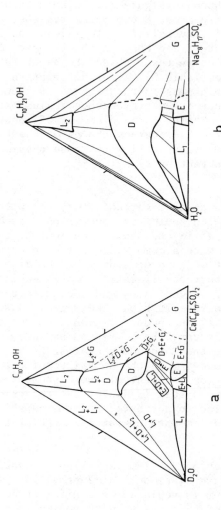

Figure 2a. Phase diagram for the system calcium octylsulphate–decanol–heavy water at 300K. L_1 and L_2 are isotropic aqueous and decanolic solutions, respectively; D and E, are, respectively, lamell–ar and hexagonal liquid crystalline phases; G is crystalline solid substance; $L_1+L_2, D+E+L_1$ etc. are two– and three–phase zones.

Figure 2b. Phase diagram for the system sodium octylsulphate–decanol–water[29] at 293K. Phase notations as in Figure 2a.

Table I. The water deuteron (^{2}H) quadrupolar splittings (Δ^2H) for some representative samples in the calcium octylsulphate-decanol-^{2}H$_2$O system at 300K.

Ca(C$_8$H$_{17}$SO$_4$)$_2$	C$_{10}$H$_{21}$OH	^{2}H$_2$O	Δ^2H	Phase
wt%	wt%	wt%	kHz	
58.02	–	41.98	0.81	E
55.77	–	44.23	0.84	E
50.0	8.0	42.0	0.81	E
49.6	28.0	22.4	1.80	D
45.2	26.0	28.8	1.50	D
43.1	30.8	26.0	1.31	D

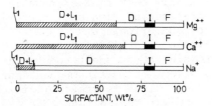

Figure 3. Region of existence of different phases for the system di-2-ethylhexyl sulphosuccinate-heavy water at 300K with different counterions. L$_1$, isotropic solution; D, lamellar; F, reserved hexagonal; and I, isotropic cubic, liquid crystalline phases.

a small region of cubic liquid crystalline phase, region I (~17-23-wt% ^2H_2O; two-phase regions are included) and a large region of lamellar liquid crystalline phase, region D (~23-89wt% ^2H_2O), followed by a two-phase region, D+L_1 (89-99wt% ^2H_2O). The two-phase regions F+I and I+D are small; this applies also to the corresponding calcium and magnesium systems. ^2H NMR produced quadrupolar splittings for samples in the D and F regions (Table II) and a singlet for the I phase.

The formation of phases and their extensions with Ca^{2+} and Mg^{2+} as counterions were established by following the ^2H NMR as stated earlier; phase extensions are included in Figure 3. The figure shows that the appearance of phases in both the calcium and the magnesium systems follows the same sequence as in the sodium system. The quadrupolar splittings obtained in the D and F phases for these two systems are approximately the same as in the D and F phases of the sodium system (Table II). Like the I phase in the sodium system, the I phases in both the calcium and magnesium systems produced a singlet in the ^2H NMR spectrum. It is obvious from Figure 3 that the location and extension of the F and I phases have hardly been affected by the counterion substitution. The same holds true for the D phase on the water-poor side (23wt% ^2H_2O). There is, on the other hand, a dramatic difference on the water-rich side of the D phase. For the calcium system, the D phase cannot incorporate more than 35wt% ^2H_2O and for the magnesium system not more than 40wt% ^2H_2O, figures to be compared with about 89wt% for the sodium system. The large two-phase region, D+L_1, extends for the calcium and magnesium systems upto almost pure water, i.e. the solubilities of the calcium and magnesium surfactants in water are very small.

It has been shown in these examples that the lamellar liquid crystalline phase has a very much smaller capability to swell and take up water in the presence of divalent counterions than in the presence of monovalent counterions; other phase extensions, such as of the hexagonal liquid crystalline or of the cubic liquid crystalline phases, remain rather unaltered by the counterion substitution. Phase diagrams of multicomponent surfactant systems are of technical and theoretical interest. It is therefore important to understand the various factors that influence the features of a phase diagram. The theoretical understanding of such complex systems has now progressed so far that an almost quantitative interpretation of the phase diagram is possible[10-12]. Long-range electrostatic interactions are found to be of great importance for the stability of the liquid crystalline phases. These interactions were computed for monovalent counterions using the Poisson-Boltzmann (PB) equation. It has been demonstrated that the capability of the lamellar liquid crystalline phase to swell and incorporate water between charged amphiphile lamellae results from a repulsion between the amphiphile layers. The PB approach predicts

Table II. The water deuteron (^{2}H) quadrupolar splittings (Δ^2H) for samples in the liquid crystalline phases of the metal ion di-2-ethylhexyl sulphosuccinate-^{2}H$_2$O systems with Na$^+$, Ca^{2+}, and Mg^{2+} as counterions. Sample composition: lamellar phase, D, 30wt% ^{2}H$_2$O; and reversed hexagonal phase, F, 10wt% ^{2}H$_2$O. Temperature 300K.

Counterion	Phase	$\dfrac{\Delta^2\mathrm{H}}{\mathrm{kHz}}$
Na$^+$	D	1.8
Na$^+$	F	3.6
Ca^{2+}	D	2.0
Ca^{2+}	F	3.5
Mg^{2+}	D	1.8
Mg^{2+}	F	4.0

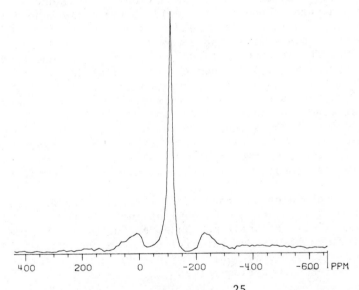

Figure 4. The quadrupolar splitting of ^{25}Mg in the lamellar liquid crystalline phase of magnesium di-2-ethylhexyl sulpho-succinate-^{2}H$_2$O system at 300K. Sample composition, 30wt% ^{2}H$_2$O. NMR parameters: pulse-duration 20μs (90° pulse), acquisition time 5.12ms, total number of pulses 4.5x10^5, (total time required is about 40 min.)

a substantially smaller interlamellar repulsion with divalent
counterions than with monovalent counterions. This leads to a reduction of the predicted swelling, a finding which corresponds to the
fact that the stability of electrostatically stabilized systems is
drastically reduced on replacing monovalent ions by divalent ions[33].
This predicted reduction of the lamellar liquid crystals is, however, not as large as was found experimentally. Recently, Monte
Carlo simulations[13] showed that the PB equation overstimates
the repulsive force by a factor of five for divalent ions under
conditions relevant to the lamellar liquid crystalline phase.
This should lead to a much smaller swelling as is indeed found
experimentally in our systems.

NMR Quadrupolar Splittings of ^{25}Mg Counterions

in the Liquid Crystalline Phase.

To our knowledge Figure 4 shows the first observation of a
quadrupolar splitting of a divalent counterion. This is the splitting of the ^{25}Mg counterion in the lamellar liquid crystalline
phase of the system magnesium di-2-ethylhexylsulphosuccinate-^{2}H$_2$O.
^{25}Mg with a natural abundance of 10.13% has a nuclear spin I=5/2.
Hence ^{25}Mg in an anisotropic liquid crystalline phase should yield
five absorption peaks in its spectrum with the peak intensity ratio
of 5:8:9:8:5. The spectrum recorded on the Nicolet superconducting
(8.5T) spectrometer at a resonance frequency of 22.15 MHz shows
only one splitting and a central peak. A quadrupolar splitting
value of 2.25 kHz, defined as the peak-to-peak distance between
the central signal and a satellite, was obtained. The spectrum
further shows that the signals are very broad, the intensity
of the outermost signals is probably spread over a large range
thus reducing the peak amplitudes to values too small to detect under
our experimental conditions. Further studies are in progress.

One immediate consequence of this finding is to pursue a
study of competition between monovalent and divalent counterions
in lyotropic liquid crystalline phases by following the quadrupolar splittings of both competitive ions. In the past[34], such
studies were performed by monitoring the quadrupolar splittings
of the monovalent counterion only.

REFERENCES

1.   (a) Physical Chemistry 2, Chemical Center. (b) Food Technology,
      Chemical Center. (c) Physical Chemistry 1, Chemical Center.
2.   V. Luzzati, in "Biological Membranes", D. Chapman, Editor,
      Chap. 3, Academic Press, London and New York, 1968

3. P.A. Winsor, Chem. Rev., 68, 1 (1968).
4. P. Ekwall, L. Mandell, and K. Fontell, Mol. Cryst. Liq.
    Cryst., 8, 157 (1969).
5. P. Ekwall, in "Advances in Liquid Crystals", G.H. Brown,
    Editor, Vol. 1, Chap. 1, Academic Press, London and New York,
    1975.
6. G. J. T. Tiddy, Phys. Rev., 57, 1 (1980).
7. K. Fontell, Prog. Chem. Fats other Lipids, 16, 145 (1978).
8. P. Ekwall, L. Mandell, and K. Fontell, J. Colloid Interface
    Sci., 31, 508 (1969).
9. K. Fontell, (1982) Unpublished Work.
10. B. Jönsson, and H. Wennerström, J. Colloid Interface Sci.,
    80, 482 (1981).
11. B. Jönsson, "The Thermodynamics of Ionic Amphiphile-Water
    Systems, A Theoretical Analysis", Ph. D. Thesis, Lund
    University, Lund, 1981.
12. B. Jönsson, G. Gunnarsson, and H. Wennerström, in "Solution
    Behaviour of Surfactants - Theoretical and Applied Aspects",
    K. L. Mittal, and E. J. Fendler, Editors, Vol. 1, pp.317-341,
    Plenum Press, New York, 1982.
13. H. Wennerström, B. Jönsson, and P, Linse, J. Chem. Phys.,
    76, 4665 (1982)
14. A. Khan, K. Fontell, G. Lindblom, and B. Lindman, J. Phys,
    Chem., 86, 4266 (1982).
15. H. Wennerström, G. Lindblom, and B. Lindman, Chem. Scr.,
    6, 97 (1974).
16. H. Wennerström, N-O. Persson,and B. Lindman, in "Colloidal
    Dispersions and Micellar Behaviour", K. L. Mitall, Editor,
    Am. Chem. Soc. Symp. Ser., 9, 253 (1975).
17. A. D. Buckingham, and K. D. McLauchlan, in "Progress in NMR
    Spectroscopy", J. W. Emsley, J. Feeney, and L. H. Sutcliffe,
    Editors, Vol. 2, p. 63, Pergamon, Oxford, 1978.
18. N-O. Persson, B. Lindman, and G. J. T. Tiddy, J. Colloid
    Interface Sci., 58, 461 (1975).
19. A. Khan, O. Söderman, and G. Lindblom, J. Colloid Interface
    Sci., 78, 217 (1980).
20. K. Beyer, J. Colloid Interface Sci., 86, 73 (1981).
21. A. Khan, K. Fontell, and G. Lindblom, J. Phys. Chem., 86,
    383 (1982)
22. P-O. Eriksson, A. Khan, and G. Lindblom, J. Phys. Chem., 86,
    387 (1982).
23. J. Ulmius, H. Wennerström, G. Lindblom, and G. Arvidson,
    Biochemistry, 16, 5742 (1979).
24. A. Khan, L. Rilfors, Å. Wieslander, and G. Lindblom, Eur. J.
    Biochem., 116, 215 (1981).
25 A. Khan, G. Arvidson, K. Fontell, and G. Lindblom, to be
    published.
26. A. Khan. and B. Lindman, to be published.

27.   L. Mandell, P.Ekwall, K. Fontell, and H. Lehtinen, Acta
      Polytechn. Scand., 74, I-III (1968).
28.   P. Ekwall, L. Mandell, and K. Fontell, Acta Chem. Scand.,
      22, 373 (1968). ibid. 22, 1543 (1968).
29.   P. Ekwall, L. Mandell, and K. Fontell, Mol. Cryst. Liq. Cryst.,
      8. 157 (1968), Figure 17a.
30.   J. Rogers, and P.A. Winsor, Nature, 216, 477 (1967).
31.   P. Ekwall, L. Mandell, and K. Fontell, J. Colloid Interf. Sci.,
      33, 215 (1970).
32.   R. R. Balmbra, J. S. Clunie, and J. F. Goodman, Proc. Roy.
      Soc. (London), Ser. A, 285, 534 (1965).
33.   P.C. Hiemenz, "Principles of Colloid and Surface Chemistry",
      Marcel Dekker, New York, 1977.
34.   B. Lindman, G. Lindblom, H. Wennerström, N-O. Persson, H.
      Gustavsson, and A. Khan, in "Magnetic Resonance in Colloid
      and Interface Science", J.P. Fraissard, and A. Resing, Editors,
      p 307, Reidel Publishing Co., 1980.

A NEW OPTICALLY ISOTROPIC PHASE IN THE DILUTE REGION OF THE SODIUM OCTANOATE - DECANOL - WATER SYSTEM

William J. Benton and Clarence A. Miller

Department of Chemical Engineering
Rice University
Houston, Texas   77251

A phase diagram is presented for the dilute region of the sodium octanoate-n-decanol-water system at 30°C based on studies using polarized light screening and optical microscopy. It includes surfactant concentrations up to 10 wt%. An optically isotropic phase is found which has not been previously reported for this system. It contains some 3 to 4 wt% surfactant and some 2 to 5 wt% alcohol and for a given water content, has an alcohol-to-surfactant ratio greater than that of the lamellar liquid crystalline phase. Two-phase and three-phase regions involving this new phase are seen at nearby compositions. Also of interest is that the lamellar phase extends well into the dilute region studied and occurs at water contents exceeding 90 wt%.

INTRODUCTION

In recent years several authors have reiewed the complex array of structures formed by self-assembly of surface-active molecules and the phase behavior in binary and ternary systems resulting form these structures[1-3]. The phase diagrams for such systems are of technological interest in such diverse fields as detergency, pharmacy, food processing, and enhanced oil recovery.

One of the earliest ternary systems studied was that containing sodium octanoate, n-decanol, and water. Due to the extensive data available from various techniques, it has become a "model" system for behavior of anionic surfactant-alcohol-water mixtures (4). Even so, the emphasis in previous studies has been placed on understanding the behavior of liquid crystalline phases present at relatively high surfactant and alcohol concentrations. A recent paper on X-ray studies, for instance, included only one sample with a water content above 80% by weight[5].

Our interest, in contrast, has been in dilute systems with water contents exceeding about 85% by weight. This range is of interest in enhanced oil recovery, where the injected fluids are often composed of 3-12 wt% of an anionic surfactant such as a petroleum sulfonate and a few percent of a short-chain alcohol in sodium chloride brine.

Our previous studies in such systems using both petroleum sulfonates and pure anionic surfactants have identified some general patterns of phase behavior[6-9]. As salinity increases, for example, the basic phase sequence is an aqueous micellar solution, a stable dispersion of lamellar liquid crystal in this solution, a single lamellar phase, and an optically isotropic phase which scatters light and exhibits streaming birefringence[9]. The same sequence has been seen in other cases when the surfactant-alcohol mixture is made more oil soluble by other means, e.g., by increasing the concentration of n-hexanol in a system containing sodium dodecyl sulfate, n-hexanol, and sodium chloride brine[8].

Although our results suggested that the entire sequence was quite general, a review of the literature indicated that the optically isotropic phase with streaming birefringence had not been reported for the sodium octanoate-decanol-water "model" system. Accordingly, we conducted an investigation of phase behavior in the dilute region of this system using optical techniques. This study, which is described below, has demonstrated that such a phase does indeed exist in the model system. Its generality in anionic surfactant-alcohol-water (or brine) systems is thus established.

MATERIALS AND SAMPLE PREPARATION

Sodium octanoate was obtained from two sources (K & K Chemicals and British Drug Houses). In the latter case the compound was used as received. In the former the compound was recrystallized from warm ethanol and dried in vacuo after filtration. N-decanol (Fluka-puriss) was used as received. The water was deionized and double distilled with a pH of 7.4.

Weighed amounts of sodium octanoate were dissolved in water with some warming. N-decanol was added to make 10.0 gram solutions in 13 mm. I.D test-tubes with Teflon insert screw caps. Solutions were heated to at least 60°C and mixed thoroughly by vortex mixing and ultrasound. Once the samples were mixed they were allowed to cool slowly to room temperature. After 24 hours the process of heating and mixing was repeated. Samples were then transferred to a constant temperature environment maintained at 30.0°C ± 0.5°C. They were again mixed thoroughly and left to reach equilibrium.

TECHNIQUES, CRITERIA FOR EQUILIBRIUM AND NOMENCLATURE

The difficulties of investigating the dilute regions in ternary systems have been pointed out in some of the original studies[4]. In order to overcome some of these difficulties, we have used improved optics for observing behavior of multiphase systems in test tubes and developed a novel closed cell technique for studying weakly birefringent phases with the optical microscope over long periods of time.

Phase volumes were measured and optical properties noted when samples in test tubes were placed between polarizers positioned with their optic axes 90° with respect to each other. Under these optical conditions isotropy, anisotropy and scattering are easily differentiated. Streaming or flow birefringence of samples was noted by disturbing the test tubes. Phase volume measurements and observations were made at selected time intervals. Equilibrium was considered to have been attained only when no further change in phase volumes or birefringent textures was observed. Some samples were remixed at 30°C either by hand rotation or by vortex mixer to confirm that they would return again to the same state.
Selected birefringent samples were observed by optical microscopy with polarizing and/or Hoffman Modulation Contrast (HMC) optics. Samples imbibed by capillary action into optical rectangular capillaries having pathwidths of either 100 or 200 μm. The capillaries were then sealed and observed for periods of time from hours to several weeks. Further details of these techniques and criteria for equilibrium are given elsewhere[6,10].

Table I. Nomenclature of Phases and Structures.

This Paper	Refs.	Definition
I_1	L_1	Water-rich isotropic phase
I_2	L_2	Alcohol-rich isotropic phase
L	D	Homogeneous lamellar phase
S	--	Dispersion of $I_1+L(L_1+D)$
C	--	Optically isotropic phase with streaming birefringence

Table I describes the differences in phase nomenclature between this paper, and the earlier papers on phase behavior of this system[4,5].

RESULTS

The results of this study are shown in the partial phase diagram of Figure 1 for the dilute region of sodium octanoate-n-decanol-water. The phase boundaries are from measurements and observations taken after 90 days at 30.0°C ± 0.5°C. Figure 1 is based on over four hundred individual samples which were prepared and studied.

There are several features of interest in this diagram. At low sodium octanoate and n-decanol concentrations an isotropic aqueous solution I_1 exists. This region is very narrow, however, as n-decanol is practically immiscible in water for these concentrations of sodium octanoate which are all below the CMC of 6.3 wt%[11]. Above the I_1 region the phase behavior becomes more complicated. A two-phase region (I_1+I_2) exists for all surfactant concentrations below about 2.1 wt%, where I_2 is an alcohol-rich phase which is less dense than the aqueous I_1 phase. This region falls below the "limiting association concentration" or l.a.c. of 2.1 wt% sodium octanoate[11], i.e., there are no additional phases formed due to the tendency for association or self-assembly of surfactant and alcohol molecules.

As the concentration of sodium octanoate is increased above the l.a.c., a third optically isotropic phase C forms as a middle, third phase between the I_1 and I_2 phases. Bordering this three-phase region of I_1+C+I_2 are an I_1+C two-phase region at less than about 5.0 wt% n-decanol and a $C+I_2$ two-phase region above about 5.0 wt% n-decanol. The volume of the I_2 phase is relatively small in all our samples because its alcohol content is much higher than the maximum of 10 wt% considered here.

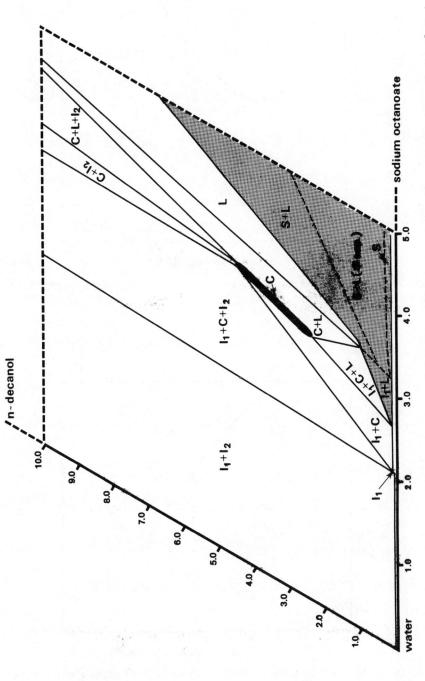

Figure 1. Partial phase diagram for the dilute corner of the sodium octanoate-n-decanol-water system T = 30°C.

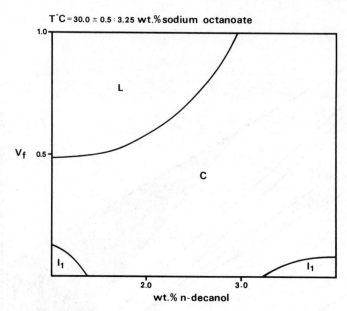

Figure 2a.   Volume fraction versus n-decanol concentration for
3.25% sodium octanoate.   T = 30°C.

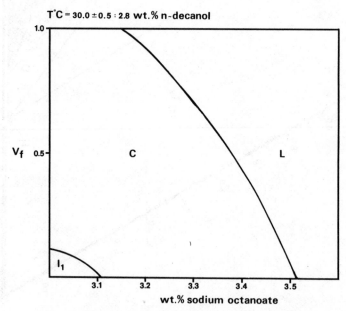

Figure 2b.   Volume fraction versus sodium octanoate concentration
for 2.8 wt.% n-decanol.   T = 30°C.

The narrow blackened region indicated by C in Figure 1 is the single-phase region of the optically isotropic phase mentioned above. It scatters blue light when placed between polarizers, an indication that aggregation occurs on a scale somewhat below the wavelength of visible light. When samples of this phase are disturbed, they exhibit streaming birefrigence with a fast relaxation rate to the optically isotropic state. Some comments on the structure of this phase are made below. In any case the range of compositions over which the pure C phase exists is small.

At higher surfactant concentrations are found a region where the C phase coexists with a lamellar liquid crystalline phase L and a single phase region of the L phase. Figure 2 shows volume fraction diagrams which describe how the transition from (I_1+C) to C to (C+L) occurs as

  (a) alcohol concentration decreases at constant surfactant concentration and
  (b) surfactant concentration increases at constant alcohol concentration.

In the latter case the transition to the single L phase is also shown. Figure 3 is a photograph in polarized light of the series of samples from which Figure 2a was plotted.

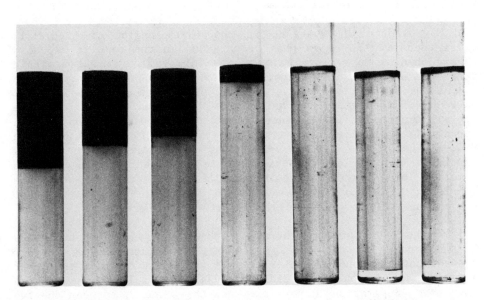

Figure 3. Part of series of test tubes in polarized light used in constructing Figure 2a. Negative Print.

In polarized light the liquid crystalline L phase has a fine-textured and multi-colored birefringence. As discussed elsewhere this texture differs from those of the two-phase mixtures of L and I_1 which are shaded and designated by S+L and S in Figure 1 and which were observed in our previous studies of anionic surfactant-alcohol-bine systems[7,9]. It is noteworthy that the L texture was seen even at water contents as high as about 95 wt.%. At surfactant concentrations above about 3 wt.% and at alcohol concentrations above the solubility limit in I_1 but below about 1 wt.%, equilibration was very slow. At these compositions dispersions of L in I_1 exist.

The birefringent L and (S+L) samples were sealed in closed rectangular campillary cells and observed with the polarizing microscope. For the L phase the initial texture was fairly bright with general orientation along the length of the cell due to the flow in this direction during the filling process. With time, reorientation occurred and numerous dark regions formed with the bilayers perpendicular to the optic axis. Separating these regions were "oily streaks" which are characteristic of lamellar phases.

After two or three days, careful focusing at higher magnification showed that the previously dark regions exhibited weak birefringence with a texture consisting of a highly ordered square grid pattern. Figure 4 shows this texture as it appears when the focus is on the upper surface of the rectangular cell. This texture is known to be the parbolic focal conic (PFC) texture which occurs in lamellar phases[10,12-14].

Microscopic observations in the (S+L) region of Figure 1, where there was no gross phase separation in the test tubes, also showed changes in texture with time. This region is not a single phase but a stable dispersion. Domains of the L texture and of an S texture with liquid crystalline spherulites dispersed in an aqueous solution were observed to grow with time. The ratio of the overall amounts of L and S decreased as sample composition shifted further from the boundary with the L phase. Figure 5, taken with HMC optics, shows and L domain during its period of growth. Particles of the lamellar phase can be seen near the edge of the domain. This behavior and that described above for the L phase are consistent with our observations with microscopy in other anionic surfactant systems which will be reported elsewhere.

At lower alcohol concentrations phase separation between L and S does occur in test tubes, as shown in Figure 1. Microscopy showed that the upper phase is a lamellar liquid crystal, the lower a stable dispersion of liquid crystalline particles in an aqueous solution. Separation is very slow and the boundary between the two regions was diffuse even after 90 days, an indication that equilibrium was not reached.

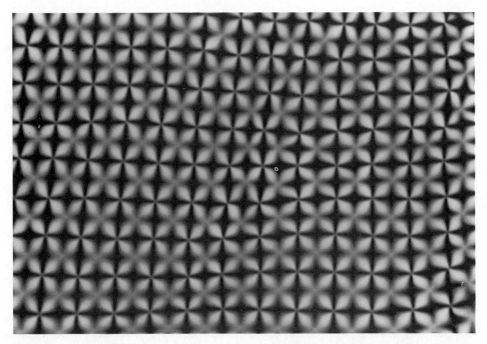

Figure 4.   Parabolic focal conic texture of lamellar phase.
Polarized light 4 wt.% sodium octanoate, 6.5 wt.% n-decanol.   PFC
point-to-point = av. 25 μm.

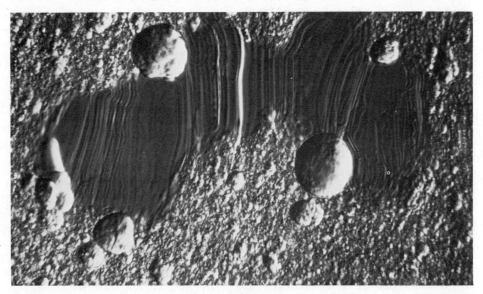

Figure 5.   Lamellar domain in stage of "growth" from dispersed
lamellar aggregates.   HMC optics 5.0 wt.% sodium octanoate, 6.0
wt.% n-decanol.   Large aggregate to left of center = 44 μm.

DISCUSSION

As indicated above the present work has focused on the dilute region of the phase diagram to a greater extent than have previous studies. One interesting finding of this detailed look at the dilute region is that the lamellar phase L can exist at high water contents. As indicated previously, the L texture was seen with polarized light in test tube samples for compositions up to about 95 wt % water. Only samples containing up to about 92 wt% water were examined by optical microscopy, but these all confirmed the existence of a single lamellar phase in the L region of Figure 1.

A second finding of this study is the existence of the optically isotropic C phase which contains some 90 to 95 wt% water and which has not been previously reported for this "model" system. That C is a true phase is indicated by the existence of various two-phase and three-phase regions involving C in Figure 1 in positions which are consistent with those of the well known I_1, I_2 and L phases.

The internal structure of the C phase is not known and awaits elucidation by further studies using techniques such as NMR and scattering of light, X-rays, and neutrons. Optically isotropic phases which scatter light and exhibit streaming birefringence have been reported in various binary and ternary systems in various parts of the phase diagram. Called "viscous isotropic" of "cubic" phases they have been found, for example, at compositions between those of the aqueous micellar solution and the hexagonal phase, between the hexagonal and lamellar phases, and between the lamellar and reverse hexagonal phases[3,15]. Although X-ray studies have provided useful information on their structure, they remain poorly understood.

Moreover, cubic phases have not been reported at the high water contents of our C phase. The closest is that containing 60-65 wt% water in the potassium oleate-n-decanol-water system[16]. X-ray examination of this phase indicated a face-centered-cubic structure.

In view of the absence of the hexagonal phase in Figure 1, it seems unlikely that the surfactant-alcohol aggregates in the C phase are rod-like. They may very well, however, be disk-like, i.e., bilayers of finite lateral extent. In monoglyceride-water systems of much lower water content than considered here, Lindblom, et. al[17] proposed that such finite bilayers fuse to form a regular, three-dimensional network. Whether the aggregates fuse in our dilute systems or whether they are distinct remains to be seen.

The results presented here and our previous work on petroleum sulfonates and other pure anionic surfactants[6-10] strongly suggest that lamellar phases of high water content and existence of the C phase are general characteristics of systems containing anionic surfactant, alcohol, and water or brine. The transition from the lamellar to the C phase occurs with increasing oil solubility of the surfactant-alcohol mixture, a condition which can be brought about by adding salt or, as in the present case, by increasing the concentration of an oil-soluble alcohol. Increasing temperature evidently can also lead to disruption of the lamellar structure and formation of a C phase[9].

Fontell[18] has described phase behavior in the dilute region of the Aerosol OT- water-sodium chloride system which is consistent with our results. With only some 5% of the surfactant present, addition of salt causes transformation to occur from the lamellar phase to an optically istropic phase he calls L_2. We have confirmed these results in our laboratory[19]. It seems likely that the optically isotropic phase, which can contain up to about 95% brine, is the same as our C phase.*

Friman, et. al[5] recently presented some modifications of the sodium octanoate-n-decanol-water phase diagram. The main change is that the "B" and "C" regions (Ekwall's nomenclature) of the original diagram[4] have been eliminated in favor of an extended lameller phase and a larger two phase region between this phase and the aqueous micellar solution. Extensive X-ray data were presented for various compositions with less than about 80 wt% water.

The I_1 and (I_1+I_2) regions of the modified phase diagram are as found here. But between about 2 wt% and 6 wt% surfactant and below about 15 wt% n-decanol, dotted lines on their diagram indicate that the phase behavior was not extensively studied. This is, of course, precisely the composition region of interest in the present work.

It should also be noted that the phase diagram of Friman, et al[5] is at 20°C while ours is at 30°C. It seems unlikely however, based on our work with other anionic systems[9] that this relatively small temperature difference would cause significant changes in the phase diagram.

Finally, it is useful to present the data of Figure 1 in a somewhat simpler manner. One method of doing this is to use "field" variables which are continuous when crossing interfaces between phases in equilibrium[20]. The pertinent field variables here are chemical potentials. A schematic phase diagram based on this approach is shown in Figure 6. Here two-phase regions are represented by lines and three-phase regions by points. This

---

* We thank Dr. Fontell for calling this paper to our attention and suggesting the analogy.

diagram differs from an earlier one of Skoulios[21] in that changes in phase behavior based on this work and on that of reference[5] have been incorporated.

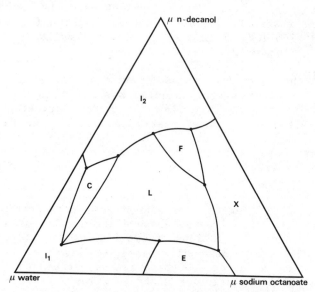

Figure 6.   Schematic representation in field space of phase behavior of the system sodium octanoate-n-decanol-water system.

SUMMARY

A careful study of phase behavior in the water-rich region of the sodium octanoate-n-decanol-water phase diagram has been made using polarized light screening and optical microscopy.   The lamellar liquid crystalline phase was found to extend to water concentrations exceeding 95 wt%, and an optically isotropic phase was discovered which has not been previously reported in this system.   This phase contains over 90 wt% water and has an alcohol-to-surfactant ration greater than that of the lamellar phase.   It scatters light between polarizers to an extent detectable visually and exhitibs streaming birefringence on being disturbed.   Its structure is unknown but likely consists of surfactant-alcohol bilayers of finite lateral extend either as individual particles or fused into some three-dimensional structure.

ACKNOWLEDGMENTS

This work was supported by grants from Amoco Production Company, Gulf Research and Development Company, Exxon Production Research Company, and Shell Development Company.

REFERENCES

1. P. Ekwall, Adv. Liq. Cyst. 1, 1 (1975).
2. K. Shinoda and S. Friberg, Adv. Colloid Interface Sci. 4, 281 (1975).
3. G.J.T. Tiddy, Physics Report 57, 11 (1980).
4. L. Mandell and P. Ekwall, Acta Polytechn. Scand. Chem. Met. Series 74, 1 (1968).
5. R. Friman, I. Danielsson, and P. Stenius, J. Colloid Interface Sci. 86(2) 501 (1982).
6. W.J. Benton, C.A. Miller, and T. Fort, Jr., Society of Petroleum Engineers Preprint 7579 presented at the SPE 53rd Ann. Tech. Conf., Houston, TX (1978).
7. C.A. Miller, S. Mukherjee, W.J. Benton, S. Qutubuddin, and T. Fort, Jr., AIChE Symp. Ser. 78(212), 28 (1982).
8. W.J. Benton, J. Natoli, S. Qutubuddin, S. Mukherjee, C.A. Miller, and T. Fort, Jr., Soc. Pet. Eng. J. 22, 53 (1982).
9. W.J. Benton and C.A. Miller, "Lyotropic Liquid Crystalline Phases and Dispersions in Dilute Anionic Surfactant-Alcohol-Brine Systems. I. Patterns of Phase Behavior." In press ,J. Phys. Chem.
10. W.J. Benton, E.W. Toor, C.A. Miller, and T. Fort, Jr., J. Phys. (Paris) 40, 107 (1979).
11. P. Ekwall, H. Eikrem, and L. Mandell, Acta Chem. Scand. 17, 111 (1963).
12. Ch.S. Rosenblatt, R. Pindak, N.A. Clark, and R.B. Meyer, J. Phys. (Paris) 38, 1105 (1977).
13. S.A. Asher and P.S. Pershan, J. Phys. (Paris) 40, 161 (1979).
14. F. Candau and F. Ballet, in "Microemulsions", I.D. Robb, Editor,p. 49, Plenum Press, New York, 1982.
15. K. Fontell, Mol. Cryst. Liq. Cryst. 63, 59 (1981).
16. P. Ekwall, L. Mandell, and K. Fontell, J. Colloid Interface Sci. 31, 508 (1969).
17. G. Lindblom, K. Landsolm, L. Johansson, K. Fontell, and S. Karsen, J. Amer. Chem. Soc. 101, 5466 (1979).
18. K. Fontell in "Colloidal Dispersion and Micellar Behavior", K.L. Mittal, Editor, p.270, ACS Symposium Series No. 9, (1975).
19. W.J. Benton, O. Ghosh, and C.A. Miller, unpublished results.
20. R.B. Griffiths, J. Chem. Phys. 60, 195 (1974).
21. A. Skoulios, Ann. Phys. 3, 421 (1978).

NMR AND POLARIZED EMISSION STUDIES OF CUBIC PHASES

AND MODEL MEMBRANES

Per-Olof Eriksson, Lennart B.-Å. Johansson and
Göran Lindblom
Department of Physical Chemistry, University of Umeå
S-901 87 Umeå
Sweden

Membrane lipids may form several different liquid
crystalline phases depending on composition and temperature. In recent years studies have been focussed on the
structure and molecular dynamics in the viscous isotropic or cubic liquid crystalline phases. The structural
investigations performed with low angle X-ray diffraction usually do not give a complete picture of the
structure and additional methods are needed. Two such
methods are provided by NMR and time-resolved polarized
emission briefly reviewed here. Lipid lateral diffusion
coefficients for several different lamellar and cubic
phases are summarized and the luminescence method for
studies of aggregate geometry is described. It is shown
that measurements of diffusional motion and quadrupole
relaxation times can give information about dynamics as
well as structure of the cubic phases.

INTRODUCTION

All biological membranes contain a number of different lipids. It is generally accepted that these lipids form a two-dimensional fluid matrix,[1] where other membrane components like proteins are incorporated. However, not all of these lipids form a bilayer structure with water but, as was shown already in the 1960's by Luzzati and coworkers,[2] other phase structures like reversed hexagonal (H_{II}) and cubic phases may form. There might be many reasons for the existence of such non-lamellar forming lipids within the membrane and in recent years an increasing number of works on this subject have appeared in the literature.[3-14] Thus, for example, we[9] have suggested that these lipids forming non-lamellar structures are needed to keep an optimal molecular packing in the bilayer membrane of the bacteria Acholeplasma laidlawii and that membrane fusion may involve such lipids,[8,9,11,12] The so called lipidic particles observed with freeze-fracture electron microscope technique have been suggested to consist of cubic liquid crystals.[4] The proposal[3] that the principal structural elements of the tight junction strands are due to the membrane lipids is very intriguing.

In spite of a lot of speculation, non-lamellar structures formed by membrane lipids are no doubt an essential area of membrane biophysical chemistry. A large part of our membrane research program contains investigations of the phase equilibria, structure and dynamics of different lipids. In particular we have focussed our efforts on the study of cubic liquid crystalline phases and the development of spectroscopic methods for such investigations. Cubic liquid crystalline phases are formed by lipids from both mamalian membranes[11] and bacterial membranes[8-14] as well as from chloroplast membranes.[5]

Hitherto we have mainly used various NMR methods for our investigations of structure and dynamics of cubic phases and only recently we have started to use time-resolved emissiontechniques also. In this communication we will first briefly review the achievments obtained in recent NMR studies and in the second part the use of luminescence methods will be delineated.

LIPID LATERAL DIFFUSION

Lamellar Phases and Model Membranes

Translational diffusion coefficients of amphiphile systems can be conveniently studied by the NMR-technique developed by Stejskal and Tanner.[18] We have used this method for about ten years to determine the lipid lateral diffusion for a number of lamellar liquid crystalline systems and model membranes.[6-8,10,11,15-17,19] So that the lateral diffusion coefficient, D_L, could be directly

measured and the static dipole coupling reduced to zero, the lamellar sample was macroscopically aligned at the so called magic angle
in the magnetic field. For further details of the method the reader
is referred to refs. 6 and 15. In table I lateral diffusion coefficients for several different lamellar systems have been summarized.
Since the table speaks for itself, we shall put emphasis only on a
few points. Monovalent ions like Na^+ or Li^+ seem to have a negligible effect on the lateral diffusion coefficient of the lipid octanoate. Also the hydrocarbon chain length seems to have a small effect compared to the very large change in the diffusion coefficients
observed for different polar head groups. Thus, for example the
translational diffusion of octyl sulphate is about ten times smaller
than that of octanoate whereas lysolecithin diffuses somewhat slower
than the corresponding lecithin although the latter has two hydrocarbon chains and the lysocompound only one; i.e. the difference in
molecular weight is almost a factor of two. On the other hand, di-
glucosyldiglyceride with oleoyl alkylchains shows a lateral diffusion, which is about ten times faster than that of the corresponding
lecithin. The most probable explanation of these observations is
that the lateral diffusion coefficient is strongly dependent on the
packing properties in the bilayer. To settle this issue we are currently studying this problem for simple amphiphilic systems.

As can be seen from Table I, cholesterol has a negligible effect on the lipid lateral diffusion and this is extensively discussed in ref. 16, and we will therefore not further elaborate on this
matter here.

Cubic Liquid Crystalline Phases

The lipid lateral diffusion coefficient has also been determined for a number of cubic liquid crystalline phases. Such data
are given in Table II. It can be seen from Tables I and II that for
most of the systems studied, the diffusion coefficients differ by
about a factor of 2-3 or is, at least for most systems, of the same
order of magnitude for lamellar and cubic phases. However, there
are exceptions, where the lipid diffusion coefficients in some cubic phases are extremely small compared with the diffusion in lamellar phases.

It has previously been shown that amphiphile diffusion data
can provide information about the structure of the cubic phases. As
the experimental diffusion time corresponds to displacements over
collodial dimensions, the NMR diffusion method can be used for determinations of any restriction for translational motion in the
system. For example, a diffusion time of 50 ms and a diffusion coefficient of 10^{-11} m^2 s^{-1} correspond to a root mean square distance
of 1700 nm, which is much larger than the dimension of a spherical
micelle. Thus, from a comparison of measured diffusion coefficients

Table I. Lateral Diffusion Coefficients of the Amphiphile for some Lamellar Liquid Crystalline Systems.

Sample	Compn. wt %	Temp. °C	$D_L \cdot 10^{11}$ m² s⁻¹	Ref.
Octylammoniumchloride Water	72 28	24	33	15
Sodium diethylhexyl sulfosuccinate Water	71 29	24 35	1.0 1.7	10
Monoolein Water	82 12	22 35	1.1 1.6	10
Monooctanoin Water	90 10	24	2.9	15
Sodium octanoate Decanol Water	22 40 38	24	21	15
Lithium octanoate Decanol Water	22 40 38	24	20	15
Sodium octylsulphate Decanol Water	30.9 35.8 33.3	24	3.5	15
Diglucosyldiglyceride (DGDG) Water	87 13	45	3.9	8

Sample	Compn. wt %	Temp. °C	$D_L \cdot 10^{11}$ m² s^{-1}	Ref.
Dioleoyllecithin (DOL) Water	80 20	35	0.5	16
Palmitoyloleyllecithin (POL) Water	80 20	45	0.9	16
Egg yolk lecithin (EYL) Water	80 20	35	0.6	16
Dioleoyllecithin Cholesterol Water	75 5 20	35	0.5	16
Dioleoyllecithin Cholesterol Water	65 15 20	35	0.7	16
Dioleoyllecithin Cholesterol Water	50 30 20	35	0.7	16
Dioleoyllecithin Sodium cholate Water	70 10 20	35	0.5	7
Lysooleoyllecithin Water	84 16	26	0.2	11

Table II. Amphiphile Diffusion Coefficients for some Cubic Liquid Crystalline Phases.

D is the measured coefficient, D_L the lateral diffusion coefficient for the corresponding lamellar phase and D_L^{cub} and $D_{||}^{cub}$ are calculated coefficients assuming lamellar or rod structural units, respectively.

Sample	Compn. wt %	Temp. °C	$D \cdot 10^{11}$ $m^2 s^{-1}$	$D_L^{cub} \cdot 10^{11}$ $m^2 s^{-1}$	$D_{||}^{cub} \cdot 10^{11}$ $m^2 s^{-1}$	$D_L \cdot 10^{11}$ $m^2 s^{-1}$	Ref.
Sodium diethylhexylsulfosuccinate Water	72.4 27.6	24	0.36	0.54	1.1	1.1	10
		35	0.59	0.89	1.8	1.6	10
Monoolein Water	88 12	43	1.5	2.3	4.5	2.3*	10
		57	2.6	3.9	7.8	3.9*	10
Potassium octanoate Water	60 40	24	8.8	13.2	26.4	–	15
Sodium Octanoate Octane Water	39.4 4.3 56.3	24	<0.1	–	–	21	15
Dodecyltrimethylammonium chloride Water	50.1 49.9	47	<0.2	–	–	–	6
Dodecyltrimethylammonium chloride Water	84.0 16.0	26	0.8	1.1	2.3	–	6
		47	2.0	3.0	6.0	–	6

* Extrapolated

Sample	Compn. wt %	Temp. °C	$D \cdot 10^{11}$ m² s⁻¹	$D_L^{cub} \cdot 10^{11}$ m² s⁻¹	$D_{\parallel}^{cub} \cdot 10^{11}$ m² s⁻¹	$D_L \cdot 10^{11}$ m² s⁻¹	Ref.
Cetyltrimethylammonium fluoride Water	75 25	59	2.5	3.8	7.5	–	6
		75	4.8	7.2	14.4	–	6
Dioleoyllecithin Water	97 3	60	0.5	0.8	1.5	1.5	16
Dioleoyllecithin Sodium cholate Water	45 30 25	35	0.1	0.15	0.3	0.5	7
Monoglucosyldiglyceride Water	97 3	35	<0.1	–	–	–	17
Lysooleoyllecithin Water	79 21	26	0.14	0.28	0.42	0.2	11

of different phases of an equilibrium system, this makes the NMR
diffusion technique useful for structural investigations of cubic
liquid crystalline phases. There exist two fundamentally different
types of cubic mesophases, viz., structures with continuous regions
of both water and hydrocarbon chains and structures composed of
closed aggregates of "oil-in-water" or "water-in-oil" type. The NMR
pulsed field gradient diffusion-technique provides a possibility of
differentiating between these basic alternatives, and may in addition give information about the structural units of a particular
phase.

The lipid molecule is free to move within the aggregates, but
it is highly unlikely[15] that it will pass over an aqueous region into an adjacent aggregate. Therefore, although the local molecular
diffusion probably is the same, a cubic phase with discontinuous
hydrocarbon regions but with continuous water regions or a phase
with discontinuous water spheres but with continuous hydrocarbon regions will have a measured lipid diffusion coefficient well below
that of a lamellar sample, [6,10,15] On the other hand, in a cubic
structure consisting of continuous hydrocarbon and continuous water
regions, the amphiphile molecule can move over macroscopical distances without passing through an aqueous region. The macroscopic
diffusion coefficient will then be of the same order of magnitude
as that of lamellar or hexagonal phases.

A quantitative comparison between the diffusion coefficients
measured for macroscopically aligned lamellar samples and various
cubic phases can also be made using a simple model.[6,10,15] This is
based wholly on the geometry of the structural units building up
the cubic phases and the assumption that the geometry of the aggregate does not influence the local lipid diffusion. Three such geometries are possible, namely, spherical-, lamellar- and rod-like
aggregates.

Consider a local diffusion tensor, where the component representing restricted diffusion in a certain direction is set to zero.
The measured diffusion coefficient, D, is a mean value of the local diffusion tensor $\tilde{D}$, yielding

$$D = \overline{(1/3)\mathrm{Tr}\ \tilde{D}} \tag{1}$$

where the bar denotes an average over all sites in a unit cell of
the mesophase. Depending on the number of available directions for
lipid diffusion within the unit aggregate, three different cases
are distinguished.

a)   For a spherical aggregate unit, the lipid diffusion is restricted in all three directions of the local coordinate system,
leading to

$$D = (1/3)\text{Tr}\ \tilde{D}^{\text{cub}}_{\text{sphere}} \approx 0 \tag{2}$$

where $\tilde{D}^{\text{cub}}_{\text{sphere}}$ is the local diffusion tensor at the surface of the spherical aggregate.

b)   A structure composed of long rods permits local lipid diffusion in only one direction along the rod, since lipid diffusion around the rod is considered to be restricted as measured by the NMR method (cf. spherical aggregates). Denoting the diffusion coefficient along the rod by $D^{\text{cub}}_{\parallel}$ one obtains

$$D = (1/3)D^{\text{cub}}_{\parallel} \tag{3}$$

c)   If the lipid diffusion occurs on a lamellar unit (i.e. two-dimensional diffusion), the lateral diffusion coefficient in the cubic phase, D^{cub}_{L} is given by

$$D = (2/3)D^{\text{cub}}_{L}$$

Obviously the aggregate structure will have an appreciable influence on the magnitude of the measured diffusion coefficient of a cubic liquid crystalline phase. Furthermore, by knowing the local diffusion coefficient, conclusions about the geometry of the structural unit can be drawn. Note, however, that for a cubic phase composed of rod-like aggregates, these rods have to be sufficiently long to give effectively one-dimensional lateral diffusion. Otherwise, the calculated $D^{\text{cub}}_{\parallel}$ (cf eq. (3)) will be larger than the measured lateral diffusion, D_{L}.

Table II includes, besides the measured diffusion coefficients of cubic phases, also calculated local diffusion coefficients for various assumed structural units and the corresponding determined lateral diffusion coefficient D_{L}.

From the table it can be seen that for many systems there is strikingly good agreement between calculated $D^{\text{cub}}_{\parallel}$ and measured D_{L} indicating that these cubic phases are built up of rod-like networks as proposed by Luzzati and coworkers.[20] However, cubic phases composed of globular aggregates show very small diffusion coefficients.

RELAXATION TIMES OF QUADRUPOLAR NUCLEI IN CUBIC PHASES

For a nucleus with spin greater than one-half, the NMR spectrum is dominated by the interaction between the nuclear quadrupole moment and the electric field gradient at the nucleus.[21] Molecular reorientation modulates this interaction and the observed relaxation times are determined by the strength of the interaction and by the time scale at which it is modulated. Usually the quadrupole interaction provides the dominant mechanism of spin relaxation. This simplifies a great deal the interpretation of the relaxation times in terms of molecular motion compared to e.g. spin 1/2 nuclei for a system that is already rather complicated. The most popular candidates for this type of investigations are the spin one nuclei ^2H and ^{14}N.

For these nuclei the T_1 and T_2 relaxation times can be written[21]

$$T_1^{-1} = (3\pi^2/40)\chi^2[2\tilde{J}(\omega_0) + 8\tilde{J}(2\omega_0)]$$ (4a)

$$T_2^{-1} = (3\pi^2/40)\chi^2[3\tilde{J}(0) + 5\tilde{J}(\omega_0) + 2\tilde{J}(2\omega_0)]$$ (4b)

where χ is the quadrupolar coupling constant and $\tilde{J}(\omega)$ is a reduced spectral density function,[21] which characterizes the molecular motion. Assuming that the molecular motion can be described by one time constant, τ_c, the correlation time $J(\omega)$ becomes

$$\tilde{J}(\omega) = 2\tau_c[1 + (\omega\tau_c)^2]^{-1}$$ (5)

It has, however, been shown previously[6,22-24] that one correlation time is not sufficient to describe the experimental data obtained for lipid systems composed of aggregates where the molecules motion is locally anisotropic. A suitable model, taking this particular property into account, comprises two modes of motion occuring at different time scales. There is one fast motion (characterized by τ_c^f) consisting of conformational and orientational fluctuations at a time scale of about 10^{-10} s. These fluctuations result in a partial averaging of the quadrupole interaction from χ to $|S\cdot\chi|$, where S is the local order parameter.[23] The other motion is a slow one which in turn average out the residual quadrupole interaction. There are two different kinds of molecular motions that may be involved. Either the slow correlation time τ_c^s is determined by lipid lateral diffusion around the curved aggregate surface or by the rotational diffusion of the whole aggregate. For cubic phases composed of e.g. rod-like networks the contribution from the aggregate rotation can, however, be neglected.

If the fast (f) and the slow (s) modes of motion occur at considerably different time scales and are statistically uncorrelated, which is often the case for the systems in question, the reduced spectral density is given by[23]

$$\tilde{J}(\omega) = S^2\tilde{J}_s(\omega) + (1 - S^2)\tilde{J}_f(\omega)$$ (6)

Equations (4) and (6) then give, assuming $\omega_0\tau_c^s \gg 1$ and $\omega_0\tau_c^f \ll 1$

$$T_1^{-1} = (3\pi^2/2)\chi^2(1 - S^2)\tau_c^f \tag{7}$$

$$T_2^{-1} - T_1^{-1} = (9\pi^2/20)(\chi S)^2\tau_c^s \tag{8}$$

The product (χS) is obtained[23] from the quadrupole splitting $\Delta = (3/4)|\chi S|$. The slow correlation time τ_c^s is obtained from equation (8) and the fast correlation time τ_c^f may be obtained from equation (7) provided the quadrupole coupling constant is known. It is interesting to note that this model can also be used under certain conditions in time-resolved light spectroscopy. Thus the phenomenological relaxation part of the polarized emission, as discussed in the next section, can be given an analogous treatment.

The correlation time of the slow motion, τ_c^s, can be interpreted[6] in terms of the lateral diffusion of the lipid molecule along the aggregate surface. For a particle diffusing on a spherical surface[21]

$$\tau_c^r = R^2(6D_L)^{-1} \tag{9}$$

where R is the radius of the sphere. Thus for a cubic phase consisting of close-packed spheres, which do not rotate, an estimate of the spherical radius can be obtained (see Table III). As can be inferred from Table III the size of the globular aggregate is quite reasonable and agrees well with the picture that the radius of a globular aggregate should not be greater than the full all-trans length of the lipid molecule.

The relation (9) can also be used for the bicontinuous cubic phase structures. Although not strictly correct, the R-value determined in this way can be taken as a measure, or at least a rough estimate, of the dimension of the unit cell of the cubic phase. The underlying reason for this is that for the residual quadrupolar interaction to be averaged out completely, the lipid molecules have to take on all possible orientations in the cubic unit cell. To put it in another way, the amphiphile has to diffuse through the whole unit cell in a time which is short compared with the inverse of the residual interaction.

EMISSION ANISOTROPY OF LIPID SYSTEMS

The isotropic lipid systems (i.e. systems that are not optically birefringent) are most conveniently studied by light spectroscopic methods. In particular measurements of time-resolved emission can be expected to be a useful method for studies of both dynamics and molecular ordering in isotropic phases containing macromolecules or aggregates of amphiphiles. Typical example of such systems are

Table III. Correlation Times for the Fast (τ_c^f) and the Slow (τ_c^s) Motions of the Polar Head Group of the Lipid in some Cubic Phases. R is Calculated from Equation (9) (see text).

Sample	Compn. (wt %)	Temp (°C)	$10^{10} \tau_c^f$ (s)	$10^8 \cdot \tau_c^s$ (s)	R sphere (nm)	R rod (nm)	R lamellar (nm)
Dodecyltrimethylammonium chloride	49.3	23	3.7	4.8	2.5	–	–
Water	50.7	43	1.1	1.6	2.1	–	–
Dodecyltrimethylammonium chloride	84	23	6.0	7.6	–	3.1	2.2
Water	16	40	4.3	3.6	–	3.0	2.1
		50	3.6	2.6	–	3.1	2.2
		60	3.2	2.1	–	3.3	2.3
Cetyltrimethylammonium fluoride	81	55	6.8	6.0	–	4.8	3.4
Water	19	65	4.8	4.7	–	5.3	3.7
		75	3.2	3.7		5.7	4.0
Lysooleoyllecithin	79	24	–	80		3.5	3.5
Water	21						

solutions of micelles, vesicles or proteins and the liquid crystalline phases with cubic symmetry. The physicochemical properties of
the system is usually obtained in the time-resolved experiment from
a measurement of the emission anisotropy defined by equation (10)

$$r(t) = [F_{||}(t) - F_{\perp}(t)][F_{||}(t) + 2F_{\perp}(t)]^{-1} \qquad (10)$$

where $F_{||}(t)$ and $F_{\perp}(t)$ are the emission intensities polarized paral-
led and perpendicular relative to the polarization of the excitation
beam. From equation (10) it can be seen that the timedependent anisotropic part of r is located to the nominator, which for an ordinary
isotropic fluid, after sufficiently long time, must be equal to zero
due to the rapid molecular motion. Hence the relaxation of the emission anisotropy provides information about the rotational motion of
the lumiphores in an isotropic fluid. However, for a system which is
locally anisotropic but macroscopically isotropic like e.g. a micellar solution, the rapid motion of the molecules in the aggregates is
usually anisotropic and cannot consequently average r(t) to zero
within the lifetime of the probe molecule. The theoretical treatment of r(t) from time-resolved fluorescence experiments on lipid
systems has been delt with by several authors.[25-28] It is found that
the dynamics of the probe molecule is obtained through a rather
rough physical model of the molecular motion and that the molecular
ordering is described by measured order parameters,[26-28] which are
model independent. In this communication we will mainly consider
how the emission anisotropy is affected by the geometry of the aggregates building up the amphiphilic system. This is very similar to
the treatment of static parameters in NMR, like quadrupole splittings, in different lyotropic mesophases as e.g. lamellar and hexagonal phases. In the time-resolved emission experiment the measured
quantity r(t) depends on the rotational motion that occurs during
the emission lifetime, τ_e.

The time dependence of the emission anisotropy is contained in angular correlation functions[25-28] as

$$r(t) = (2/5)(Tr\tilde{P}Tr\tilde{Q})^{-1} \sum_{n=-2}^{2} <P_n(\Omega_0)Q_{-n}(\Omega_t)>(-1)^n \qquad (11)$$

P_n and Q_n are the second order irreducible tensor components generated from the absorption and emission transition dipoles, respectively. Ω_0 and Ω_t specify the Euler angular coordinates of the chromophore at the time of absorption and emission, respectively. Tr
stands for the trace of the tensors $\tilde{P}$ and $\tilde{Q}$.

The timedependence of r(t) is thus obtained from an evaluation
of the correlation function. For lipid aggregates this can be performed in practice only by assuming a specific physical model for
the molecular motion, since the local anisotropy of the aggregates

leads to that the correlation function contains an infinite number of exponential functions.[29] By using the strong collision model one obtains

$$r_f(t) = [r(0) - r_f(t_\infty)]e^{-t/\tau_f} + r_f(t_\infty) \tag{12}$$

$$r(0) = (2/5)D_{oo}^{(2)}(\delta) \tag{13}$$

$$r_f(t_\infty) = (2/5)(Tr\tilde{P}Tr\tilde{Q})^{-1} \sum_{mm'} P_m^M Q_{m'}^M \langle D_{om}^{(2)*}(\Omega_{MN}) \rangle \langle D_{om'}^{(2)*}(\Omega_{MN}) \rangle \tag{14}$$

Here the space averaged Wigner matrix elements $\langle D_{oq}^{(2)*}(\Omega_{MN}) \rangle$ are order parameters in the uniaxial system characterizing the lumiphore orientation (M) relative to a locally fixed coordinate system (N) at the surface of the aggregate (cf. Figure 1). Ω denotes the eularian angles. Hence, if the emission lifetime $\tau_e \approx \tau_f$ then r(t) will only depend on the fast local motions and the orientation of the chromophores. It should be noted here that it has been assumed that there are no interactions between chromophores i.e. energy migration, eximer formation etc. do not occur.

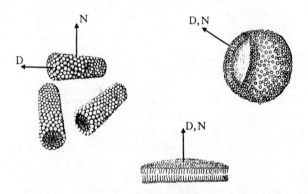

Figure 1. A schematic picture of different aggregate geometries: cylindrical rodlike micelles (left), lipid bilayer (middle) and single shell lipic vesicle (right). The director and the normal to the surface of the aggregates are denoted by D and N, respectively.

   In the analysis above of the time-resolved decay of $r(t)$ only
the rapid motion of the lumiphores around the normal to the aggregate surface is effective. On the other hand, if the emission lifetime is long enough, the translational diffusion around the aggregate or the rotation of the aggregate itself will affect $r(t)$. This
can be accounted for by modifying the model of the molecular motion
to include a slow motion in analogy with the model used above for
the NMR relaxation times. Then the geometry of the aggregate will
influence the measured emission anisotropy at long times. Several experimental observations[6,30-34] indicate that for phases in equilibrium the local environment is similar in lipid aggregates with rod-
like and lamellar geometry. Therefore the local orientational distribution of the solubilized chromophores can be assumed to be independent of the aggregate shape. This leads to that information about the
geometry of the aggregate can be obtained from a measurement of $r(t)$
in appropriate phases in equilibrium as discussed below. Assuming
that the fast and slow motions are independent of each other one
obtains

$$r(t) = r_f(t_\infty) \int_{\Omega_{ND}\Omega_{N'D}} F(\Omega_{ND}) G(\Omega_{ND}|\Omega_{N'D}, t) \times$$

$$\sum_q D_{qo}^{(2)^*}(\Omega_{ND}) D_{-qo}^{(2)^*}(\Omega_{N'D})(-1)^q d\Omega_{ND} d\Omega_{N'D} \qquad (15)$$

Here the slow motion of the N coordinate cystem (cf. the NMR case)
is described relative to a coordinate system, D, of the symmetry
axis of the lipid aggregate (cf. Figure 1). It can be assumed that
the conditional probability in equation (15) can be written

$$G(\Omega_{ND}|\Omega_{N'D}, t) = \{\delta(\Omega_{N'D}-\Omega_{ND}) - F(\Omega_{N'D})\}e^{-t/\tau_s} + F(\Omega_{N'D}) \qquad (16)$$

where δ denotes the Dirac delta function and τ_s is the correlation
time of the slow motion. For spherical aggregates $F = 1/8 \pi^2$, which
with equations (15) and (16) give

$$r_{s,sph}(t) = r_f(t_\infty)e^{-t/\tau_{s,sph}} \qquad (17)$$

The steady state emission anisotropy defined by

$$\bar{r}_s = 1/\tau_e \int_0^\infty r(t)e^{-t/\tau_e} dt \qquad (18)$$

becomes for spherical aggregates

$$\bar{r}_{s,sph} = r_f(t_\infty)(1 + \tau_e/\tau_{s,sph})^{-1} \qquad (19)$$

For cylindrical aggregates with a negligible rotational motion
around their short axes during τ_e

Table IV. The Dependence of Emission Anisotropy on Aggregate Geometry and Emission Lifetime.

Aggregate geometry	Timescale $\tau_e \approx \tau_f \ll \tau_s$		Timescale $\tau_e \approx \tau_s \gg \tau_f$	
	$r_f(t_\infty) \equiv r_f$	$\bar{r}_f$	$r_s(t_\infty) \equiv r_s$	$\bar{r}_s$
Spherical	$r_{f,sph} = r_{f,cyl} =$ $= r_{f,pla}$	$\bar{r}_{f,sph} = \bar{r}_{f,cyl} =$ $= \bar{r}_{f,pla}$	0	$\bar{r}_{s,sph} = \bar{r}_{f,sph} \dfrac{1}{1+\dfrac{\tau_e}{\tau_{s,sph}}}$
Cylindrical	$r_{f,sph} = r_{f,cyl} =$ $= r_{f,pla}$	$\bar{r}_{f,sph} = \bar{r}_{f,cyl} =$ $= \bar{r}_{f,pla}$	$r_{s,cyl} = \dfrac{1}{4} r_{s,pla}$	$\bar{r}_{s,cyl} = \bar{r}_{f,cyl} \dfrac{1}{1+\dfrac{\tau_e}{\tau_{s,cyl}}}$
Planar	$r_{f,sph} = r_{f,cyl} =$ $= r_{f,pla}$	$\bar{r}_{f,sph} = \bar{r}_{f,cyl} =$ $= \bar{r}_{f,pla}$	$r_{s,pla}$	$\bar{r}_{s,pla} = \bar{r}_{f,pla}$

τ_e = emission lifetime; τ_s and τ_f are the slow and the fast correlation times, respectively.

$$f(\Omega_{ND}) = f(\alpha,\beta,\gamma) = \delta(0,\pi/2,0) \text{ and}$$

$$r_{s,cyl}(t) = r_f(t_\infty)(1/4)(3e^{-t/\tau_{s,cyl}} + 1) \tag{20}$$

The steady state anisotropy

$$\bar{r}_{s,cyl} = r_f(t_\infty)(1/4)(4 + \tau_e/\tau_{s,cyl})(1 + \tau_e/\tau_{s,cyl})^{-1} \tag{21}$$

The effect of aggregate geometry and emission lifetime on the emission anisotropy is presented in Table IV. The table shows that $r_s(t_\infty)$ strongly depends on the aggregate shape. It can therefore be suggested that measurements of time-resolved emission would be a useful complementary method for studies of liquid crystalline phase structures, in particular cubic phases.

REFERENCES

1. S.J.Singer and G.L. Nicolson, Science., 175, 720 (1972).
2. V. Luzzati, A. Tardieu, Y.Gulik-Krzywicki, E. Rivas and F. Reis-Husson, Nature (London) 220, 485 (1968).
3. B. Kachar and T.S. Reese, ibid., 296, 464 (1982).
4. S.W. Hui and L.T. Boni, ibid., 296, 175 (1982).
5. E. Selstam, G. Lindblom, I. Brentel and M. Ryberg,(1982) in: Biochemistry and Metabolism of Plant Lipids (Wintermans,J.F.G.M. and Kuiper, P.J.C. eds). Dev. Plant Biol. vol.8 p 389.,Elsevier Biomedical, Amsterdam, New York.
6. P.-O. Eriksson, A.Khan and G. Lindblom, J. Phys. Chem.,86, 387 (1982)
7. J. Ulmius, G. Lindblom, H. Wennerström, L.B.-Å. Johansson, K. Fontell, O. Söderman, and G. Arvidson, Biochemistry, 21, 1553 (1982).
8. Å. Wieslander, L. Rilfors, L.B.-Å. Johansson, and G. Lindblom, ibid., 20, 730 (1981).
9. Å. Wieslander, A. Christiansson, L. Rilfors, and G. Lindblom, ibid., 19, 3650 (1980).
10. G. Lindblom, K. Larsson, L.B.-Å. Johansson, K. Fontell, and S. Forsén, J.Am. Chem. Soc. 101, 5465 (1979).
11. P.-O. Eriksson, G. Arvidson, and G. Lindblom, to be published.
12. P.R. Cullis and B. De Kruijff, Biochim. Biophys. Acta., 559, 399 (1979).
13. A. Stier, S.A. Finch, and B. Bösterling, FEBS Lett. 91, 109 (1978).
14. L. Rilfors, A. Kahn, I. Brentel, Å. Wieslander and G. Lindblom, FEBS Lett. 149, 293 (1982).
15. G. Lindblom and H. Wennerström, Biophys. Chem. 6, 167 (1977).
16. G. Lindblom, L.B.-Å. Johansson, and G. Arvidson, Biochemistry 20, 2204 (1981).

17.   A. Khan, A.J. Verkleij, P.-O. Eriksson, Å. Wieslander, and
      G. Lindblom, to be published.
18.   E.O. Stejskal and J.E. Tanner, J. Chem.Phys. $\underline{49}$, 288 (1965).
19.   H. Wennerström and G. Lindblom, Q. Rev. Biophys., $\underline{10}$, 67 (1977).
20.   V. Luzzati and P.A. Spegt, Nature (London) $\underline{215}$, 701 (1967).
21.   A. Abragam, "The Principles of Nuclear Magnetism", Oxford Uni-
      versity Press, London 1961.
22.   J. Charvolin and P. Rigny, J. Chem. Phys. $\underline{58}$, 3999 (1973).
23.   H. Wennerström, G. Lindblom, and B. Lindman, Chem. Scr. $\underline{6}$, 97
      (1974).
24.   B. Halle and H. Wennerström, J. Chem. Phys. $\underline{75}$, 1928 (1981).
25.   K. Kinosita, S. Kawato, and A. Ikegami, Biophys. J. $\underline{20}$, 289
      (1977).
26.   C. Zannoni, Mol. Phys. $\underline{38}$, 1813 (1979); ibid $\underline{42}$, 1303 (1981).
27.   L.B.-Å. Johansson and G. Lindblom, Q. Rev. Boophys. $\underline{13}$, 63
      (1980).
28.   A. Szabo, J. Chem. Phys. $\underline{72}$ 4620 (1980).
29.   P.L. Nordio and P. Busolin, J. Chem. Phys. $\underline{55}$, 5485 (1971);
      H. Wennerström, from pure symmetry arguments it can be shown
      that an infinite number of exponentials are needed, personal
      communication.
30.   H. Wennerström and J. Ulmius, J. Magn. Reson. $\underline{23}$, 431 (1976).
31.   J. Ulmius and H. Wennerström, ibid., $\underline{28}$, 309 (1977).
32.   U. Henriksson, L. Ödberg and J.-C. Eriksson, J. Phys. Chem.
      $\underline{81}$, 76 (1977).
33.   O. Söderman, G. Lindblom, L.B.-Å. Johansson, and K. Fontell,
      Mol. Cryst. Liquid Cryst. $\underline{59}$, 121 (1980).
34.   L.B.-Å. Johansson, O. Söderman, K. Fontell, and G. Lindblom,
      J. Phys. Chem. $\underline{85}$, 3694 (1981).

THE USE OF FREEZE-FRACTURE AND FREEZE-ETCHING ELECTRON MICROSCOPY

FOR PHASE ANALYSIS AND STRUCTURE DETERMINATION OF LIPID SYSTEMS

T. Gulik-Krzywicki and L.P. Aggerbeck

C.N.R.S., Centre de Genetique Moleculaire
91190 Gif-sur Yvette, France

K. Larsson

Chemical Center, University of Lund
P.B. 740, S-220 07 Lund, Sweden

Recent improvements including the development of
novel ultra-rapid cryofixation procedures and the use
of high resolution shadowing techniques suggest that
freeze-etching electron microscopy may be particularly
useful for studying phase diagrams and the structure of
individual phases in lipid containing systems. Use of
the freeze-etching technique, however, requires careful
evaluation of the effects of cryofixation, fracture and
etching upon the sample structure. The combined use of
X-ray diffraction before and low temperature X-ray diffraction after freezing prior to fracture, etching and
replication of samples whose structures are known provides a sensitive assessment of the effects of cryofixation upon lipid phases. Investigation of a number of
different lipid phases by the combined use of X-ray
diffraction, low temperature X-ray diffraction and
freeze-etching electron microscopy revealed the cryofixation procedures best adapted for studying each lipid phase and led to a better understanding of the
cryofixation methods. We discuss here the results obtained by using this combined X-ray diffraction - freeze-
etching approach and describe in more detail the most
recent data on the structure of lipid-water cubic phases.

INTRODUCTION

Electron microscopy, which is one of the most direct and
most powerful techniques for obtaining structural information, has
been applied only sparingly in the study of surfactants. Examination of untreated samples is virtually impossible due to the dama-
ging effects of the high vacuum and electron beam and because of
the very weak contrast provided by the surfactant molecules. Generally pretreatments (chemical fixation and staining) to permit
electron microscopic observation modify or completely destroy the
sample structure.[1] One procedure which seems to avoid some of the
pitfalls of the other techniques is cryofixation. Frozen hydrated samples may thus be examined directly[2] or by freeze fracture
electron microscopy. Ideally, when cryofixation is used, the
sample should be quenched rapidly enough to avoid structural alterations due to temperature induced structural transitions, change of the partial specific volumes, ice crystal formation, etc.
Practically, this goal may be difficult to achieve and one must
assess the extent of preservation of the sample structure after
cryofixation. Among the methods that might be used for such an
assessment, X-ray, neutron or electron diffraction, which provide
detailed structural information, are probably the most straightforward. However, spectroscopic (NMR, ESR etc.) or other techniques
could be applied. We have developed an approach based on the combined use of ambient and low temperature X-ray diffraction and
freeze-etching electron microscopy to study to effect of quenching
upon sample structure. This approach was applied first to study a
variety of lipid-water phases of known structure cryofixed by different procedures.[3,4] More recently the same approach was used
to study solutions of low density serum lipoproteins.[5] The results of these studies provided some understanding of the factors
which affect sample structure during quenching and permitted the
elaboration of a more general approach for structural investigations by freeze etching electron microscopy. In this article we
will describe this general approach and its use in the study of
some lipid water cubic phases.

FREEZE-ETCHING ELECTRON MICROSCOPY

Freeze-etching electron microscopy has become, at least
for biological membranes, a routine method for studying structure.
Well defined procedures giving highly reproducible results have
been devised and applied to a variety of different materials.[6]
The technique consists of four main steps. Each step, for which
there are many technical variations, must be optimized to insure
that the structure observed in the electron microscope is an
exact representation of the ultrastructure of the original sample.

Cryofixation of the Sample

The goal of this step is to quench the sample rapidly enough so that, after fixing, the structure is the same as it was at the initial temperature. Many factors, however, such as ice crystal formation, temperature induced structural transitions, changes in partial specific volume, etc., have the potential to alter the sample structure during quenching. One approach to minimize such perturbations is to pre-treat the sample by chemical fixation or by adding cryoprotectants.[7] For surfactants and some other materials, such pre-treatment itself will alter the sample structure. In these cases, optimal quenching must be achieved by one of the ultrarapid cryofixation methods such as sandwich,[4] propane jet,[8] spray[9] or copper block freezing.[10] In the first two methods, the sample is squeezed between two thin metal (copper or gold) plates and then quenched by rapid plunging into liquid propane or by projecting, at high speed, a stream of liquid propane onto the metal surfaces. The rate of cooling, about 13×10^3 °K/s in both cases, can be increased by applying high pressure to the sample.[11] The spray freezing technique employs an artist's air brush and filtered compressed air to spray small droplets (5-15 μm diameter) of sample containing solution into liquid propane. The rate of cooling for this technique is unknown but may be surmised to be of the same order of magnitude as the other rapid quenching methods. Projection of the sample against a very clean copper surface maintained at liquid helium temperature forms the basis for the copper block technique (cooling rate about 6×10^3 °K/s.) The relative advantages and disadvantages of these procedures have been discussed recently by Costello et al.[12]

Fracture of the Cryofixed Sample

Quenched samples are fractured at low temperatures (less than 173 °K) under high vacuum (better than 10^{-6} Torr) by the action of a liquid nitrogen or liquid helium cooled knife or by separation of the two metal plates of "sandwiched" samples. Deposition of residual water vapor resulting in contamination of the exposed surface may be reduced by operating under a higher, cleaner vacuum. The sample may or may not be deformed during the fracture process. Plastic deformation[13] may be reduced by fracturing at lower temperature. Finally, parts of the sample situated below the fracture plane may be exposed subsequently by sublimation (freeze-etching) of the solvent prior to replication. Some samples may be very sensitive to this procedure and undergo structural changes[14].

Replication of the Fracture - Exposed Surfaces

The replication of surfaces exposed by fracture (freeze fracture) or by fracture and etching (freeze etching) is performed

by evaporation of heavy metals. Platinum-carbon is commonly used
but a finer grain size is obtained with tungsten or tungsten-tan-
talum alloys.[15,16] The metal deposit, which has a mean thickness
of 1-2 nm, is too fragile to manipulate or observe in the electron
microscope. To strengthen the replica, a 10-20 nm thick layer of
carbon is evaporated onto the surface. The carbon layer, at this
thickness does not contribute appreciably to the final electron
microscope image. Since evaporation of the metals during the replication step requires high temperatures, heating and possible
distortion of the exposed surface must be carefully avoided. This
is particularly true when metals other than platinum are used.
Surface heating can be minimized by working at the optimal conditions for a given source of evaporation and by using a multistep
evaporation procedure allowing cooling of the partially shadowed
surface between steps.[16] The best revelation of the various aspects
of the sample ultrastructure frequently needs to be determined
empirically by using unidirectional or circular (obtained by rotating the sample) shadowing or combinations thereof at different
shadowing angles.

Cleaning and Observation of the Replica

        After replication, the sample is thawed and then solu-
bilized or digested with an appropriate solvent system. After
elimination of the sample, the replica is washed with distilled
water, deposited on an electron microscope grid and dried prior
to observation in the electron microscope. Incomplete elimination
of sample material giving a "dirty" replica can, on occasion, raise
a problem. Finally, it must be stressed that since the replica is
an object having three dimensions, stereo electron microscopy is
invaluable for properly determining the sample ultrastrucure.

        Further, for ordered phases, optical diffraction of the
electron micrographs may provide information about the lattice
parameters and symmetry of the sample ultrastructure revealed by
different fracture planes.

X-RAY DIFFRACTION

        The X-ray scattering instrument and the experimental
procedures are those routinely used in our laboratory.[3-5] The
sample is placed on an appropriate holder which is mounted on a
small metal support (O) fixed on a glass capillary (p), whose
position in the X-ray beam is controlled by a micromanipulator
(Figure 1). The X-ray diffraction pattern of the sample is recorded with linear position sensitive detectors.[17] This pattern is
compared with that obtained from the sample in a sealed specimen

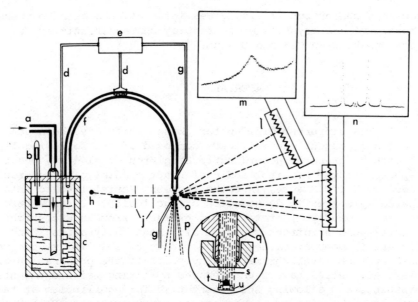

Figure 1. Schematic drawing of the low temperature X-ray diffraction set-up : (a) liquid nitrogen input ; (b) level indicator ; (c) constant level Dewar ; (d) heaters ; (e) heater control ; (f) flexible Dewar transfer tube ; (g) thermocouples ; (h) X-ray source ; (i) focusing mirror ; (j) slits ; (k) beam attenuator ; (l) linear position-sensitive detectors ; (m) wide-angle output from multichannel analyser ; (n) low-angle output from multichannel analyser ; (o) sample support cup ; (p) glass rod ; see insert for (q) cryostat output tip ; (r) heated cap to eliminate frost ; (s) laminar flow of gaseous nitrogen ; (t) sample ; (u) sample holder

holder, to ensure that no changes occur during preparation of the specimen for freezing. If no changes occur, then the sample is immediately frozen and replaced on the same small metal support (o) under a laminar stream of cold dry nitrogen gas (133°K).After recording the low temperature spectrum, the sample is transferred to liquid nitrogen for storage prior to subsequent analysis by freeze-etching electron microscopy .Comparisons of X-ray diffraction patterns from samples before and after freezing permit the assessment of the structural changes induced by the freezing process. Low angle diffraction was used to follow changes in the long range organization of samples, specifically, the change of dimensions of lattices and their disordering (as judged from the broadness of X-ray reflections). Wide angle X-ray diffraction was used mainly to follow the perturbations in the packing of the lipid hydrocarbon chains but it also gives information about the presence or absence of ice crystals which display characteristic strong reflections between 0.3 and 0.4 nm.[18]

MATERIALS

The different lipid-water phases, which have been used in our past and present studies, were composed of the following lipids : Egg lecithin (prepared according to Singleton et al [19]), wheat germ phosphatidyl- inositol and beef heart cardiolipin (generously provided by Dr. M. Faure, Institut Pasteur, Paris), lysolecithin (prepared by enzymatic degradation of egg lecithin), cholesterol (purchased from Sigma), total lipid extracts from an Escherichia coli fatty acid auxotroph grown on different fatty acids[20] and sunflower oil monoglycerides[21]. The purity of all lipids used was checked by thin layer chromatography prior to the preparation of samples. All phases have been prepared by mixing weighed amounts of constituents and allowing sufficient time for equilibrium at temperatures corresponding to the existence of pure phases. Cytochrome c (Sigma, type IV) -Sunflower oil monoglycerides- water cubic phase was prepared by equilibration of monoglycerides with a concentrated solution of cytochrome c (30% by weight) at 313°K.

RESULTS AND DISCUSSION

Preserving Structure During Cryofixation : Some Important Factors

Our previous combined X-ray diffraction-freeze-etching electron microscope studies have shown that the extent of conservation, after cryofixation, of the initial sample structure depends strongly upon the composition of the sample, its structure and, in most cases, the quenching procedure. The lipid phases containing less than 10%, by weight, water appear to be the least sensitive to the rate of cooling. Among these, the phases displaying an ordered conformation of the hydrocarbon chains (L_β phase) showed no changes in the low angle X-ray scattering pattern, before and after quenching, indicating complete preservation of the lattice order and dimensions. Broadening of the wide angle X-ray reflections for these samples after cryofixation reflects some disordering of the initial packing of the hydrocarbon chains. In contrast, phases which initially have disordered hydrocarbon chains (L_α phase) become more ordered and more closely packed after quenching. This ordering is reflected by the narrowing and shifting to a wider angle of the 0.45 nm broad X-ray diffraction band (a band characteristic of highly disordered hydrocarbon chains).[22] The ordering, however, is very limited since there is no change in the lattice dimensions.

That such ordering is limited is also suggested by similar behavior
for phases having structures incompatible with the existence of
highly ordered hydrocarbon chains such as the two dimensional egg
lecithin-water hexagonal phase (H_α) shown in Figure 2. This phase
also can be quenched at relatively slow cooling rates without any
appreciable change in lattice dimensions. The laser optical diffraction pattern of the electron micrograph of the cross fractured,
hexagonally ordered water cylinders shows a hexagonal lattice whose
dimensions match those observed by X-ray diffraction before quenching. This last observation also indicates that no major structural perturbations occured during fracture, replication and washing
of the replica.

For phases containing 10 to 30% water, insufficiently
rapid cooling rates result in an increase in lattice dimensions
and segregation of different lipid species into separate domains
after cryofixation.

When the water content of a phase exceeds 30% by
weight, only ultrarapid cryofixation methods conserve, with varying
degrees of success, the initial sample structure. Slower rates of
quenching lead to the formation of ice crystals resulting in disordering of the lattice concomittant with an important decrease
in lattice dimensions. Some samples may undergo phase transitions
toward phases existing either at lower temperatures or lower water
contents. In extreme cases, a highly ordered phase, such as a
cubic phase, may be broken down into an amorphous like material.
The addition of cryoprotectants, such as glycerol, may partially
relieve the pertubations due to quenching.[3]

For the study of surfactants in solution, several
partial conclusions may be drawn. Knowledge of the phase diagram
for various water contents may be useful for predicting how difficult adequate quenching (that is, quenching with minimal structural perturbation) might be. Since, in some cases, even the most
rapid cooling rates presently in use do not provide quenching
without structural perturbations, comparison of structural parameters (lattice dimensions, lattice order, ice crystal size, etc.)
before and after cryofixation are of crucial importance. One very
powerful method which permits such a comparison is X-ray diffraction which may be performed before and after quenching using the
same sample that will undergo analysis by electron microscopy.

Structural Analysis of Some Lipid-Water Cubic Phases

A large number of cubic phases, situated in different
places in binary amphiphilic lipid-water phase diagrams, have been
reported.[23] They are all optically isotropic, highly viscous and
show X-ray diffraction spectra in the low angle region, which are

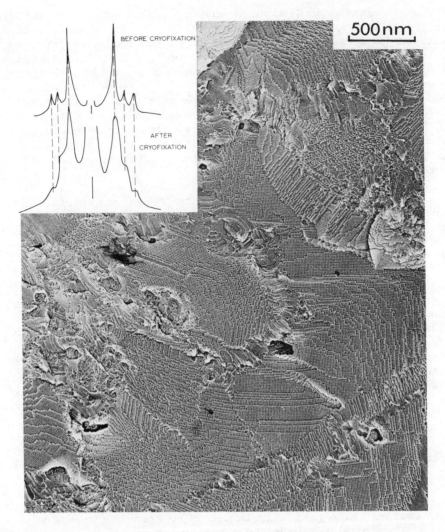

Figure 2. Freeze-fracture electron micrograph of egg lecithin-water reversed hexagonal phase (H_α). Shadowing with W-Ta (shadows are white) is unidirectional from lower right to the upper left. The structure of the phase is shown at the bottom left corner (hexagonally packed water cylinders, surrounded by lipid polar head groups and separated by disordered lipid hydrocarbon chains[22]). Two laser optical diffraction pictures, shown at the bottom right corner, were obtained from the areas indicated by arrows. Notice that fracture planes propagating in between the cylinders show striations, whose period corresponds to the distance between the cylinders (according to X-ray diffraction results). The cross fractured cylinders show the hexagonal packing on the optical diffraction patterns obtained from the corresponding area.

compatible with three dimensional crystalline cubic lattices and, at wide angles, a diffuse 4.5 Å broad band, characteristic of highly disordered hydrocarbon chains. Some of these cubic phases have been extensively studied by X-ray diffraction and NMR spectroscopy. The main conclusions of these studies (summarized by Fontell[23]) were that, in spite of still prevailing opinion, the structures of most of these cubic phases are not formed of independent globular entities, closely packed in cubic lattices. Instead, more complex models of the molecular organizations were proposed. In most cases, however, the X-ray diffraction data are insufficient to derive, unambigously, molecular models of the structures. Since freeze-etching electron microscopy can provide structural information complementary to that obtained by X-ray diffraction, we undertook the present study with the goal of providing independent structural information which may help in discerning the correct molecular models for some of these phases.

Monoglyceride-water cubic phase. This phase is situated between the lamellar and reversed hexagonal phases in many monoglyceride-water systems. A detailed molecular model of its structure was recently proposed on the basis of extensive X-ray diffraction and NMR studies .[24,25] This model is composed of two independent three-dimensional systems of continuous water channels, isolated by hegaxonally shaped lipid bilayer faces.

The first step, in our study, was to determine the conditions which provide the best conservation, after cryofixation, of the initial sample structure. As assessed by low temperature X-ray diffraction, this phase was quite difficult to quench in its original state. Even after ultrarapid cryofixation with glycerol as a cryoprotectant, there is only a very broad X-ray diffusion peak in the angular domain where sharp cubic reflections were present before quenching (inset, Figure 3). Freeze-etching electron microscopy of this sample showed only rare, isolated, small domains of ordered structure resembling the cubic phase (arrow, Figure 3). Markedly less structural perturbation, after quenching, occurs when the sample contains water dissolved cytochrome c (without glycerol). The low temperature X-ray diffraction pattern, in this case, exhibits only slightly broadened reflexions as compared to that at 313 °K. Freeze fracture electron microscopy reveals very large domains of highly ordered crystallites (Figure 4). The analysis of electron micrographs obtained at higher magnification (Figure 5), shows clearly that this ordered structure is composed of globular elements arranged in a body centered cubic lattice. As revealed by the fracture plane descending through the sample (arrows, Figure 5), each plane of globules displaying square symmetry is shifted by one-half the lattice dimension with respect to those immediately above and below. The interglobular distance coincides exactly with the cubic lattice dimensions (13.0±0.2 nm) determined from the X-ray diffraction data obtained before and

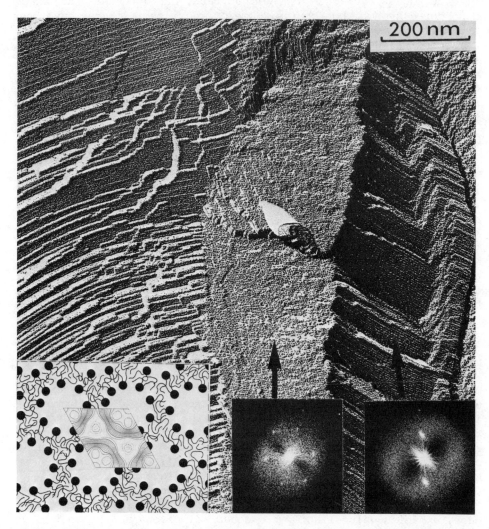

Figure 3. Freeze-fracture electron micrograph of a sample of
sunflower oil monoglycerides in water which originally was organized in a cubic phase. Shadowing with W-Ta (shadows are white) is
unidirectional from the bottom left toward the top right. The X-
ray diffraction patterns of the sample before and after cryofixation (inset) show that the structure of the sample did not with-
stand the quenching process. It is, however, possible to find, in
some extremely rare areas of the replica, an organization compatible with a cubic structure (arrow).

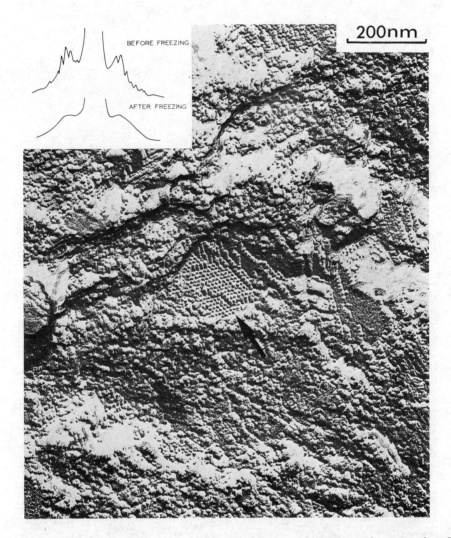

Figure 4. Low magnification freeze-fracture electron micrograph of
sunflower oil monoglycerides-cytochrome c - water cubic phase.
Rotary shadowed with W-Ta (shadows are white). X-ray diffraction
patterns of the sample before and after cryofixation (inset) show
a good preservation of the sample structure in the quenched state.
The freeze-fracture replica shows extended domains of different,
highly ordered, fracture planes, each compatible with a body centered cubic symmetry. In some places, very small domains of a reversed hexagonal phase can also be seen.

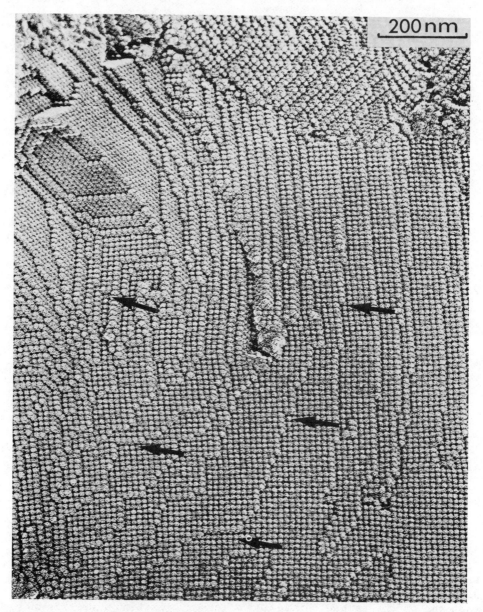

Figure 5. Freeze-fracture electron micrograph of sunflower oil
monoglycerides-cytochrome c-water cubic phase. Rotary shadowed with
W-Ta but contrast is reversed (shadows are black). Notice the
different fracture planes through the phase. Each plane displaying
a square lattice is shifted by one-half the lattice dimension as
compared to those immediately below and above (arrows) indicating
that the structure is body centered.

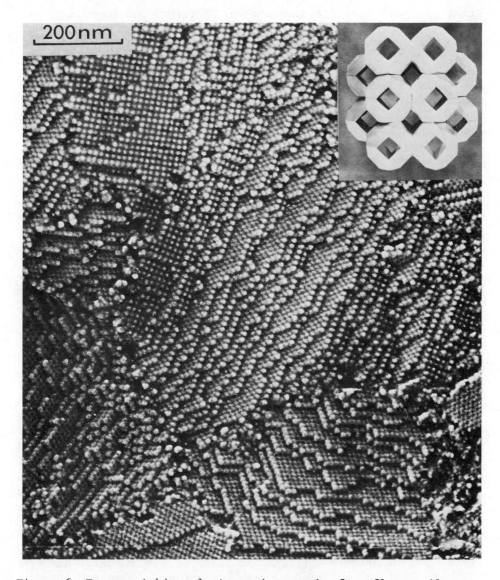

Figure 6. Freeze-etching electron micrograph of sunflower oil
monoglycerides -cytochrome c-water cubic phase. Rotary shadowed
with W-Ta but contrast is reversed (shadows are black). Partial
sublimation of the ice (30 seconds at 173 °K, under 5.10^{-7} Torr)
leads to some changes in fracture face morphology, mainly due to
an enlargement of the water channels, which are clearly seen at
the bottom left part of the electron micrograph. Inset. Photograph
of a model of the tetrakaidekahedral polyhedron proposed as the
building block of the body centered cubic structure in monoglyceride water systems. [24,25]

after quenching with the same sample. Therefore, up to this point we may conclude that the sample is composed of globular elements (probably interconnected) arranged in a body centered cubic lattice. The information is insufficient, however, to suggest a precise molecular model although it is compatible with that proposed on the basis of X-ray diffraction data (inset Figure 6). This model is composed of two, independent, three dimensional water channels, isolated by hexagonally shaped lipid bilayer faces forming individually repeating units. The relatively limited resolution, 1.0-2.0 nm, obtained with our best shadowing procedure may explain the presence of globular rather than hexagonally shaped units in the electron micrographs (Figure 5). Good visualization of the water channels is not obtained by freeze-fracture alone. However, etching of freshly fractured surfaces should result in sublimation of some of the water frozen within the phase leaving holes where the water channels were situated. Comparison of freeze-fracture (Figure 5) and freeze-etching (Figure 6) electron micrographs shows that etching does, indeed, result in the production of holes in the fractured surfaces. The dimensions and distribution of these etching induced holes are compatible with the above model for the structure of the phase. Continued etching leads, eventually, to complete breakdown of the initial structure probably due to the fragility of the exposed lipid bilayers. Additional evidence supporting the previously proposed model might be obtained from a similar study of the cubic phase prepared with water soluble proteins larger than cytochrome c. In this case, one might expect larger dimensions for the cubic lattice which would allow better visualization of the individual structural units on the fractured and etched surfaces.

The limited etching of fractured surfaces also is useful in the study of the transition between the sunflower oil monoglyceride lamellar and cubic phases. Replicas of samples quenched while undergoing this phase transition display many structures other than pure cubic or lamellar organization and probably represent intermediate steps in the structural transition. The stacked arrangement of smooth (characteristic of a pure lamellar phase) and rough lamellae may correspond to an early stage of reorganization of the lipid lamellae (Figure 7A). Stacked lamellae showing large number of holes (probably enlarged by etching) which are arranged in square or hexagonal lattices (Figure 7B) may correspond to the later stages of the transition. Hexagonally arranged globules (Figure 7C) are also frequently seen.

This type of periodic arrangement of holes or globules might be characteristic of intermediate structures such as the rhombohedral R phase[26] or the tetragonal T phase[27] occurring during the lamellar to cubic phase transition of this and other lipid-water systems. For these systems the lamellar to cubic phase

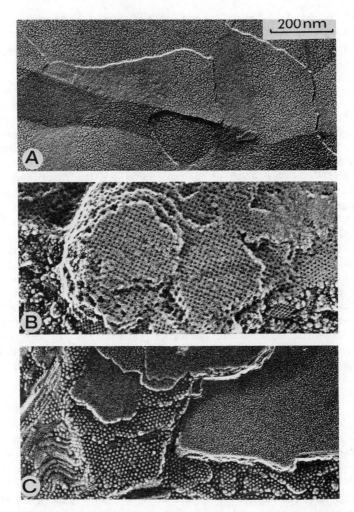

Figure 7. Freeze-etching electron micrograph of, probably, the
early stages of the transition between the lamellar and the cubic
phases in sunflower oil monoglycerides-cytochrome c -water system.
The micrographs were obtained from the replicas of etched (1 minute
at 173°K, under 5.10^{-7} Torr vacuum) surfaces of the sample in which
the formation of cubic phase was not complete. A. Stacked lamellae,
parts of which are very smooth (typical of corresponding pure lamellar phase) and parts of which are rough. Unidirectional shadowing
with W-Ta from the bottom toward the top. B. Large holes in stacked
lamellae, some of which form square lattices whose dimensions are
very close to that of the final cubic lattice dimension. Rotary
shadowing with W-Ta. C. Stacked lamellae, parts of which are smooth
and parts of which are arranged in small globules (particles) forming hexagonal lattices with interglobular distances approximating
the dimensions of the cubic phase. Rotary shadowing with W-Ta. The
contrast is reversed for all 3 photographs (shadows are black).

transition might induce an important reorganization of the stacked lamellae and the production of holes protruding through the lamellae or the formation of globules is not surprising. The latter, called "lipidic particles" have been the topic of much discussion and speculation recently.[28-30]

Egg lecithin and lysolecithin cubic phases. The egg lecithin cubic phase, like the monoglyceride-water cubic phase, is situated between the lamellar and reverse hexagonal phases. Its X-ray diffraction pattern is of the same type as those given by strontium soaps at elevated temperatures, for which a detailed molecular model has been derived on the basis of the large number of X-ray reflections.[26,32] The good agreement between the measured and calculated intensities of the reflections suggests that the model has a high degree of credibility. The model is composed of identical rod-like elements of finite length, linked three by three into two distinct three dimensional networks, unconnected but mutually interwoven (inset, Figure 8). Low temperature X-ray diffraction (inset, Figure 8) showed only minimal changes in the lattice parameters (as compared to the initial diffraction pattern) after cryofixation indicating that this phase is easier to quench than the sunflower oil monoglyceride-water cubic phase. In contrast to this latter phase, the egg lecithin cubic phase does not show any globular elements but rather parallel striations. By optical diffraction, the most frequently occurring fracture planes display rectangular symmetry (inset, Figure 8). The precise paths of the fracture planes relative to the proposed structural model is not clear from these preliminary results.

The lysolecithin cubic phase is distinguished from the other cubic phases by its position in the phase diagram encompassed by the hexagonal and the micellar phases and by the spacings ratios of the small angle reflections.[32] As for the egg lecithin cubic phase, the lysolecithin cubic phase was relatively easy to quench.

By electron microscopy, this phase reveals some globular elements in the fracture planes which display square symmetry (Figure 9). The morphology and dimensions of these elements are in good agreement with the tentative model proposed for the structure of this phase.[32] This model is composed of a system of rods of finite length, joined three by three at one end and four by four at the other, and a body centered arrangement of spherical elements (inset, Figure 9). The detailed analysis of this phase by freeze etching electron microscopy is still underway.

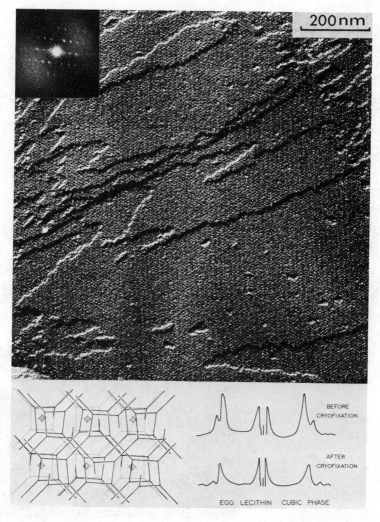

Figure 8. Freeze-fracture electron micrograph of the egg lecithin-
water cubic phase. Shadowing with W-Ta is unidirectional from the
bottom toward the top. The contrast is reversed (shadows are black).
The inset at the bottom left corner shows a schematic drawing of
this cubic phase structure, as deduced from previous X-ray diffraction data[26,31]. The structure is formed from identical rod like
elements of finite length, linked three by three into two distinct
three dimensional networks, which are unconnected but mutually
interwoven. The inset at the bottom right corner shows the X-ray
diffraction patten of the sample before and after freezing. It
indicates that the cubic structure is preserved in the quenched
state with just a small decrease in the lattice dimension. The
inset at the top left corner shows the optical diffraction pattern
from the central part of the electron micrograph.

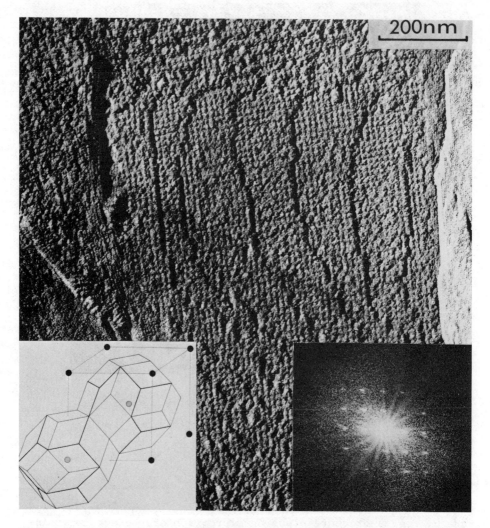

Figure 9. Freeze-fracture electron micrograph of the lysolecithin-
water cubic phase. Shadowing with W-Ta is unidirectional from the
left toward the right. The contrast is reversed (shadows are black).
The inset at the bottom left corner shows a schematic drawing of
the tentatively proposed structure, as deduced from X-ray diffraction data[32]. The structure consists of a system of rods of finite
length, all identical and crystallographically equivalent, joined
three by three at one end and four by four at the other, and a body-
centered arrangement of spherical elements. The inset at the bottom
right corner shows the optical diffraction pattern corresponding
to the central part of the electron micrograph. It shows a well
developed square lattice.

The Role of Freeze-Fracture Electron
Microscopy in Structural Analysis

Given the preceding results concerning some lipid-water phases, we can now summarize the relationship of freeze fracture and freeze-etching electron microscopy to X-ray diffraction for the structural analysis of lipid phases. In particular, we can attempt to show what unique information can be provided by the electron microscopic technique.

First, the extent of this information depends upon the nature of the sample. For pure, highly ordered phases, providing rich X-ray diffraction data, electron microscopy can reveal the overall (general) morphology of the sample and shape and dimensions of some of the structural elements. As previously illustrated this information may permit one to determine which of several molecular models (all compatible with the X-ray diffraction data) is the correct one. Further, the combination of freeze-etching with freeze fracture may give information about the location of the solvent within the structure.

For less well ordered phases, providing more limited X-ray diffraction data, the information provided by electron microscopy is of greater importance. For samples composed of a mixture of ordered phases, the freeze-etching data can indicate the number of different phases present and their structure. Further, the data may help to correctly assign X-ray reflections in, often, very complex diffraction patterns.

For phases composed of both ordered and disordered domains or disordered domains alone, freeze-etching electron microscopy permits the morphological analysis of each individual domain.

In the near future, one may expect further technological developments that will result in more rapid quenching with even less perturbation, finer grain size after shadowing giving better resolution and more detailed stereological analysis. There is, thus, good reason to expect that freeze-fracture and etching electron microscopy combined with low temperature X-ray diffraction should provide an excellent approach for investigating the molecular organization of surfactant solutions.

ACKNOWLEDGEMENTS

The authors gratefully acknowledge the constant encourage-
ment, many useful discussions and critical comments about the
manuscript from V. Luzzati, help with the optical diffraction
apparatus from J.L. Ranck, excellent technical assistance from
J.C. Dedieu and secretarial assistance from B. Gascard. This work
was supported by the Délégation Générale à la Recherche Scientifi-
que et Technique contracts number 79-7-097 and 80-7-0386.

REFERENCES

1. H.P. Zingsheim and H. Plattner, in "Methods in Membrane
   Biology," E.D. Korn, Editor, Vol. 7, p. 1, Plenum Publishing
   Corporation, New York, 1976.
2. J. Lepault, F.P. Booy and J. Dubochet, (1982) J. Microsc. in
   press.
3. M.J. Costello and T. Gulik-Krzywicki, Biochim. Biophys. Acta,
   455, 412   (1976).
4. T. Gulik-Krzywicki and M.J. Costello, J. Microsc., 112, 103
   (1978).
5. L.P. Aggerbeck and T. Gulik-Krzywicki, J. Microsc., 126, 243
   (1982).
6. E.L. Benedetti and P. Favard, Editors, "Freeze-Etching, Technique and Applications," Société Française de Microscopie Electronique, Paris, 1973
7. F. Franks, J. Microsc., 111, 3 (1977).
8. M. Mueller, N. Meister and H. Moor, Mikroscopie (Wein), 36,
   129, (1980).
9. L. Bachmann and W.W. Schmitt-Fumian, in "Freeze-Etching Technique and Applications", E.L. Benedetti and P. Favard, Editors,
   p. 73, Société Française de Microscopie Electronique, Paris,
   1973.
10. J.E. Heuser, T.S. Reese, M.J. Denis, Y. Jan, L. Jan and L. Evans,
    J. Cell. Biol., 81, 275 (1979).
11. H. Moore, in "Proceedings of Electron Microscopy Society of
    America," G.W. Bailey, Editor, P. 334, Claitor's Publishing
    Division, Baton Rouge, 1977.
12. M.J. Costello, R. Fetter and M. Höchli, J. Microsc., 125,
    125, (1982).
13. V.B. Sleytr and A.W. Robards, J. Microsc., 110, 1 (1977).

14. J. Lepault and J. Dubochet, J. Ultrastruct. Res., 72, 223
    (1980).
15. R. Abermann, M.M. Salpeter and L. Bachman, in "Principles and
    Techniques of Electron Microscopy", M.A. Hayat, Editor, Vol. 2,
    p. 197, Van Nostrand Reinhold, New York, 1972.
16. A. Gulik, L.P. Aggerbeck, J.C. Dedieu and T. Gulik -Krzywicki,
    J. Microsc., 125, 207 (1982).
17. A. Gabriel and Y. Dupont, Rev. Sci. Instrum., 43, 1600,
    (1972).
18. L.G. Dowell, S.W. Moline and A.P. Rinfret, Biochim. Biophys.
    Acta, 59, 158 (1962).
19. W.S. Singleton, M.S. Gray, H.L. Brown and J.L. White, J. Am.
    Oil Chem. Soc., 42, 53 (1965).
20. E. Shechter, L. Letellier and T. Gulik-Krzywicki, Eur. J.
    Biochem., 49, 61, (1974).
21. J.V. Boyd, N. Krog and P. Sherman, in "Theory and Practice of
    Emulsion Technology", A.L. Smith, Editor, p. 61, Academic Press,
    1976.
22. V. Luzzati, in "Biological Membranes", D. Chapman, Editor,
    p. 71, Academic Press, New York, 1968.
23. K. Fontell, Mol. Cryst. Liq. Cryst., 63, 59 (1981).
24. G. Lindblom, K. Larsson, L. Johansson, K. Fontell and S. Forsen,
    J. Am. Chem. Soc., 101, 5465, (1979).
25. K. Larsson, K. Fontell and N. Krog, Chem. Phys. Lipids, 27,
    321, (1980).
26. V. Luzzati, T. Gulik-Krzywicki and A. Tardieu, Nature, 218,
    1031, (1968).
27. V. Luzzati, A. Tardieu and T. Gulik-Krzywicki, Nature, 217,
    1028, (1968).
28. S.W. Hui and L.T. Boni, Nature, 296, 175, (1982).
29. W.P. Williams, A. Sen, A.P.R., Brain, P.J. Quinn and M.J.
    Dickens, Nature, 296, 175, (1982).
30. A.J. Verkleij, C. Monbers, J. Bijvelt-Leunissen and P.J.J.Th.
    Ververgaert, Nature, 279, 162, (1979).
31. V. Luzzati and D.A. Spegt, Nature, 215, 701, (1967).
32. A. Tardieu and V. Luzzati, Biochim. Biophys. Acta, 219, 11, (1970).

THE STRUCTURE OF PLASMA LIPOPROTEINS: EVALUATION BY X-RAY AND NEUTRON SMALL-ANGLE SCATTERING

P. Laggner

European Molecular Biology Laboratory
Hamburg Outstation at DESY
Hamburg, Federal Republic of Germany;
and
Institut für Röntgenfeinstrukturforschung
Austrian Academy of Sciences
Research Centre Graz
Graz, Austria

The methods of X-ray and neutron small-angle scattering on dilute solutions have been used to obtain structural information on plasma, High Density (HDL) and Low Density Lipoproteins (LDL). The structural principle emerging is remarkably simple and can be related in a logical way to the molecular properties of the lipid constituents. Both LDL and HDL are quasi-spherical particles with the apolar lipids (cholesteryl esters and triglycerides) arranged radially in the core, surrounded by a surface layer of polar consti-tuents. In HDL, where the amphipathic components dominate, the particle sizes are determined by the degree of interdigitation between the cholesteryl esters and the phospholipid hydrocarbon chains from the surface. In LDL, the cholesteryl esters and tri-glycerides form a separate domain in the core which undergoes a reversible, thermotropic order-disorder transition at physiological temperatures. Below the transition, the core arrangement is reminiscent of the liquid-crystalline state of pure cholesteryl esters. The surface is formed by a monolayer of polar lipids and the protein, the latter being located, on average, above the level of the polar phospholipid headgroups. While the scattering methods do not show any temperature

dependent changes in the surface structure of LDL,
e.s.r. studies using spin labels inserted into the
outer monolayer revealed that the fine-structure of
the surface is sensitive to the transition in the
particle core.

INTRODUCTION

In simple physical terms, the lipoproteins of blood plasma
can be regarded as microemulsion particles by which the water-
insoluble lipids are transported in the circulation. They constitute a continuous, multimodal spectrum of particle sizes, ranging
from about 100 $\overset{o}{A}$ to 10,000 $\overset{o}{A}$, and consist to different proportions,
of the following components held together by noncovalent bonds:

- proteins

- phospholipids

- cholesterol

- cholesteryl esters

- triglycerides

Ultracentrifugal fractionation has formed the basis of the
widely-applied definition of density classes introduced by the group
of Gofman[1]. Figure 1 shows a diagramatic representation of the lipoprotein classes and subclasses as they appear in ultracentrifugal
separation, together with their size and composition ranges.
Although primarily operational, this classification has also some
physiological relevance since the individual classes play different
roles in lipid transport and metabolism[2,3].

Up to now, eight different specific proteins have been isolated
and characterized as parts of lipoproteins. For several of these
apoproteins, there exists already a documented primary structure
(Table I). Some additional polypeptides have been reported as apoproteins, but their nature and specificities are not yet sufficiently
documented. These apoproteins are unevenly distributed among the
lipoprotein density classes, also reflecting their different physiological and biochemical specificities. According to this fact,
Alaupovic and coworkers[13] have introduced an alternative classification, whereby lipoprotein families are defined by their protein
moieties, such that e.g. lipoprotein A (LpA) is the group of particles
containing only apoproteins A. This implies that a family is not
necessarily monodisperse and hence for the purpose of structural
description this classification is inconvenient unless an additional
defining constraint is introduced, as, e.g., LpA from HDL$_3$ indicating

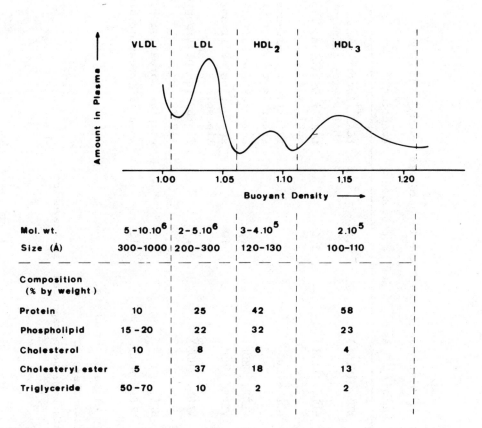

	VLDL	LDL	HDL$_2$	HDL$_3$
Mol. wt.	$5-10.10^6$	$2-5.10^6$	$3-4.10^5$	2.10^5
Size (Å)	300-1000	200-300	120-130	100-110
Composition (% by weight)				
Protein	10	25	42	58
Phospholipid	15-20	22	32	23
Cholesterol	10	8	6	4
Cholesteryl ester	5	37	18	13
Triglyceride	50-70	10	2	2

Figure 1. Lipoprotein classification: Particle mass, size, and chemical composition of the major classes.

particles of the HDL$_3$ density subclass containing only apo-A proteins. Note that even then two different polypeptides, Apo-AI and Apo-AII, can be contained in this category.

The greatest stimulus for the widespread interest in plasma lipoproteins originates from their physiological implications. It is

Table I. Apoprotein Distribution in Human Lipoprotein Density Classes

Apoprotein	Mol. Wt.	Density Class +	Ref. for Sequence	Function
A – I	28.300	HDL, (VLDL)	4,5	LCAT[++] activation; cholesterol transport.
A – II	17.400	HDL, (VLDL)	5,6	LCAT inhibition; lipase activation.
B	8000-575.000[+++]	LDL, VLDL	–	intestinal lipid absorption; specific receptor binding.
C – I	6.625	VLDL, (HDL)	8,9	cofactor for LCAT.
C – II	8.837	VLDL, (HDL)	10	activation of lipoprotein lipase.
C – III	8.764	VLDL, (HDL)	11	inhibition of lipoprotein lipase.
D	22.000	HDL$_3$	–	complex formation with LCAT.
E	35.000	VLDL, HDL	12	specific receptor binding.

+   Minor component of density class in brackets.

++   Lecithin-cholesterol acyltransferase.

+++   Different molecular weights have been reported, see Reynolds[7].

now generally accepted that variations in plasma lipoprotein concentrations and in the relative proportions of density classes are
related, by as yet poorly understood ways, to the incidence of
atherosclerosis. In the recent past, increasing evidence has been
produced to support the view that lipoproteins are much more than
just lipid transport containers as they also play specific regula-
tory roles in lipid metabolism. Most excitingly, this has been demonstrated by the work of Goldstein, Brown and coworkers[14] revealing a specific mechanism for recognition and internalization
of LDL through a plasma-membrane bound receptor. This mechanism
involves also a regulation of the cell's own cholesterol synthesis
as well as the synthesis of receptors. In order to understand the
molecular basis for this and other biochemical events involving
lipoproteins, it is obviously required to know their molecular
architecture.

As a consequence of this strong interest, virtually the whole
arsenal of physico-chemical methods - with the important exception
of crystallography - has been applied to obtain information on different aspects of lipoprotein structure (for reviews, see[15-17]).
These aspects can be grouped into four categories:

- Morphology and internal molecular organization

- Structure-forming role of the components

- Interactions between components and with aqueous
  environment

- Dynamic state of components

In addition to their chemical complexity, lipoproteins have two
properties which make a complete structural description a difficult and as yet unsolved problem: firstly, as they are part of an
apparently continuous spectrum, particles within a given subclass
obtained by any separation technique are to some degree heterogeneous; secondly, mobility of components within the particles and
exchange between particles are the rule rather than the exception.
Both properties necessarily limit the degree of resolution by
which the structures may be described. While microheterogeneity is
merely a nuisance and might be overcome by future developments in
separation techniques, the structural dynamics are, as in biological membranes, of functional importance and have to be included in
a complete description. Very likely, these properties can be held
responsible for the failure of any attempts to crystallize carried
out so far.

From what has been said above, it is clear that a structural
description can provide merely a time-average picture. A substantial
part of the present knowledge in this respect has come from X-ray
small-angle scattering, and to a lesser degree from the related method of neutron small-angle scattering. In the following the pertinent results obtained in the author's laboratory on this subject

will be reviewed.
The discussion will largely omit methodological details (for which
reference is made to the original literature, as well as to recent
reviews[18,19]). In addition, the results of spin – label e.s.r.-
spectroscopy on the thermotropic transition in LDL particles will
also be included to illustrate the importance of combining different
methods in lipoprotein structure research.

HIGH DENSITY LIPOPROTEINS

Of all lipoproteins, HDL contain the largest proportion of
amphiphilic constituents (protein, phospholipids and cholesterol)
which are able to stabilize lipid/water interfaces. Consequently,
this density class represents the smallest lipoprotein particles,
ranging in molecular weight from about 2×10^5 to 5×10^5. Con-
ventionally, two major subfractions, HDL_2 and HDL_3 are isolated from
this class, but it has to be borne in mind that a considerably
larger number of subfractions with different compositions can be
resolved by suitable techniques[20]. It is not entirely clear,
whether these subfractions are indeed present in vivo or are
artifacts caused by separation.

Detailed X-ray small-angle scattering studies have been carried
out on three selected subfractions in the author's laboratory[21-23].
They are representative of the whole range of HDL particle sizes.
The bulk chemical and physical data of these subfractions are listed
in Table II.

Table II. Chemical and physical data of the HDL subfractions which
have been studied by X-ray small-angle scattering in the author's lab.

	LpA from HDL_3 (LpA$_3$)	LpA from HDL_2 (LpA$_2$)	LpC from HDL_2
Mol. Wt $(\times 10^{-5})$	2.0 ± 0.1	3.6	4.6
$\bar{V}$ at $4^\circ C$	0.859	0.903	0.905
R_f (Å)*	42	47	N.D.
Composition (% by weight):			
Protein	58.0	42.0	35.0
Phospholipid	23.1	31.9	31.7
Cholesterol	3.9	5.5	4.7
Cholesteryl Esters	13.0	18.0	21.1
Triglycerides	2.0	2.6	7.5

* Radius of gyration of the excluded volume.

The X-ray scattering curves i(h) and the corresponding electron
pair-distance distribution functions p(r) are shown in Figure 2.
The distance distribution functions p(r) were obtained from the experimental scattering curves by Fourier-transformation according to

$$p(r) = \frac{1}{2\pi^2} \cdot {}_0\!\int^\infty i(h).hr.\sin(hr)dh \qquad (1)$$

and correspond to the distribution of distances r between any two
volume elements within one particle, weighted by the product of
their respective electron densities. As the most obvious result,
the maximum particle diameter can be determined from the point
where p(r) finally approaches zero. The striking similarity of the
scattering curves as well as of the p(r) functions reveals that the
particles of the HDL class must be built according to a common
structural principle differing mainly in their sizes. The well developed secondary maxima and minima in the scattering curves and

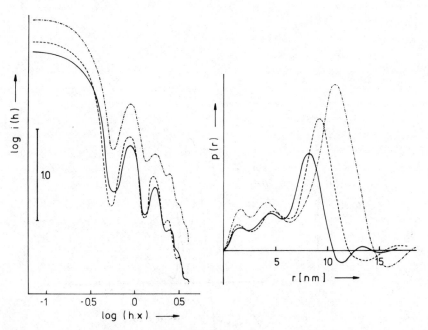

Figure 2. X-ray small-angle scattering curves i(h) and distance
distribution functions p(r) of HDL subfractions. ——— ... LpA
from HDL$_3$, - - - - ... LpA from HDL$_2$, -·-·- ... "LpC" from HDL$_2$.
The angular scale (h=(4π/λ).sinθ, where λ... wavelength and 2θ ...
scattering angle) has been multiplied by a factor x such that the
positions of the maxima coincide. The value of x is 1.0, 0.85, and
0.76 for HDL$_3$, HDL$_2$ and LpC, respectively, in good agreement with
the maximum particle diameters in the p(r) functions.

the steep decay of the p(r) function towards the maximum diameter
indicate a compact, highly isotropic and symmetrical structure.
Further support for this notion comes from the radius of gyration R_f
of the excluded volume found by contrast variation measurements in
different sucrose solutions[18]. For LpA₃ and LpA₂, the R_f values are
42 Å and 47 Å, respectively. They correspond to theoretical diameters
of spheres (as the particles with the smallest maximum diameter for a
given R_f) of 108 Å and 121 Å, respectively. These values are within
the error margin of the maximum particle diameters found experimentally (Figure 2). It is therefore justified to consider the
particles, at low resolution, as spherical. In this special case,
the particle structure is described by a radial electron density
distribution which can be determined by Fourier transformation of
the scattering amplitudes, given by the square roots of the intensities. The results of this approach are shown in Figure 3.

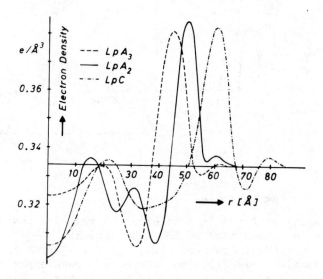

Figure 3.   Radial electron density distributions of three subfractions
of HDL.

On the basis of this evidence and considering the known molecular parameters of the individual molecular components, the following
structural model has been proposed[18] (Figure 4).

In this model, the particle core is formed by a spherical micellar arrangement of the cholesteryl esters into which the hydrocarbon chains of the phospholipids and the unesterified cholesterol are
interdigitating. The smallest core radius of about 37 Å corresponds
well to the extended length of a cholesteryl ester (e.g. oleate)
molecule, and in this case full interdigitation of the phospholipid
hydrocarbon chains is postulated. This agrees with the fact that
LpA from HDL$_3$ is the smallest particle in this density class. As the
particles increase in size, the overlap length for interdigitation
decreases, reaching the lowest value in LpC from HDL$_2$. In this latter case the core radius of about 51 Å correspondes to the full

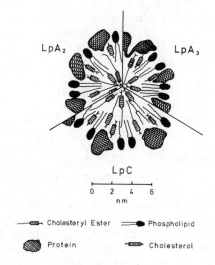

Figure 4.   Schematic cross-sectional model for the molecular
arrangement in HDL subfractions.

length of a cholesteryl ester plus the length of an all-trans
$(CH_2)_{11}$ - chain, leaving about the length of six methylene groups
for overlap. Again, this appears to be a limiting case, since the
HDL spectrum normally does not contain larger particles. According
to this scheme, the interdigitation between core and surface lipids
limits the particle sizes in the HDL class.

With regard to the surface structure, the fairly constant
thickness of 15 - 20 Å of the high - electron - density shell
corresponds well to the cylindrical diameter of an α -helix. Since
the major apolipoproteins of the HDL class contain substantial
amounts of α -helix when interacting with phospholipids[17], it is
postulated that they form a two-dimensional cage-like arrangement
at the particle surface into which the phospholipid hydrocarbon
chains are intercalated. Unfortunately, an individual assignment of
protein and polar lipid head-groups is not possible on the basis of
electron densities since they are too similar. A potential solution
to this problem could come from selectively deuterated lipoproteins
and neutron small-angle scattering as has been done with LDL (see
below). Comparison of the available surface areas with the surface
requirements of the constituents has shown that only in the case of
LpA from HDL_3 the particle surface can be fully covered. In the two
larger particles, about 30% of the surface would be uncovered. It
is suggested therefore, that in the latter cases protein intrudes
partially into the core regions to prevent unfavourable hydrophobic
contacts between water and hydrocarbon moieties. According to these
concepts, HDL particles could accommodate additional phospholipids or
cholesterol without changing their absolute amounts of core lipids
and their diameters. This has been shown to occur in vitro[24], and
it is tempting to speculate that such mechanisms are also realized
in vivo, where HDL has been shown to reduce the cholesterol content
of arterial tissue cells[25,26], which has been suggested to be of
importance in the prevention of atherosclerosis.

LOW DENSITY LIPOPROTEINS.

In its chemical composition (Figure 1) LDL differs from all other
lipoproteins by the relatively highest proportion of cholesteryl
esters and by the exclusive presence of apolipoprotein B. X-ray
small-angle scattering studies[23,27,28] have provided strong evidence
for a quasi-spherical particle symmetry, along similar lines of
argument as described above for HDL. The maximum particle diameter
was found to be between 230 and 250 Å, and the radius of gyration of
the excluded volume R_f was found to be 90 Å (by X-ray[28]) and 77 Å
(by neutrons[29]). The comparison of these values as well as the
observation of well-resolved scattering maxima and minima has
justified the approach to evaluate a radial electron-density
distribution by Fourier transformation of the amplitudes. The
results of this treatment are shown in Figure 5.

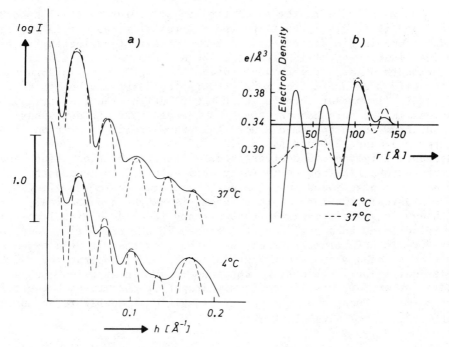

Figure 5. X-ray small-angle scattering curves and radial electron density distributions of LDL at 4 and 37° C, respectively. Broken lines in the scattering pattern represent theoretical scattering curves of ideal spheres with radial density distributions as shown on the right hand side.

The observation of a pronounced temperature dependence in the scattering patterns has provided an important clue to the structural interpretation. Deckelbaum et al.[30] were the first to point out that isolated cholesteryl esters show very similar behaviour in the temperature sensitivity of the 37 Å diffraction band suggesting that these lipids are organized in a similar way within the LDL particle. The observation of a reversible disappearance of the two inner maxima in the electron density distribution, around 30 and 65 Å, indicated that the electron-dense cholesterol ring system is centered at these radii. The temperature-sensitive core extends to radii of about 75 Å, i.e. twice the length of a cholesteryl ester molecule, and the volume of this core (1.8×10^5 Å^3) is in very good agreement with the calculated volume for the apolar lipids (cholesteryl esters plus triglycerides) of an LDL particle. This has lead to a model in which the apolar lipids form two concentrical shells in the particle core, surrounded by the polar constituents, phospholipids, cholesterol and protein (Figure 6).

The detailed arrangement of the apolar lipids is still hypothetical and it is likely that the spatial restrictions within the core induce certain differences to the known phases of cholesteryl esters in the liquid-crystalline state[31]. Here again, neutron scattering experiments on selectively deuterated LDL samples may yield further information, as shown from model studies by Burks and Engelman[32]. It is also important to note that the relative amount of triglycerides in the core has an effect on the thermotropic behaviour[33]. In LDL from normolipidemic donors the thermotropic transition is centered around 25°C and has a width of about 10°C. LDL with increased triglyceride contents show a lower transition point, and other factors like degree of fatty acid saturation[34] and particle size[35] were also found to influence the thermotropic behaviour. Any physiological relevance of this effect is as yet hypothetical, although it seems remarkable that dietary induced hypercholesteremia in swine leads to a significant increase in the transition temperature[36]. In this context it may be noteworthy that it is a misconception to assume a uniform, physiological body temperature, and that under quite normal conditions the temperature in arteries such as the radial artery or dorsalis pedis may drop to temperatures between 20 and 25°C[37]. Thus, even the normal transition temperature of LDL falls within the physiological range.

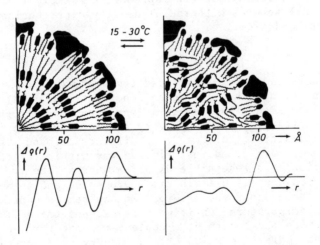

Figure 6. Top: Cross-sectional model of LDL. Below: Radial electron density distributions below and above the transition.

Regarding the surface structure of LDL, the same problems of insufficient contrast between protein and phospholipid head-groups make it difficult to obtain their respective locations from X-ray data, as indicated above for HDL. It can only be concluded that both protein and polar headgroups are located within the outer high electron density shell centered around 105 Å. Moreover, from X-ray data, no temperature changes in this region could be detected. Experiments on partially trypsin-digested LDL[38] have essentially confirmed this view.

To further elucidate the surface structure of LDL we have performed neutron small-angle scattering studies on selectively deuterated samples[29]. The basic idea of such experiments is to enhance the scattering contrast of one component over the other by replacing hydrogen by deuterium atoms within this component. Since the scattering length density for neutrons is vastly different between hydrogen and deuterium, this approach results in an essentially isomorphous contrast enhancement[39]. To this end we have replaced the endogenous LDL phospholipids by $N(CD_3)_3$-phosphatidylcholine. Direct subtraction of the scattering amplitudes of deuterated and native LDL has revealed the location of the phospholipid head groups in a spherical shell around 103 Å. By comparison to the radius of gyration of the entire outer shell (containing protein and polar head-groups) of 110 Å, a radius of gyration of 111 Å is calculated for the protein alone. This indicates that the protein lies, on average, about 8 Å above the level of the phospholipid head-groups.

These parameters can be used in further model considerations concerning the arrangment at the particle surface. The radius of 103 Å for the polar head-goups is in excellent agreement with the view that the polar lipids form a monolayer surrounding a core of 75 Å radius. Assuming average surface areas of 60 $Å^2$ and 38 $Å^2$ for phospholipids and cholesterol, respectively, a total area of 6.9 x 10^4 $Å^2$ is calculated for these constituents which is almost exactly that of the surface of a 75 Å radius sphere. Thus complete coverage is provided at this level. On the other hand, at the radius of the head-groups, the theoretical surface area of 1.33 x 10^5 $Å^2$ is only covered to about 50%. If the rest is to be covered completely by the available protein (volume estimated to 6.3 x 10^5 $Å^3$), the average thickness of the protein layer would only be about 10 Å, which appears unrealistic. In any case, however, the protein has to be spread rather thinly over the surface to prevent extensive hydrophobic contacts. This is difficult to be reconciled with the suggestion made on the basis of freeze-etching electron microscopy, that the protein of LDL is organized in a small number (possibly four) globular particles at the surface[40].

Although by the above techniques no temperature-dependent changes in the surface structure of LDL could be detected, it remained uncertain whether such changes exist but fall below the low-resolution level. To answer this question we have undertaken an e.s.r. study using spin labels inserted into the surface monolayer[41]. Two alternative types of spin labels have been used: doxylstearic acids carrying the oxazolidine–N–oxyl rings in different positions along the chain, and a maleimide analog spin label covalently attached to the protein moiety. The results are shown in Figures 7 and 8.

The results can be summarized as follows:

(1) The stearic acid spin labels show reversible changes in their mobilities between 25 and 30°C, corresponding to the temperature of the core transition. This effect is more pronounced for 12- and 16-doxylstearic acid than for the most strongly immobilized 5-doxyl analog. The polarity, reflected by the average hyperfine coupling

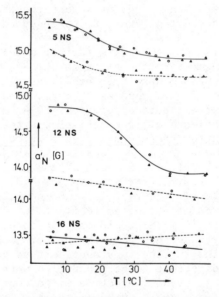

Figure 7. Temperature dependence of the average hyperfine coupling constant of stearic acid spin labels in LDL (full lines) and in liposomes of egg yolk phosphatidylcholine and cholesterol (molar ratio 2:1; broken lines). 5 NS...5-doxyl, 12 NS...12-doxyl, 16 NS...16 doxylstearic acid.

constant, changes concomitantly, with the exception of the region
probed by the 16-doxal analog, residing most closely to the apolar
core. This latter finding agrees with the notion that at the interface between core and surface monolayer (near radii of 75 Å), an
all-hydrocarbon environment prevails, whereas towards the surface
the influence of protein-lipid interaction becomes effective.

(2) Below the transition the spectral characteristics are dominated
by lipid-protein interaction, while above 30° C the situation
approaches that of the protein-free model system of lecithin/chol-
esterol (2:1).

(3) The spectral parameters from the protein-bound spin label indicate reversible changes in the temperature range where the core lipid transition is observed by XSAS.

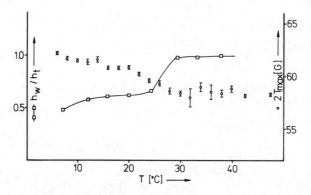

Figure 8. Temperature dependence of the spectral parameters relating to the mobility of a maleimide-analog spin label bound to
the protein moiety of LDL. $2T_{max}$ ... maximum hyperfine splitting;
h_w/h_t ...line height ratio of "weakly" to "tightly immobilized"
e.s.r. signals.

Although the molecular details of the surface structure are
still largely obscure, it has been demonstrated, nevertheless, that
the surface structure is sensitive and reflects the cholesteryl
ester transition within the particle core. In view of the variations
in transition-temperature occuring, and the wide range of temperatures to which LDL is exposed during circulation in blood, it
appears likely that the transition has physiological implications
which are so far unrevealed.

ACKNOWLEDGEMENTS

The original work cited in this article has been supported by
the Oesterreichischer Fonds zur Foerderung der wissenschaftlichen
Forschung under grants no. 2737, 3524 and 4102.

REFERENCES

1. O. DeLalla and J.W. Gofman, Methods Biochem. Anal., 1, 459 (1954)
2. V.P. Skipsky, in "Blood Lipids and Lipoproteins. Quantitation,
    Composition and Metabolism", G.J. Nelson, Editor, p. 474, Wiley
    Interscience, New York, 1972.
3. R. J. Havel, Ann. N. Y. Acad. Sci. 348, 16 (1980).
4. H.N. Baker, A.M.Gotto, Jr. and R.L. Jackson, J. Biol. Chem.,
    250, 2725 (1975).
5. J.C.Osborne, Jr. and H.B. Brewer, Jr., Advan. Protein Chem.,
    31, 253 (1977).
6. H.B. Brewer, Jr., S.E. Lux, R. Ronan and K.M. John, Proc. Natl.
    Acad. Sci., USA, 69, 1304 (1972).
7. J.A. Reynolds, Ann. N. Y. Acad. Sci., 348, 174 (1980).
8. R.S. Shulman, P.N. Herbert, K. Wehrli and D.S. Fredrickson,
    J. Biol. Chem. 250, 182 (1975).
9. R.L. Jackson, J.T. Sparrow, H.N. Baker, J.D. Morrisett,
    O. Daunton and A.M. Gotto, Jr., J. Biol. Chem.., 249, 5308 (1974).
10. R.L. Jackson, H.N. Baker, E.B. Gilliam and A.M. Gotto, Jr.,
    Proc. Natl. Acad. Sci., USA, 74, 1942 (1977).
11. H.B. Brewer, R. Shulman, P.N. Herbert, R. Ronan and K. Wehrli,
    J. Biol. Chem. 249, 4975 (1974).
12. S.C. Rall, H.W. Weisgraber and R.W. Mahley, J. Biol. Chem. 257,
    4171 (1982).
13. P. Alaupovic, in "Protides of the Biological Fluids", H. Peeters,
    Editor, p. 9, Pergamon Press, Oxford, 1972.
14. M.S. Brown, P.T. Kovanen and J.L. Goldstein, Ann. N. Y. Acad.
    Sci., 348, 48 (1980).
15. P. Laggner, in "Low Density Lipoproteins", C. Day and R.S. Levy,
    Editors, p. 49, Plenum Press, New York, 1976.
16. P. Laggner, in "High Density Lipoproteins", C. Day, Editor,
    p. 43, Marcel Dekker, New York, 1980.
17. J.D. Morrisett, R.L. Jackson and A.M. Gotto, Jr., Biochem.
    Biophys. Acta, 472, 93 (1977).
18. P. Laggner and K.W. Mueller, Quart. Rev. Biophys. 11, 3 (1978).

19. P. Laggner, in "Small Angle X-ray Scattering", O. Glatter and
    O. Kratky, Editors, p. 329, Academic Press, London, 1982.
20. G. Kostner, in "High Density Lipoproteins", C. Day, Editors,
    p. 1, Marcel Dekker, New York, 1981.
21. P. Laggner, K. Mueller, O. Kratky, G. Kostner and A. Holasek,
    FEBS Letters, 33, 77 (1973).
22. K. Mueller, P. Laggner, O. Kratky, G. Kostner, A. Holasek and
    O. Glatter, FEBS Letters, 40, 213 (1974).
23. P. Laggner, K. Mueller and O. Kratky, J. Colloid Interface Sci.,
    55, 102 (1976).
24. A.R. Tall and P.H.R. Green, J. Biol. Chem., 256, 2053 (1981).
25. G.J. Miller and E. Miller, in "High Density Lipoproteins",
    C. Day, Editor, p. 435, Marcel Dekker, New York, 1981.
26. G. Bjonders, O. Wiklund, S.-O. Olofsson, A. Gustafson and
    S.U. Bjorkerud, in "High Density Lipoproteins", C. Day, Editor,
    p. 436, Marcel Dekker, New York, 1981.
27. P. Laggner, G. Degovics, K.W. Mueller, O. Glatter, O. Kratky
    and A. Holasek, Hoppe-Seyler's Z. Physiol. Chem. 385, 771 (1977).
28. K. Mueller, P. Laggner, O. Glatter and G. Kostner, Eur. J.
    Biochem., 82, 73 (1978).
29. P. Laggner, G. Kostner, U. Rakusch and D. Worcester, J. Biol.
    Chem., 256, 11832 (1981).
30. R.J. Deckelbaum, G.G. Shipley, D.M. Small, R.S. Lees and P.K.
    George, Science, 190, 392 (1975).
31. B.M. Craven and G.D. DeTitta, J. Chem. Soc. Perkin Trans., 2,
    814 (1976).
32. C. Burks and D.M. Engelman, Proc. Natl. Acad. Sci. USA, 78,
    6863 (1981).
33. R.J. Deckelbaum, G.G. Shipley and D.M. Small, J. Biol. Chem.,
    252, 744 (1977).
34. T. Kirchhausen, S.M. Untracht, G.M. Fless and A.M. Scanu,
    Atherosclerosis, 33, 59 (1979).
35. G. Juergens, G.M.J. Knipping, P. Zipper, R. Kayushina,
    G. Degovics and P. Laggner, Biochemistry 20, 3231 (1981).
36. H.J. Pownall, R.L. Jackson, R.I. Roth, A.M. Gotto, J. Patsch
    and F.A. Kumerow, J. Lipid Res., 21, 1108 (1980).
37. H.C. Bazett, L. Love, M. Newton, L. Eisenberg, R. Day and
    R. Forster II, J. Appl. Physiol. 1, 3 (1948).
38. P. Laggner, S. Goldstein and M.J. Chapman, Biochem. Biophys.
    Res. Commun., 82, 1332 (1978).
39. B. Jacrot, Rep. Progr. Phys. 39, 911 (1976).
40. T. Gulik-Krzywicki, M. Yates and L. Aggerbeck, J. Mol. Biol.
    131, 457 (1979).
41. P. Laggner and G. Kostner, Eur. J. Biochem. 84, 227 (1978).

Part II
Structure, Dynamics and Characterization
of Micelles

THE PACKING OF AMPHIPHILE CHAINS IN MICELLES AND BILAYERS

David W. R. Gruen and Emma H. B. de Lacey

Department of Applied Mathematics
Research School of Physical Sciences
A.N.U., P.O. Box 4, Canberra, A.C.T., Australia

A simple model for the packing of amphiphile chains in spherical and rod-shaped micelles and in bilayers is presented. The model assumes that each aggregate has a hydrophobic core in which neither headgroups nor water can exist.

All possible conformations — *trans*, *gauche*$^+$ and *gauche*$^-$ at each bond — of a single fully saturated C_{12} chain attached to a headgroup are generated and assigned Boltzmann statistical weights on the basis of their energy. This energy has three contributions. The first is the internal energy of the chain, determined by the number of *gauche* bonds in the chain. The second is a hydrophobic free energy; this term is zero unless the chain conformation has segments sitting outside the hydrophobic core of the aggregate. The third contribution to the energy arises from the action of a lateral surface pressure on segmental areas of the chain. This term models the constraints imposed by the surrounding amphiphiles and the geometry of the aggregate. The lateral pressure is allowed to vary as a function of distance from the hydrophobic core center; its correct profile is determined, after several iterations, as that profile which leads to the hydrophobic core of the aggregate being packed at liquid hydrocarbon density throughout.

As an illustration of the information that may be gleaned from the model, consider a spherical micelle of aggregation number 53 at 25°C. This micelle has a

hydrophobic core radius slightly less than a fully
extended C_{12} chain. The free energy cost of forcing
the chains to pack into this micelle (from a reservoir
where they have the freedom enjoyed by bulk alkane
chains) is only 0.8 kT per chain. On average, there
are 3.4 *gauche* bonds per chain; in liquid alkane the
same chains average 3.6 *gauche* bonds (a theoretical
estimate based on Flory's random-coil model). From
the headgroup to the terminal methyl group, the chain
segments are, on average, progressively closer to the
micelle center. Even so, the CH_3 group sits on average
10.0 Å from the micelle center (and hence only 6.1 Å
from the hydrophobic core surface). All chain segments
spend an appreciable time in contact with the micelle
surface — the CH_3 group is partly outside the hydrophobic core 20% of the time.

Similar information is available for all the
aggregates studied.

1.   INTRODUCTION

This paper considers the behaviour of the amphiphile chains in
micellar aggregates and in hexagonal and lamellar phases. A
theoretical model is presented which attempts to give a detailed
and reasonably accurate view of the conformational state of these
chains with the aim of elucidating the role of aggregate geometry.

From a theoretical viewpoint, if either temperature or
aggregate geometry preclude an array of parallel frozen chains, the
small difference in energy between *trans* and *gauche* conformers
(approximately 0.8 kT at room temperature) suggests, a priori, that
there will be considerable configurational disorder in the chains.
This view is supported by the ample experimental evidence suggesting
a liquid-like nature of the hydrophobic interior of micelles[1] and
of hexagonal[2] and lamellar[2,3] phases. (In the case of lamellar
phases, we are concerned only with the high temperature, liquid
crystalline phase.)

From the experience of liquid-state physics we expect that the
combined action of short-range intermolecular repulsive forces and
longer range van der Waals attractions will determine the density
of chain packing in the liquid hydrophobic core within narrow
limits of the density of bulk n-alkane phases. This expectation is
supported by measurements of the partial molar volumes and
compressibilities of n-alkanes dissolved in micelles[4] and of the
partial molar volumes of the lipid chains in bilayers.[5] In both
studies the measured values are very close to the values for the
same quantities in bulk liquid n-alkanes.

Although still a controversial subject, there is also strong experimental evidence suggesting minimal water penetration into the hydrophobic core of micelles.[1] In the case of lipid bilayers, very strong evidence for a negligible amount of water in the hydrophobic region comes from neutron diffraction experiments[6] and from capacitance measurements on "black" lipid films.[7,8] (Assuming the dielectric constant of the hydrophobic core has the value $\varepsilon \simeq 2.1$, characteristic of bulk liquid n-alkane, thicknesses of the core deduced from capacitance measurements are in excellent agreement with independent measurements which do not rely on an estimate of this dielectric constant. Even small amounts of water in the hydrophobic core would increase its dielectric constant markedly and ruin the agreement.) If measurements of the volume fraction of water in bulk n-alkane phases[9] are used as a guide, one should expect approximately one water molecule for each 2×10^4 CH_2 groups in the interior of amphiphilic aggregates.

In the light of the above discussion, our model has the following features. Each aggregate has a hydrophobic core in which neither headgroups nor water can exist. The amphiphilic chains are constrained to fill the core in such a way that, when an ensemble average is taken over all chain conformations, the density of chain packing throughout the core is the same as liquid n-alkane. Thus, we assume that the volume of a CH_2 group is 27 $\overset{\circ}{A}{}^3$ and of a CH_3 group 54 $\overset{\circ}{A}{}^3$ (reference 10). The chains may also, to a limited extent, exist outside the core but they are then subject to a 'hydrophobic free energy' cost. In other words, we allow a region of the aggregate where the chains mix with headgroups and water: between the hydrophobic core and the 'pure' aqueous phase.

The paper is arranged as follows. In section 2, the details of the model are presented. On a first reading, or whenever the reader finds himself lost in these details, he should consult section 3, PROTOCOL FOR GENERATING THE RESULTS, where an overview of the whole calculation is presented. The fourth section explains the presence of all the terms in the exponent of the Boltzmann statistical weights and derives free energy expressions for the chain. In section 5, results are presented and explained. This is followed by a DISCUSSION section and the paper concludes with a SUMMARY.

The following symbols will be used throughout. The superscript i refers to the ith chain conformation. The subscript j refers to the segment number in the chain (j = 1-11 for CH_2 groups and j = 12 for the CH_3 group). The term 'radius' will be used for all three aggregate shapes to mean the distance from the aggregate's center to any point along a line which runs normal to the hydrophobic core surface. r_j is the radius of segment j and R denotes the hydrophobic core radius (which, in the case of a bilayer, is the hydrophobic core half-thickness).

2. THE MODEL

The model involves explicit generation of all conformations of a single amphiphilic molecule. The amphiphile has the chemical formula $G-(CH_2)_{11}CH_3$ where G is an unspecified headgroup. All the C-C bonds and the G-C bond have length 1.53 Å. The angle between successive C-C bond vectors is assumed to be $\theta = 112°$. Each H_2C-CH_2 bond can exist in three conformations — one trans (t) and two gauche (g^+, g^-) states. In the all-trans state, the carbon atoms lie in a planar zig-zag. At each bond, the gauche states are formed from the trans state by rotation of the C-C bond by $\phi = \pm120°$. Sequences g^+g^- and g^-g^+ on successive bonds are excluded. Flory[11] discusses these assumptions in detail.

Three orientations of the G-C bond were allowed. They are the same as those used previously by one of us[12,13] and can most easily be described by imagining that the headgroup consists of two CH_2 groups. Call the outer imaginary C-C bond the first imaginary bond and inner one (which was the G-C bond) the second imaginary bond. The outer CH_2 group is fixed with the plane spanned by its two C-H vectors parallel to the surface of the hydrophobic core of the aggregate. The three orientations of the G-C bond are the orientations of the second imaginary bond which result from the bond sequences (i) tt, (ii) tg^+, (iii) g^-t in the first and second imaginary H_2C-CH_2 bonds. Having served their purpose, the imaginary groups will be considered no further. The statistical weights for the three orientations of the G-C bond were chosen 1/2, 1 and 1 respectively. Without other contributions to the chain energy, this leads to a small order for the first CH_2 group in the chain.

Three consecutive segments of length 1.27 Å are chosen on the z-axis (see Figure 1) from the hydrophobic core surface to a distance of 3×1.27 Å = 3.81 Å into the aqueous phase. (The number 1.27 Å was chosen simply because, for an all-trans chain with $\theta = 112°$, this is the distance between successive CH_2 groups resolved along the axis of the chain.) Three random points are chosen (one from each segment on the z-axis) as positions of the headgroup and a single chain conformation is generated (the same conformation for each random headgroup position). Three new random headgroup positions, one in each segment, are chosen, and another chain conformation is generated. This process is repeated until all possible chain conformations have been generated. Therefore, for each aggregate studied, each iteration of the problem involves the generation of all possible chain conformations three times. As will be clear from the results, inclusion in the model of amphiphiles whose headgroups sit more than 3.81 Å from the hydrophobic core would make negligible difference (see Figure 10).

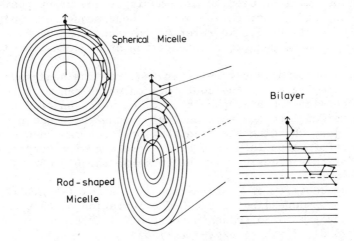

Figure 1.   Schematic drawings of a spherical micelle, an infinite
rod-shaped or cylindrical micelle (equivalent, for our purpose,
to a single cylinder in a hexagonal phase) and a bilayer
(isolated or part of a lamellar phase).   In each case a possible
chain conformation is shown.   The arrow defines the z-axis of
the aggregate.   In the case of the rod-shaped micelle, the x-axis
is the axis of the rod and the y-axis completes a set of
cartesian axes.   The seven symmetrical regions within the hydrophobic core of the aggregate are shown.   The relative volumes of
the regions, from the center to the surface, are chosen as
1/10:2/5:1/2:1:1:1:1 in the case of all spheres;
1/3:1/3:2/3:2/3:1:1:1 in the case of all rods;   and all equal in
the case of bilayers.   The smaller volume regions near the center
of the curved aggregates are introduced so they don't have too
large radii which would defeat the aim of packing the aggregates
at constant density throughout.

     Both spherical micelles and bilayers have cylindrical
symmetry around the z-axis of Figure 1.   However, for rod-shaped
micelles, a chain conformation which extends along the x-axis is
different from one with the same internal bond sequence which
extends along the y-axis.   This situation is dealt with as follows.
Define α as the angle between the x-axis and the projection of the
G-C vector on the x-y plane.   For rod-shaped micelles, a random value
of α between 0 and 2π is chosen for each internal bond sequence.   Before Boltzmann weightings are applied, the chain conformations are
randomly oriented in the x-y plane.

For the rest of this section and the next, the term "chain conformation" will be used to mean a single headgroup position and orientation, and a particular internal bond sequence. Only in the FREE ENERGY section will we refer to the fact that we have three copies of the same headgroup orientation and internal bond sequence.

Chain conformations were generated from the headgroup to the terminal methyl group. Any conformation which had more than five segments outside the hydrophobic core, or which entered the core and then re-emerged into the surrounding aqueous phase was excluded. (Clearly, just changing the headgroup position can change an allowed conformation into a forbidden one and vice versa.) The reasons for this decision will be found in the DISCUSSION section.

The probability of conformation i is $a^i \exp(-E^i/kT)/Z$ where a^i is 1/2 or 1 depending on the G-C orientation of this conformation (see earlier), k is Boltzmann's constant, T is temperature (set at 298 K or 25°C), $Z = \sum_i a^i \exp(-E^i/kT)$ and

$$E^i = E^i_{int} + \sum_j \omega^i_j + \sum_j \pi(r_j) A^i_j \ . \tag{1}$$

E^i contains more than the energy of chain conformation i as will be clear from the description of the terms, which follows. The reason for the latter two terms, and their form, is explained in the FREE ENERGY section. As previously discussed[13], Equation (1) ignores any orientation dependence of the van der Waals attractions between the chains. E^i_{int} is the internal energy of the chain, assumed equal to $500.n_g$ cal/mol where n_g is the number of *gauche* bonds in the chain.[11] The variable j in the second and third terms of Equation (1) refers to the segment number in the chain (j = 1-11 for the CH_2 groups, j = 12 for the CH_3 group). The second term represents the hydrophobic free energy cost of exposing hydrocarbon to water.

$$\omega^i_j = (1 - \phi_{HC}(r_j)) \ f^i_j \ F_{HC-W} \tag{2}$$

$\phi_{HC}(r_j)$ is the volume fraction of hydrocarbon at r_j. For r < R, $\phi_{HC}(r) = 1$. For r > R, the form of $\phi_{HC}(r)$ from the previous iteration was used [$\phi_{HC}(r)$ falls off approximately exponentially for increasing r — see Figure 10]. f^i_j is the fractional increase in chain-water contact as a result of this chain segment. f^i_j is defined simply as

$$f^i_j = (r_j - R)/_{1.86 \text{ Å}} \qquad R < r_j < R + 1.86 \text{ Å}$$

$$= 1 \qquad\qquad r_j > R + 1.86 \text{ Å} \tag{3}$$

1.86 Å is the radius of a sphere of volume 27 Å^3. (Of course, CH_2 groups are not particularly spherical. To take the shape of the group into account would complicate Equation (3) considerably without making much quantitative difference.) When $r_j = R$, the segment is partly in contact with water but it is assumed not to *increase* chain-water contact. Given that the hydrophobic core has radius R, if this segment was not there another would be. $F_{HC-W} = 884$ cal/mol is the measured free energy cost of transferring CH_2 groups from bulk n-alkane to water.[14] As defined, it is the value for pure water; the presence of salt changes this free energy only slightly.[14] The definition for allowed conformations (see earlier) implies that the CH_3 group is always 'inside' the hydrophobic core (i.e. $r_{12} \leq R$).

The third term in Equation (1) models the constraints which surrounding amphiphiles in, and the geometry of, the aggregate impose on the chain under consideration. Each segmental chain area A_j^i is acted on by the lateral pressure at a distance r_j from the aggregate's center ($\pi(r_j)$). For $r > R$, $\pi(r) = 0$, while for $r < R$, the variation of $\pi(r)$ with r is discussed in the next section. A_j^i is defined as follows. Let V be the volume of the chain; $V = 27 \times 11 + 54 = 351$ Å^3. Assume for the ith chain conformation that k segments sit inside the hydrophobic core. Define $\delta_j = 1$ if $r_j < R$ and $|r_{j-1} - r_j| < 1.27/2$ Å; $\delta_j = 0$ otherwise. (r_0 is the radius of the headgroup). Let $m = \Sigma_j \delta_j$; $\ell^i = R - \min (r_1,\ldots,r_{12})$; $A_0^i = V/(1.27 \text{ Å} \times 12) = 23 \text{ Å}^2$ and $A^i = V/\ell^i$.[1] Then

$$A_j^i = \begin{cases} 0 & \text{if } r_j > R \\ CA_0^i/k & \text{if } m = 0 \text{ and } r_j \leq R \\ C(A_0^i/k + \delta_j(A^i - A_0^i)/m) & \text{if } m > 0 \text{ and } r_j \leq R \end{cases} \qquad (4)$$

The average area per chain (A_{AV}) is defined as the volume weighted mean area per chain in the hydrophobic core. For a sphere with a hydrophobic core which contains N chains and has radius R we have

$$N A_{AV} = \int_0^R 4\pi r^2 \, (4\pi r^2 dr / [4\pi R^3/3]) \qquad (5)$$

and $NV = 4\pi R^3/3$. This analysis, and similar ones for the cylinder and bilayer lead to

$$A_{AV} = 9V/5R \qquad \text{spheres}$$

$$= 4V/3R \qquad \text{cylinders} \qquad\qquad (6)$$

$$= V/R \qquad\quad \text{bilayers} \ .$$

The constant C in Equation (4) is determined uniquely for each aggregate by applying Equation (6) and requiring

$$A_{AV} = < \sum_j A^i_j > \ . \qquad\qquad (7)$$

By design, $C \approx 1$. The definition of segmental chain areas has the following desirable properties

(i)   The total area of conformation i is $\sum_j A^i_j = CV/\ell^i$ which is a simple and physically appealing result. Thus, the all-*trans* conformer lying perpendicular to the hydrophobic core surface has the smallest area; a conformer which stays near the hydrophobic core surface has the largest area.

(ii)   Within any given chain conformation, those segments which lie one under another have a small area while those which "stick out" from the rest of the chain have a large segmental area.

(iii) The ensemble average area per chain is fixed at its correct value, as determined by the geometry of the aggregate.

3.   PROTOCOL FOR GENERATING THE RESULTS

Before any calculations are begun, the size and shape of the hydrophobic core of an aggregate are chosen. Several symmetrical regions in the hydrophobic core are then defined (see Figure 1). A process of iteration is begun. Before each iteration, the profile of $\pi(r).C$ (the lateral pressure multiplied by the constant in Equation (4)) for $r < R$ and the hydrocarbon volume fraction profile ($\phi_{HC}(r)$) for $r > R$ are set. (For the first iteration, they are guessed; for subsequent iterations, the results of previous iterations are a guide.) The lateral pressure profile, $\pi(r)$, is simply a set of seven numbers; the values of $\pi(r)$ for the seven symmetrical regions (see Figure 2). E^i for each conformation (Equation (1)) is now well-defined and all allowed conformations are generated and assigned Boltzmann statistical weights ($a^i \exp[-E^i/kT]$ where k is Boltzmann's constant and T is temperature; set at 298 K or 25°C). After the iteration, the partition function for the chain ($Z = \sum_i a^i \exp(-E^i/kT)$ is evaluated and all required ensemble averages are calculated. In particular, the ensemble average profile $\phi_{HC}(r)$ for $r > R$ and the average density of chain packing (assuming a CH_2 group has a

volume of 27 $\overset{\circ}{A}^3$ and a CH_3 group, 54 $\overset{\circ}{A}^3$) in the seven symmetric
regions, are evaluated. The profile of $\pi(r) \cdot C$ for the next
iteration can be guessed by considering the following. If the
density in a region is too high, increasing the value of $\pi(r) \cdot C$
for r values in this region will lower its average density because
conformations with few, or no, segments in this region are favoured
over conformations with many segments in the region. Thus,
successive iterations improve the evenness of chain packing. A
"final" result is obtained when (i) the packing density in all seven
regions is within 1% of the liquid hydrocarbon value and (ii) there
is close agreement between the assumed $\phi_{HC}(r)$ profile before the
iteration and the calculated ensemble average $\phi_{HC}(r)$ profile after
the iteration. After the final iteration, the value of C can be
determined by applying Equation (7); then the $\pi(r)$ profile is
known.

An important problem remains: there is not a unique $\pi(r)$
profile which ensures even packing of the seven symmetrical regions
of the aggregate. This can easily be illustrated by considering a
bilayer with a hydrophobic core thickness equal to the fully
extended chain length. This aggregate can certainly be packed at
constant density either with fluid chains which rarely extend far
beyond the mid-plane, or with rigid all-*trans* chains which extend
the full width of the bilayer and interdigitate with chains from
the opposing monolayer. The latter packing arrangement would be
approached as $\pi(r) \to \infty$ for all r < R.

Thus, a further constraint is needed to provide unique answers.
For all aggregates studied, the final result has a uniformly packed
hydrophobic core *at the smallest possible free energy cost to the
chains*. The free energy of the chains is defined by Equation (8)
in the next section. For all aggregate dimensions we went to some
trouble to satisfy this constraint. However, we have no guarantee
that in each case we found the absolute minimum of the chain free
energy. This problem will be discussed in detail in a forthcoming
paper.[15]

4. FREE ENERGY

In the equation for E^i (Equation (1)), the first term, E^i_{int},
is the internal energy, and the other two terms model external
constraints on the chain. The hydrophobic effect involves both
energetic and entropic changes to water surrounding a hydrocarbon
chain (Reference 14, Chapter 4). When we evaluate a partition
function (Z), we do not sum over the possible configurations of
water. Hence, when evaluating statistical weights, it is closer
to the truth to include not the energy, but the hydrophobic free
energy for chain conformations outside the hydrophobic core.

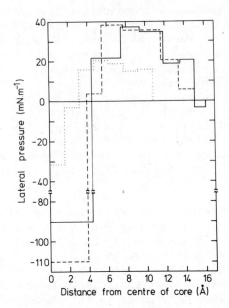

Figure 2. Lateral pressure ($\pi(r)$) profiles for a bilayer with
$R = 10.8$ Å (....); a rod-shaped micelle with $R = 15.0$ Å (----)
and a spherical micelle with $R = 16.1$ Å (———). In all aggregates,
$\pi(r) = 0$ for $r > R$. The lateral pressure profile imposes constant
density of chain packing in the hydrophobic core at the smallest
possible free energy cost to the chains.

(*This* assumes that the hydrophobic free energy is the same for
different chain conformations which, presumably, is not exact
either.) Further, during any given iteration, the profile $\pi(r)$
remains fixed while the areas A_j^i vary with conformation i. Thus,
we evaluate a "constant surface pressure" ensemble and the term
$\Sigma_j \, \pi(r_j)A_j^i$ must be included in E^i.

We now assume that an amphiphile conformation is defined by
an internal bond sequence and a G-C bond orientation and that each
amphiphile conformation is generated at most three times, corresponding to different headgroup positions. (This assumption is
necessary; otherwise the calculated entropy of the chains would
depend strongly on how many conformations with the same internal
bond sequence and head group orientation are generated.)

The ensemble average internal energy of the chains is
$U = \langle E_{int}^i \rangle$. The entropy is $S = -k \Sigma_\ell \, p_\ell \ln p_\ell$ where $p_\ell = a^i \, \Sigma_\alpha \, \exp(-E^i/kT)/Z$. a^i, as defined previously, depends on the
C-C orientation, ℓ is used as a subscript for different amphiphile
conformations and Σ_α is a sum over repeats of amphiphile conformation
ℓ. Then the free energy per chain is

$$F = U - TS = \langle E_{int} \rangle + kT \sum_\ell p_\ell \ln p_\ell \; . \tag{8}$$

We wish to compare this free energy with the free energy of the
chains in the rotational isomeric state or random-coil model (F_{RC}),
used to describe the behaviour of bulk liquid n-alkane.[11] We
choose not to include any contribution from the initial orientation
of the chain, and hence assume that the random-coil model chain
has three initial orientations with the same statistical weights as
our model aggregate chain. Then

$$F_{RC} = -kT \, [\ln(2.5 \, Z_{RC}) + 1/5 \, \ln 2] \tag{9}$$

where $Z_{RC} = \Sigma_i \exp(-E_{int}^i/kT)$. E_{int}^i is the internal energy of
conformation i and the sum is over *all* t, g^+, g^- conformers in the
chain, excluding g^+g^- and g^-g^+ sequences (i.e. in the random-coil
model there are no constraints, external to the chain, which
exclude any conformations).

5. RESULTS

The model is applied to spherical micelles with R values of
10.8 Å, 13.6 Å, 15.0 Å, 16.1 Å and 17.1 Å (the hydrophobic cores of
these micelles contain, respectively, 15, 30, 40, 50 and 60 chains);
rod-shaped micelles with R values of 6.0 Å, 9.0 Å, 10.8 Å, 13.6 Å,
15.0 Å and 16.1 Å; and bilayers with R values of 3.5 Å, 5.0 Å,
7.8 Å, 8.8 Å, 9.8 Å, 10.8 Å and 13.6 Å. The reason for the
different ranges for the three different aggregate shapes should
become clear from what follows. The fully extended length of a C_{12}
chain is approximately 16.7 Å.[14] The consequences of having a
spherical hydrophobic core with a radius somewhat larger than this
are discussed in the next section.

The graphs in Figure 3 display only one contribution to the
free energy of packing amphiphiles in aggregates. Other important
contributions come from the interactions of the headgroups and from
the chain-water interface. For all shapes, the experimentally
observed aggregates tend to have a larger value of R (and hence a
smaller area per amphiphile) than the minima of the curves in
Figure 3.

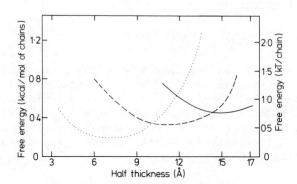

Figure 3. Free energy cost of packing chains into bilayers (....),
rod-shaped micelles (----) and spherical micelles (——) as a
function of aggregate size. Half thickness is the value of R.
The graphs display $F-F_{RC}$ (see Equations (8) and (9) in the text),
the free energy cost of transferring the chains from a (model)
bulk n-alkane environment to the aggregate. For comparison, the
free energy gained when a C_{12} chain is transferred from water to
bulk n-alkane is approximately 20 kT[14]. The number of chain
conformations which are inside the hydrophobic core decreases from
bilayers to cylinders to spheres; this explains the different
depths of the minima in the curves.

     The most striking results to emerge from Figures 3 and 4 are
that, on being transferred from a (model) bulk n-alkane environment
to a spherical micelle with a radius equal to a fully extended
chain, there is a free energy cost of only ~0.8 kT per chain and a
loss of only ~0.2 *gauche* bonds per chain. In this micelle, one or
two chains (of the fifty to sixty in the aggregate) must be fully
straightened in order to fill the volume at the center of the
sphere; but no more! *On average*, the chains are only slightly
straighter than in their bulk state.

     The sizes of the aggregates studied in Figures 5-10 are chosen
as possible equilibrium sizes. Figures 5-7 show three effects of
changing the aggregate geometry. From Figure 5 it is clear that,

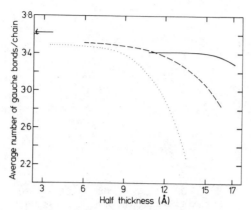

Figure 4. The ensemble average number of *gauche* bonds per chain
for bilayers (....), rod-shaped micelles (----) and spherical
micelles (———) as a function of aggregate size. The ensemble
average number of *gauche* bonds in the random-coil state
characteristic of bulk n-alkane is shown by the arrow.

for all aggregates, spatial delocalization is smallest near the
headgroup and increases all along the chain. There is a large
free energy cost associated with moving the headgroup either into
or out from the hydrophobic core surface (the model, of course,
prohibits the former movement). The headgroup, therefore, has a
narrow spatial distribution, while, as the free end of the chain
is approached, the chain segments become progressively less well
localised. For the spherical micelle, the volume available to the
chains increases sharply with distance from the center. On taking
an ensemble average, although the CH_3 group is closer to the center
than any other group, it sits a full 10.0 Å from the center and
only 6.1 Å from the hydrophobic core surface (Figure 5). Consistent
with this is the proportion of time (or, more strictly, the
probability that) each segment is at least partly outside the
hydrophobic core (Figure 6). For the spherical micelle, it is
clear that without any water penetration into the hydrophobic core,
all segments of the chain spend a reasonable amount of time in a
partly hydrophilic environment. This statement applies, to a

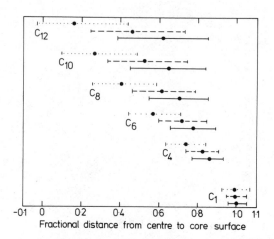

Figure 5. Distribution of chain segments in three aggregates: a bilayer with R = 10.8 Å (....); a rod-shaped micelle with R = 15.0 Å (----); and a spherical micelle with R = 16.1 Å (———). The jth dot shows the ensemble average value of r_j, i.e. $\langle r_j \rangle$. The length of the error bars on the left and right of the jth dot are, respectively, $\langle r_j^2 - \langle r_j^2 \rangle \rangle^{1/2}$ for all r_j less than $\langle r_j \rangle$, and $\langle r_j^2 - \langle r_j^2 \rangle \rangle^{1/2}$ for all r_j greater than $\langle r_j \rangle$. Thus, the difference in length of the two error bars is a measure of the skewness of the distribution of r_j. Just for this figure, r_j values for the bilayer are negative when the jth segment is sitting on the opposite side of the midplane from its headgroup.

lesser extent, for the rod-shaped micelle, while for the bilayer there is almost no contact with the surface beyond C_5. Figure (7) displays the distribution of terminal methyl groups in the same aggregates.

    Figure 8 shows the very different behaviour of the order parameter as a function of distance and as a function of carbon atom number in a spherical micelle. The general features of the profile as a function of distance may be understood as follows. Chain segments outside the hydrophobic core are subject to a 'hydrophobic free energy cost' and hence there is an incentive for the chain to straighten to allow more segments into the core. Hence the small positive order parameter outside the core. Inside the core, the constant density constraint is an important factor in determining the chain order. A third of the core volume is taken up by an annulus 2 Å thick at the surface. In this annulus,

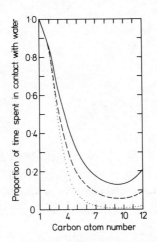

Figure 6. Probability of each segment in the chain being at least
partly in contact with water for the aggregates studied in Figure 5.
To be deemed at least partly in contact with water, $r_j > R - 1.86$ Å
for CH_2 groups and $r_j > R - 2.345$ Å for the CH_3 group. (1.86 Å
and 2.345 Å are the radii of the groups if they are assumed
spherical and have volumes 27 Å^3 and 54 Å^3.)

     It should be noted that numbers taken from this Figure cannot
be used as a direct measure of the polarity of the environment of
a chain segment. At least partly in contact with water, as defined
above, usually means mostly surrounded by other chains.

to fill space, the chains have a *slight* tendency to lie parallel
to the surface. Nearer the center of the micelle, one is sampling
only those chains which have reached the inner parts of the micelle
and which therefore have higher than average order. It should be
noted, however, that the order parameters are never very large.

     While it is true that only fully extended chains sample the
very center of the micelle, it is also true that only one or two
chains can do so at any given time. These 'center-sampling' chains

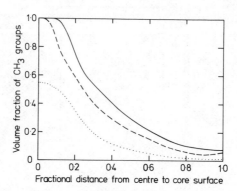

Figure 7. Distribution of terminal methyl groups in the three
aggregates studied in Figure 5.

can diffuse rapidly around the micelle surface and can also exchange
rapidly with other chains. The average order parameter for segments
in the innermost 2% of the core volume — a sphere with radius 4.4 Å —
is 0.13 (see Figure 8). While this is partly due to our definition
of the CH_3 order parameter, it is mainly due to the fact that most
of the chain segments in this innermost sphere are disordered. A
"degree of order approaching that in a crystal"[16] is not indicated.

The order parameters along the chain, for segments near the
headgroup, are determined largely by the probability distribution
of the G–C orientation. We have assumed a distribution which leads
to a small positive order parameter for C_1. Further along the
chain, the order parameter profile depends on the aggregate
geometry. For the bilayer, there is a well–known order parameter
'plateau'[2,3] (reproduced by our model[15]) but this is completely
absent for the spherical micelle (Figure 8). The cylindrical
micelle[15] (or hexagonal phase[2,17]) shows intermediate behaviour.
All the experiments, and all our model calculations show a
substantial reduction in order as the free end of the chain is

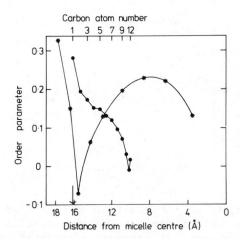

Figure 8. Order parameters as a function of distance from the
micelle center (*) and as a function of carbon atom number in the
chain (●) for a spherical micelle with R = 16.1 Å (marked by an
arrow). Define the jth chain segment vector, for CH_2 groups, as
the vector perpendicular to the plane spanned by the two C-H vectors,
and for the CH_3 group, as the vector perpendicular to the plane
spanned by the three H groups, i.e. the terminal C-C bond vector.
The jth segmental order parameter, S_j, is defined as $S_j =
\frac{3}{2} <\cos^2\theta_j>-1/2$ where θ_j is the angle between the radial vector $(\vec{r}_j)$
and the jth chain segment vector. The segmental order parameters
are plotted at their mean positions in the micelle. The order
parameter as a function of distance into the micelle is evaluated
for the seven spherical annuli of Figure 1 as well as for two
spherical annuli outside the hydrophobic core. For a given
spherical annulus, the displayed order parameter (*) is the volume
weighted ensemble average order parameter for all segments in the
annulus. This order parameter is plotted at the volume weighted
mean position in the annulus (e.g. for the innermost annulus,
with radius 4.4 Å, the star is plotted at $2^{-1/3} \times 4.4$ Å = 3.5 Å).
The lines are drawn as an aid to the eye.

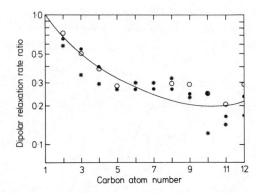

Figure 9.   Comparison of the predictions of the model with experiments which measure the ratios of the contributions of paramagnetic
ions to the nuclear magnetic resonance relaxation rates of ^{13}C nuclei
in sodium dodecylsulfate micelles (SDS)[18].   The experimental data
is the ratio: [the contribution of the paramagnetic ions to the
relaxation rate of the jth ^{13}C nucleus/the same quantity for the 1st
^{13}C nucleus in the chain].   The experimental data is taken from
Figure 11 of reference 18.   ($\bullet$), 0.068 molar SDS with Mn^{++}, ratios
of $1/T_2$ values.   (0), 0.165 molar SDS with Mn^{++}, ratios of $1/T_2$
values.   (*), 0.165 molar SDS with Co^{++}, ratios of $1/T_1$ values.
Under all these conditions the micelles are approximately spherical
and the experiment measures the quantity $<1/d_j^6>/<1/d_1^6>$ where d_j is
the distance from the jth ^{13}C nucleus to the paramagnetic ion.   The
paramagnetic ions are assumed to be uniformly distributed with their
centers on a sphere of radius D from the micelle center, in which
case $<1/d_j^6> = <(D^2+r_j^2)/(D+r_j)^4(D-r_j)^4>$.   (Our notation;   analysis
from reference 18.)   The values derived from the model (———) are
plotted for the spherical micelle with R = 16.1 Å and D – R = 7 Å.
Had data been plotted with D – R = 5 Å (which is probably too
small[18]), the curve would be a similar shape, but lowered by about
0.05 for segments 3-12.   Alternatively, had the data been plotted
for the R = 17.1 Å spherical micelle (with D – R = 7 Å), the curve
would look very similar to the one shown;   the main change would be
to the ratios for the carbon atoms 8-12 which would all be about
0.017 smaller.

approached[*]. This can be understood as follows. There is always
an entropic advantage in disordering. The energy cost of doing so
drops as the free end of the chain is approached, because
introduction of disorder near the middle of the chain perturbs a
larger part of the chain than introduction of disorder near the
free end. By contrast, Dill and Flory[16] predict a slowly increasing
order parameter all along the chain for both spheres and cylinders.

Figure (9) demonstrates that predictions from the model are in
good agreement with experiments which measure the distance of each
chain segment from the aqueous phase. From Figure 10, it is clear
that the hydrophobic effect is sufficiently strong that, had allowance been made for headgroups to sit more than 3.81 Å from the hydrophobic core surface, they would have contributed only negligibly to
the properties of the micelle. Exactly fifty chains fit *inside* the
hydrophobic core of the R = 16.1 Å spherical micelle. Its aggregation
number is 53 which demonstrates, as does Figure 10, that very little
hydrocarbon sits outside the core.

6. DISCUSSION

None of the parameters in the model are determined using
experiments on amphiphilic aggregates. Experimental results which
are explicitly used are: the dimensions and angles of the C-C
bond[11]; the energy difference between *trans* and *gauche* isomers[11];
and the free energy cost of transferring a hydrocarbon chain from
n-alkane to water[14]. Once the size and shape of a hydrophobic core
are chosen, all other parameters in the model (the surface pressures
in the seven symmetrical regions and the constant in Equation (4))
are determined.

The most important short-range effect acting on the chains in
an amphiphilic aggregate is the excluded-volume effect whereby a
chain cannot pass through itself or its neighbours. At first
thought, it seems that a single chain calculation would be
particularly inappropriate for studying systems where this is an
important effect. However, this is not so. The (single chain)
rotational isomeric state or random-coil model[11] has been very
successful in predicting properties of bulk liquid n-alkane even
down to length scales of less than 10 Å[11,19-21]. This augurs well
for the use of a similar model to describe the hydrophobic core of
amphiphilic aggregates where the short-range intermolecular forces
between chains are the same as those in bulk liquid n-alkane. Our
model can be roughly described in the following terms. One end of
an (admittedly very short) 'random coil chain' is anchored near
the hydrophobic core surface and the relative probabilities of the

---

[*] It should be noted, however, that for the hexagonal phase, many
   of the resonance lines are assigned *assuming* this behaviour[2,17].

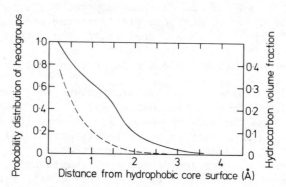

Figure 10. Distribution of headgroups (———, left hand scale),
and fraction of available volume filled by the hydrocarbon chains
(————, right hand scale) as a function of distance from the
hydrophobic core surface of a spherical micelle with R = 16.1 Å.
Both quantities were evaluated in nine segments of width 1.27/3 Å
from R to R + 3.81 Å and curves were drawn through the resulting
points. The headgroup distribution was set to 1.0 in the segment
nearest the hydrophobic core. For all aggregates studied, both
the headgroup distribution and hydrocarbon volume fraction outside
the core behaved similarly to the curves shown here.

chain conformations are changed in such a way as to pack the
aggregate with constant liquid alkane density throughout at the
smallest possible cost in free energy. The random-coil chains are
perturbed as little as is possible consistent with setting up a
state of constant density in the aggregate interior.

    Thus far, there has been little comment on the interaction of
the headgroups and the role of the 'hydrophobic effect'. Without
the 'hydrophobic effect', the aggregates being discussed would not
exist at all. In our model, we *assume* the existence of an aggregate
of a particular size and then evaluate the consequences for the
chains. We cannot *predict* equilibrium sizes solely from the free
energies displayed in Figure 3.

Once we have assumed a hydrophobic core of a particular size
and shape, we need not consider energy terms which are the same for
all chain conformations. Chain-water contact at the hydrophobic
core surface leads to such a term. (Clearly, chain conformations
which exist *outside* the hydrophobic core increase chain-water
contact and account is taken of this in Equation (1)). Interaction
between the headgroups can be split into two terms which are,
respectively, independent and dependent on the chain conformation.
(The latter term depends on the z-coordinate of the headgroup and
on the area of the chain conformation at the hydrophobic core
surface.)

In our model, *all* interactions with surrounding amphiphiles
which are dependent on chain conformation are modelled by the third
term in Equation (1). Because of this, $\pi(r)$ must not be regarded
as an accurate picture of the *actual* lateral pressure profile in an
aggregate. However, this does not detract from its usefulness —
it plays the crucial role of imposing the constraint of constant
density on the ensemble of chain conformations. It should also be
noted that once the lateral pressure has played this role, it is
not required further in evaluation of any of the results. Finally,
on this point, the overall lateral pressure imposed by the chains
in any aggregate can be evaluated from the thermodynamic identity
$\pi = -(\frac{\partial F}{\partial A})$ where F is free energy (see Figure 3) and A is area.

We now consider a further aspect of the free energy curves in
Figure 3. Near the minima of all curves, the chains have very
little order (average order parameters, as defined for Figure 8,
of ≤ 0.1) and have almost as many *gauche* bonds as in the random-
coil (see Figure 4). Here the assumptions of our model seem most
reasonable. However, when aggregates are formed which require
considerable straightening of the chains two effects combine to
reduce the free energy cost below the value calculated by our model.
(The effects are described in reference 22 in the footnote on
p. 112 and the legend to Figure 4 on p. 113.) Hence, we have most
confidence in our free energy estimates for all the spheres studied
and for most rods, but for bilayers with $R \gtrsim 11$ Å, the actual free
energy cost to the chains may be considerably smaller than
calculated.

The results displayed in Figure 3 demonstrate that there is
a small free energy cost (to the chains) of forming spherical
micelles with a hydrophobic core radius equal to a fully extended
chain (16.7 Å). Assuming the volumes of CH_2 and CH_3 groups
already quoted, the hydrophobic core of this micelle contains ~56
chains. We now wish to examine the possibility of forming spherical
micelles with a larger hydrophobic core radius than a fully extended

chain.[*] For the sake of clarity, we will not include in the
discussion the small amount of hydrocarbon which sits outside the
hydrophobic core and increases aggregation numbers by ~5% (see
last paragraph in the Results section). Consider a spherical
hydrophobic core of radius 19.05 Å which, ignoring fluctuations in
its shape, must have a 2.35 Å radius hole at its center with volume
~54 Å^3. The hydrophobic core of this micelle contains ~82 chains.
There are two important contributions to the free energy cost of
forming the hole: the work done to form the hole, acting against
the internal pressure in the micelle; and the van der Waals
energy loss of atoms near the hole.

The *largest possible* pressure which could be acting in the
micelle is the Laplace pressure arising from a surface with the
tension of an n-alkane/water interface (~50 mN·m^{-1}) at the
hydrophobic core surface (R = 19.05 Å). (There is good reason
to believe that the internal pressure in a micelle is much smaller
than this, but as we shall see, even this pressure leads to a
negligible free energy cost.) The free energy contribution is
PV ≃ 2γV/R = 408 cal/mol of micelles or 5 cal/mol of chains in the
hydrophobic core.

As bulk liquid n-alkane chains are conformationally disordered[20], the proportion of *gauche* bonds should be not too
different in liquid and gaseous n-alkane at the same temperature.
Hence, measurement of ΔH (vapourization) of liquid hydrocarbons as
a function of chain length should give a good measure of the
van der Waals energy between the chains in bulk liquid n-alkane,
and hence also in the micelle. Such measurement yields (at 25°C[24]),

$$\Delta H \text{ (vapourization)} - RT = 1137\ n_{CH_3} + 1169\ n_{CH_2} \text{ cal/mol}$$

of alkane

(10)

$$r^2 = 0.9996$$

(ΔU = ΔH - PΔV ≃ ΔH - RT, where U, H, P, V, R, T are energy,
enthalpy, pressure, volume, the gas constant and temperature,
respectively.)

In order to estimate the van der Waals energy cost of forming
the hole, we require some preliminary analysis. Let U_{jk} be the
van der Waals energy of CH_2 group j as a result of the presence of
CH_2 group k. (Of course, $U_{jk} = U_{kj}$.) Assuming pairwise additivity,
the total van der Waals energy U_{VDW} in a sample is

---

[*] We are grateful to Drs. Gunnar Karlström, Mats Almgren and
John Nagle for illuminating discussions on this possibility.

$$U_{VDW} = \frac{1}{2} \sum_{\substack{j,k \\ j \neq k}} U_{jk} \tag{11}$$

(The factor $\frac{1}{2}$ corrects for the double counting of the energies.)

The hole in our micelle has a volume of two CH_2 groups. We wish therefore to consider a hypothetical molecule consisting of two bonded CH_2 groups (labelled 1 and 2) vapourizing from a bulk n-alkane phase. It is instructive to consider the process in two steps: (i) vapourization of the hypothetical molecule leaving a 54 Å^3 hole, (ii) rearrangement of the remaining molecules to fill the hole. The first step has a van der Waals energy *cost* (U_1) of

$$U_1 = \frac{1}{2} \sum_{\substack{j=1,2 \\ k>2}} (U_{jk} + U_{kj}) \tag{12}$$

As should be clear from the definition of U_{jk}, we identify the first term in Equation (12) as the loss of van der Waals energy of the j groups (comprising the vapourizing molecule), and the second term as the loss of van der Waals energy of the k groups (comprising the remaining molecules). Two conclusions emerge from this identification. Firstly, on allowing rearrangement of remaining molecules, there is no loss of van der Waals energy for these k groups and hence the van der Waals energy of vapourization ($\Delta U_{VAPOURIZATION}$) is simply the first term in Equation (12).

$$\Delta U_{VAPOURIZATION} = \frac{1}{2} \sum_{\substack{j=1,2 \\ k>2}} U_{jk} \tag{13}$$

Secondly, formation of the hole in the micelle involves only surrounding groups (as there was never any molecule which vapourized from the hole). The van der Waals energy cost of forming the hole (U_{HOLE}) is then simply the second term in Equation (12):

$$U_{HOLE} = \frac{1}{2} \sum_{\substack{j=1,2 \\ k>2}} U_{kj} \tag{14}$$

Since $U_{jk} = U_{kj}$, we have from Equations (13) and (14):

$$U_{HOLE} = \Delta U_{VAPOURIZATION} \quad . \tag{15}$$

The loss in van der Waals energy on creation of a hole with the volume of 2 CH_2 groups can now be estimated (using Equations (10)

and (15)) as 1169 × 2/82 ≃ 29 cal/mol of chains in the
hydrophobic core of the micelle. In fact, the groups in the
immediate vicinity of the hole are CH_3 groups and hence a better
estimate is probably obtained by the same calculation for a single
CH_3 group (which again has volume 54 $Å^3$). Using Equations (10)
and (15) leads to an estimate of 1137/82 ≃ 14 cal/mol of chains.
Thus, if a spherical hydrophobic core containing 82 chains is to
be formed, the free energy cost associated with the hole in this
micelle is probably in the range 14 to 29 + 5 = 34 cal/mol of
chains, which is very small indeed.

A spherical hydrophobic core radius of 20.4 Å will contain
~101 chains and must have a hole with volume ~4 × 54 $Å^3$ at its
center. A generous estimate for the free energy cost of forming
this hole (based on the above) is (408 × 4 + 1169 × 8)/101 ≃ 109
cal/mol of chains in this hydrophobic core. Probably a more
realistic estimate (assuming a pressure much closer to atmospheric
in the micelle and the van der Waals calculation for CH_3 groups)
is 1137 × 4/101 = 45 cal/mol of chains.

The critical reader may wonder how we can argue throughout
this paper for a constant liquid hydrocarbon packing density, while
in the previous few paragraphs we discuss the possibility of a
vacuum at the center of spherical micelles. Firstly, the average
change in density as a result of these holes is very small indeed
(for the micelle with a hydrophobic core containing 82 chains, it
is 0.2%; for the micelle with a core containing 101 chains, it is
0.6%). Secondly, there is something very special about these
holes in that, at minimal free energy cost, they result in large
increases in the number of chains in the hydrophobic core, and
hence substantial reductions in the area per amphiphile at the
hydrophobic core surface.

The calculations we have presented in the last few paragraphs
are clearly crude in some respects. Nevertheless, they establish
an important point. As we have seen from Figure 3, there is
minimal free energy cost in packing the amphiphile chains into
spheres with a hydrophobic core radius equal to (or slightly
greater than) the fully extended chain length. (We will present
the results for larger spheres in a forthcoming paper.[15]) This
result, together with the calculations presented above, suggest
that spherical micelles can exist with aggregation numbers
substantially larger (of the order of 50% or more) than those
deduced from assuming a sphere with radius equal to the fully
extended chain length.

Clearly, as chemical changes are made which favour larger
aggregates with smaller areas per amphiphile (e.g. the addition
of salt or an increase in amphiphile concentration), aggregates

must eventually become profoundly non-spherical. Our point is simply that non-sphericity need not exist in micelles with a hydrophobic core radius slightly larger than the fully extended chain length.

We now turn our attention to Figure 10. We do not claim to have very accurately modelled the interfacial region, where chains, headgroups and water mix. In the model, the possibility of such a region was included because we felt that the alternative, of completely segregating the chains from the headgroups and water, was more undesirable.

Apart from the simplicity of Equation (2), there are at least two effects which make the graphs in Figure 10 inaccurate. Firstly, the presence of hydrocarbon in the annulus in which most of the headgroups sit, will lower the effective dielectric constant in this annulus; this has an adverse effect on the headgroups. The free energy cost, to the aggregate, of chains sitting in this region is therefore larger than that determined in Equation (2). Secondly, for ionic amphiphiles, both the electric potential arising directly from other headgroups and the repulsion from an 'image headgroup' in the hydrophobic core, are reduced when the headgroup sits further from the core surface.[*] A calculation which evaluated the effect of the electric potential for a case chosen to model sodium dodecylsulphate just above the cmc in the absence of added salt[23], gave an average headgroup protrusion of 1.96 Å. However, the chains are not all-*trans*, as assumed in Reference 23 and they have more configurational freedom inside the core than outside it. Thus, there is a larger free energy cost of pulling them a given distance out of the core than assumed in Reference 23. This, together with the 'lowering dielectric constant' effect will make the average headgroup protrusion less than 1.96 Å (the value derived from Figure 10 is 0.93 Å). In summary, although the results displayed in Figure 10 are not quantitatively accurate, they should not be too far from the truth.

---

[*] The reason for forbidding chain conformations which entered the core and then re-emerged into the surrounding aqueous phase was that these conformations are, in reality, *much* less likely than the model would make them. They are largely suppressed because, as well as requiring space which is partly filled by headgroups and counterions, they would contribute to the first unfavourable effect, while contributing nothing to the second favourable effect.

7. SUMMARY

(i) It should be emphasised that the aggregates we are
discussing are highly dynamic entities. The average time required
for a monomer to leave a micelle is of the order of 10^{-6} -10^{-5}s
(reference 23) and it may protrude more than 4 Å from the micelle
surface and then rejoin the micelle many times before it leaves.[23]
(Although we have suggested earlier that the average headgroup
protrusion derived by Aniansson[23] is too large, we agree with the
qualitative picture painted by him.) In studying the kinetics of
micelles, these events are very important. However, from the point
of view of the *average* equilibrium properties of an aggregate,
they contribute only negligibly. The *vast* majority of time any
given monomer is associated with an aggregate, it sits with almost
all of its chain inside the hydrophobic core of the aggregate.

(ii) For a bilayer to have a fluid hydrophobic core, the
half thickness of the core must be considerably less than the
length of a fully extended chain. For a cylinder, our calculations
suggest that the free energy cost of packing the hydrophobic core
at constant density is small until the core radius is within about
1 Å of the fully extended chain length. For spherical micelles,
the free energy cost to the chains is small for core radii up to
and even slightly larger than the fully extended chain length
(from Figure 3).

(iii) As a consequence of (ii), the possibility of small
holes in the center of spherical micelles must be considered.
Calculations of the free energy cost of such holes suggest that
spherical micelles may be formed with aggregation numbers
substantially larger (of the order of 50% or more) than the number
of amphiphiles which can pack into a sphere with a radius equal to
a fully extended chain.

(iv) On being transferred from a (model) bulk n-alkane
environment to an aggregate, there is an average loss of less than
0.3 *gauche* bonds per chain for all bilayers with R < 10 Å; for
all cylinders with R < 13 Å and for all spherical micelles with
R < 17 Å (from Figure 4).

(v) For a given aggregate shape, the equilibrium aggregate
dimensions depend not only on the chains but also on several
factors external to the chains. For possible equilibrium sizes
(see Figure 5 for details) we find the following. For all
aggregate shapes, the extent of spatial freedom of movement is
smallest for groups near the headgroup and increases all along
the chain. Also, the chains show a tendency to order in a
direction perpendicular to the hydrophobic core surface (i.e. they
have a positive order parameter, as defined in Figure 8). Although
the order parameter profiles along the chain change with aggregate

shape, a common feature is a substantial drop in order as the free end of the chain is approached.

(vi) For the spherical micelle (and to a lesser extent, for the cylindrical micelle), it is clear that without any water penetration into the hydrophobic core, all segments of the chain spend a reasonable amount of time in a partly hydrophilic environment (from Figure 6).

ACKNOWLEDGEMENTS

One of us (D.W.R.G.) is very grateful to the members of Divisions of Physical Chemistry 1 and 2, University of Lund Chemical Center and to Professor John F. Nagle, Department of Biological Sciences, Carnegie-Mellon University, for their hospitality and helpful comments during the writing of much of this paper.

REFERENCES

1. B. Lindman and H. Wennerström, Topics in Current Chemistry, 87, 1 (1980).
2. B. Mely, J. Charvolin and P. Keller, Chem. Phys. Lipids, 15, 161 (1975).
3. J. Seelig, Q. Rev. Biophys., 10, 353 (1977).
4. E. Vikingstad and H. Høiland, J. Colloid Interface Sci., 64, 510 (1978).
5. J. F. Nagle and D. A. Wilkinson, Biophys. J., 23, 159 (1978).
6. G. Büldt, H. U. Gally, A. Seelig, J. Seelig and G. Zaccai, Nature, Lond., 271, 182 (1978).
7. R. Fettiplace, L. G. M. Gordon, S. B. Hladky, J. Requena, H. P. Zingsheim and D. A. Haydon, Methods Membrane Biol., 4, 1 (1975).
8. J. P. Dilger, L. R. Fisher and D. A. Haydon, Chem. Phys. Lipids, 30, 159 (1982).
9. P. Schatzberg, J. Phys. Chem., 67, 776 (1963).
10. F. Reiss-Husson and V. Luzzati, J. Phys. Chem., 68, 3504 (1964).
11. P. J. Flory, "Statistical Mechanics of Chain Molecules", Wiley-Interscience, New York, 1969.
12. D. W. R. Gruen, Biochim. Biophys. Acta., 595, 161 (1980).
13. D. W. R. Gruen, J. Colloid Interface Sci., 84, 281 (1981).
14. C. Tanford, "The Hydrophobic Effect: Formation of Micelles and Biological Membranes", 2nd ed. Wiley-Interscience, New York, 1980.
15. D. W. R. Gruen, in preparation (1983).
16. K. A. Dill and P. J. Flory, Proc. Natl. Acad. Sci. USA, 78, 676 (1981).

17. T. Klason and U. Henriksson in "Solution Behavior of Surfactants: Theoretical and Applied Aspects,"K.L. Mittal and E.J. Fendler, Editors,Vol.1,pp.417-429,Plenum Press,New York, 1982.

18. B. Cabane, J. Physique, 42, 847 (1981).

19. D. Y. Yoon and P. J. Flory, J. Chem. Phys., 69, 2536 (1978).

20. P. J. Flory, Faraday Discuss. Chem. Soc., 68, 14 (1979).

21. M. Vacatello, G. Avitabile, P. Corradini and A. Tuzi, J. Chem. Phys., 73, 548 (1980).

22. D. W. R. Gruen, Chem. Phys. Lipids, 30, 105 (1982).

23. G. E. A. Aniansson, J. Phys. Chem., 82, 2805 (1978).

24. F. W. Billmeyer, J. Appl. Phys., 28, 1114 (1957).

MOLECULAR ORGANIZATION IN AMPHIPHILIC AGGREGATES

Ken A. Dill

Department of Chemistry
University of Florida
Gainesville, Florida 32611

Molecular organization within the hydrocarbon
cores of micelles, bilayers, and other amphiphilic
aggregates depends on i) intramolecular structure: the
continuity and flexibility of the chains, and ii) the
balance of intermolecular forces. Attractive exchange
interactions are responsible for the differential solubilities of polar heads and hydrocarbon tails in water.
But dispersion forces among aliphatic hydrocarbons are
isotropic; steric repulsive forces are required to
induce the anisotropy responsible for internal structure.

The hydrocarbon core within an amphiphilic
aggregate is better represented as an interfacial phase
of matter than as a bulk phase. For example, physical
properties vary with depth from the interface; they
can be predicted from the surface density of the chains,
at constant bulk density; and present evidence strongly
supports the view that water interacts through contact
(ubiquitous for small micelles) at the core interface,
being otherwise excluded from the hydrocarbon interior.

Present address:
Departments of Pharmaceutical Chemistry and Pharmacy
University of California,
San Francisco, California 94143

INTRODUCTION

Irving Langmuir was the first to provide clear evidence for
the physical character of the forces of aggregation of amphiphilic
molecules. His classic work in 1917 established the principle that
the unique molecular organization in surfactant monolayers results
from the differential solubility of the polar heads and the hydrocarbon tails in water[1]. The polar heads readily associate with
water while the nonpolar tails collectively avoid it. On the basis
of Langmuir's principle, Perrin interpreted the structure of soap
bubbles[2], Gorter and Grendel proposed the now familiar lipid
bilayer structure of biomembranes[3], and Hartley described the
structures of micelles[4].

Molecular organization within amphiphilic aggregates, as
within other condensed phases, is determined by intermolecular
forces and intramolecular structure. A surfactant chain molecule
has a polar or ionic head group attached to a continuous "flexible"
hydrocarbon tail. Flexibility is an intramolecular property. It
may refer to the near equality of the internal energies of different
rotational isomeric states of the chains in equilibrium, or of the
energy of activation between states. The latter represents a
dynamic property; the former meaning is used exclusively here.
Models of flexibility may adopt the approximation that rotational
isomers occur as discrete states[5]; or the further approximation
that chain bonds are oriented along the principal axes of a spatial
lattice[6].

Langmuir's principle describes a consequence of intermolecular
attractive forces. These are balanced by intermolecular steric
repulsive forces; the combination gives rise to molecular organization. Here we consider only the time-averaged structures of pure
unswollen non-crystalline phases of the hydrocarbon cores of amphiphilic aggregates.

ATTRACTIVE FORCES

Attractive forces are responsible for the aggregation of amphiphiles in water. Two molecular species will mix if their mutual
attractive free energy is lower than the sum of the self attractions
of the individual species[7]. This "exchange energy", the cost of
replacing a like neighbor with an unlike neighbor, determines
whether mixing or aggregation will be preferred in solutions in
which nearest neighbor interactions are dominant. Because the
attraction of water for itself is large, mixing with hydrocarbons
is disfavored.

In condensed phases of aliphatic hydrocarbons, attractive forces are due to dipolar London dispersion interactions. These attractions cause cohesion within the system by reduction of the average intermolecular separation; they have little tendency to align the molecules. Changes in internal attractive energy, ΔE, can be expressed in terms of the relative alignment, S, of the molecules and of their volume, V:

$$\Delta E = \left(\frac{\partial E}{\partial V}\right)_{S,T} \Delta V + \left(\frac{\partial E}{\partial S}\right)_{V,T} \Delta S$$

$$+ \frac{1}{2}\left[\left(\frac{\partial^2 E}{\partial V^2}\right)(\Delta V)^2 + 2\left(\frac{\partial^2 E}{\partial V \partial S}\right)(\Delta V \Delta S) + \left(\frac{\partial^2 E}{\partial S^2}\right)(\Delta S)^2\right] + \ldots$$

The terms higher than the first order in this series are expected to be small[8], therefore

$$\Delta E \cong \Delta E_{\text{isotropic, S const}} + \Delta E_{\text{anisotropic, V const}}$$

where

$$\Delta E_{\text{iso}} = \left(\frac{\partial E}{\partial V}\right)_{S,T} \Delta V$$

$$\Delta E_{\text{aniso}} = \left(\frac{\partial E}{\partial S}\right)_{V,T} \Delta S$$

The relative magnitudes of these energies can be estimated from theory[8-10]; for aliphatic hydrocarbons, ΔE_{aniso} is less than 4% of of the magnitude of ΔE[9,10]. Thus even for major changes in hydrocarbon organization represented by sublimation, vaporization, or melting, the van der Waals attractive forces are responsible predominantly for isotropic changes in density; the constraints they impose on molecular orientation are insignificant. Therefore, phase changes in amphiphilic aggregates, which are accompanied by changes in both order and density, are characterized predominantly by a van der Waals energy difference, ΔE_{iso}, rather than by

$$\Delta E_{\text{alignment}} = \left(\frac{\partial^2 E}{\partial V \partial S}\right)(\Delta V \Delta S),$$

which has been adopted in some theoretical treatments[11-13]. These "alignment potentials" are of questionable molecular origin. Therefore they are regarded as phenomenological representations of combinations of these and steric forces[11,14,15], the latter of which, however, find more suitable molecular representation through steric packing models discussed in the following sections.

REPULSIVE FORCES

Attractive forces underlying Langmuir's differential solubility principle increase the concentration of amphiphilic mole-
clues to the limit imposed by steric repulsive forces. The balance
of these forces holds the density to within a narrow range, which
is about the same as for other liquids and solids[16,17]. As in
other condensed phases[18], the attractive forces provide general
cohesion; the steric repulsive forces impose constraints on the
molecular configurations. Steric exclusion of volume may be
represented by the occupancy of sites on a spatial lattice by
chain segments[5,19,20] or as "effective" pressures[12,13,21] or
forces[19] acting on single chains.

WATER DISTRIBUTION IN AMPHIPHILIC AGGREGATES

The large difference in solubility in water of polar heads
and hydrocarbon tails dictates that water penetration into the
hydrophobic cores of amphiphilic aggregates should be negligible.
This assertion of the Langmuir principle is supported by experiments which measure bulk properties of hydrophobic cores. Micellar
hydrocarbons have approximately the same partial molar volume[17],
compressibility[17], and amorphous structure[22] as bulk, unhydrated
n-alkanes. Neutron scattering contrast variation experiments and
measurements of aggregation number and radius of gyration of lithium
dodecylsulfate micelles show that core volumes do not contain
significant concentrations of water[23]. Other neutron scattering
experiments indicate that lipid bilayers[24] and other membranes[25]
are similarly devoid of water.

The assertion that water should not penetrate the hydrophobic
cores does not imply that the fraction of hydrocarbon in contact
with water is small. On the contrary, more than half the surface
of the hydrocarbon cores of small micelles is unprotected by head
groups. The large interfacial surface of the core must be the site
of substantial water contact. Except for micellar systems with
large oxyethylene head groups which reduce such contact[26,27], proton NMR chemical shift and relaxation measurements indicate some
chain exposure to water[28]. Measurements of the rates of hydrolysis
of alpha-methylene groups of alkyl sulfates above and below their
critical micelle concentrations[29] and of the dependence of partial
molar volumes on chain length in homologous series of surfactants[30]
indicate some hydration of the first 1 - 4 methylene groups from
the polar head. A similar conclusion has been drawn from ^{19}F-NMR
relaxation rate[31] and chemical shift[32] experiments with perfluoro-
carbon micelles.

Many water-insoluble spectroscopic probes[33-40] are exposed to
water upon solubilization within micelles. From these experiments

it cannot be concluded that water penetrates the cores, however, for that conclusion would require the assumption that the probe molecules themselves are completely buried within the hydrophobic cores. On the contrary, it is likely that many of these probes reside at the core interface adjacent to external water[33,34,39]. The probes may prefer the interface for several reasons. A substantial fraction of the hydrocarbon core is interfacial (see below). The free energy of solubilizing a probe depends not on the absolute attraction of the probe for water, but on the exchange free energy: the difference in energy due to exchanging an exposed hydrocarbon segment at the interface for a probe molecule at the interface. According to the well-known consequence of the exchange principle that intermolecular attractions are greatest for molecules of similar structure[7], few molecules should prefer the micellar core interior more than aliphatic hydrocarbons. Thus, even small polarizability of the solute may lead to significant preference for the interface.

The same caveat must be applied to probes which are covalently bound to the hydrocarbon chains, including carbonyl groups and fluorine atoms[38,41]. Partially fluorinated chains do not obey ideal mixing laws[42,43]; such groups may prefer the interface. These mixing problems may be avoided by using solute probes which are chemically identical to the amphiphilic chains. For example, solubilities of alkanes in alkyl chain micelles are approximately the same as in bulk hydrocarbon[44]; this provides additional evidence against water penetration into hydrophobic cores. Ideal mixing may also be expected from completely perfluorinated micellar chains[31,32]. Muller and Simsohn[32] have derived a parameter from ^{19}F-NMR chemical shift experiments which characterizes the environment of the CF_2 groups in perfluorooctanoate micelles. Their data, shown in Figure 1, indicate that no CF_2 group is in a completely water-free environment. However, this can be suitably interpreted as representing the probability of residence of the various chain segments at the interfacial surface. The line in Figure 1 is the prediction of that probability according to the interphase theory[20], (see below), which assumes that no water penetrates the core. Because such a large fraction of the micellar core is at the interface, there is substantial probability that even terminal chain ends will occur there. For some probes[38,45], the relative importance of non-ideal mixing behavior and of interfacial distribution has not yet been determined; NMR ring current shift methods[39,46] may help resolve these issues. Present evidence thus firmly supports the view that the hydrocarbon cores are largely devoid of water, but that there is substantial hydrocarbon interaction with water at the interface.

A further consequence of Langmuir's principle is that the area of the interface, defined as the boundary between the hydrocarbon core and its polar surroundings, should be minimal. Al-

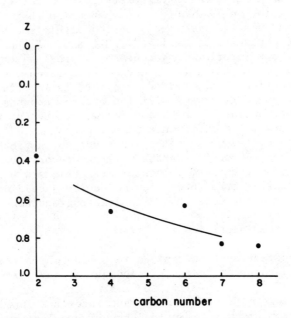

Figure 1. Water contact of CF_2 groups in perfluorooctanoate micelles ($Z = 1$ in bulk fluorocarbon; $Z = 0$ in bulk water), as a function of chain position numbered from the head group. The points represent the data of Muller and Simsohn[32]. The line is the prediction, according to the interphase theory[20], of the probability that a given chain segment resides at the interface of micelles presumed to be spherical. Chains are assumed to have 3 segments centered at carbons 3,5,7. The first segment is assigned $Z = 0.52$, representing the estimated C_3 value (averaged from C_2 and C_4), and which also thus represents the approximate interfacial character of methanol[33,34].

though the polar head region may be penetrated by solvent[47], in the ensemble average over a large number of micelles and in time average over their structural fluctuations, the interfacial surface is expected to be reasonably sharp and geometrically smooth. Neutron scattering[23], particularly at large wavevectors[48], and micellization electrostatic self-potential experiments[49] are consistent with a sharp interface. Theoretical models which predict a rough core interface[50,51] generally neglect the severe penalties which must be paid for water-hydrocarbon contact. One of these, a

Monte Carlo model[51], exacts no penalty for solvent penetration and supposes that the only energies important for micellar structure are those of the interactions among heads and tails. However, these energies alone, in the absence of artificially applied constraints, are apparently insufficient to sustain the micelle against dis-aggregation. Such energies are inadequate characterizations of am-phiphilic aggregates in water. The limiting extreme of surface roughness is represented by the radial chain model in which the amphiphilic tails are all-trans and aligned along the radii of the micelle, so that water may penetrate to the core center[52,53]. This model substantially overestimates radial alignment of the chains[52,53] and the concentration of terminal methyls near the micellar center[23]. In addition to these shortcomings, such models are difficult to reconcile with the bulk properties cited above.

A simple model can be constructed to compute the fraction of hydrocarbon chain segments which lie at the interfacial surface of an amphiphilic aggregate. Suppose the core, defined by a sharp interface, is exclusively filled by chain segments. All chain seg-ments occupy equal volumes of space, represented by sites on a lat-tice. The sites are considered to occur in spatial layers, the first of which is at the interface. If the aggregate is planar (monolayer or bilayer), layers are planar; if the aggregate is non-planar (micelle or vesicle), layers are concentric with the inter-face. In order that each site may be occupied by a chain segment in any orientation, sites must be isodiametric with all dimensions approximately equal to the chain width. On that basis, sites con-tain segments comprising about 3.6 methylene groups[19,20,53].

Consider a spherical micelle. The total number, N_t, of chain segments in the core is

$$N_t = 4\pi r^3/3$$

where r is the radius in units of lattice site lengths. The number of segments, N_1 in the layer at the interfacial surface is[20]:

$$N_1 = 4\pi(3r(r-1)+1)/3$$

Thus for micelles with chains of length n+1 = r = 4 (4 x 3.6 ≅ 14 methylene groups), the fraction of hydrocarbon segments at the in-terface is N_1/N_t = 0.58.

The number of fully hydrated methylenes per chain can be com-puted as follows. The total number of chains is $J_1 = N_t/(n+1)$. The number of segments at the surface which are not protected by head groups is $N_1 - J_1$, so the number of unprotected methylene groups at the surface (per chain) is $(3.6)(N_1 - J_1)/J_1$. This supposes, for simplicity, that each head group covers one segment, and thus has an area of 1.65 x (4.6 Å)^2. This represents approximately the area of a phosphatidylcholine or trimethylammonium group, and simply serves the purpose of illustration; introduction of arbitrary head group area is straightforward. The factor of 1.65 arises from the greater radial distance from the core center of the head groups than the first segments. Thus the number, n_w, of fully wetted methylene groups per micellar chain is

$$n_w = (\sigma^{-1}-1)(3.6 \ f) = [(N_1/N_t)(n+1)-1](3.6 \ f)$$

where $\sigma = J_1/N_1$, the surface density of chains, and the fraction, f, of each interfacial segment in contact with interfacial water is estimated to be in the range 1/6 - 1/2 (1/6 is the minimum fractional contact admitted by the cubic lattice; 1/2 represents a segment solvated half by water and half by the core. These values are chosen arbitrarily as approximate bounds on our ignorance of interfacial structure). On this basis it is predicted that 0.8 - 2.4 methylene groups are fully "wetted", in general concurrence with the experimental values of 1 - 4[28-31,54]. Similar calculations show that less hydrocarbon contact with water should occur for cylindrical micelles and lamellar structures. This simple model shows that even the minimum equilibrium-averaged interfacial area of a micellar hydrocarbon core is quite large; roughness[50,55] need not be invoked to account for substantial water contact.

INTERNAL ORDER

Although the balance of intermolecular attractive and repulsive forces holds the bulk density (number of molecules per unit volume) of a condensed phase to within a narrow range, the surface density of amphiphiles in aggregates (number of molecules per unit interfacial area) varies over a wide range with changes in thermodynamic parameters including temperature, concentration, and lateral pressure (applied to monolayers). Surface density, like bulk density, is determined by a balance of "opposing forces"[56,57]. The progression from small spherical micelles to rodlike micelles and microemulsions to condensed monolayers and bilayers is characterized by increasing surface density of the chains[56]. Phase changes in amphiphilic aggregates are associated with large changes in surface density, and relatively much smaller changes in bulk density[16,58]. This primary dependence on surface density is a consequence of the large surface-to-volume ratios of amphiphilic aggregates, which bear closer resemblance to interfacial than to bulk systems. Among the first to recognize this, Wolfgang Ostwald remarked that: "Colloid particles are not phases according to Gibbs' own definition, because the amount of energy and entropy located in their surfaces is *not* to be neglected when compared with the amount of energy and entropy located in the interior of the particle" (italics his)[59].

While the amphiphilic aggregate geometry may change with surface density, the simplest physical model approximation is that the
internal chain configurations do not. The primary model of this
type supposes that the interior of an amphiphilic aggregate has
approximately the structure of some bulk phase of matter; the micellar core is often modelled as disordered amorphous hydrocarbon[56,57]. This "oil-droplet" representation of the micellar
interior has had moderate success. It qualitatively predicts the
solubilities of small hydrophobic molecules in the micellar cores,
chemical potentials from which sizes, shapes and critical micellar
concentrations can be computed, and some of the mechanical properties of amphiphilic aggregates[56,57].

Inadequacies of the bulk amorphous hydrocarbon model are indicated by observations that configurational order within amphiphilic aggregates increases with increasing surface density of the
chains[60-62]. Chains are less ordered in reverse micelles than in
normal ones[63], more ordered in cylindrical than in spherical
micelles[64], and less ordered in micelles than in bilayers, as
measured by solubility experiments[65], and by accessibility of the
hydrocarbon chains to attack by external molecules[66].

An additional weakness of bulk phase models is the prediction
that configurational order throughout the core of an amphiphilic
aggregate should be everywhere the same. That the order varies
with depth from the interfacial surface of an aggregate is shown
by ^{2}H-NMR, electron spin resonance, and fluorescence experiments,
primarily in bilayers[67-70], by ^{13}C-NMR experiments in micelles[52,53],
by neutron scattering experiments of the distribution of chain ends
with depth in bilayers[71] and micelles[23], and by neutron scattering
measurements of the distribution with depth of the concentration of
dissolved hydrophobic solute molecules in bilayers[72] and micelles[33,34,39,46]. Chain segment mobilities also vary with chain
segment position in bilayers and micelles[73]. These observations
support the conjecture that amphiphilic aggregates closely resemble
interfacial systems in which properties vary over molecular dimensions along the interfacial normal spatial coordinate.

The interfacial character of amphiphilic aggregates can be
described on the basis of a lattice model[19,20]. The purpose of
the lattice is to allow systematic enumeration of the intramolecular configurations of all the chains within the hydrophobic core,
subject to i) the volumetric density constraint imposed by the
balance of isotropic attractive and steric repulsive forces, ii)
specification of the surface density of the chains, the chain
lengths, and the aggregate geometry, and iii) the exclusion of
water from the core interior. Interphase theory accounts for the
steric constraints on molecular configurations as follows. Given
that all J_1 chains have first segments in the first lattice layer
containing N_1 sites, the fraction of sites in that layer filled by

first segments is $\sigma = J_1/N_1$. The assumption that no solvent penetrates the core requires that the remaining sites in the first layer,
$R_1 = N_1 - J_1$, must be occupied by bonds oriented laterally in the
plane of the layer. All chains which do not terminate in the first
layer must enter the second layer. The number of chains per unit
area which enter layer i from i-1 can be described by a "flux", J_i.
Flux is conserved from one layer to the next, except where chain
terminations serve to decrease it. In each layer, as in the first,
the number of segments oriented laterally is determined by the
difference between the total number of sites in the layer, and the
flux from the previous layer. This model neglects backtracks, chain
segments which return toward the interface, an approximation which
is least reliable for systems of low surface density. The present
model[19,20] also neglects the small bending energy between intramolecular pairs of neighboring bonds. The statistical mechanical
principle that all configurations of equal energy have equal probability of occurrence is satisfied by the requirement that for
any chain which has entered layer i, the probability that its next
segment is lateral is given by the overall incidence of lateral
bonds in that layer, determined as described above. Thus the
relative probabilities for forward or lateral placements are the
same for any bond of any chain, provided it is within a given layer.

The interphase theory predicts that in the ensemble average
over all possible chain configurations within the core, the chain
segments should be statistically distributed. The model is consistent with the diffuse 4.5 - 4.6 Å alkane-like x-ray diffraction
patterns characteristic of systems with no long range order[22,58].
Predictions of the widths of segment distributions along the interfacial normal[19,20] are in reasonable agreement with neutron scattering experiments with specifically deuterated methylene and
methyl groups in bilayers[71] and micelles[23]. Such distributions, in
bilayers at least, broaden with chain position toward the termini.

In addition to its spatial distribution, each bond has an
orientational distribution. With increasing depth within planar
bilayers, the flux of chains decreases as chains terminate. The
area per chain (for chains which enter layer i), the reciprocal of
the flux, increases with depth. The requirement to fill space de-
mands that a greater fraction of chain segments be oriented laterally near mid-bilayer. Thus the distribution of chain termini is
seen to be the source of the increasing orientational disorder toward
the chain ends. This orientational "disorder gradient" has been
measured by ^{2}H-NMR in a variety of systems including those of pure
lecithins[67,68], single chain surfactants[60,68], unsaturated amphiphiles[67-69], mixtures of lecithins with cholesterol, peptides and
proteins[67,68,74,75] and in biological membranes[67,68,70]. It is
thus a general physical property of planar surfactant systems. The
orientational order parameter, S_1, for first segments gives an
approximate measure of the surface density, σ, of the chains[19,20]

$$S_1 = \frac{3}{2} <\cos^2\theta> - \frac{1}{2} \cong \frac{1}{2}(3\sigma-1)$$

where θ is the angle between the chain axis and the interfacial normal.

Closely related to the disorder gradient along the chains is the disorder gradient as a function of spatial position from the interface[13,19,20]; the latter is not directly measurable. This distinction is most important for nonplanar aggregates. Theory predicts for example, that there should be sharply increasing orientational order toward the core center of spherical micelles, but because so few chains reach that restricted region of space, the disorder should be virtually independent of positon along the chain[20]. Recent ^{13}C-NMR measurements of a configurational property related to both the orientational and positional distributions of chain segments in spherical micelles[52] are consistent with this prediction[53], but ^{2}H-NMR experiments on perdeuterated potassium laurate in the hexagonal phase are not[60]. Recent work in our lab suggests that this discrepancy is due to the neglect of intra-molecular chain stiffness in this model.

The outer region of spherical micellar cores, with its sub-stantial chain disorder, should provide a favorable site for mixing solute. Similarly, small hydrophobic solutes should concentrate in the planar center between half-bilayers where chain disorder is greatest. Neutron scattering from deuterated hexane dissolved in bilayers[72] supports this prediction. Interphase theory also pre-dicts slightly greater orientational order in spherical micelles than in bulk alkanes, in accord with the increased trans/gauche ratio of amphiphiles above their cmc[63], and with the slightly lower solubilities of alkanes in micelles than in bulk hydrocarbons[44].

It may be concluded that Langmuir's Principle reliably predicts the general structures of amphiphilic aggregates and rationalizes the near complete exclusion of water from their hydrocarbon cores. However the molecular configurations therein are not wholly spec-ified by that principle alone. The additional principle required, now well-known from broad-ranging studies of the liquid state, is that steric repulsive forces impose constraints which dictate molecular organization. The consequence of these principles combined is that molecular configurations within these systems do not resemble those within ordinary bulk phases of matter, but are instead interfacial in character.

ACKNOWLEDGEMENTS

I thank R. Cantor and R. Orwoll for helpful discussions.

REFERENCES

1. I. Langmuir, J. Am. Chem. Soc., 39, 1848 (1917).
2. J. Perrin, Ann. Phys., 9, 160 (1918).
3. E. Gorter and F. Grendel, J. Exp. Med.,41, 439 (1925).
4. G. S. Hartley, "Aqueous Solutions of Paraffin Chain Salts",
    Hermann, Paris, 1936.
5. P. J. Flory, "Statistical Mechanics of Chain Molecules",
    Interscience, New York, 1969.
6. P. J. Flory, "Principles of Polymer Chemistry", Cornell Univ-
    ersity Press, New York, 1953.
7. J. H. Hildebrand and R. L. Scott, "The Solubility of Nonelec-
    trolytes", 3rd ed., Reinhold, New York, 1950.
8. W. M. Gelbart and A. Gelbart, Mol. Phys., 33, 1387 (1977).
9. L. Salem, J. Chem. Phys., 37, 2100 (1962).
10. M. Warner, J. Chem. Phys., 73, 5874 (1980).
11. S. Marcelja, J. Chem. Phys., 60, 3599 (1974)
12. S. Marcelja, Biochim. Biophys. Acta, 367, 165 (1974).
13. D. W. R. Gruen, Biochim. Biophys. Acta, 595, 161 (1980)
14. J. N. Israelachvili, S. Marcelja and R. G. Horn, Quart. Rev.
    Biophys., 13, 121 (1980).
15. D. W. R. Gruen, Chem. Phys. Lip., 30, 105 (1982).
16. J. F. Nagle and D. A. Wilkinson, Biophys, J., 23, 159 (1978).
17. T. S. Brun, H. Hoiland and E. Vikingstad, J. Colloid Interface
    Sci., 63, 89 (1978).
18. J. A. Barker and D. Henderson, Ann. Rev. Phys. Chem., 23, 439
    (1972).
19. K. A. Dill and P. J. Flory, Proc. Nat'l. Acad. Sci., 77, 3115
    (1980).
20. K. A. Dill and P. J. Flory, Proc. Nat'l. Acad. Sci., 78, 676
    (1981).
21. D. W. R. Gruen, J. Colloid Interface Sci., 84, 281 (1981).
22. F. Reiss-Husson and V. Luzzati, J. Phys. Chem., 68, 3504
    (1964).
23. D. Bendedouch, S. H. Chen and W. C. Koehler, (1982), J. Phys.
    Chem., in press.
24. G. Zaccai, J. K. Blasie and B. P. Schoenborn, Proc. Nat'l.
    Acad. Sci., 72, 376 (1975).
25. D. L. Worcester and N. P. Franks, J. Mol. Biol., 100, 359
    (1976).
26. C. J. Clemett, J. Chem. Soc., A, 2251 (1970).
27. F. Podo, A. Ray and G. Nemethy, J. Am. Chem. Soc., 95, 6164
    (1973).
28. J. Clifford, Farad. Soc. Trans., 61, 1276 (1965).
29. J. L. Kurz, J. Phys. Chem., 66, 2239 (1962).
30. J. M. Corkill, J. F. Goodman and T. Walker, Farad. Soc., Trans.,
    63, 768 (1966).
31. J. Ulmius and B. Lindman, J. Phys. Chem., 85, 4131 (1981).
32. N. Muller and H. Simsohn, J. Phys. Chem., 75, 942 (1971).

33. P. Mukerjee, J. R. Cardinal and N. R. Desai, in "Micellization, Solubilization, and Microemulsions", Vol. 1, p. 241, K. L. Mittal, Editor, Plenum Press, New York, 1977.

34. P. Mukerjee, in "Solution Chemistry of Surfactants", Vol. 1, p. 153, K. L. Mittal, Editor, Plenum Press, New York, 1979.

35. W. Reed, M. J. Politi and J. H. Fendler, J. Am. Chem. Soc., 103, 4591 (1981).

36. M. A. J. Rodgers, J. Phys. Chem., 85, 3372 (1981).

37. F. M. Menger and J. M. Bonicamp, J. Am. Chem. Soc., 103, 2140 (1981).

38. F. M. Menger, J. M. Jerkunica and J. C. Johnston,, J. Am. Chem. Soc., 100, 4676 (1978),and references therein.

39. K. N. Ganesh, P. Mitra and D. Balasubramanian, (1982), these proceedings.

40. B. Lindman and H. Wennerstrom, Topics in Current Chemistry, 87, 1 (1980).

41. N. Muller and R. H. Birkhahn, J. Phys. Chem., 71, 957 (1967).

42. N. Muller, J. Magn. Res., 28, 203 (1977).

43. P. Mukerjee and K. H. Mysels, in "Colloidal Dispersions and Micellar Behavior", K. L. Mittal, Editor, ACS Symp. Ser. No. 9, p. 239, American Chemical Society, Washington, DC, 1975.

44. A. Wishnia, J. Phys. Chem., 67, 2079 (1963).

45. F. M. Menger, (1982), these proceedings.

46. K. A. Zachariasse and G. Kozankiewicz, (1982), these proceedings.

47. D. Stigter and K. J. Mysels, J. Phys. Chem., 59, 45 (1955).

48. B. Cabane, R. Duplessix, T. Zemb, (1982), these proceedings.

49. D. Stigter, J. Phys. Chem., 78, 2480 (1974).

50. F. M. Menger, Acc. Chem. Res., 12, 111 (1979).

51. S. W. Haan and L. R. Pratt, Chem. Phys. Lett., 79, 436 (1981).

52. B. Cabane, J. de Phys. (Paris), 42, 847 (1981).

53. K. A. Dill, J. Phys. Chem., 86, 1498 (1982).

54. B. Halle and G. Carlstrom, J. Phys. Chem., 85, 2142 (1981).

55. G. Aniannson, J. Phys. Chem., 82, 2805 (1978).

56. C. Tanford, "The Hydrophobic Effect", 2nd ed., Wiley-Interscience, New York, 1980.

57. J. N. Israelachvili, D. J. Mitchell and B. W. Ninham, J. Chem. Soc. Faraday Trans. II, 72, 1525 (1976).

58. M. J. Janiak, D. M. Small and G. G. Shipley, Biochem., 15, 4575 (1976).

59. W. Ostwald, Farad. Soc. Trans., 31, 79 (1935).

60. B. Mely, J. Charvolin and P. Keller, Chem. Phys. Lipids, 15, 161 (1975).

61. E. Oldfield, M. Glaser, R. Griffin, R. Haberkorn, R. Jacobs, M. Meadows, D. Rice, R. Skarjune and D. Worcester, in "Biomolecular Structure and Function", p. 55, P. F. Agris, Editor, Academic Press, New York, 1978.

62. B. J. Forrest and L. W. Reeves, Chem. Rev., 81, 1 (1981).

63. J. B. Rosenholm, T. Drakenberg and B. Lindman, J. Colloid Interface Sci., 63, 538 (1978).

64. K. Kalyanasundaram and J. K. Thomas, J. Phys. Chem., 80, 1462 (1976).
65. S. A. Simon, W. L. Stone and P. Busto-Latorre, Biochim. Biophys. Acta, 468, 378 (1977).
66. M. F. Czarniecki and R. Breslow, J. Am. Chem. Soc., 101, 3675 (1979).
67. J. Seelig, Quart. Rev. Biophys., 10, 353 (1977).
68. J. Seelig and A. Seelig, Quart. Rev. Biophys., 13, 19 (1980).
69. J. Seelig and J. L. Browning, FEBS Lett., 92, 41 (1978).
70. G. W. Stockton, K. G. Johnson, K. W. Butler, A. P. Tulloch, Y. Boulanger, I.C.P. Smith, J. H. Davis and M. Bloom, Nature, 269, 267 (1977).
71. G. Zaccai, G. Buldt, A. Seelig and J. Seelig, J. Mol. Biol., 134, 693 (1979).
72. S. H. White, G. I. King and J. F. Cain, Nature, 290, 161 (1981).
73. M. F. Brown, J. Magn. Res., 35, 203 (1979).
74. E. Oldfield, M. Meadows, D. Rice and R. Jacobs, Biochem., 17, 2727 (1978).
75. D. Rice and E. Oldfield, Biochem., 18, 3272 (1979).

THE NATURE OF THE SURFACTANT-BLOCK MODEL OF MICELLE STRUCTURE

Peter Fromherz

Department of Biophysics
University Ulm
D-79 Ulm-Eselsberg, W. Germany

The molecular packing of surfactants in micellar aggregates in water is unknown. Usually a spherical hydrocarbon core is assumed to be formed with chaotic packing, i.e., with negligible near order, as formalized by mean-field theories. True molecular pictures of a micelle, however, must include not only segmental distribution functions but pair-correlation functions of the segments as well. The surfactant-block model is an attempt to propose on a semiempirical basis a micelle structure with explicit molecular regularity, with a parallel correlation of the hydrocarbon chains up to a width equal to the chain length, with preferential orthogonal assembly of these molecular blocks. Its architecture is based on the features of the liquid-crystalline lipid bilayer modified in a minimal way to account for the well established features of the droplet-like micelle. The implications of the model in its present qualitative issue are not in contradiction with the experimental data. As the model is more detailed than the chaotic model, more detailed experiments are required in order to prove or to disprove it. A quantitative theoretical treatment of the model is to be attained only by a distinct extension of approaches presently available. The concept of block assembly is particularly useful with respect to non-bilayer defects in lipid membranes, leading to the process of membrane fusion.

1. INTRODUCTION

The Surfactant-Block Model of micelle structure was proposed in 1981 [1-3]. It claims that the Droplet Model of micelle structure[4-6] is to be interpreted in terms of a molecular packing following the main features of surfactant packing in liquid crystalline bimolecular layers. This packing rule of restricted parallel correlation leads to a certain assignment of spatial distribution functions for the surfactant segments. Experimental data which would disprove the model in its present state have not been reported. As some misinterpretations of the model, however, became apparent[7,8] the nature of the Surfactant-Block Model is clarified here as well as its implications with respect to various experimental data.

2. DROPLET AND BILAYER

Two micelle models have competed in classical times of micelle research:
- The Bilayer Model (Figure 1a) was designed intuitively on the basis of the intrinsic shape of the surfactant molecules[9-11]. It was considered - unfortunately - to be a solid.[12].
- The Droplet Model (Figure 1b) was designed on the basis of energetic reasoning about the stability of micelles.[4-6,13,14] It was considered to be comparable to a bulk liquid.[12]

The Bilayer Model is a genuine molecular model. The Droplet Model becomes a molecular model only when the nature of the liquid is specified, i.e., when its molecular spatial distribution functions and its pair-correlation functions are pointed out.[15]

Experimental information on molecular structure was not available for many years.[16] Consequently the droplet won the competition as it referred exclusively to the thermodynamical data available. The micelle community got accustomed to substituting the repressed molecular structure by fancy cartoons.[4,14,17] The bilayer was kept alive only by a few chemists who could not forget about the natural shape of the assembling molecules.[18]

Two developments changed the view on micelle structure in recent years:
- A wide range of experimental techniques was applied to micelles, techniques which refer directly to structural issues such as the conformation and distribution of hydrocarbon and the nature and extent of hydrocarbon/water contact.[19-26]
- The infinite planar lipid bilayer was characterized so extensively both experimentally and theoretically that a generally accepted view was attained.[27-29] So the simplest assembly of amphiphiles, the infinite planar bilayer, is available as a reference.

Examining the novel structural data on micelles and considering the features of the infinite lipid bilayer, it appeared that
most of the data could be rationalized consistently by the Bilayer
Model of micelle structure, if the cylindrical bilayer fragment
(Figure 1a) is considered to be punched out from the infinite
planar bilayer it its liquid-crystalline state.[1-3]

So two perspicuous, though basically different models, do
account together for all features of the actual micelle: The Droplet Model describes the energetic features, the Bilayer Model describes the structural features. Droplet and Bilayer are antithetic
though complementary models of micelle structure.

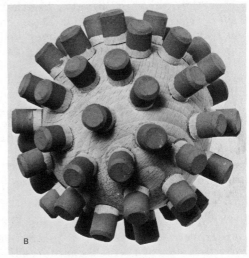

Figure 1. Liquid-space filling models of two classical micelle
structures: a. The Bilayer Model and b. The Droplet Model.
Note the identical hydrocarbon/water interface of the cylindrical
bilayer-fragment and of the packing-free droplet. The models refer
to micelles of dodecylsulfate or dedecyltrimethylammonium of aggregation number 64. The dark zones indicate the polar heads. The two
antithetical models form together a complementary description of
micelle structure.

3. THE DESIGN OF THE SURFACTANT-BLOCK MODEL

It is the conventional paradigm for designing micelles to take the energetic model of the complementary pair, the Droplet, and to cast this spherical mould with surfactant such that a molecular model is gained.[23,24,30,31] This procedure is not adequate: An external shape of the isometric micelle is not established down to sufficient resolution.[26,32] The sphere being thermodynamically sufficient[6,13] must be taken as a structural constraint only on average. Its molecular extrapolation must not be trusted with the same confidence. Micelle assembly is self-assembly of free molecules. The imposition of an external constraint is convenient but arbitrary.

I propose to start micelle design with the structural model of the complementary pair, i.e., with the Bilayer Model. The segmental distribution function and the segmental pair-correlation functions of the bilayer are to be taken as a reference to construct the unknown assembly of a droplet-like micelle. The goal is to transform the bilayer-fragment such that the energetic features of a droplet are attained, retaining the local structure of hydrocarbon and the nature of hydrocarbon/water contact as far as possible.

The hydrocarbon/water contact area is identical for the bilayer-fragment and for the ideal compact droplet (Figure 1).[1] The energetic hurdle of the bilayer model is its extreme headgroup repulsion[33,34] and its tendency to unlimited growth.[12]

A drop of energy is attained by rotating a sheet of molecules across the cylindrical fragment by a right angle. This operation leads to stabilization just if the hydrocarbon fragment is isometric, i.e., if the diameter equals twice the hydrocarbon chain length. This move is the magic motif of Rubik's cube.[35] Here, however, the motif cannot be reiterated. Molecules are impenetrable. A partitioning into smaller blocks is required. The range of parallel correlation is to be restricted to one chain length.

This is the rule of the Surfactant-Block Model of micelle structure: An assembly of molecules with parallel unstaggered correlation up to a width corresponding to the chain length with orthogonal correlation of the assembly of these blocks. Geometry and energetics resemble on the average closely to the thermodynamical model of the droplet. Hydrocarbon conformation and hydrocarbon/water contact of the structural model, the bilayer fragment, are retained.

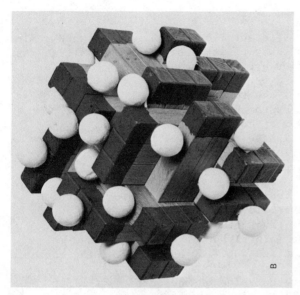

Figure 2. Liquid-space-filling representation of the Surfactant-Block Model referring to a micelle of sodium-dodecyl-sulfate or dodecyl-trimethylammonium-chloride. For octanoate, a parallel correlated block consists of two molecules; for cetyltrimethylammonium-chloride of four molecules; as the width of one block is determined by the length of the chains. In the 64-aggregate, eight edge positions are unoccupied. The wooden models represent the architecture of the micelle with parallel correlation. The wooden sticks do not represent space-filling models with crystal van der Waals radii with all-trans configuration of the hydrocarbon chains. They represent the average volume occupied by a surfactant molecule in a liquid-crystalline bilayer, accounting for gauche conformations and voids. The relation to the usual CPK-models is indicated by the two molecules inserted in the front of the object in Figure 2a. The headgroup repulsion within blocks is lowered by the association of counterions which neutralize one to two charges per block of three as indicated by the white spheres in Figure 2b.

The procedure of designing the Surfactant-Block Model out of the Bilayer does not mean, of course, that actual micellar assembly follows this path. The path of actual self-assembly from monomers to submicellar aggregates to isometric micelles and beyond is described elsewhere.[3]

The features of the Surfactant-Block Model are assigned semi-empirically guided by the pathway of its design: The known features of the infinite planar bilayer are modulated by the two transformations applied:
- The first transformation is the restriction to an isometric cylinder. It creates a large hydrocarbon/water contact and it restricts the lateral freedom of configurational change.
- The second transformation is the rotation of blocks. It changes intermolecular chain interactions and modifies headgroup interactions providing the energetic basis for stability of limited aggregates.

Unchanged by the transformations are the main dynamical features of the infinite planar bilayer i.e. the rotation around C-C bonds with the formation of gauche-trans-gauche kinks, the rotational diffusion of the chains, the lateral hopping of the aligned molecules, little radial fluctuations and little curling of the chains.

A wood-paint-glue representation of the Surfactant-Block Model is shown in Figure 2. The sticks do not represent all-trans configurations of the paraffin chains, as clearly stated in previous papers.[1-3] So the model does not attribute a solid state to the micelle as imputed.[7,8] The sticks represent the average volume occupied by a surfactant in a liquid-crystalline bilayer. The width of the sticks is thus distinctly larger than the width of usual space-filling models with van der Waals radii as taken from crystals. The packing of liquid-space-filling models accounts for kink formation and voids, which is difficult to achieve with a packing of solid-space-filling models.[23,36] The relation of the two types of models is indicated in Figure 2a by two Corey-Pauling-Koltun-models (CPK) incorporated into the wood-paint-glue object at an appropriate scale.

The energetic features of the Surfactant-Block Model resemble those which have to be assigned to a strict (packing-free) Droplet Model.[6,13]

- The hydrocarbon/water contact for a cuboid hydrocarbon body is only a few percent larger than for a smooth spherical body of identical volume. On the other hand packing of a spherical mould may be difficult to achieve without roughness such that an enhanced hydrocarbon/water contact is obtained in that case too.

- The headgroup repulsion energy seems to be too high, as the compact headgroup plane of the bilayer is not completely partitioned. However, on one hand, the loose association of counterions to charged micelles[37,38] compensates one to two of the three charges within a block in SDS as indicated by the white spheres in Figure 2b. On the other hand, for any real cast of a spherical mould random accumulations of headgroups do occur also in order to avoid extreme conformational constraints.[23,30] Quantitative estimation of headgroup repulsion has not been included in any theory available.[6,13,14,30,31] Estimation of negligible head-head interaction based on negligible pK-shifts of amines[39] is unsound, as it has been shown that for amines effects of electrical potential are compensated by the reduced local polarity in micelle surfaces.[40]
- Packing energy is composed of intramolecular conformation energy, of intermolecular van der Waals energy and of the configurational entropy.
  The Surfactant-Block Packing may accomodate a large number of gauche conformations but they are not a requirement whereas a cast of a spherical micelle requires a large number of gauche and presumably cis conformations.[30]
  Van der Waals attraction of poly-methylene chains induces parallel correlation of chains.[41]
  The packing entropy of the Surfactant-Block Model should be relatively high in contrast to the first impression of a high "order": Configurational correlation is an intrinsic feature of any micelle packing. Parallel correlation implies a relative independent rearrangement of each molecule on its lattice site, whereas any cast of a spherical mould implies a less visible though highly cooperative configurational interaction of the molecules[30], which of course is not accounted for in a mean-field approach of packing.[31]

To emphasize: The Surfactant-Block Model is a semiempirical model at the present moment. It is not a model where molecular features are computed by statistical mechanics on the basis of intra- and intermolecular potentials with packing constraints as imposed by an internal lattice or an external shape.

At the present moment the Surfactant-Block Model interprets the micelle through the infinite planar bilayer as modified by the minimal molecular transformation accounting for limited micelle size.

The Surfactant-Block Model comprises the features of the two antithetical classical models: of McBain's Bilayer and of Hartley's Droplet. Their complementarity is resolved into a synthesis.

4. IMPLICATIONS OF THE MODEL

A model cannot be proven. Every model is a good model when any attempt to falsify it is unsuccessful. The Surfactant-Block Model provides a long list of implications. These may be compared with experimental data. The implications of the model are divided into four sets referring to (i) general features, (ii) the nature of the hydrocarbon, (iii) the hydrocarbon/water contact and (iv) the micelle surface. Every statement is accompanied with a comment.

General Implications

Surfactant self-assembly is driven by minimization of hydrocarbon/water contact with the constraint of low headgroup repulsion and negligible conformation energy.

These are the implications of Hartley's droplet as supported by cmc-data.[6,13]

The ratio of chain length to chain width determines the regularity of aggregation numbers.

Typical numbers are 32,72 and 128 for octanoate, SDS and CTAC.[1,3]

The average radius of the hydrocarbon may exceed the chain length without arbitrary vacuoles in the center or arbitrary ellipsoidal deformations.

The cuboid isometric shape is a natural consequence of packing. It resolves a classical antinomy concerning spherical shape and aggregation number. [6,42,43]

The equilibrium shape is isometric but non-spherical.

Available diffraction data using X-rays[32] and neutrons[44] are of insufficient resolution. Recent data are not explained in terms of a compact sphere.[26]

Transition from an isometric micelle to a rod-like micelle is allowed. Due to the distinct packing of the saturated cube it may occur as a sharp transition.

Models of the rod-like packing have been described.[3] The transition is blocked by high headgroup repulsion.[13,14]

The Hydrocarbon

The hydrocarbon chains form a compact body. Water penetrates in the sense of usual molecular solubility.

This is the classical droplet concept. It is consistent with the low hydration numbers.[45] It is in contrast to the idea of aqueous fjords or the picture of a loose porous cluster. 20,46,47,48

The local motion of the chains in the assembly is diffusive not harmonic as in a solid.

The time constant of local motion lies certainly in the picosecond range[49], as even in compact proteins diffusive motion proceeds on this time scale.[50]

Lateral motion of monomers occurs in the micelle by lateral diffusion within parallel correlated zones.

This motion is related to the lateral diffusion in bilayers NMR-data have been reported.[49]

The chains are not in all-trans configuration though they are in a fairly extended state as perturbed mainly by g-t-g kinks.

Various authors have reported chain extension in micelle formation.19,51,52 Absolute numbers of gauge rotamers are not available.

The local order of the chains decreases towards the chain ends.

This assignment is based on the analogy to the bilayer, as the geometrical reason for the order profile[53] holds here as well. The limited size of parallel correlated blocks and the orthogonal assembly of the blocks may modify this analogy, as has to be studied by simulations.

Large deflections of the chain director within a block do occur rarely.

This motion is allowed in infinite bilayers.[27] It should be hindered in any micelle packing which goes beyond a mean-field treatment.[31]

The total directional disorder of a chain segment is composed of the local disorder of the chain on a lattice site and of the disorder of the overall chain director due the orthogonal assembly of the blocks.

In the comparison of measured order parameters with models it should be clarified whether both types of disorder are probed. This is important for a comparison of "disorder" in micelles and in planar bilayers.

Curling of molecules is not forbidden, but is assumed to be insignificant for the average micelle packing.

Rare curling may be important, however, for certain chemical reactions.

The packing allows a first order calculation of segmental distribution functions.

Comparison with recent data[26] is under investigation. Definite evaluation requires involved simulation as the model does not allow a mean-field approach[31] or the imposition of a simple lattice.[30,53]

Even in the liquid-crystalline state of the assembly, the lattice may form pockets of defined size.

Stretching and jolting of short and long chains in mixed micelles[54] may be rationalized.

About fifty percent of molecules are positioned on tangential sites of the assembly.

Paramagnetic quenching profiles [24,25] may be described in detail.

The chains are assembled locally on an orthogonal – not a radial – lattice.

The order parameter in micelles decreases towards the chain end[21,49]. A radial lattice implies an increase[30] whereas an orthogonal lattice as in bilayers implies a decrease.[53] The data[21,49] together with the calculations[30,53] suggest that micelles could be described better by a bilayer-like lattice.

The Hydrocarbon/Water Contact

The average water contact of hydrocarbon is reduced in micellisation by about 80 % for the stem, more for the terminal methyl – due to its peculiar shape -, less for the few methylenes behind the headgroup due to adjacent empty sites. The α-methylene may stay wetted permanently.

The drastic decrease of wetted area is a typical feature of the droplet model.[6,13] The F[19] NMR data reported[7] are not in disagreement with the Surfactant-Block Model as claimed by the authors.

The hydrocarbon surface between the headgroups corresponds to two fully wetted anonymous methylene groups per monomer. About one anonymous methylene is wetted in addition due to vacancies being present on edge sites. Without considering the

Even in the compact smooth droplet the hydrocarbon/water contact between the headgroups corresponds to two (anonymous) methylene groups[5,6]. So even in the case of the ideal classical Hartley model three methylenes are wetted at least α-methylene, two to three methylenes are being wetted on the average.

considering the α-methylene to be wetted permanently.[6],[13] So with some roughness the sphere- irrespective of packing- and the Surfactant-Block Model imply an identical wetting, being in agreement with data on specific volume[55] and proton T_1 measurements[56] as well with hydration numbers.[45] The extreme low wetting reported recently[22] is difficult to rationalize even by a compact sphere if an appropriate specific volume is considered.

By lateral diffusion fast exchange between radial and tangential sites occurs, i.e., on the order of 1 ns all molecules become laterally wetted.

For all probe molecules a distribution between radial and tangential sites has to be considered. Even for rather hydrophobic probes, contact with water or ions occurs within the time scales of fluorescence and NMR.[20],[47],[48],[57-59]

The Surface

The surface provides "fatty" patches as binding sites for aromatic dyes as illustrated previously.[3]

Many of the conventional probes, e.g., pyrene, may be considered to be adsorbed on the surface without perturbing much the micelle structure.[19],[48],[59],[60] Large chromophores such as porphyrins may lead to a local rearrangement of the block-assembly.

Little roughness occurs by radial fluctuations of the chains. Discrete roughness is produced by empty sites.

Calculations reported on fluctuation probability[61] are based on a continous structureless core. Considering the nature of the radial and tangential sites the probability of a fluctuation by the first methylene is about 3 kT. Paramagnetic shielding data indicate low local radial fluctuations.[24]

The local accumulations of charged headgroups induce binding of counterions.

Counterions bound to micelles [37,38] compensate one to two charges per block by association to these blocks. The headgroup accumulations in the block-assembly could even be the cause for the low dissociation degree of micelles which is not understood up to now.

Large headgroups are accomodated by chain bends near the heads.[1]

For extreme sizes of headgroups the block-concept must break down. It is not claimed that parallel correlation dominates all micelles.

Short amphiphiles may bind to fatty patches or they may substitute surfactants on empty sites, thereby lowering the aggregation number.

For a coiled cluster-like packing, stabilization of micelles would be expected by filling unfavourable gaps. Experimental data[62] indicate decrease of aggregation number for low amounts of alcohol solubilizate.

Monomer association to micelles may occur onto open tangential sites by a diffusion - controlled process.

This is in agreement with well known kinetic data.[63]

All the implications of the Surfactant-Block Model mentioned above originate from a consideration of the known features of the infinite planar bilayer, modifying them by the two operations of size limitation as the restriction to an isometric cylinder and the rotation of blocks. Some assignments are definite, some have to be confirmed by quantitative evaluation or by a statistical mechanical treatment of the model. It is apparent that no experimental data are available which would falsify the model in its present qualitative issue.

To be clear: Within the accuracy of the experiments many of the data may not falsify other micelle models either. Most of the data available are not selective. It is doubtful whether there exists any single experiment which disproves one of the competing models, as the compact coiled model in its various issues[2,3,30,31] the porous cluster[36,46] and the Surfactant-Block Model[1,3]. Micelle characterization must be a multimethod characterization: The whole set of data has to be compared with the whole set of predictions

of the models with some intuitive weigthing of the various assignments of experiments and models.

The point to be made here is not to discuss an incorrectness
of other models but to emphasize the amazing ease of explanation
of all experimental data on micelles by the Surfactant-Block Model.
Considering the extreme simplicity and transparency of the model
this finding is surprising indeed.

5. NON-BILAYER ARCHITECTURE

The crucial issue of the Surfactant-Block Model of micelle
structure is the idea that narrowest convex forms are assembled
not by appropriate coiling of amphiphiles, but by the assembly of
rotated blocks of parallel correlated amphiphiles. The average occupation of wedge-like spaces by surfactants is not due to actual
wedge-shaped molecules, but to block-assembly. The classification
of amphiphiles into wedge-type, inverted-wedge type and cylinder
like[64] remains valid and useful. Its molecular basis with respect
to the formation of membrane defects[65] and membrane fusion may be
understood, however, consistently by the Block-Architecture Principle for any convex and concave amphiphile assembly. This issue
is described elsewhere.[3,66]

6. SUMMARY

The difficulty in accepting the unconventional Surfactant-
Block Model of micelle structure may lie in a certain tradition.
The micelle community is accustomed to considering the micelle as a
structureless spherical chaos. Hartley described micelle as liquids[4,12] and liquids were considered to be chaotic as compared to
solids. However, within the last ten years all physics of liquids
has endeavoured to introduce local structure into bulk liquids
within the range of a few molecular diameters, by measuring and
calculating pair-correlation functions.[15] Why should this approach stop just in the particular case where the size of the presumed fluid approaches the range of a few molecular diameters?
The novel model proposes a certain type of segmental distribution
function and a certain type of segmental pair-correlation function
connected with some molecular dynamics. The Surfactant-Block Model
does not attack the Droplet Model, it fills the Droplet with adequate molecular structure.

The message to the experimentalist is to measure correlation
functions which may be obtained by diffraction techniques or by
spectroscopic methods. The message to the theoretician is to improve approaches which are based on the mean-field scheme[31] -

which implies chaos from the very beginning - or to improve approaches which impose a certain arbitrary lattice from the very beginning[30], i.e. to simulate true self-assembly of surfactants. The Surfactant-Block Model may help to avoid unsound constraints in such future theories.

REFERENCES

1. P. Fromherz, Chem. Phys. Lett., 77, 460 (1981)
2. P. Fromherz, Nachr. Chem. Techn. Lab., 29, 537 (1981)
3. P. Fromherz, Ber. Bunsenges. Phys. Chem., 85, 891 (1981)
4. G.S. Hartley, "Aqueous Solutions of Paraffinic Chain Salts", Hermann, Paris, 1936
5. D. Stigter, J. Phys. Chem., 68, 3603 (1964)
6. C. Tanford, J. Phys. Chem., 76, 3020 (1972)
7. J. Ulmius and B. Lindman, J. Phys. Chem., 85, 4131 (1981)
8. B. Lindman, M.C. Puyal, N. Kamenka, B. Brun and G. Gunnarsson, J. Phys. Chem., 86, 1702 (1982)
9. J.W. McBain, J. Chem. Educ., 6, 2121 (1929)
10. W. Philippoff, Kolloid-Z., 96, 255 (1941)
11. J.W. McBain, "Colloid Science", p. 255, Heath, Boston 1950
12. G.S. Hartley, Quart. Rev. Chem. Soc., 2, 152 (1948)
13. C. Tanford, J. Phys. Chem., 78, 2469 (1974)
14. I.N. Israelachvili, D.J. Mitchell, and B.W. Nisham, J. Chem Soc., Faradaday Trans. II, 7, 1525 (1976)
15. J.P. Hansen and I.R. McDonald, "Theory of Simple Liquids", Academic Press, London, 1976
16. P. Mukerjee, Adv. Colloid Sci., 1, 243 (1967)
17. B. Lindman, H. Wennerström, H. Gustavsson, N. Kamenka and B. Brun, Pure Appl. Chem., 52, 1307 (1980)
18. K. Shinoda, J. Phys. Chem., 85, 3311 (1981)
19. K. Kalyanasundaram and J.K. Thomas, J. Phys. Chem., 80, 1462 (1976)
20. F.M. Menger, J.M. Jerkunica and J.C. Johnston, J. Amer. Chem. Soc., 100, 4676 (1978)
21. U. Henriksson and L. Ödberg, Colloid Polym. Sci., 254, 35 (1976)
22. B. Halle and G. Carlström, J. Phys. Chem., 85, 2142 (1981)
23. H. Wennerström and B. Lindman, Phys. Reports, 52, 1 (1979)
24. B. Cabane, J. Physique, 42, 847 (1981)
25. Th. Zemb and C. Chachaty, Chem. Phys. Lett., 88, 68 (1982)
26. B. Cabane, R. Duplessix and T. Zemb, These proceedings
27. J. Seelig and A. Seelig, Quart. Rev. Biophys., 13, 19 (1980)
28. J.F. Nagle, Ann. Rev. Phys. Chem., 31, 157 (1980)
29. S. Marcelja, Biochim. Biophys. Acta., 367, 165 (1974)
30. K.A. Dill and P.J. Flory, Proc. Natl. Acad. Sci. USA, 78, 676 (1981)
31. D.W.R. Gruen, J. Coll. Interface Sci., 84, 281 (1981)

32. F. Reiss-Husson and V. Luzzati, J. Phys. Chem., 68, 3504 (1964)
33. M.E. Hobbs, J. Phys. Chem., 55, 675 (1951)
34. J.T.G. Overbeek and D. Stigter, Rec. Trav. Chim., 75, 1263 · (1963)
35. D.R. Hofstadter, Sci. Amer., 244, (3), 14 (1981)
36. K.A. Zachariasse, B. Kozankiewicz and W. Kühnle, These proceedings
37. D. Stigter, R.J. Williams and K.J. Mysels, J. Phys. Chem., 59, 330 (1959)
38. H. Gustavsson and B. Lindman, J. Amer. Chem. Soc., 100, 4647 (1978)
39. F.M. Menger, These proceedings
40. M.S. Fernandez, P. Fromherz, J. Phys. Chem., 81, 1755 (1977)
41. W. Pechhold, private communication 1981
42. H.V. Tartar, J. Phys. Chem., 59, 1195 (1955)
43. H. Schott, J. Pharm. Sci., 60, 1594 (1971)
44. J.B. Hayter, Ber. Bunsenges. Phys. Chem., 85, 887 (1981)
45. H. Wennerström and B. Lindman, J. Phys. Chem., 83, 2931 (1979)
46. F.M. Menger, Accounts Chem. Res., 12, 111 (1979)
47. F.M. Menger, H. Yoshinaga, K.S. Venkatasubban and A.R. Das, J. Org. Chem., 46, 415 (1981)
48. D.G. Whitten, I.C. Russell, T.K. Foreman, R. H. Schmehl, J. Bonilha, A.M. Braun and W. Sobol, J. Am. Chem. Soc., in press
49. T. Ahlnäs, O. Söderman, H. Walderhaug and B. Lindman, These proceedings
50. M. Karplus and J.A. McCammon, CRC Critical Reviews in Biochemistry, 9, 293 (1981)
51. B.O. Persson, T. Drakenberg and B. Lindman, J. Phys. Chem., 80, 2124 (1976)
52. J.B. Rosenholm, K. Larsson and N. Dinh-Nguyen, Colloid Polym. Sci., 255, 1098 (1977)
53. K.A. Dill and P.J. Flory, Proc. Natl. Acad. Sci. USA, 77, 3115 (1980)
54. J. Ulmius, B. Lindman, G. Lindblom and T. Drakenberg, J. Colloid. Interface Sci., 65, 88 (1978)
55. J. Clifford, Trans. Faraday Soc., 61, 1276 (1965)
56. J.M. Corkhill, J.F. Goodman and T. Walker, Trans. Faraday Soc., 63, 768 (1967)
57. N. Mueller and R.H. Birkhahn, J. Phys. Chem., 71, 957 (1967)
58. P. de Mayo, J. Miranda and J.B. Stothers, Tetrahedron Lett., 22, 509 (1981)
59. B. Kozankiewicz, W. Kühnle, P. Nguyen Van and K.A. Zachariasse, in "Fast reactions in Solution Discussion Group Meeting", W. Knoche, Editor, p. 119, The Chemical Society, London and the Max-Planck-Gesellschaft, Göttingen 1980
60. W. Reed, M.J. Politi and J.H. Fendler, J. Amer. Chem. Soc., 103, 4591 (1981)
61. G.E.A. Aniansson, J. Phys. Chem., 82, 2805 (1978)

62. R. Zana, S. Yiv, C. Straziella and P. Lianos, J. Colloid.
     Interface Sci., 80, 208 (1981)
63. G.E.A. Aniansson, S.N. Wall, M. Almgren, H. Hoffmann,
     I. Kielmann, W. Ulbricht, R. Zana, J. Lang and C. Tondre,
     J. Phys. Chem., 80, 905 (1976)
64. D.H. Haydon and J. Taylor, J. Theoret. Biol., 4, 281 (1963)
65. B. de Kruijff, P.R. Cullis and A.J. Verkleij, Trends. Biol.
     Sci., 5, 79 (1980)
66. P. Fromherz, Biophys. Struct. Mech., 7, 297 (1981)

STRUCTURE IN MICELLAR SOLUTIONS: A MONTE CARLO STUDY

P. Linse[*] and B. Jönsson[+]

[*]Division of Physical Chemistry 1
[+]Division of Physical Chemistry 2
Chemical Center, P.O. Box 740
S-220 07 Lund, Sweden

A model system of a solution consisting of small ionic micelles with monovalent or divalent counterions has been investigated. The purposes of this study were to determine: 1) how is the intermicellar structure affected by amphiphile concentration and by the valency of the counterions; 2) how satisfactory is the cell model compared to the isotropic model; and 3) how well does the Poisson-Boltzmann equation describe these systems. The study was done using the Monte Carlo simulation technique and by solving the Poisson-Boltzmann equation for a spherical cell.

INTRODUCTION

Electrostatic interactions have a considerable influence on the structure in a solution of charged aggregates. Two examples are micellar solutions of ionic amphiphiles[1] and colloid dispersions[2]. In these solutions, the pair interactions between different types of charged particles are of different orders of magnitude. This fact makes it fruitful to apply the cell model[3] in order to describe a micellar solution.

In the cell model, the total solution is divided into spherical cells, each one containing one micelle and an amount of electrolyte corresponding to the particular system. By this approximation, the influence of the electrostatic interaction on the micelle-ion and ion-ion correlations is taken into account, while quantities depending on micelle-micelle correlations cannot be treated.

337

    A common way to obtain the distribution of the ions around a
micelle in a cell and other mechanical and thermodynamical quantities is by means of the Poisson-Boltzmann (PB) equation[4-8]. The
essence of the approximations used in deriving the PB equation[9]
are: i) the ions are treated as point charges, and ii) the correlation between the ions is neglected.

    We have obtained exact numerical solutions of an isotropic
model and of the cell model corresponding to the same system by
using a computer based method, the Monte Carlo (MC) simulation
technique[10-11]. The non-liniarized PB equation in the cell model
has also been solved.

    The purpose of this contribution is threefold. First, we
have obtained information on the intermicellar structure from the
isotropic model, which is not available from the cell model. Secondly, the validity of the cell model is tested against the isotropic model and finally, the validity of the PB approximation in
the cell model is checked.

MICELLAR SOLUTION

    The model used mimics two different micellar solutions of so-
dium n-octanoate (NaO), and calcium n-heptylsulfate (CaHS), respectively. The critical micelle concentration (cmc) for NaO and CaHS at
room temperature in the absence of salt is ~0.4M[1,12]. In this model,
the micelle is assumed to have a fixed aggregation number and is
treated as a hard sphere with a uniform surface charge density. Since
we focus on the long-range electrostatic aspects of the micellar solution, the deviation of the micellar geometry from a sphere and a
small polydispersity in the micellar radius are not likely to affect
the results to any great extent.

    The presence of free amphiphile is neglected in the calcula-
tions. We also neglect the increase of the aggregation number at
amphiphile concentration above cmc. These two simplifications balance each other to some extent regarding the effect on the electrostatic repulsion between micelles. Although, a direct comparison
with experimental results must be postponed until the free amphiphiles, variable aggregation number and perhaps, added salt also
are included in the calculations. Thus, we have two different charged
species, a negatively charged micelle and positively charged ions.

    Figure 1 shows schematically the cell and the isotropic model
and that in the latter one the direct micelle-micelle interaction
is explicity treated.

    The interaction between two particles i and j (i and j can
be a micelle or an ion) consists of two parts: 1) An infinitely
repulsive potential at a center-to-center distance, r, closer
than contact, $r<R_i+R_j$; 2) a long range Coulombic potential with
a relative permittivity of 78.3. All calculations are done at a
system temperature of 298 K. Further details of the numerical solutions of the PB equation and the MC simulations can be found in
Ref. 8 and 13, where the models are also described in more detail.

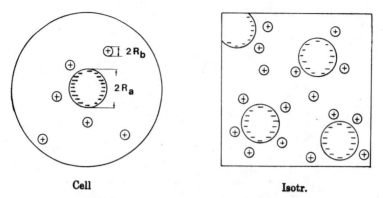

Figure 1. The drawings show the cell model and the isotropic model. The micellar radius, R_a, is 10Å and the ionic radius, R_b, is 1.0Å. The aggregation number of the micelles is assumed to be 12.

MICELLAR STRUCTURE

Concentration Dependence

The micelle-micelle radial distribution function, $g_{MM}(r)$, has been calculated for monovalent counterions at amphiphile concentrations $C_A = 0.025 M$ and $C_A = 0.10 M$. The results are shown in Figure 2. $g_{MM}(r)$ expresses the probability of finding a micelle at a distance r from another micelle[14].

It is seen in Figure 2 that at the low concentration, the micelles are very well separated from each other and it is unlikely that two micellar surfaces come closer to each other than 10-15Å, (r=30-35Å). g_{MM} shows a broad maximum at 85Å, which is close to the shortest distance in a face centered cubic (fcc) lattice, 92.8Å, at this concentration. Thus, the electrostatic repulsive forces keep the micelles apart, but they become uncorrelated already at $r \gtrsim 100$Å.

By increasing the amphiphile concentration, the micelles become more likely to come closer to each other, although the probability of direct contact is very small. The maximum in g_{MM} at the high concentration, ~ 55 Å, is also close to the shortest distance in a fcc lattice, which at this concentration is 58.5Å. The maximum is now more pronounced, which also demonstrates that the micellar solution is more structured at $C_A = 0.10 M$. The structure

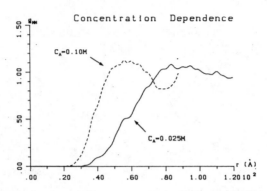

Figure 2. The micelle—micelle radial distribution function with
monovalent counterions at amphiphile concentrations C_A = 0.025 M
and 0.10 M, respectively.

obtained is due to the electrostatic interaction and not to the
hard core of the micelles. The volume fraction of micelles is
only 2.1 percent at C_A = 0.10 M.

The small fluctuations in the curves are not significant and
are rather due to numerical uncertainties.

Counterion Dependence

Figure 3 shows how the valency of the counterions influences
the micelle—micelle radial distribution function at C_A = 0.10 M.
In the case of divalent counterions, the micelles may come into
surface contact with each other with a high probability even if
the micellar charge is −12 e . Of course, the explanation for this
is the higher degree of accumulated counterions close to the micellar surfaces (see below), which screen the otherwise strong
micelle—micelle repulsive force. After the steep rise at r = 20Å,
the $g_{MM}(r)$ is approximately constant at r $\gtrsim$ 35 Å and at larger distances, the micelles become uncorrelated.

Thus the degree of structure in these isotropic model systems
varies to a high degree with the valency of the counterions.

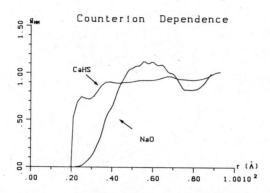

Figure 3.   The micelle-micelle radial distribution function at
C_A = 0.10 M with mono- and divalent counterions.

THE VALIDITY OF THE CELL MODEL AND
THE POISSON-BOLTZMANN EQUATION

        In order to estimate the validity of the cell model and also
the PB equation for systems of small micelles, we have chosen to
compare the electrostatic energy, E, the degree of accumulated
counterions, β, and the osmotic coefficient, ϕ. These three quantities are calculated by MC simulation in the isotropic model (MC
Isotr.), by MC simulation in the cell model (MC Cell), and by the
PB equation in the cell model (PB). The comparison is made for
both monovalent and divalent counterions over a range of amphiphile
concentrations.
        The electrostatic energy is the excess energy arising from
the interaction among the charged particles. The degree of accumulated counterions is the ratio of the charge of all the counterions within 3Å from the micellar surface and the micellar
charge[8]. The osmotic coefficient is calculated as the derivative
of the free energy with respect to the volume[13,15]. Figures 4a
and 4b show the electrostatic energy as a function of the amphiphile concentration for monovalent and divalent counterions, respectively. The isotropic and the cell model give almost identical
results and the PB equation overestimates the electrostatic energy

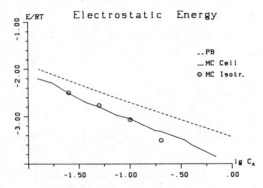

Figure 4a. The electrostatic energy in RT per mole amphiphile
calculated by three different ways (see text) as a function of
$\log_{10} C_A$ for the case of monovalent counterions.

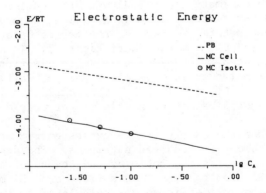

Figure 4b. Same as Figure 4a, but for the case of divalent counterions.

by ~ 0.4 RT/(mole amph.) and ~ 1.0 RT/(mole amph.), respectively.

The distribution of counterions close to the charged aggregate is of interest in many spectroscopic applications[16-18]. Here, the isotropic and the cell model agree both for monovalent and divalent counterions (Figures 5a and 5b). The PB equation is satisfactory in the case of monovalent counterions, but underestimates the degree of accumulated divalent counterions.

As Figure 6a and 6b show, the isotropic and the cell model give different osmotic coefficients and the difference is considerable for divalent counterions.

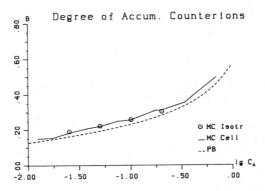

Figure 5a. The degree of accumulated counterions (see text for definition), as a function of $\log_{10} C_A$ with monovalent counterions.

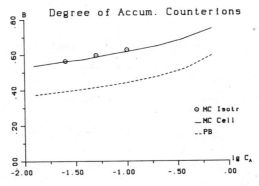

Figure 5b. Same as Figure 5a, but for the case of divalent counterions.

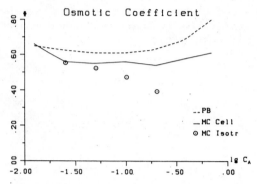

Figure 6a. The osmotic coefficient as a function of $\log_{10} C_A$ for the case of monovalent counterions.

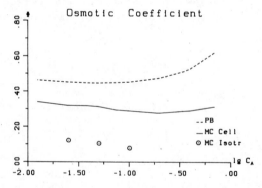

Figure 6b. Same as Figure 6a, but for the case of divalent counterions.

The results for the PB equation also deviate from the exact solution of the cell model and the deviation is larger for divalent ions than for monovalent ones. The PB equation overestimates the osmotic coefficient by 10-20 percent for monovalent and 50-200 percent for divalent counterions.

To summarize, the cell model seems to be satisfactory for the electrostatic energy and the counterion distribution, but overestimates the osmotic coefficient, especially for divalent counterions. The PB equation seems to be a good approximation for monovalent counterions, but it breaks down for divalent ones.

The breakdown of the PB equation can be illustrated by examining the two approximations introduced in the cell model[9] to obtain the PB equation. In models of micellar systems, it appears that the approximation of ions as point charges, when considering interaction among them, is not a serious one[8,19]. The other and more important approximation in deriving the PB equation is thus the neglect of the ion-ion correlation. Since one allows for the correlation between the ions in the MC simulations, the ions can accumulate closer to the micellar surface than they do according to a mean field theory such as the PB equation. The reason is that the repulsion between the ions is lowered through the correlations. Since the direct repulsion between the ions increases with the square of the valency, the correlation effect becomes more important at higher valencies.

There is a qualitative consistency in the deviations of β and ϕ between the results of the PB equation and the MC simulations in the cell model. By introducing the ion-ion correlation, the higher degree of accumulated counterions close to the micellar surface must be balanced by a lower ion concentration close to the cell boundary, i.e. a lower osmotic coefficient will be obtained.

CONCLUSIONS

By performing MC simulations for the isotropic model, it is possible to obtain structural information about the micellar solution and to test the validity of the cell model applied to systems of small micelles.

The valency of the counterions affects the micellar structure to a high degree in these model systems. By including free amphiphiles in the simulations, it will be possible to make direct comparisons of intermicellar structure using X-ray and neutron scattering data. Another application of structural information is to explain possible fluorescence quenching mechanisms[20].

In the second part of this work, it is shown that the PB equation is a satisfactory approximation *only* for monovalent counterions in this model system. The cell model introduced to facilitate the use of the PB equation does not affect the electrostatic energy or the degree of accumulated counterions, but does increase the osmotic coefficient in the concentration region investigated.

REFERENCES

1. H. Wennerström and B. Lindman, Phys. Reports, $\underline{52}$, 1 (1979).
2. A.L. Leob, J. Th. G. Overbeek and P.H. Wiersema,"The Electrical Double Layer around a Spherical Colloid Particle", M.I.T. Press, Cambridge, Mass., 1961.
3. T.L. Hill, "Statistical Mechanics", Addison-Wesley, Reading, Mass., 1960.
4. G.M. Bell and A.J. Dunning, Trans. Faraday Soc., 66, 500 (1970).
5. M. Mille and G.J. Vanderkooi, J. Colloid Interface Sci., $\underline{59}$, 211 (1977).
6. S.W. Westra and J.C. Leyte, Ber. Bunsenges. Phys. Chem., $\underline{83}$, 672 (1979).
7. G. Gunnarson, B. Jönsson and H. Wennerström, J. Phys. Chem., $\underline{84}$, 3114 (1980).
8. P. Linse, G. Gunnarsson and B. Jönsson, J. Phys. Chem., $\underline{86}$, 413 (1982).
9. B. Jönsson, H. Wennerström and B. Halle, J. Phys. Chem., $\underline{84}$, 2179 (1980).
10. N.A. Metropolis, A.W. Rosenbluth, M.N. Rosenbluth, A.H. Teller and E. Teller, J. Chem. Phys., $\underline{21}$, 1087 (1953).
11. W.W. Wood in "Physics of Simple Liquids", H.N. Temperley, J.S. Rowlinson, G.S. Ruchbrooke, Eds., Chapter 5, North Holland Publishing Co., Amsterdam, 1968.
12. B. Lindman and H. Wennerström, Top. Curr. Chem., $\underline{87}$, 1 (1980).
13. P. Linse and B. Jönsson, J. Chem. Phys. (to be published).
14. D.A. McQuarrie, "Statistical Mechanics", Chapter 13, Harper and Row, New York, 1973.
15. H. Wennerström, B. Jönsson and P. Linse, J. Chem. Phys., $\underline{76}$. 4665 (1982).
16. J.J. van der Klink, L.H. Zuiderweg and J.C. Leyte, J. Chem. Phys., $\underline{60}$, 2391 (1974).
17. C.F. Anderson, M.T. Record, Jr. and P.A. Hart, Biophys. Chem. 7, 301 (1978).
18. J.K. Thomas, F. Grieser and M. Wong, Ber. Bunsenges. Phys. Chem., $\underline{82}$, 937 (1978).
19. C.W. Outhwaite, L.B. Bhuiyan and S. Levine, J.C.S. Faraday II, 76, 1388 (1980).
20. M. Almgren, G. Gunnarsson and P. Linse, Chem. Phys. Lett., $\underline{85}$, 451 (1982).

MULTI-METHOD CHARACTERIZATION OF MICELLES

Fredric M. Menger

Department of Chemistry
Emory University
Atlanta, Georgia 30322

The "Menger micelle" possesses striking predictive and rationalizing power. Many examples are presented.

Models are devised when reality is too complicated to analyze. A model can be extremely useful as long as one never forgets that reality is being simulated, not identified. Good models permit us to rationalize observed phenomena and to predict phenomena not yet observed. In the case of micelles, models are required because micelles are elusive, transient, fluctuating, and disordered assemblages. Three micelle models have drawn particular attention, namely those of Hartley, Menger, and Fromherz. Their properties are discussed briefly below.

A.   Hartley

The classical Hartley micelle, accepted for decades[1], features a hydrocarbon-like core surrounded by a polar layer of head groups, counterions, and water. The model is often referred to as an "oil droplet with an ionic coat" or as a "two-phase" system. Its main characteristics include (a) a radial organization of surfactant chains; (b) a smooth micelle surface; (c) little water penetration into the hydrocarbon core; (d) minimal hydrocarbon-water contact at the micelle surface. A schematic representation depicts these properties:

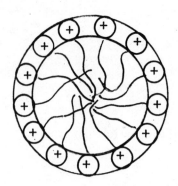

Hartley, it turns out, never meant others to accept literally what has become known as the "Hartley micelle". He wrote[1] in 1936:

"The symmetrical asterisk form...has no physical basis and is drawn for no other reason than that the human mind is an organizing instrument and finds unorganized processes uncongenial. It is easier to speak a sentence than a jumble of words."

Yet, for the most part, the world (with the Lund group being a notable exception) has regarded the "oil droplet in an ionic coat" as central dogma in micellar chemistry. For example, solubilization (the most important micellar property) is traditionally viewed as a type of partitioning into a hydrocarbon phase. We will see below that this is misleading if not outright incorrect.

B.   Menger

    A 1978 Accounts of Chemical Research article[2] discusses the failings of the Hartley model and proposes another which I will henceforth call the "Menger model". In essence, the article shows (see Figure 5 in ref. 2) that it is impossible to occupy a sphere with 58 fully extended surfactant molecules in a more-or-less radial arrangement without leaving huge cavities. If the chains are bent to help fill the cavities with hydrocarbon units (see Figure 6 in ref. 2), then one creates a "Menger micelle" with the following features:

    a.   a disorganized collection of surfactant molecules many of which are severely bent.

    b.   a rough micelle surface with small water-filled pockets.

    c.   the presence of hydrocarbon chains ("fatty patches") at the micelle surface in direct contact with water.

    d.   a central hydrocarbon core (as in the Hartley model) but comprising a smaller percentage of the total micellar volume.

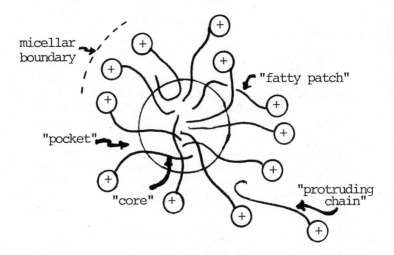

The dynamic nature of the micelle model should be stressed.
At one instant a chain could be bent, or a pocket filled with
water, or a terminal methyl group exposed to the micelle surface.
An instant later the chain could be fully extended, the cavity no
longer present, and the terminal methyl group buried in the core.
It is these fluctuations which make the micelle structure so elusive and difficult to define. On a time-averaged basis, however,
the differences between the Hartley and Menger micelles are substantial. As stated above, a water-insoluble guest molecule
would enter a Hartley micelle much as it would partition from
water to heptane. With a Menger micelle, the guest molecule binds
hydrophobically by one of three mechanisms:

    <u>a</u>.   displacement of water in a pocket at the micelle sur-
face.

    <u>b</u>.   hydrophobic binding to a "fatty patch".

    <u>c</u>.   association to one or more extended chains which, owing
to thermal motion, protrude from the micelle like a pin-cushion.

It should be clear from this analysis that <u>extensive hydrocarbon-
water contact</u> constitutes the fundamental difference between the
Menger micelle and the traditional Hartley picture portrayed
widely in the literature.

Almost all discussions of micelle models end up on the subject
of water-content. It is important in this regard to differentiate
"dynamic" from "static" (time-averaged) wetting. Failure to do

so has led to confusion in the past. Clefts exist transiently within a micelle, fill up with water, and thereby expose inner methylenes to the aqueous medium. Looping of the chains near the clefts contributes to the wetting. This effect, coupled to the presence of fatty patches and protruding chains, causes all carbons of the chains to be exposed transiently to water. The point here is that it is misleading to speak of, for example, "water penetration to only the β-carbon". It would be far better to deduce the total static water content of a micelle and realize that all chain carbons become exposed to this water. The percentage of time a given carbon contacts water depends on the micellar disorder which unfortunately has not yet been defined. Recent unpublished experiments (described later in the article) indicate, however, extensive wetting even of terminal carbons of the chains.

Lindman and Brun[3] have estimated from diffusion data that sodium octanoate micelles contain 9 waters for each surfactant molecule. This agrees with a hydration number of 9 found for SDS by Mukerjee[4]. Much of this micellar water, perhaps 4-5 molecules, is needed to solvate the ions; the balance of the water is engaged in hydrocarbon-water contact. Now a rough micelle surface (containing head groups, counterions, segments of hydrocarbon chains, and 9 waters per surfactant) necessarily occupies a substantial volume of space. Assume for the moment that the Hartley 2-phase model is correct and that the hydrocarbon core has a radius equal to the extended hydrocarbon chain (16.6 Å for SDS). Since the diameter of the head group is 4.6 Å, then the Stern layer volume would be roughly 50% of the total micelle volume. In the Menger micelle, the percent of the micelle volume containing static water is probably larger than 50% because the outer reaches are rough, ill-defined, and laced with hydrocarbon. In summary, the Menger micelle possesses an enormous hydrocarbon-water contact. There is no Stern layer but rather a highly complex Stern region in which most of the interesting micellar chemistry occurs.

One final and somewhat puzzling point should be mentioned. The radii of micelles appear to be somewhat larger than the length of the extended surfactant molecule. This could mean one of two things: (1) There is a hole in the center of the micelle devoid of matter[5]; (2) The micelle contains more water (i.e. is more porous) than even we imagine. In regard to the latter possibility, we pointed out in the Accounts article[2] first describing the Menger micelle that a porous, totally disorganized "brushheap" micelle has a radius only 20% greater than a more organized and compact configuration.

C. Fromherz

Fromherz constructed a "surfactant-block" model of micelles

consisting of doublet and triplet subunits[6]. These surfactant
subunits have parallel chains and assemble orthogonally to form
the micelles. The model was physically constructed from wooden
rods and therefore exaggerates, in my opinion, the rigidity of
the hydrocarbon tails. Work by others demonstrates the presence
of folding within micelles[7,8]. The most attractive feature of
the model is the large degree of hydrocarbon-water contact which
is necessary, as we see below, to explain micellar chemistry.
The Fromherz micelle is more organized, and the Menger micelle
less organized, than the Hartley model. At present, there are no
data proving or disproving the Fromherz micelle although it is
difficult to imagine how the model could accommodate severely
branched surfactants which are known to form micelles readily.

As stated above, a useful model must rationalize and predict.
It is the purpose of this article to demonstrate that the Menger
micelle has the power to do both. Since each experimental
approach has its particular assumptions and thus cannot by itself
settle the question of micelle structure, I have purposely chosen
a wide variety of methods. The resulting data taken together
seem to point strongly to the micelle structure that we and others
have advanced in recent years.

Prediction: Oxygen should be less soluble in micelles than
in nonane.

The Menger micelle is a porous structure containing water
throughout a substantial fraction of the volume. Since oxygen
solubility in water is about 2% of that in nonane, the gas
solubility inside a micelle should be less than that in nonane if
the "oil droplet" concept is incorrect. This is found to be the
case[9]:

Medium	Solubility (m_f)
Water	0.35×10^{-4}
Nonane	21×10^{-4}
SDS	10×10^{-4}

Adherents of the Hartley micelle have invoked, in order to rationalize reduced solubility in "dry" micelles, a huge (400 atmosphere!)
Laplace pressure on the micelle interior[10]. But barring this possibility, the best explanation is that either partial ordering some-
how affects solubility or that the micelle is in fact not entirely
dry hydrocarbon.

Prediction: A water-insoluble compound adsorbed in a micelle
should display a reactivity characteristic of that in a protic
solvent.

Scores of examples, of which I will give but one, bear out this prediction. p-Chlorobenzhydryl chloride solvolyzes with an extremely solvent-sensitive rate:[11]

Medium	k_{obs}, min^{-1}
30% dioxane-H$_2$O	6.0
40% dioxane-H$_2$O	1.1
60% dioxane-H$_2$O	0.028
80% dioxane-H$_2$O	0.00045
0.02 M HTAC	1.0

Thus, solvolysis is over 10^3 times faster in 30% dioxane-H$_2$O than in 80% dioxane-H$_2$O. The rate constants for solvolysis inside micelles can be extracted from equations we developed years ago[12]. As is seen from the table, the reaction rate in 0.02M hexadecyl-trimethylammonium chloride (well above its CMC) resembles that in 40% dioxane-H$_2$O. The point here is that the guest clearly locates itself not in an "oil droplet" but in a highly aqueous micellar medium. This is particularly striking because p-chlorobenzhydryl chloride is water-insoluble and will completely partition into heptane when shaken in a water-heptane mixture. Why should a hydrophobic compound bind tightly to micelles in a water-rich environment? The Hartley micelle gives no simple answer to this dilemma. The Menger micelle, on the other hand, has water-filled cavities and fatty patches which can bind compounds through hydro-phobic forces not unlike those that gave rise to the micelle in the first place.

   Prediction: Since micelles have "fatty patches" near the surface, organic counterions should bind more effectively than inorganic counterions of the same charge.

Work of Whitten and coworkers (major proponents of porous cluster micelles) demonstrates impressively the difference between organic and inorganic counterions[13]. It was shown that the binding capacity of SDS micelles is much larger for methyl viologen than for Cu^{++}:

$$Cu^{++} \qquad\qquad CH_3-N\bigcirc-\bigcirc N-CH_3{}^{++}$$

Capacity: 1 Cu^{++}/6 SDS   1 MV^{++}/3 SDS

The difference is particularly noteworthy when one considers that MV^{++} is so much larger than Cu^{++}. Evidently the Cu^{++} binds to the micelle through Coulombic forces alone, whereas the MV^{++} has <u>both</u> Coulombic and hydrophobic forces working to its advantage. Presumably, the MV^{++} attaches to the micelle at fatty patches and in shallow cavities not far from the anionic head groups.

Prediction: Association constants between ionic and hydrophobic species should be enhanced in micelles.

If a hydrophobic species is solubilized deep within a Hartley oil droplet while an ionic species is solubilized in the Stern layer, then association between the two should be impaired (if not prevented) by micellization. On the other hand, association of the two species should be enhanced in a Menger micelle because both can reside comfortably near the surface:

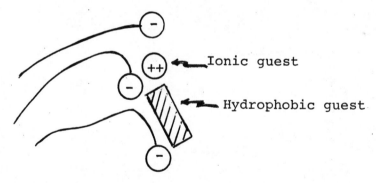

Martens and Verhoeven[14] recently showed that the association constant for charge-transfer complexation between hydrophobic pyrene and ionic paraquat increases from 0.93 in methanol to 485 in 0.1M SDS. A large increase such as this is not as readily explained by a "two-phase" Hartley micelle as it is by a Menger micelle having a complex surface comprised of water, hydrocarbon, head groups, and counterions.

   Prediction: Even terminal carbons of surfactant chains
should be exposed to water within a Menger micelle.

A Menger micelle is a rather disorganized assemblage. Chains
fold back on themselves within the micelle while others lie, for
a fleeting moment, almost entirely at the surface. Still other
chains line the cavities that penetrate the micelle. The net
result is that all the carbons of a chain are exposed to water
for a finite portion of their residence time within a micelle.
This conclusion would not follow from a Hartley oil droplet model.
We recently attempted to exploit this distinction by determining
the average polarity experienced by the termini of surfactant
chains inside micelles[15]. The problem was, however, to find an
"inoffensive" probe which would not sterically disrupt the normal
micelle structure nor "drag" water into a normally dry region.

   Most probes are, of course, large multi-functional systems
which create a lingering doubt as to how they affect the environment upon which they are reporting. We found that the acetylenic
proton in $RC\equiv CH$ molecules has a chemical shift which responds to
the medium polarity. This was a fortunate observation because the
acetylenic unit is small and relatively non-polar. Hence, we
synthesized surfactants terminating with a triple bond (and suitably deuterated to preclude splitting and overlap complications).
Two examples are shown, along with relevant chemical shift data,
in the table below.

	δ, PPM
$\underline{H}C\equiv C-CD_2-(CH_2)_{10}NMe_3^+$	2.16
$\underline{H}C\equiv C-CD_2-(CH_2)_{10}OSO_3^-$	2.10
Cyclohexane	1.66
Acetone	2.04
Methanol	2.14
33% H_2O-Methanol	2.19

Clearly, the acetylenic surfactants in the micellar state display
chemical shifts much closer to those in protic solvents than in
cyclohexane. Although it is best not to overinterpret such numbers, there is no doubt that the chain termini are experiencing
water a substantial percentage of the time. An oil droplet model
simply cannot accommodate our data.

   Prediction: In general, a micellar parameter should fall
intermediate between that in heptane and that in water.

An excellent example of "intermediate behavior rule" was discussed recently by Turro and Okubo[16]. It turns out that the viscosity of water remains constant with pressure (up to 2 kbar) whereas the viscosity of liquid hydrocarbons undergoes a sharp upward curvature. The microviscosity inside a micelle, on the other hand, behaves <u>neither</u> like that of water <u>nor</u> hydrocarbon as shown in the graph below:

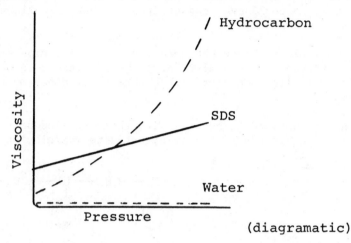

(diagramatic)

It is striking how commonly one observes intermediate behavior in micellar systems. If a probe binds to a micelle at a fatty patch or in a surface cavity, then the guest would have hydrocarbon on one side and water on the other, thereby explaining a large body of micellar data.

    <u>Prediction</u>: Branching of surfactant chains should enhance hydrocarbon solubilization.

Small saturated hydrocarbons should be able to enter the micelle core and thus experience a rather dry environment. Since the micelle core is far less than half the micelle volume[2], the binding capacity of the core is not large. By placing alkyl substituents on the main surfactant chain, one can help occupy the water-filled cavities with side-chains and thus expand the core. Work by Shinoda <u>et al.</u>[17] illustrates the effect beautifully. They compared the relative solubilization capacity of cyclohexane in branched and straight-chained surfactants of an equal number of carbons:

Rel. Capacity

$C_{10}H_{21}CH(C_6H_{13})COO^-$ 3

$C_{17}H_{35}COO^-$ 1

Much more cyclohexane can be solubilized in the micelles comprised of the branched surfactant. One would predict, similarly,
that a cosurfactant (e.g. n-pentanol) added to a surfactant system
would effectively fill the cavities and thus enhance solubilization of saturated hydrocarbons which seem to prefer completely
nonpolar environments. Hoskins and King have shown this to be
the case[18]. I may add parenthetically that it is by no means
clear how one could even construct a Fromherz micelle with the
branched surfactant.

   Prediction: Since chain-to-chain relationships are not
highly ordered in a micellar "brush-heap", it should be easy to
simulate micellar behavior with a variety of other multi-chained
systems.

This has been found to be true with the so-called polysoaps of
Cordes and coworkers[19]. Polyvinylpyridine was alkylated with
dodecyl bromide:

PVP + $C_{12}H_{25}Br$ $\longrightarrow$

The resulting polysoap binds long-chain esters and catalyzes their
alkaline hydrolysis -- very similar to the action of cationic
micelles. Clearly, chain organization is not critical to "micelle-
like" behavior. Another example comes from the hexapus system
which we recently synthesized and studied[20]:

Hexapus

Hexapus solubilizes a host of water-insoluble guests and inhibits the alkaline hydrolysis of long-chain esters (as do fatty acid micelles). With its aromatic cap and widely spread chains, hexapus can hardly be likened to a classical micellar aggregate.

Many additional predictions and examples could be cited if space and the readers' patience permitted. Naturally, no single example constitutes "proof" of the Menger micelle in the usual sense of the word. But models are not supposed to be proved; they are supposed to be useful in rationalizeing and predicting. I hope that this article serves to persuade a few that the Menger micelle can do both.[21]

ACKNOWLEDGMENTS

Support of the National Science Foundation, the Petroleum Research Corporation, and the North Atlantic Treaty Organization is gratefully acknowledged.

REFERENCES

1. G. S. Hartley, "Aqueous Solutions of Paraffin-chain Salts. A Study of Micelle Formation", Hermann and Co., Paris, 1936.
2. F. M. Menger, Acc. Chem. Res., 12, 111 (1979).
3. B. Lindman and B. Brun, J. Colloid Interface Sci., 42, 388 (1973).
4 P. Mukerjee, J. Colloid Sci., 19, 722 (1964).
5. This intriguing possibility was suggested by B. Lindman, private communication.
6. P. Fromherz, Ber. Bunsenges. Phys. Chem., 85, 891 (1981).
7. K. A. Dill and P. J. Flory, Proc. Natl. Acad. Sci. USA, 78, 676 (1981).
8. D. W. R. Gruen, J. Colloid Interface Sci., 84, 281 (1981).
9. F. M. Menger, J. Phys. Chem., 83, 893 (1979).
10. I. B. C. Matheson and A. D. King, Jr., J. Colloid Interface Sci., 66, 464 (1978).
11. F. M. Menger, H. Yoshinaga, K. S. Venkatasubban and A. R. Das, J. Org. Chem., 46, 415 (1981).
12. F. M. Menger and C. E. Portnoy, Jr. Am. Chem. Soc., 89, 4698 (1967).
13. D. G. Whitten, J. C. Russell, T. K. Foreman, R. H. Schmehl, J. Bonilha, A. M. Braun, and W. Sobol, in "Proceedings of the 26th OHOLO Conference, Zichron, Israel," B. S. Green, Y. Ashani, and D. Chipman, Editors, Elsevier, Amsterdam, 1982.
14. F. M. Martens and J. W. Verhoeven, J. Phys. Chem., 85, 1773 (1981).
15. F. M. Menger and J. F. Chow, unpublished results.
16. N. J. Turro and T. Okubo, J. Am. Chem. Soc., 103, 7224 (1981).
17. H. Sagitani, T. Suzuki, M. Nagai, and K. Shinoda, J. Colloid

Interface Sci., 87, 11 (1982).

18. J. C. Hoskins and A. D. King, Jr., J. Colloid Interface Sci., 82, 260 (1981).

19. T. Dodulfor, J. A. Hamilton, and E. H. Cordes, J. Org. Chem., 39, 2281 (1974).

20. F. M. Menger, M. Takeshita and J. F. Chow, J. Am.Chem. Soc., 103, 5938 (1981).

21. Small-angle neutron scattering measurements are being applied with increasing frequency to micellar systems (see for example L. J. Magid, R. Triolo, J. S. Johnson, Jr., and W. C. Koehler, J. Phys. Chem., 86, 164 (1982)). These studies will lead to valuable insights into micellar structure only if the assumptions involved in the complicated analyses are clearly delineated. Thus, it is important to know how conclusions are affected by assumptions regarding sphericity, monodispersity, surface roughness, long-range ordering effects, micelle dimensions, critical scattering, etc., or, in other words, by the nature of the working model used to "prove" a structural model. Note that many experiments described in this paper, such as the p-chlorobenzhydryl solvolyses, require no assumptions about shape, size, and monodispersity.

TRACER SELF-DIFFUSION STUDIES OF SURFACTANT ASSOCIATION

[1]N. Kamenka, [1]M. Puyal, [1]B. Brun, [1]G. Haouche and
[2]B. Lindman

[1]Laboratoire de "Physicochimie des Systèmes Polyphasés"-
USTL, Place E. Bataillon, 34060 Montpellier-Cedex,France
[2]Dept. of Physical Chemistry 1, Chemical Center,
Lund University, P.O.B. 740, S-220 07 Lund, Sweden

Self-diffusion studies were performed to obtain an
insight into the association phenomena in aqueous solutions of sodium p-octylbenzene sulfonate (SOBS). The
open-ended capillary tube method, using radioactive labelling of all the kinetic entities of the solution, has
been employed so the self-diffusion coefficients of amphiphile ions, sodium counterions, solubilized decanol
molecules confined to micellar aggregates and water molecules were studied as a function of surfactant concentration. SOBS micelles can be considered to remain spherical throughout the concentration range studied with a hydrodynamic radius corresponding to the length of the extended surfactant ion and a hydration corresponding to about
one layer of water molecules. This result is in accordance with the low "hydration numbers" obtained from water
molecule self-diffusion coefficients; there is no marked
water penetration deep into the SOBS micelles.
From surfactant ion self-diffusion, the concentration of free surfactant ions has been calculated as a
function of surfactant concentration. The self-diffusion technique provides the direct demonstration that the concentration of free amphiphile ion decreases substantially above the cmc in agreement with a recently developed
theory based on the Poisson-Boltzmann equation.
The association of ionic surfactant is compared
with the aggregation occuring in aqueous solutions of
a nonionic surfactant (octaethyleneglycol dodecyl
ether) and a zwitterionic surfactant (n-dodecylbetaine).

359

INTRODUCTION

The aim of the present work was to obtain a description of the translational motion of the kinetic entities in aqueous solutions of three surfactants, sodium octylbenzene sulfonate (SOBS), n-dodecylbetaine (NDB) and octaethyleneglycol n-dodecylether ($C_{12}(EO)_8$) in order to study surfactant association as a function of concentration.

As the mobility of a molecule or an ion in a solution depends on its interaction with other species in the solution, it is obvious that aggregation or interaction phenomena affect the self-diffusion of this molecule or ion. In the case of micellar solutions, since the micelle has a much lower translational mobility than the free molecule or ion in the bulk solution, the association of a species with the micelle leads to a marked change in the self-diffusion coefficient. For example, a solubilizate with very low aqueous solubility will be confined to the micelle and thus will have a very low self-diffusion coefficient corresponding to the self-diffusion of the micelle itself; investigations of the diffusion coefficient of the micelles provide information on micellar size and shape.

For a certain species which occurs in different kinetic entities, the observed self-diffusion coefficient D is given by the simple equation (1):

$$D = \Sigma \, p_i \, D_i \qquad\qquad\qquad (1)$$

where p_i is the fraction of the molecules or ions in the complex i, and D_i is the self-diffusion coefficient of this complex. In our tracer experiments, the diffusion time is very long, of the order of a few days and Equation (1) may be applied without any limitations, the lifetime of the species studied at a site i always being much shorter than the measuring time.

It seems reasonable to describe the micellar system in terms of a two-site model where the molecules or ions belong to free and micellized entities, i.e. diffuse freely in the intermicellar solution and with the micelle. Equation (1) can be schematically written as:

$$D^X = p_f^X \, D_f^X + p_m^X \, D_m = D_f^X - p_m^X \, (D_f^X - D_m) \qquad\qquad (2)$$

where X represents the amphiphile, the counterion, the water molecule or the solubilizate which diffuses with the observed self-diffusion coefficient D^X. p_m^X is the fraction of micellized X, i.e. micellized amphiphile, bound counterions, hydration water molecules, solubilizate within the micelles; all of them diffuse with D_m, the self-diffusion coefficient of the whole micelle. p_f^X and D_f^X are, respectively, the fraction and the self-diffusion coefficient of free X located in the intermicellar solution. When D_m and D_f^X are known, one can calculate p_m^X values and also the concentration of micellized X at each concentration studied; thus, the aggregates can be quantitatively described.

Two remarks are in order here:
--Firstly, D_m is much lower than D^X and the errors inherent in the solubilization of an organic compound within the micelle are negligible when p_m^X values are calculated. The obtained values of D_m, directly related to the size of the micelle, will be discussed below.

-- Secondly, a much more difficult problem is to deduce the various D_f^X values at different concentrations of surfactant. The diffusion of the free species in aqueous surfactant solutions is not identical to the diffusion of these species at infinite dilution in water for several reasons. The translational mobilities of the free entities are lowered due to ion-ion and ion-solvent interactions and to the obstruction effect of the micelles which act as an excluded volume for certain molecules or ions. For ionic surfactant solutions, the excluded volume depends markedly on the molecular charge; it will not be the same for free amphiphile, water or counterion diffusing in the intermicellar solution.

We attempt here to describe quantitatively the micelle composition for sodium n-octylbenzene sulfonate and present a discussion of the possibilities and limitations of the self-diffusion coefficient method.

As the magnitude of the repulsive forces between amphiphile head groups influences the micelle size and is the limiting factor for the size of the micelle, it seems interesting to apply the same method to two other types of surfactants with chain length similar to that of SOBS, one non ionic and one zwitterionic. Some results were recently obtained on the diffusion in n-dodecylbetaine (NDB) and octaethyleneglycol dodecylether ($C_{12}(EO)_8$) solutions as a function of surfactant concentration.

EXPERIMENTAL

The experimental technique used was the open ended capillary tube method using radioactive labelling as described elsewhere[1,2,3]. Each component is successively labelled to obtain its self-diffusion coefficient at each composition of the solution. The time of diffusion is of the order of a few days, the relative error is about 2%, and the self-diffusion coefficients obtained are averages of six measurements.

Diffusion in SOBS and $C_{12}(EO)_8$ solutions was studied at $33^{\circ}C \pm 0.5^{\circ}C$. As NDB has been used in our laboratory to formulate micro-emulsion systems and as pseudo-ternary phase diagrams have been determined at $25^{\circ}C$, so the self-diffusion experiments in NDB aqueous solutions were carried out at this temperature.

The total radioactivity content in the capillaries before and after diffusion was measured directly for γ emitters by means of a Packard auto-gamma spectrometer; while for β emitters, the capillary contents were transferred to an Instagel Packard scintillator and radioactivity measurements were done on a Packard liquid scintillation spectrometer. SOBS was synthesized according to the synthesis proposed by Gray[4]. ^{35}S-labelled SOBS was obtained analogously in a microsynthesis. ^{35}S labelled sulfuric acid, necessary in the first step of the reaction, was provided by CEA, Department of Radioelements, Gif-sur Yvette, France. $C_{12}(EO)_8$ was obtained from Nikko Chemicals, Co. Ltd., Tokyo. NDB and ^{14}C labelled NDB were obtained according to the synthesis described by Tori and Nakagawa[5]. ^{14}C labelled decanol was obtained from Paris-Labo, Paris, ^{22}Na labelled sodium chloride, from Amersham, Radiochemical Center, Buckinghamshire, England, and ^{3}H labelled water and ^{14}C labelled $C_{12}(EO)_8$ from C.E.A.

ASSOCIATION OF SODIUM p-OCTYLBENZENESULFONATE[6]

A: Micelle Self-Diffusion

D_m, the self-diffusion coefficient of the SOBS micelles, was obtained at various concentrations of the surfactant by solubilizing small amounts of ^{14}C-decanol. This alcohol has a very low solubility in water and its partitioning between micellar aggregates and intermicellar solution is shown to be greater than 99%[7]. To minimize the effect on micellar shape and size possibly due to the added solubilizate, all the measurements are made at very low decanol concentration; the molar ratio alcohol to aggregated amphiphile is about 0.04. We observe a slow decrease of D_m as the concentration of SOBS increases. Following the theory of Mazo[8], since a plot of $1/D_m$ versus the concentration of micellized amphiphile (obtained from the self-diffusion of the amphiphile and Equation (2)) is approximately linear (Figure (1)), the decrease in D_m may be attributed to electrostatic intermicellar repulsion. In the concentration range studied, the SOBS micelles can be considered to remain approximately spherical.

The extrapolated value of the self-diffusion coefficient of the micelles at zero micelle concentration is equal to $0.123 \times 10^{-9}m^2s^{-1}$. The hydrodynamic radius calculated from the Stokes-Einstein equation is equal to 24Å. Because of hydration (and because of surface

roughness and maybe also small deviations from the spherical shape)
the radius obtained from self-diffusion measurements is somewhat
larger than the length of the extended amphiphilic molecule.

B: Amphiphile Self-Diffusion

The experimental amphiphile self-diffusion coefficients, D^a,
are plotted in Figure (2) versus the total surfactant concentration. D^a decreases markedly when the surfactant concentration is
greater than the cmc which is equal to 1.28×10^{-2} mol $\cdot$ kg^{-1} at
33°C.

Knowing the experimental D^a and D^m values, we are able from
Equation (2) to calculate the concentration of free and micellized
amphiphile if the self-diffusion coefficient of free amphiphile
can be estimated.

$$m_f^a = m_t - m_m^a = \frac{D^\alpha - D_m}{D_f^a - D_m} \, m_t \tag{3}$$

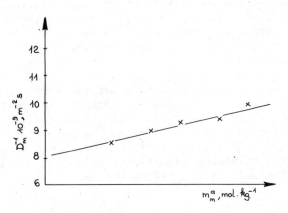

Figure 1. The inverse micelle self-diffusion coefficient as a
function of micellized surfactant concentration, m_m^a, for aqueous solutions of sodium octylbenzenesulfonate. Temperature $\cdot$33°C.

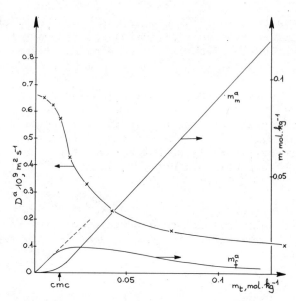

Figure 2. Amphiphile self-diffusion coefficient and deduced concentrations of free, m_f^a, and micellized, m_m^a, amphiphile as a function of total surfactant concentration for solutions of sodium
octylbenzenesulfonate. Temperature: 33°C.

Calculations of m_m^a and m_f^a were performed, taking the infinite dilution D^a value and correcting it for interionic interactions in an
iterative procedure. On the other hand, we think that it is not
appropriate to correct the free amphiphile diffusion for an excluded
volume effect due to the micelles[6]. Actually, several different
hypotheses made for the estimation of D_f^a give qualitatively the
same behaviour of m_f^a. In Figure (2), we present the deduced free
and micellized surfactant concentrations obtained from interpolated
self-diffusion values (D_m and D^a were not obtained at exactly the
same m_t concentrations).

The m_f^a values are identical to m_t at low surfactant concentrations. Around the cmc, a broad maximum occurs and then, with
increasing surfactant concentration, the concentration of free amphiphile decreases to values well below the cmc; for example,
m_f^a = cmc/2 when m_t = 0.1 mol · kg^{-1}. The concentration of micellized
amphiphile starts to become significant below the cmc and increases
approximately linearly with m_t after the cmc.

From the kinetic studies, Aniansson et al.[9] have deduced a decrease in the monomer concentration above the cmc and a theoretical
analysis based on the Poisson–Boltzmann equation by Gunnarsson

et al.[10] has rationalized this. The behaviour can be attributed to changes in the counterion distribution since an enhanced concentration of the free counterions would stabilize the micellar state by levelling out the uneven counterion distribution. Cutler et al.[11] have demonstrated a decrease in the monomer activity as the surfactant concentration increases, while the self-diffusion measurements give a direct demonstration for a decreased concentration of free amphiphile above the cmc.

C: Counterion Self-Diffusion

The experimental self-diffusion coefficients for the counterions are plotted as a function of surfactant concentration in Figure (3) and D^{Na} can be seen to decrease markedly above the cmc.

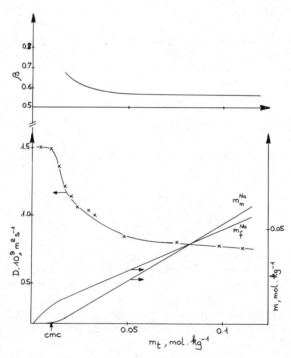

·Figure 3. Sodium counterion self-diffusion coefficients, concentrations of free (m_f^{Na}) and micellar (m_m^{Na}) counterions versus the concentration of sodium octylbenzenesulfonate. Upper part shows β, the degree of counterion binding. Temperature: 33°C.

Recently some theoretical work has shown that the counterion distribution is sharp enough to make a two-state model a suitable approximation[10,12,13]. From Equation (2), the concentration of bound counterions and β, the degree of association of sodium ions to the micelles (defined as the ratio of bound counterion concentration to micellized amphiphile concentration) can be calculated if D^{Na} is known.

$$m_m^{Na} = m_t^{Na} - m_f^{Na} = \frac{D_f^{Na} - D^{Na}}{D_f^{Na} - D_m}\, m_t \quad \text{and} \quad \beta = \frac{m_m^{Na}}{m_m^a} \tag{4}$$

It is obvious that the micelles constitute an excluded volume for the counterions and this affects the diffusion of free counterion by an obstruction effect. The latter effect was deduced according to Wang[14]. The values of the free and micellized counterion concentrations as well as the degree of counterion association β thus obtained are plotted in Figure (3). The increase in m_f^{Na} above the cmc and the slower increase of m_m^{Na} than of m_m^a are as expected.

β is quite constant (at 0.55 to 0.60) over a wide range of concentrations but at a low micelle concentration, the degree of counterion binding is somewhat higher and decreases as surfactant concentration increases. Despite the fact that the error in β is large around the cmc, the reduction in β is significant. This behaviour is predicted by the calculations based on the Poisson-Boltzmann theory[10].

D: Water Self-Diffusion

Two factors affect the self-diffusion coefficient of water molecules in surfactant solutions:
  - an obstruction effect due to the presence of the micelles
  - an hydration effect arising from the water molecules moving with the micelles.
In reality, it is difficult to make a distinction between bulk water molecules and "bound" water molecules as solvent molecules are close to a solute even in the absence of strong attraction and furthermore the concept of a single hydration number is a simplification depending on the experimental approach. However, self-diffusion measurements and simple calculations using a two-site model give an estimation of the average hydration number and therefore help in deducing whether or not there is a deep water penetration into micelles.

In Figure (4) we have plotted the observed values of the water self-diffusion coefficient, D^W, versus the total SOBS concentration. Firstly, and this is a general observation in our studies of water

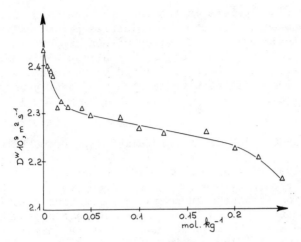

Figure 4. Concentration dependence of the water self-diffusion co-efficient in solutions of sodium octylbenzenesulfonate. Temperature: 33°C.

diffusion in various surfactant solutions, D^W decreases much more rapidly below than above the cmc. As the mobility of water molecules hydrating free amphiphile ions is greater than the mobility of water molecules bound to a micelle ($D_f^a \gg D_m$), this behaviour indicates that upon micellization, the hydrocarbon chains of the amphiphile are almost completely withdrawn from water contact.

It is possible to quantify the hydration effect using Equation (2) which can be written as

$$D^W = D_f^W - (nm_m^a/55.5)(D_f^W - D_m) \qquad (5)$$

where n is the micelle hydration number per amphiphile ion. The term corresponding to non-micellar water molecules consists of three parts: free water molecules, water molecules hydrating free amphiphile ions, and water molecules hydrating free sodium ions. These contributions can be calculated using the deduced concentrations of free amphiphile ions and counterions and D_f^W is then empirically corrected on the basis of the pre-cmc data. The obstruction effect was evaluated[14] using micellar concentrations deduced above.

All hydration numbers in the range of concentration studied were
quite low, between 5 and 15 water molecules per micellized amphiphile
ion. The self-diffusion study gives direct support for the generally
accepted picture[15-17] of a micelle with a liquid-like hydrocarbon
interior and with water contact limited to the surface of the micelle. It is inconsistent with appreciable water penetration deep
into the micelle.

ASSOCIATION OF OCTAETHYLENEGLYCOL DODECYLETHER

In the phase diagram of the nonionic surfactant $C_{12}(EO)_8$ and
water, the isotropic L_1 phase is very extensive. In order to compare aggregation phenomena in nonionic surfactant solutions with
the above results obtained for an ionic surfactant, we have chosen
to study aqueous $C_{12}(EO)_8$ solutions at 33°C, a temperature significantly lower than the lower critical point of this compound (i.e.
74°C). The self-diffusion coefficient of the surfactant molecules
is plotted as a function of the surfactant concentration in Figure
(5). A rapid decrease of the self-diffusion coefficient is observed to start when the surfactant concentration reaches the cmc
value (i.e. $6.5 \cdot 10^{-5}$ mol·kg^{-1} at 33°C[19]). Over a large range of
concentrations, D decreases slowly.

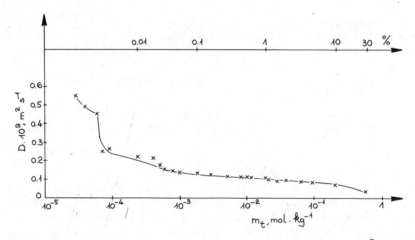

Figure 5. The observed self-diffusion coefficient at 33°C of
$C_{12}(EO)_8$ versus the $C_{12}(EO)_8$ concentration. Concentration values
are given on a logarithmic scale, upper scale in percent by weight
and lower scale in molality.

D_m has been calculated from these values according to Equation (2), assuming the amount of free amphiphile to be equal to the cmc. As the cmc is very low, the contribution of free monomer is appreciable only for the lowest surfactant concentrations. For example, D_m evaluated from Equation (2) for a 2% (by weight) surfactant concentration is 1% lower than D. Above this concentration the measured self-diffusion coefficient equals the micelle self-diffusion within the experimental error. If there is no change with concentration in the size of the micelles, the decrease in D with increasing surfactant concentration may be attributed to micelle-micelle interactions (mainly an obstruction effect).

At infinite dilution, we can calculate the micellar radius with the Stokes-Einstein equation and we obtain a value equal to 30Å. Exactly the same value has been derived from quasielastic light-scattering experiments on the same solutions[20]. As the surfactant concentration increases, the apparent radius obtained from tracer diffusion remians quite small. Taking into account the hydration of the polar head, the micellar aggregation number is deduced to be 55[21].

For more concentrated solutions the self-diffusion coefficient decreases (i.e. the apparent radius increases). This may be attributed to interactions between the micelles as shown by Nilsson et al.[21] from NMR self-diffusion and proton relaxation studies. In the whole concentration range studied, the $C_{12}(EO)_8$ aggregates appear to remain small and spherical at 33°C; other experiments are in progress in our laboratory for more concentrated solutions. This behaviour can be expected from geometrical considerations and is in line with the phase equilibria. The micelles formed by a nonionic surfactant with a long ethyleneoxide chain are preferentially spherical while for a nonionic surfactant with a short ethyleneoxide chain, a micellar growth occurs as found in the case of $C_{12}(EO)_5$[21] and $C_{12}(EO)_6$. A direct determination of the aggregation number of $C_{12}(EO)_6$ gives a value of 320[22].

ASSOCIATION OF N-DODECYLBETAINE

The n-dodecylbetaine (NDB), $C_{12}H_{25}N^+(CH_3)_2CH_2COO^-$, is soluble in water in all proportions. The cmc has been measured by two different techniques, i.e. from the variation of the refractive index and the density of the solutions, and is equal to $2 \cdot 10^{-3}$ mol $\cdot$ kg^{-1} [23]. The cmc value of the zwitterionic surfactant falls between the values of the nonionic and ionic surfactants of the same hydrocarbon chain length. The heating of a 50% NDB solution does not produce any turbidity and, furthermore, the solutions stay clear as the salinity increases.

The self-diffusion coefficients (Figure 6) for the more concentrated solutions is quite a bit lower than those observed in the ionic and nonionic surfactant solutions and, furthermore, the viscosity of the NDB solutions is quite large (>100 cP). The variation of the self-diffusion coefficient of the micelles, obtained by applying Equation (2), does not correlate with a two-site model assuming a constant micellar size. Instead, the zwitterionic surfactant displays a growth into large micelles. For zwitterionic surfactants, the value of the aggregation number is strongly dependent on the distance between the charged groups. When the two charges may neutralize each other, the micelle aggregation number is quite large[22]. Aggregate and water self-diffusion measurements are in progress in our laboratory.

CONCLUSION

The results obtained in this study, by the self-diffusion tracer method, indicate that in the concentration range studied the micelles remain small and roughly spherical. The decrease in free amphiphile concentration as well as the variation in counterion binding with surfactant concentration are in agreement with the recently developed theory based on the Poisson-Boltzmann equation. The results are inconsistent with a strong water penetration into the micelles.

The nonionic surfactant micelles appear to remain small and spherical while the aggregation of the zwitterionic compound depends markedly on the surfactant concentration. For the latter, the cmc seems to be less well-defined and the self-diffusion results suggest that the micelle size increases with concentration.

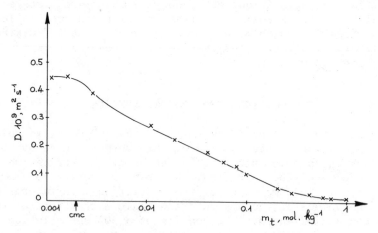

Figure 6. The observed self-diffusion coefficients at 25°C of n-dodecylbetaine (NDB) versus the NDB concentration. Concentration values are given on a logarithmic scale.

REFERENCES

1. B. Lindman and B. Brun, J. Colloid Interface Sci., 42, 388 (1973).
2. N. Kamenka, B. Lindman and B. Brun, Colloid Polym. Sci., 252, 144 (1974).
3. B. Lindman, N. Kamenka, T.M. Kathopoulis, B. Brun and P.G. Nilsson, J. Phys. Chem., 84, 2485 (1980).
4. F.W. Gray, J.F. Gerechet and I.J. Kreus, J. Org. Chem., 20, 511 (1955).
5. K. Tori and T. Nakagawa, Kolloid-Z.Z. Polym., 1, 188 (1963).
6. B. Lindman, M.C. Puyal, N. Kamenka, B. Brun and G. Gunnarsson, J. Phys. Chem., 86, 1702 (1982).
7. P. Stilbs, J. Colloid Interface Sci., 87, 385 (1982).
8. R.M. Mazo, J. Chem. Phys., 43, 2873 (1965).
9. E.A.G. Aniansson, S.N. Wall, M. Almgren, H. Hoffman, I. Kielman, W. Ulbricht, R. Zana, J. Lang and C. Tondre, J. Phys. Chem., 80, 905 (1979).
10. G. Gunnarsson, B. Jönsson and H. Wennerström, J. Phys. Chem., 84, 3114 (1980).
11. S.G. Cutler, P. Mears and D.G. Hall, J. Chem. Soc., Faraday Trans. 1, 74, 1758 (1978).
12. B. Jönsson, H. Wennerström and B. Halle, J. Phys. Chem., 84, 2179 (1980).
13. P. Linse, G. Gunnarsson and B. Jönsson, J. Phys. Chem., 86, 413 (1982).
14. J.H. Wang, J. Am. Chem. Soc., 76, 4755 (1954).
15. B. Lindman and H. Wennerström, Topics in Current Chemistry, 87, 1 (1980).
16. B. Cabane, J. Phys. (Orsay, Fr.), 42, 847 (1981).
17. J. Ulmius and B. Lindman, J. Phys. Chem., 85, 4131 (1981).
18. G.J.T. Tiddy, Physics Reports, 57, 1 (1980).
19. K. Meguro, Y. Tagasawa, N. Kawahashi, Y. Tabata and M. Ulno, J. Colloid Interface Sci., 83, 50 (1981).
20. J. Appel (1982), personal communication.
21. P.G. Nilsson, H. Wennerström and B. Lindman, to be published.
22. P. Lianos and R. Zana, J. Colloid Interface Sci., 84, 100 (1981).
23. G. Haouche (1981), D.E.A. Montpellier, France.

AN INTRODUCTION TO NEUTRON SCATTERING ON SURFACTANT

MICELLES IN WATER

B. Cabane

Physique des Solides[*], Université Paris-Sud
91 405 Orsay, France

R. Duplessix[+]

Institut Laue Langevin, 156 X, 38042 Grenoble Cedex
France

T. Zemb

Département de Physico-Chimie, CEN-Saclay
91191 Gif sur Yvette, Cedex, France

The structure of surfactant micelles can be investigated
through neutron scattering. The usefulness of such data is
discussed in terms of their resolution. Spatial resolution,
i.e. the ability to resolve fine details of the structure, is
achieved by extending the data to high scattering vectors.
Chemical resolution, i.e. the ability to discriminate between
the various chemical groups of the surfactant molecules, is
achieved through isotopic labeling. An experiment illustrating these criteria is presented for the surfactant sodium
dodecylsulfate (SDS).

* Laboratoire associe au C.N.R.S.

+ Present address: CRM, 6 rue Boussingault, 67083 Strasbourg, France.

INTRODUCTION

What are the requirements of an experiment devoted to determining the structure of particles in solution ? It should measure
distances, and it should do so with as high a resolution as possible. A well known example in this respect is the work of crystallo-
graphers on biological particles such as viruses and lipopro
teins [1], [2]. Surfactant micelles are no exception to this rule,but
here the situation has remained at an earlier stage of development:
most of the available information has been obtained through classical techniques of physical chemistry, and concerns such characteristics as the fluidity of the hydrocarbon core of micelles, or the
polarity of their surface layers [3], [4]. Now the structural work, i.e.
the measurement of distances, can begin. Most techniques of physical chemistry are inadequate for this purpose, because they use the
molecule and not the micelle as a frame of reference. For example,
a measurement which would imply some contact between water and the
γ carbons of the surfactant chains will not tell us where in the
micelle those γ carbons and water molecules are located : such contact could happen in the middle of the micellar "core", or near
the surface, or even in water, beyond the surface of the micelle.
In other words, distances should be measured in spatial coordinates,
not in molecules coordinates - a distance measured along a chain
will not help, if the configuration of the chain is not known a
priori.

One class of experiments which does meet these conditions is
the NMR measurement of dipolar interactions between nuclei within
the micelle and paramagnetic ions adsorbed on the surface of the
micelle [9-12]. Then the charged surface of the micelle is used as a
frame of reference; the measured interactions vary as the inverse
sixth power of the distances, yielding a very good "spatial" resolution; also, most ^{13}C nuclei in a surfactant chain can be resolved,
leading to a good "chemical" resolution. These experiments were
very useful in disproving some of the most common models for micelle
structure; yet they did not enable us to discriminate between some
of the recently proposed theoretical models [13], [14].

The classical method for measuring distances is the study of
interference patterns in the scattering of radiation; this is how
the structures of biological particles were determined. Previous
scattering experiments on micelles have suffered from a lack of
resolution. Spatial and chemical resolution are equally important
here. By spatial resolution we mean the ability to resolve fine
details of the structure - such as the thickness of the water/hydro-
carbon interface, or that of the charged surface. By chemical resolution we mean the ability to identify those atoms or nuclei
between which distances are being measured - for example specific
methylene or methyl groups within the hydrocarbon core. As a

result of this lack of resolution, the only structural parameters
which have been determined in an unambiguous way concern the $_{5-8}$
micelle as a whole : aggregation number, basic shape, charge   .
Yet the acquisition of high resolution structural data is a required step for the future design of specific chemical reactions
using the micellar surface as a catalyser. We have taken a systematic approach of this problem through neutron scattering. Our aim
is to improve over previous scattering experiments both in spatial
and chemical resolution. Firstly, in order to resolve the various
chemical groups of the surfactant molecules, we take advantage of
the fact that neutrons are scattered by the nuclei in the sample.
This allows the use of isotopic labelling; in particular, hydrogen
can be replaced by deuterium, which has a very different cross-
section for the scattering of neutrons. In this work, we have used
molecules of the surfactant sodium dodecylsulfate (SDS) which have
been labeled in 4 different ways :

$C_{12}H_{25}SO_4Na$ (fully protonated, or "p25")

$C_{12}D_{25}SO_4Na$ (fully deuterated, or "d_{25}")

$(CD_3)(C_{11}H_{22})SO_4Na$ (methyl deuterated or "12d_3")

$(C_9H_{19})(CD_2)(C_2H_4)\ SO_4Na$ (γ deuterated, or "3d_2")

By using either H_2O or D_2O as a solvent for these molecules,
we generate 8 micellar solutions.We have checked that the micelles
formed in each of them share the same basic parameters (aggregation number, charge). They differ, however, in the way they scatter
neutrons, according to the distribution of H vs. D in and around
the micelles. Hence the 8 scattering curves are expected to provide
access to the internal structure of the SDS micelle : any model for
this internal structure will have to fit these 8 curves, once the
scattering properties of H vs. D are taken into account.

Then, in order to improve the spatial resolution, we extended
our data to scattering vectors much larger ($Q = 1\ \AA^{-1}$) than those
of previous neutron scattering experiments in this field. In
this range, the scattering curves of spherical micelles showed
show damped oscillations superimposed to an asymptotic behavior.
The asymptotic law is related to the structure of the water/hydro-
carbon interfaces; for sharp interfaces the law is Q^{-4}; it changes
to a much faster decrease with Q if the interface becomes diffuse
(interpenetration of water and hydrocarbon over a few $\AA$). The damped oscillations are very sensitive to the shape and size of the
micelles; in particular they are smeared out if the weight distribution of the micelles in broader than $\sigma/M_n = 0.6$.

ELASTIC SCATTERING OF NEUTRONS BY PARTICLES IN SOLUTION

Excellent reviews of this field have been published already 15-20. Here we present a simplified description, with emphasis on some typical problems which we met in the study of SDS micelles.

Inelastic effects will be ignored; in fact, they do affect the magnitude of the intensities scattered by very light molecules (H_2O); this effect can be accounted for in an empirical way.

As mentioned above, neutrons interact almost exclusively with the nuclei in the sample. For a given nucleus j, located at $\vec{R}_j$, the strength of that interaction is determined by the <u>nuclear scatte-ring length b_j</u> which appears in the potential of <u>interaction</u> :

$$U(\vec{R}) = \frac{2\pi\hbar}{m} b_j \; \delta(\vec{R} - \vec{R}_j)$$

where m is the neutron mass. The interaction results in a momentum transfer to the neutron; in the experiment, neutrons are sorted out according to the values of the scattering vector $\vec{Q}$, which is the difference of momentum between incident and scattered neutrons :

$$\vec{Q} = \vec{k} - \vec{k}_o, \quad |\vec{Q}| = \frac{4\pi}{\lambda} \sin(\Theta), \; 2\Theta = \text{scattering angle}$$

In a disordered system, it is not possible to get informations at the atomic level, and the data are limited to values of $|\vec{Q}|$ which are small compared to the inverse of typical interatomic distances. For this range of Q vectors, the molecules can be considered as a continuous medium, and the coherent scattering is controlled by the local density of scattering length :

$$\rho(\vec{R}) = \frac{1}{v} \left| \sum b_j \; \delta(\vec{R} - \vec{R}_j) \right| \tag{1}$$

where the summation is over nuclei located in a volume v which is large compared to interatomic distances, but small compared to the resolution of the data. Typical densities ρ range from $-0.5 \times 10^{10} \text{cm}^{-2}$ for protonated materials, to $7 \times 10^{10} \text{ cm}^{-2}$ for deuterated ones.

However a sample with a <u>uniform</u> density of scattering length will produce no scattering at <u>small</u> angles. Scattering is produced by <u>fluctuations</u> in $\rho(\vec{R})$ within the sample : there must be regions (particles ...) where $\rho(\vec{R})$ differs from the mean scattering length density of the total sample. These regions, studied in small angle neutron scattering, are typically 10-1000 Å large. For particles in solution, the relevant quantity will be the <u>excess density</u> of scattering length with respect to the solvent, i.e. $\overline{|\rho(\vec{R}) - \rho_s|}$. It is also useful to define the contrast between one particle and the solvent, which is the average value of the excess density of scattering length :

$$\kappa = \bar{\rho} - \rho_s = \frac{1}{V} \int_V (\rho(\vec{R}) - \rho_s) \, d^3\vec{R} \tag{2}$$

where the integration is over the volume V of the particle. Thus high contrast is achieved with protonated micelles in D_2O, or per-deuterated ones in H_2O.

1. Single particle analysis

1.1. Scattered amplitude

The angular distribution of the scattered amplitude is a 3 dimensional Fourier transform of the distribution of excess scattering density :

$$A(\vec{Q}) = \int_V (\rho(\vec{R}) - \rho_s) \exp(i\vec{Q}.\vec{R}) \, d^3\vec{R} \tag{3}$$

where the scattering vector $\vec{Q}$ is the difference between the wave vectors of incident and scattered neutrons :

$$\vec{Q} = \vec{k} - \vec{k}_o, \quad |\vec{Q}| = \frac{4\pi}{\lambda} \sin(\Theta), \quad 2\Theta = \text{scattering angle}$$

Much of the information contained in $A(\vec{Q})$ or $\rho(\vec{R})$ is lost when the scattering particle can take random orientations in the solution; what remains is the isotropic averages :

$$\rho(R) = \int_o^{4\pi} \rho(\vec{R}) \, d\omega \quad \text{and} \quad A(Q) = \int_o^{4\pi} A(\vec{Q}) \, d\omega \tag{4}$$

When the particle has a center of symmetry, these are still related to each other by a Fourier transformation :

$$Q \, A(Q) = 4\pi \int_o^{\infty} R[\rho(R) - \rho_s] \sin QR \, dR \tag{5}$$

Thus the set of reciprocal Fourier transforms relating the scattered amplitudes to the scattering densities can be summarized as follows :

$$\rho(\vec{R}) - \rho_s \xleftrightarrow{\text{3 dim F.T.}} A(\vec{Q})$$

iso. average $\downarrow$ $\qquad\qquad\qquad\qquad$ $\downarrow$ iso. average

$$R(\rho(R) - \rho_s) \xleftrightarrow{\text{sine F.T.}} QA(Q)$$

In general, we do not have access to the scattered amplitudes. Indeed, the neutron detector measures the flux of incident neutrons, i.e. the intensity :

$$I(\vec{Q}) = A(\vec{Q}) \, A^*(\vec{Q}) = |A(\vec{Q})|^2 \tag{6}$$

In this process the information on the phase of $A(\vec{Q})$ is lost. An exception is the case of spherically symmetrical particles, where $A(\vec{Q})$ is identical with $A(Q)$, which is a real number.

1.2. Scattered intensity

$$I(\vec{Q}) = \int_V P(\vec{r}) \cos \vec{Q}.\vec{r} \ d^3\vec{r} \tag{7}$$

where

$$P(\vec{r}) = \int_V [\rho(\vec{R}) - \rho_s][\rho(\vec{R} + \vec{r}) - \rho_s] \ d^3\vec{R} \tag{8}$$

is the Patterson function of the particle. Again these functions suffer an isotropic averaging when the particle can take random orientations, yielding only :

$$I(Q) = \int_0^{4\pi} I(Q) \ d\omega \quad \text{and} \quad P(r) = \int_0^{4\pi} P(\vec{r}) \ d\omega \tag{9}$$

The set of reciprocal transformations relevant to the intensities can be summarized in the following way :

$$
\begin{array}{ccc}
\text{(distribution } P(\vec{r}) & \xleftarrow{\quad \text{3dim FT} \quad} & I(\vec{Q}) \\
\text{of vectors)} & & \\
\quad \big\downarrow \text{ iso.average} & & \quad \big\downarrow \text{ iso. average} \\
\text{(distribution } rP(r) & \xleftarrow{\quad \text{sine FT} \quad} & QI(Q) \\
\text{of distances)} & &
\end{array}
$$

It is also useful to write out explicitly these relationships :

$$I(Q) = \int_0^\infty 4\pi r^2 P(r) \ \frac{\sin Q \ r}{Qr} \ dr \tag{10}$$

$$P(r) = \int_0^\infty \frac{1}{2\pi^2} \ Q^2 I(Q) \ \frac{\sin Q \ r}{Qr} \ dQ \tag{11}$$

Of course the ultimate aim of the experiment is to proceed back from the measured intensity to the density of scattering length $\rho(\vec{R})$, and from that to the distribution of protonated (CH_2, CH_3, H_2O) and deuterated (CD_2, CD_3, D_2O) groups in and around the particle. In the case of spherical, monodisperse particles such as latexes, lipoproteins and some viruses the above Fourier transformation have been used to calculate the excess density of scattering length from the data. In all other cases, the averaging over particle orientations and particle sizes prevents us from deducing the structure, i.e. $\rho(\vec{R})$, from the scattering, i.e. $I(Q)$, through a mathematical procedure. Then the classical strategy is the following.

(a) Maximize the amount of information which can be obtained from
the data. This is achieved by measuring the scattered intensity not
only as a function of the scattering vector Q, but also as a function of the scattering length density of the solvent ρ_s. This amounts
to changing the contrast of the particles with respect to the solvent, thus the name "contrast variation".
(b) Calculate from the scattering the largest possible number of
geometrical parameters characterizing the structure of the particles.
(c) Use these parameters together with some basic knowledge on the
structures of the molecules to build some structural models; then
compare the scattering curves predicted according to these models
to the experimental ones.

    1.3.   Contrast variation.

    As described in equation (3) to (8), the scattered intensity
is controlled by the difference between the local density of scattering length in the particle $\rho(\vec{R})$ and that of the solvent ρ_s.
Thus by changing ρ_s it is possible to enhance or suppress various
features of the structure of the particle. Consider for example a
particle made of 2 regions, one of density ρ_1 and the other of density ρ_2; then for $\rho_s = \rho_1$ the observed scattering will be produced
exclusively by the region with density ρ_2 and vice versa. Obviously this requires that ρ_s be changed with minimal disturbance to the
chemical and physical equilibria of the system; in neutron scattering this is conveniently achieved through isotopic substitution :
typically H_2O/D_2O mixtures are used, covering almost the whole
range of scattering length densities found in neutron scattering.
    In the general case of a particle with arbitrary $\rho(R)$, the
scattering will be determined by 3 basic functions which describe
respectively the external shape of the particle, its internal structure, and a crossed term between external shape and internal structure [17]. Therefore it will be necessary to perform experiments for
at least 3 values of ρ_s; in practice a larger number of solvent
mixtures is used, and the data are collected in the widest possible
domain of contrasts $K = \bar{\rho} - \rho_s$ and scattering vectors Q.

    Some simple <u>geometrical parameters</u> can also be calculated from
each scattering curve obtained in one particular solvent mixture.
Again, because the observed profile of scattering density depends
on ρ_s, the values of these parameters will also vary with ρ_s. The
mathematical expressions for these parameters will be given in the
next section. At this point, however, it is useful to focus on the
$Q \to 0$ limit of the scattered intensity, because its variation with
ρ_s provides a standard way to check the <u>composition of the particle</u>.
From equations (8), (9), and (10) this limit can be expressed as :

$$I(Q \to 0) = \int_0^\infty 4\pi r^2 P(r) \, dr = \left[V(\bar{\rho} - \rho_s) \right]^2 \qquad (12)$$

where V is the volume of the particle which is excluded to the solvent. Thus a plot of $I^{1/2}$ vs. $\bar{\rho} - \rho_s$ should be linear, with a cancellation of the intensity in the solvent whose density ρ_s matches $\bar{\rho}$. In this way $\bar{\rho}$ is measured; from this information the composition of the particles can be determined, if the molar volumes are known. This information is useful when the particles are aggregates of more than one type of molecule, e.g. mixed micelles; indeed the composition of such mixed micelles may differ from the overall composition of solutes in the solution, because one component may be only partly aggregated.

More generally, a micellar solution in a multi-component system may contain many types of particles, differing by their compositions as well as by their structures [28]. The contrast variation experiment is particularly useful in this case. Indeed, equation (12) can also be applied to a solution containing n particles labeled as i, j, k..., in the case where the amplitudes scattered by the various particles are not coherent with each other (this implies that interparticle interferences are neglected; they will be discussed in § 2). Then the total intensity is the sum of the intensities scattered by all the particles :

$$I\ (Q \rightarrow 0) = \sum_{i=1}^{n} \left[V_i (\bar{\rho}_i - \rho_s) \right]^2 \tag{13}$$

If all the particles of the solution share the same composition, then there will still be one solvent whose density ρ matches all ρ_i's simultaneously; this solvent will cancel the intensity scattered by the solution. However, if the solution contains a mixture of particles with different compositions, than the intensity will be a sum of independent squared terms corresponding to each type of particle, and no solvent can cancel them all. The (intensity)$^{1/2}$ vs. contrast plot will only show a minimum instead of going through zero, and the height of this minimum measures the dispersity in the composition of the particles.

Obviously this procedure will yield significant results only if the ρ_i's of different particles have different values. Figure 1 shows an example of such a contrast variation experiment for a micellar solution which contains water-soluble plymers (PEO) adsorbed on the surface of SDS micelles [28]. A priori, such solutions could contain a mixture of pure SDS micelles, isolated PEO macromolecules, and mixed PEO + SDS aggregates of various compositions. In order to separate the scattering densities of these various particles, we used perdeuterated SDS ("d_{25}", $\rho = 6.73 \times 10^{10}$ cm^{-2}) and protonated PEO ($\rho = 0.68 \times 10^{10}$ cm^{-2}) ;[25] H_2O/D_2O mixtures were used as solvents of variable ρ_s. According to the data, the square root of I $(Q \rightarrow 0)$ is a linear function of the solvent's scattering density ρ_s; therefore these particular solutions only contain a single class of particles. Then the value of ρ_s which would cancel I$(Q \rightarrow 0)$

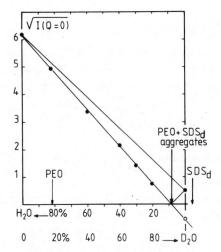

Figure 1: Contrast variation for aggregates made of surfactant micelles adsorbed polymer molecules in water.

must be equal to $\overline{\rho}$, the average scattering density of the particles; from $\overline{\rho}$, the composition of the particles can be calculated; it turns out to be identical with the overall composition of solutes in the solution (SDS, PEO). Thus in PEO + SDS solutions made with this particular composition, all the available solutes are used up in building one class of PEO + SDS aggregates; of course this may not hold for solutions of different compositions [28].

Remark that for <u>pure</u> SDS micellar solutions the contrast variation experiments are of little use; indeed , all the micelles in such a solution will share for $\overline{\rho}$ that of SDS, which can be calculated from known values of nuclear scattering lengths and molar volumes [28]. The hydration water which is associated with the micelles is not seen in a contrast variation experiment; indeed its composition is always identical with that of the solvent as a whole.

1.4.   Geometrical parameters.

The apparatus yields an experimental intensity curve K.I(Q) where I(Q) is the normalized intensity defined in § 1.2, and K can be determined by measuring the scattering of H_2O under the same conditions [21]. A number of geometrical parameters of the particles can be extracted directly from the normalized intensity curves.

(i) Total content · The limit of the normalized intensity for Q → O is given by eq (12); thus one measures :

$$KI(Q \rightarrow 0) = K [V (\overline{\rho} - \rho_s)]^2 \qquad (14)$$

where $\overline{\rho} - \rho_s$ is the contrast of the particle with respect to the
solvent, which is determined through a contrast variation experiment, and V is the volume of the particle which is inaccessible to
the solvent. Thus from a determination of K, $\overline{\rho}$, and $K.I(Q \rightarrow 0)$ one
can get the "dry volume" V and the "apparent mass" $V\overline{\rho}$; for a micelle, this will yield the aggregation number.

For these measurements of $I(Q \rightarrow 0)$ the single particle analysis can also be retained in the case of a real solution containing
a finite concentration C of particles; however it is necessary to
extrapolate the measured intensities down to vanishingly small particle concentrations, assuming that the structure of the particles
does not depend on C (this may not hold for micelles). Then an
equation similar to (14) applies, with $I(Q \rightarrow 0)$ replaced by I/C
at $(Q \rightarrow 0, C \rightarrow 0)$. At finite concentrations the reduced intensity
I/C will differ from its extrapolated limit by a factor
$S(Q \rightarrow 0, C)$ which is the compressibility of the "gas" of particles.
For particles interacting like hard spheres one can show that
$S(Q \rightarrow 0, C) = 1 - 8\phi$ where ϕ is the volume fraction of the hard
spheres in the solution. For particles interacting through long
range electrostatic potentials the factor $S(Q \rightarrow 0, C)$ is much
lower; its calculation is presented in § 2.

(ii) "Total fluctuation"   From equations (11) and (8), the measurement of the integral of the scattered intensity yields the value of the correlation function P(r) at r = 0 :

$$\frac{K}{2\pi^2} \int_0^\infty Q^2 I(Q) \, dQ = K \cdot P(r = 0) = K V \overline{[\rho(\vec{R}) - \rho_s]^2} \qquad (15)$$

where the squared fluctuation of the scattering density is averaged
over the volume of the particle. This reduces to a particularly
simple expression when this density is uniform within the particle,
i.e. $\rho(\vec{R}) = \rho$

$$K \cdot P(r = 0) = K.V (\rho - \rho_s)^2 \qquad (16)$$

or if it is a multistep distribution with n constant levels ρ_i
for volumes V_i within the particle :

$$K.P (r = 0) = K.\sum_i V_i (\rho_i - \rho_s)^2 \qquad (17)$$

In the general case Luzzati et al.[19] have shown that the ratio
$K.P (r=0)/K.I(Q \rightarrow 0)$ can be extrapolated to high contrasts, yielding a direct determination of V without having to measure K.

Remarkably, the integral of the scattered intensity is largely insensitive to the values taken by I(Q) at small Q. Thus the single particle analysis can be retained in the case of real solutions
of finite, but small, concentration. Indeed, in such solution the
interparticle interferences will occur at large distances, and therefore affect I(Q) at small Q. Hence it will not be necessary to
extrapolate towards C → 0 the value of the integral of the intensity
The same applies to the Porod limit of I(Q)., which will be discussed now.

(iii) "Total interface". For particles with sharp boundaries, the
correlation function P(r) can be expanded at small r. Since rP(r)
and QI(Q) are Fourier transforms of each other, this yields an
expansion for the high Q part of the scattering curve. In the simplest case of a particle with uniform scattering density these
expansions are :

$$P(r) = P(0) \cdot (1 - \frac{S}{4V} r)$$

$$I(Q) = 2\pi(\bar{\rho} - \rho_s)^2 S Q^{-4} + O(Q^{-6}) + \text{oscillating terms } (Q^{-3}) \qquad (18)$$
$$Q \to \infty$$

where S is the external surface of the particle, and the oscilla-
tiong terms have a pseudoperiod QD/2, D being the largest dimension
of the particle.

For a particle with n constant levels of scattering density
ρ_i separated by interfaces of areas S_{ij}, the scattering curves still
follow Porod's law (Q^{-4} behavior), but with a lower asymptotic
value :

$$\lim_{Q \to \infty} KQ^4 I(Q) = 2\pi K [(\rho_1 - \rho_2)^2 S_{12} + \ldots + (\rho_n - \rho_s)^2 S_{ns}] \qquad (19)$$

On the other hand, for a particle with diffuse boundaries, the
intensity drops much faster than Q^{-4} at large Q. In the case where
the interface is a step function smoothed by a gaussian of width σ,
the asymptotic trend is [22]

$$KI(Q) \simeq 2\pi K (\bar{\rho} - \rho_s)^2 S Q^{-4} \exp(-\sigma^2 Q^2) \qquad (20)$$

(iV) Radius of gyration

The intensity scattered at very small Q can be approximated
by the Guinier expansion :

$$I(Q) = I(Q \to 0) \cdot (1 - Q^2 R_g^2/3) \qquad (21)$$

where R_g is the radius of gyration for the distribution of excess
scattering density :

$$R_g^2 = \frac{\int_V (\rho(\vec{R}) - \rho_s) \cdot R^2 \, d^3\vec{R}}{\int_V (\rho(\vec{R}) - \rho_s) d^3\vec{R}} \tag{22}$$

and the origin of vectors R has been taken at the center of gravity of $\rho(\vec{R}) - \rho_s$. In general, this radius will depend on the value of ρ_s; this dependence takes a simple form if one uses the contrast defined in equation (2) :

$$R_g^2 = R_{gv}^2 + \frac{\alpha}{\overline{\rho} - \rho_s} - \frac{\beta}{(\overline{\rho} - \rho_s)^2} \tag{23}$$

In this expression R_{gv} is the radius of gyration for a particle of the same shape as the scattering particle, but with a uniform distribution of scattering density; this term is independent of contrast, and dominates at high contrast. The two other terms measure the inhomogeneities of the scattering density within the particle; they can be of either sign, and will dominate at low contrast.

For particles of arbitrary shape, the expansion (18) is a good approximation only when $QR_g \ll 1$. If an approximate shape is known a priori, it becomes possible to use other forms of this expansion, which minimize the higher order terms. The following expansions are valid up to $QR_g \sim 1$. For compact particles (Guinier) :

$$I(Q) = I(0 \to 0) \cdot \exp(-Q^2 R_g^2/3)$$

For random coils (Zimm) (23)

$$1/I(Q) = \left\{ 1/I(Q \to 0) \right\} \cdot (1 + Q^2 R_g^2/3)$$

Note that such expansions, when used at $QR_g \sim 1$, can yield aberrant results if the shape of the particle is not the expected one (for example if it is a long rigid rod)

Extreme care has to be excercised when applying the single particle analysis to measurements of R_g in a real solution of finite particle concentration [15]. Just as for the measurements of $I(Q \to 0)$, the relevant range of Q is one where strong interparticle interferences can be observed. However, whereas the value of $I(Q \to 0)$ can always be extrapolated to $C \to 0$, the extrapolation of R_g to $C \to 0$ may not be valid.

The "best" case is that of particles interacting through very short range potentials (hard sphere and the like). Then extrapolations to $C \to 0$ are commonly attempted. A typical case is that of uncharged polymer coils dissolved in a solvent : there an equation similar to (21) is retained for the values at finite concentration

of $I(Q)$ and R_g; both can be extrapolated linearly to vanishingly
small concentrations C[23].

A bad case for applying the single particle analysis is that
of particles interacting through long range electrostatic potentials. Then instead of the parabolic behavior expressed by equation (21), the scattering curve $I(Q)$ may show a peak at low Q.
In this case attempts to extrapolate towards $C \to 0$ will be futile,
as the peak will move but not disappear. Worst of all, the peak
may be located at Q values below the smallest Q available for the
experiment; then the experimenter may believe that he is measuring a
single particle radius, whereas he is in fact measuring the shape
of the tail of the interference peak.

A common practice is to screen out such electrostatic potentials by adding salt to the solution;for SDS micelles the addition
of 0.4 M of monovalent salt will screen out the electrostatic repulsions to the point where they are balanced by the van der Waals
attractions[24]. Of course, the structure of the micelles may no
longer be the same at such high ionic strengths[25-27].

A better way to handle this problem is to perform an exact
separation of the interparticle interference term from the observed scattering curve to obtain the single particle scattering
function. This is presented in the section 2.

1.5.   Model fitting

As indicated in equations (4) and (9), much of the structural
information is lost through the averaging over particle orientations and particle sizes. Nevertheless, if the scattering curves
are measured over a wide range of Q vectors (typically 2 decades)
and contrasts (H_2O/D_2O mixtures as solvents), they should still
contain a large amount of information. Yet from the geometrical
analysis presented in § 1.4. we can only extract a small number
of parameters ; much of the information contained in the scattering
curves is left aside in those procedures. For example, the scattering curves of some micelles show damped oscillations superimposed
on an asymptotic Q^{-4} law[28] ; such oscillations are not used in the
determination of the geometrical parameters in § 1.1.4. How can
such information be taken into account ? The easiest way is to
build a priori a structural model for the particle, calculate the
scattering from this model, and then try to fit the whole set of
experimental scattering curves. Two important choices have to be
made in this procedure. Firstly, the number of parameters used to
describe the model should be choosen to be smaller than or equal
to the smallest number of parameters which are necessary to describe the set of scattering curves[29]. Secondly, there is some

choice in the representation of the scattering curves : plots such
as I(Q) vs. Q emphasize the data at small Q, others such as
Log I vs. Log Q, IQ^4 vs. Q, IQ^4 vs. Q^4 emphasize the data at large
Q; it is wise to use them all.

First take the simplest possible model : a single sphere of
uniform scattering density $\rho(R)$, with a perfectly sharp interface.
Then $\rho(R)$ is a step function, with $\rho(R) = \rho$ for $0 < R < R_o$ and
$\rho(R) = 0$ beyond. The scattered amplitude calculated through a
Fourier transformation of $\rho(R)$ is a decreasing oscillating function:

$$A(Q) = V\rho \cdot \phi(QR) \quad \text{with} \quad \phi(x) = 3 \ (\sin x - x \cos x)x^{-3} \quad (24)$$

and the scattered intensity is the square of A(Q), i.e. a decreasing
function with pseudooscillations of period $QR_o = 2\pi$ superimposed on
a Q^{-4} law. Pictures of this scattering curve are found in the book
by Guinier and Fournet [15] as well as in other papers [1,2,16].

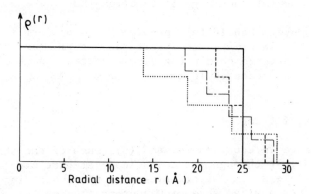

Figure 2. Scattering densities for model particles made of
concentric spherical cells (arbitrary units).

Then take a model made of many concentric sphericals shells :
here $\rho(R)$ is equal to ρ_j in each shell j, for $R_{j-1} < R < R_j$ (Figure 2).
The scattered amplitude is a sum of oscillating functions with different periods :

$$A(Q) = \sum_{\text{shells}} V_j \ (\rho_j - \rho_{j+1}) \ \phi \ (QR_j) \quad (25)$$

At low Q vectors the contributions of all shells are in phase, i.e.
they interfere constructively. Then it does not matter whether the
model contains a single sphere or many concentric ones : as shown
in Figure 3, the resolution is not sufficient to distinguish between
them :

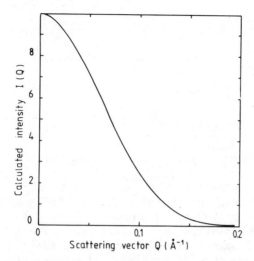

Figure 3 : Scattering curves for model particles made of concentric spherical shells. In this range of Q vectors the scattering curves corresponding to the 4 density models shown in Figure 2 are indistinguishable from each other.

Hence data taken in this range of Q vectors will only support the determination of 2 parameters : a mass or aggregation number, and an average radius. On the other hand, data taken at high Q vectors should be more interesting : in the region where $\phi(QR)$ oscillates, the contributions of the various shells will become out of phase and will interfere destructively. Then the calculated intensity will drop much lower than that of a single sphere. This loss of intensity at high Q is in agreement with equation (19), which predicts a faster decrease of the intensity for particles with diffuse boundaries.

In this range of Q vectors the intensity actually drops so fast that a plot of I(Q) vs. Q is no longer appropriate. A better choice is a plot of $I(Q).Q^4$ vs. Q : then the single sphere gives oscillations around a constant value, whereas the models made of concentric shells still give a fast decrease with Q. The curves are shown in Figure 4 . Note that we have assumed a small polydispersity of the spheres, this has the effect of damping the oscillations.

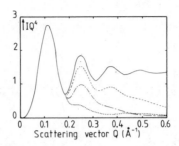

Figure 4. Scattering curves for model particles made of concentric shells, the plot of IQ^4 vs. Q emphasizes the data taken at high Q, which control the spatial resolution. Full line : single sphere of uniform density. Dashed lines : concentric shells, as depicted in Figure 2. For each type of particle, the calculated scattering curve is an average over many particles of slightly different sizes, according to a gaussian distribution of particle weights with a width σ/M_n = 0.3.

          Thus for strictly spherical particles, the data taken at "high Q" (QR $\sim 2\pi$) can discriminate between particles with sharp vs diffuse boundaries. A similar analysis would apply to other simple geometries such as disks or spherocylinders. However, the scattering from particles with an irregular surface is much more difficult to analyze. Consider a sphere with a compact inner core surrounded by an outer layer which contains <u>large</u> voids open to the solvent, and let a be the lateral size of such voids. When Q < 1/a, individual voids are not resolved; the particle will scatter like a sphere with a dense inner core and a less dense outer layer; I(Q) will decrease faster than that for a sphere with a full outer layer. On the other hand, for Q > 1/a, individual holes are resolved, and the particle will appear as having a surface area larger than the corresponding sphere; according to equation (18), the intensity scattered at Q >> 1/a will be higher than that of a sphere with a full outer layer. A similar pattern is also expected for particles with "branches" (chains) sticking out of a compact core; in fact, we have observed this type of behavior in micelles of a polyoxyethylene alkyl ether. In general, it will be possible to find more than one structure which could produce such a composite behavior.

These cautionnary remarks can be summarized as follows

(i) Scattering data must be taken over a wide range of Q vectors, ideally from $Q\ell \sim 0.1$ to $QR \sim 10$ where ℓ and R are respectively the longest and shortest dimensions of the particle

(ii) Some structures can be recognized with no ambiguity, and their parameters determined with a high precision, because their scatter ing shows some characteristic features. For example, a uniformly dense sphere produces a unique pattern of oscillations in $I(Q)$; so does a cylinder, or a flat disk

(iii) When such characteristic features are absent from the scatter- ing, the only tractable approach is to use all the geometric para- meters which can be obtained from the scattering curves, together with some basic knowledge on the shape of the molecules, to build a structural model; then check wheter this model can reproduce the whole scattering curves (a trial and error routine) and whether it can also explain the properties of the particles; finally consider whether it is unique in this respect.

2. Interparticle interferences

Scattering experiments are usually conducted on a sample con- taining a very large number of particles. In a solution where the positions and orientations of the particles are totally uncorrela- ted, the amplitudes scattered by the particles are not coherent with each other, and the intensities add up as indicated by equa- tion (13). If all the particles are identical, the measured scat- tering curve will be the same on a relative scale as that of an isolated particle. If the solution is polydisperse, the quantities which are deduced from the measured scattering curve will be average quantities; for example the calculated volume V and radius of gyra- tion R_g will be weight averages of the distributions of particle volumes and radii.

However, the neutron scattering curves obtained from solutions of direct micelles in salt free water differ from the expected curve for a single micelle; in particular, they show a prominent peak at low Q vectors, which depends on the concentration of micelles Moreover, similar peaks are observed in the scattering curves of any other charged particles in water. As shown by Hayter and Penfold [30], those peaks are produced by interferences of neutrons scattered by nuclei located on <u>different</u> particles. Such interpar- ticle interferences would be averaged out if the particles were able to take completely random positions in the solution, as in a perfect gas. However when the particles repel each other, especial- ly through long range electrostatic potentials, their positions become correlated, and these correlations yield a non trivial fac- tor in the expression of the scattered intensity. A mathematical expression of these statements lies in the general expression for the intensity scattered by a group of identical particles which are randomly oriented in the solution [15] :

$$I(Q) = \bar{N} \left\{ I_1(Q) + A^2(Q) \left[S(Q) - 1 \right] \right\} \qquad (26)$$

where :
$$I_1(Q) = \int_0^{4\pi} A(\vec{Q}) A^{*}(\vec{Q}) \, d\omega$$

is the single particle intensity, averaged over particle orientations,

$$A(Q) = \int_0^{4\pi} A(Q) \, d\omega$$

is the single particle amplitude, also averaged over orientations,

$$S(\vec{Q}) = \frac{1}{\bar{N}} < \sum_{i\,j} \exp(i\,\vec{Q}.\vec{R}_{ij}) >$$

is the structure factor for the correlations between the particles, averaged over all orientations and positions of the particles, $\vec{R}_{ij}$ is the vector joining the centers of the particles, and N is the average number of particles in the scattering volume.

When the particles are spherically symmetric, $A(\vec{Q})$ becomes identical to $A(Q)$, and equation (20) reduces to :

$$I(Q) = I_1(Q) \cdot S(Q)$$

An analytic form of the structure factor $S(Q)$ for charged particles in water has been found by Hayter and Penfold [31]. Using this form, a self consistent fitting procedure permits the extraction of $I_1(Q)$ from $I(Q)$ [32]. As an example, we present here in Figure 5 the results of this procedure for data obtained with a 2% micellar solution of SDS in water, with no salt added.

The steps followed in the use of this fitting procedure are as follows [30-32]. The program first calculates $I_1(Q)$ and $S(Q)$ on the basis of an approximate model for the structure of the micelles. The shape of the resulting theoretical scattering curve $I_1(Q) \cdot S(Q)$ is compared with that of the experimental curve; then a fitting program changes the values of some structural parameters of the model micelles in order to improve the quality of the fit; thus new functions $I_1(Q)$ and $S(Q)$ are calculated, and the whole procedure is iterated until the quality of the fit no longer improves. At this point one obtains the "best" values of the structural parameters from which the last set $I_1(Q)$, $S(Q)$ was calculated.

We still have to specify which model was used for the structure of the micelles and which were the parameters of the fit. For the calculation of $I_1(Q)$ it is convenient to consider spherical micelles made of concentric shells of uniform density; then the parameters required to calculate $I_1(Q)$ are the radii and scattering length densities for each shell and for the solvent. We restricted our-

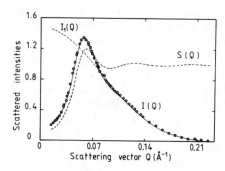

Figure 5. Scattering data from a 2% solution of SDS-p25 in D_2O ; at this concentration most SDS molecules are aggregated in globular micelles. Circles : (I sample-I background)/IH_2O. Dashed lines : intermicellar structure factor S(Q) and single micelle scattering function $I_1(Q)$, as extracted through the SQHP program. Full line : $I_1(Q)$. S(Q), arbitrary vertical scale.

selves to micelles made of two regions : a dense core (no water) containing all methyl and methylene groups exept for the methylene groups closest to the heads (α position); and surrounding this core a polar layer containing the α methylene groups, the charged sulfate groups, all the water molecules which hydrate the surface of the micelle (9 water molecules per SDS molecule), and those counterions which are condensed on the surface; the remaining counterions are randomly dispersed throughout the solvent. The scattering from such a micelle is fully determined if one specifies : (i) the scatttering lengths of the various chemical groups of a SDS molecule, i.e. CH_3, CH_2, SO_4^- , Na^+ , and eventually CD_3 , CD_2; (ii) the number N of SDS molecules in the micelle; (iii) the charge Z,

which is the number of counterions not condensed in the polar
layer. For the calculation of S(Q) we need the hard sphere fraction
ϕ of the micelles in the solution, the electrostatic screening
length which can be deduced in the Debye-Hückel approximations
from the ionic strength of the solution, and the charge Z of the
micelle. Thus for a given micellar solution $I_1(Q)$ and S(Q) can be
calculated according to one set of fixed parameters (SDS concentration C, screening length, scattering lengths b_i) and one set
of free parameters (aggregation number N, charge Z); N
and Z are the only parameters varied in the fit.

At this point it is worth mentioning that the absolute value of
the scattered intensity is not used as a criterion for the fit, i.e.
the program is free to provide theoretical scattering curves which
differ from the experimental ones by a scale factor Iexp/Itheor.
Yet for the present structural model of the micelles, the absolute
value of the intensity is also determined by N and Z ; therefore
Iexp/Itheor should be found equal to 1 within the precision for the
measurement of absolute intensities, i.e. $\pm$ 30%. Thus the value of
this scale factor can be used to assess the validity of the structural model.

For fully protonated or fully deuterated SDS micelles at high
contrast, with C = 2%, T = 25°C, and no added salt, the best fit is
obtained for the following values of the parameters :

SDS-p25/D_2O N = 73.6   Z = 28   with Iexp/Itheor = 0.91
SDS-d25/H_2O N = 86.1   Z = 24   with Iexp/Itheor = 0.89

In order to see whether the fit is meaningful, we tested it in
two ways: (i) Start with different initial values of N and Z, and
observe whether the procedure converges towards the same final values
of N and Z; we find that it does (ii) given final set of N and Z,
modify the scattering lengths b_i to predict the scattering curve observed for SDS molecules labeled in a different way, or dissolved in
a different solvent; the prediction is good for the shape of the
curves, but for solutions at low contrast the value of Iexp/Itheor
is too low. This might have been expected, as the model is too sim-
plistic to describe the micelles at low contrast.

In summary, the quality of the fit, the proper convergence of
the procedure, and the value of the intensity scale factor obtained
at high contrast indicate that the separation of $I_1(Q)$ and S(Q)
achieved by the procedure is meaningful and unique.

THE WATER/HYDROCARBON INTERFACE OF SDS MICELLES

One problem of current interest is the structure of water/hy-

drocarbon interfaces : is water/hydrocarbon contact limited to a
thin layer, or is there substantial water penetration between the
aliphatic chains of surfactant molecules ? This problem involves
the measurement of short distances, which are reflected in the high
Q region of the scattering curves. For example, sharp interfaces
will yield scattering curves which follow Porod's law, $I(Q) \alpha\ Q^{-4}$,
equation (18), while diffuse boundaries will produce a scattering
which drops much lower, equations (19) or (20). For this approach
to be used, the data obtained at high Q must be sufficiently reliable; here is a brief account of the relevant constraints.Three factors control the quality of the data obtained in the high Q region
of small angle scattering. These are the contrast between the particles and the solvent, the selection of the Q values, and the
substraction of backgrounds.
-Contrast- As explained in the introduction, high contrast between
hydrocarbon and water can be achieved by using either SDS-p25 in
D_2O or SDS -d25 in H_2O. Because H_2O produces a large incoherent
scattering it is much easier to use SDS-p25 in D_2O

-Q values are selected in 2 stages. Firstly, the wavelength of the
incident neutron beam is selected; we used the instrument DIB at
ILL, whose dispersion in wavelength is 1%. Secondly, scattered neutrons are counted according to their scattering angle. This procedure defines the values of Q, provided that multiple scattering
events can be neglected, and that inelastic scattering does not
result in a change of the average momentum transfer. We checked
that both effects are indeed negligeable in our experiment.

- The collected intensities contain a number of background contributions : incoherent scattering , mostly from H nuclei, and coherent scattering from the solvent. The usual procedure for handling
this problem is to substract from the "sample spectrum" a "background
spectrum" which is obtained from a solvent containing the same
number of H nuclei as the sample. This is not an exact procedure;
furthermore its accuracy becomes insufficient at high Q(OR $\sim 4\pi$),
where the structure-related intensity may have decreased by a factor of 10^3 from its low Q limit. This problem can easily be solved
when the structure related term follows a simple law in Q; then the
measured intensity can be plotted in a way which does not depend
on the accuracy of the background substraction. For example, the
intensity obtained with an approximate substraction of a background
can be plotted as $I(Q)$ vs. Q^{-4}. If the structural term follows
Porod's law, the total intensity will be of the form $AQ^{-4}+B$, where
B represents the residue from the background substraction; then a
straight line will be obtained, yielding a determination of A
(Porod's limit, equation (18)), and B(residue). Alternatively the
data can be plotted as $I(Q)\cdot Q^4$.vs.Q^4. If the structural term does
not follow Porod's law, the data will not follow a straight line
in these plots, and it will be necessary to find another way to
plot them.

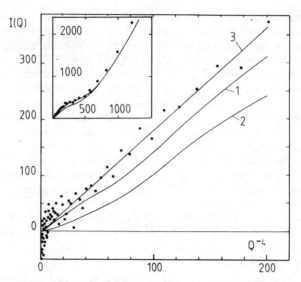

Figure 6. Asymptotic trend at high Q for SDS-p25 in D_2O; surfactant
concentration 2%; added salt NaBr 0.1 M. The scattering vectors Q
of the experiment range from 0.26 $\mathring{A}^{-1}$ at the right hand side of the
plot (distances ∿24 $\mathring{A}$) to 1 $\mathring{A}^{-1}$ at the left hand side (distances ∿
6 $\mathring{A}$). Curve 1: fit by spheres of uniform scattering density. Curve
2: fit by spheres with diffuse boundaries. Line 3: linear fit according to Porod's law. The insert shows the same data over a wider
range of Q vectors (from 0.15 $\mathring{A}^{-1}$).

        Figure 6 shows this type of plot for a micellar so-
lution of SDS; the data are represented over a wide range of Q vectors in the insert, and show the distinctive oscillations of spherical particles, equation (24). The high Q tail of the data (left
hand side of the insert, from QR = 5 to QR = 20) is expanded below
(dots); it can be approximated by a Q^{-4} decrease towards I = 0
at $Q \to \infty$. For clarity, the residue from the background substraction
has already been substracted out; otherwise it would show up as a
non-zero intercept on the I(Q) axis.

A reasonably good fit to the data can be obtained with the theoretical curve for a solution of spheres, each of which has a uniform scattering density (curve 1). Yet in the asymptotic range this curve is systematically lower than the data; according to Porod's law this indicates that SDS micelles have a larger surface area than the corresponding spheres.

Curve 2 of Figure 6 shows an attempt to fit the data by the curve for spheres with diffuse boundaries, according to equation (20), with a boundary thickness $\sigma = 2$ Å. Although the fit is acceptable at low Q (large values of Q^{-4}), it fails in the asymptotic range: compared with the data, the predicted intensities are too small by a factor of 2.

This comparison makes it possible to exclude for SDS micelles any structural model with a <u>diffuse</u> water/hydrocarbon interface, where diffuse means that the density of hydrocarbon would decrease smoothly across the interface over a range of a few Å. This does not exclude more complicated structures, such as interfaces with <u>large</u> holes, bumps or outgrowths; in order to discriminate between such structures it is necessary to study the whole scattering curve (see the next section).

That SDS micelles must have a sharp water/hydrocarbon interface had already been shown by the NMR experiment [10]; that technique has a much better resolution than the scattering experiment, but measures different distances; thus the agreement between both methods is significant.

Finally it is noteworthy that <u>reverse</u> micelles may exhibit a different behavior; Tavernier, Vonk and Gijbels have found that the asymptotic scattering from zinc di-isohexadecyl sulfonate in iso-dodecane drops faster than Porod's law; they have concluded that the zinc/hydrocarbon interface must be diffuse, with a width of 5 Å [33].

THE SHAPE OF SDS MICELLES

Now that the asymptotic trend is established, it is best to use a representation where the intensity tends towards a constant value at high Q. Such was the plot of $I(Q) \cdot Q^4$ versus Q used in Figure 3, where the scattering curves from full spheres had been compared with those from concentric shells. Figure 7 shows in this representation our data for SDS-p25 in D_2O, with no salt added.

The main feature of these data is the damped oscillations which are reminiscent of the theoretical curve for the scattering by a single sphere (curve 1). Yet the IQ^4 vs. Q^4 plot reveals that beyond $Q = 0.15$ Å^{-1} the data diverge from this theoretical curve. This discrepancy can be accounted for in two ways: either it is

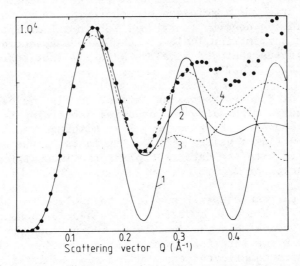

Figure 7. Scattering data for a 2% micellar solution of SDS-p25 in
D$_2$O, with no salt added (dots). Curve 1: fit by spheres with sharp
boundaries radius 20 Å. Curve 2: fit by a polydisperse solution of
spheres, average radius 20 Å; a gaussian distribution of the volumes has been assumed, with a 30% width. Curve 3: fit by monodisperse oblate ellipsoids, half diameters 23, 23 and 13.8 Å. Curve 4:
fit by polydisperse globules (see text)

assumed that the micelles are indeed spherical, but that there is
a significant dispersion in their aggregation numbers, or it is
assumed that they are globular but not spherical.

Curve 2 shows the fit obtained with the theoretical curve for
a solution of spheres with sharp boundaries, assuming a gaussian
distribution of their volumes. Remark that the intensity scales of
the fits are matched to the data near Q = 0; the first maximum of
the curves is reproduced by adjusting the average radius of the
spheres, and the depth of the first minimum by adjusting the dispersion of their volumes. Beyond this point there remain no adjustable parameters; discrepancies between the data and the fits correspond to deviations from the spherical shape.

Another indication that these micelles cannot be spherical is
provided by the value of the radius given by this fit. The precision on this value is quite good: R = 20 $\pm$ 0.3 Å; indeed, curves

for spheres with different radii would fail to match both the height
of the first peak in Figure 7 and the period of the oscillations.
This radius R corresponds to the large discontinuity in scattering
density at the hydrocarbon/water interface; in a spherical geometry,
a hydrocarbon radius of 20 Å would correspond to an aggregation
number $N = 95.7$. This is incompatible with the aggregation number
obtained from $I(Q \to 0)$ after the separation of intermicellar interferences: $N = 73.6$ (see Figure 4). Yet this last value can be checked
by measuring the integral of the scattering curve, which yields the
value of the correlation function $P(r)$ at $r = 0$. The aggregation
number obtained in this way does not depend upon the measurement
of absolute intensities; thus it is an independent measurement; its
value is $N = 72.6$, in excellent agreement with the value obtained
from $I(Q \to 0)$. Therefore the measured values of the hydrocarbon
"radius" and hydrocarbon volume cannot be reconciled within a
strictly spherical geometry.

The alternative approach, trying to fit the data with the
theoretical scattering curve of non spherical, monodisperse particles is more open-ended, in the sense that many different shapes
can yield similar scattering curves, because of the loss of structural information suffered in the isotropic average. Yet it is quite
possible to discriminate against some simple shapes. Curve 4 represents the "best fit" obtained with oblate ellipsoids; this is obviously not the correct shape; the fits obtained with prolate ellipsoids are even worse. On the other hand, acceptable fits are
obtained from globular particles whose shape is intermediate between
a sphere and a torus; if some polydispersity is allowed, such fits
can become better than those obtained with spheres (curve 4).

It has long been known that the addition of salt produces a
growth of the SDS micelles [25-27]. At first, this growth is rather
slow, until the added salt raises the Krafft point in the vicinity
of the temperature of the solution; then the growth is more rapid,
producing micelles with aggregation numbers on the order of 1000.
Accordingly, we have taken data for 2% SDS solutions with various
amounts of added NaBr.

Figure 8 shows the scattering curve obtained with 0.1 M NaBr;
in this case as well an acceptable fit is obtained with the curve
for polydisperse spheres, and a better one with that for pseudo-
toroidal globules.

Figure 9 shows the scattering curve obtained with 0.6 M NaBr
in the solution; in this case it is no longer possible to fit the
data with the curves calculated for solutions of spheres, regardless
of the radii or polydispersities which may be used. This is in
agreement with the results of light scattering experiments, which
indicate dramatic growth of the SDS micelles in this range of ionic

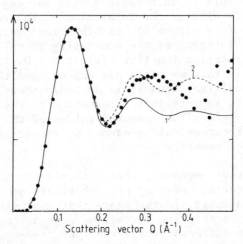

Figure 8. Scattering data for a 2% solution of SDS-p25 in D_2O at 25°C, with 0.1 M NaBr added. Curve 1: fit by a polydisperse solution of spheres, average radius 22.6 Å, width of the distribution 60%. Curve 2: fit by a polydisperse solution of pseudo-toroidal spheres.

strengths at 25°C [25-27]. More interestingly, our result excludes the possibility that this growth could happen by aggregation of spherical micelles which would preserve their individual shapes : such a "necklace" of spheres would give, in the high Q region, the same scattering curve as a single sphere.

The two simplest shapes into which a SDS micelle can grow are a cylinder and a disk. Figure 10 presents our data plotted in the appropriate representation for cylinders. The data taken in solution with 0.6 M and 0.8 M NaBr show a linear variation of Log $\{Q\ I(Q)\}$ with Q^2 : this is indeed the behavior expected for cylinders [15].

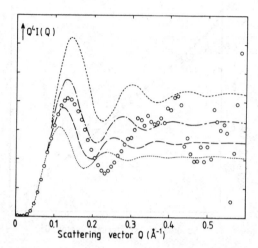

Figure 9. Scattering data with 0.6 M NaBr added. The lines are the theoretical scattering curves for solutions of spheres, with average radii ranging from 22 to 28 Å. All these curves are scaled to match the intensity scattered by the micelles at Q = 0.

The excess intensity scattered at low Q by the SDS solution with 0.8 M NaBr is presumably produced by the formation of microcrystals. We have also checked that the data cannot be fitted by the law expected for disks, i.e. Log $\{Q^2 I (Q)\}$ vs Q^2.

THE INTERNAL STRUCTURE OF SDS MICELLES

One of the goals listed in the Introduction was to resolve the locations of the various methylene or methyl groups of the SDS molecules within the hydrocarbon core of the micelle. Some structural information on this problem is contained in the scattering curves of specifically labeled micelles. However, this information has been submitted to the averaging processes discussed in § 1. For this reason, the most we can get is, for a spherical micelle, the radial distributions for each type of methylene or methyl groups.

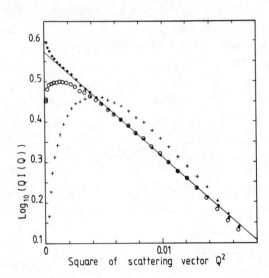

Figure 10. Scattering curves plotted in the appropriate representation for cylinders, for 2% SDS solutions at 25°C with various ionic strengths. Crosses: NaBr 0.1 M; Circles: NaBr 0.6 M; Dots: NaBr 0.8 M.

To this effect we have obtained the scattering curves for the eight micellar solutions mentioned in the Introduction. At low values of Q (below 0.15 Å^{-1}) these scattering curves are very similar to each other. This is not surprizing; indeed, the intermicellar structure factor S (Q) defined in § 1.2. should be the same in all these solutions. Also, the micelles in these eight solutions have the same outer radius, and therefore the low Q part of the single micelle scattering function should also be similar in all of them.

As in the previous sections, the spatial resolution required to discriminate between the distributions of scattering length density for these eight micelles is obtained by going to higher values of Q. As an example, we present in Figure 11 the high Q range of the scattering curves for SDS micelles with deuterated methyl groups. At high contrast, i.e. for SDS-12d$_3$ in D$_2$O, the methyl groups only contribute a small fraction of the excess scattering density $\rho(R)-\rho_s$. Accordingly, the scattering curve of SDS - 12 d$_3$ in D$_2$O does not differ much from that of the unlabeled material, SDS-p25 in D$_2$O. At

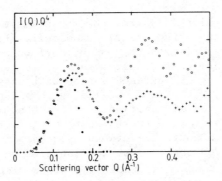

Figure 11. Scattering curves for micelles of methyl-deuterated SDS
molecules. Dots: SDS-12d$_3$ in H$_2$O (low contrast between the micelle
and the solvent). Circles: SDS-12d$_3$ in D$_2$O (high contrast).
Crosses: SDS-p25 in D$_2$O (high contrast, no labeling).

low contrast however, i.e. with H$_2$O as a solvent, the deuterated
methyl group contribute the major part of the excess scattering density (some excess scattering density also comes from the polar head
groups). Figure 11 shows that the scattering curve of SDS-12 d$_3$ in
H$_2$0 is very different from that of SDS-12 d$_3$ in D$_2$0 : the pattern of
oscillations has changed, and the Porod limit is much higher. Clearly,
the distribution of methyl groups within the micelle must be non
uniform. However, because the polar head groups contribute a fraction of the excess scattering density, the radial distribution of the
methyl groups cannot be extracted directly from this scattering curve.
Rather, it will be necessary to take the eight scattering curves together
and use them to build a model for the inner structure of the micelle.
This model should predict the radial distribution for each type of
methylene or methyl groups in the micelle ; the user should only specify the scattering lengths of these group, according to the labeling of the SDS molecule (partial or full deuteration) · Once these
scattering lengths are introduced in the model, it would yield a
distribution of excess scattering density · The scattering curve calculated from this distribution should then fit the scattering curve
observed in the corresponding conditions of labeling and contrast.
This work is in progress. At present we can only say that our set
of scattering curves contains more than enough information to build
such a model.

CONCLUSIONS

We are still a long way from the foreseeable end of this study. Yet at this stage the structure of SDS micelles has already revealed a much stronger organization than might have been expected. Much of the recent literature on micellar solutions suggests a rather fuzzy picture of micelles : ill defined shapes with many molecules half way in and half way out of the micelle, a rough surface and considerable interpenetration of water and hydrocarbon. These views now appear to be inconsistent with the scattering data obtained for SDS micelles with a high contrast between hydrocarbon and water. Indeed the scattering curves obtained at low ionic strengths show the from the spherical shape and dispersion in the aggregation numbers damped oscillations expected for dense globular particles. Deviations seem effects of comparable importance. More remarkably, the asymptotic limit of the scattering curves shows that the water/hydrocarbon interface of these micelles must be very sharp. Contact between hydrocarbon and water must be limited to a thin layer of thickness comparable to the dimensions of one CH_2 group and containing one quarter of all the CH_2 and CH_3 groups in the micelle[10]. The other CH_2 and CH_3 groups are packed at constant density within a dry core. That about on quarter of the CH_2 groups should be hydrated is consistent with the available thermodynamic and NMR data [34,35]. Of course the SDS molecules retain the possibility of frequently leaving or joining the micelle; the scattering data just imply that such events are fast and contribute little to the ensemble averaged structure.

This concept of a strict segregation between water and hydrocarbon, coupled with extensive water/hydrocarbon contact at the interface was already proposed by Hartley some 46 years ago [36] ; it might also extend to other water/hydrocarbon interfaces ; the driving forces behind it are the very strong hydrophobic forces [4]. It appears that the strength of such forces was overlooked by some of the contenders in the recent debate about the extent of water penetration within the core of the micelle [37,38].

ACKNOWLEDGEMENTS

It is a pleasure for us to thank J.B. Hayter for making available to us the procedure which calculates the intermicellar structure factor. Thanks are also due to both referees, whose criticism caused a significant improvement in the quality of this paper.

REFERENCES

1 C. Chauvin, J. Witz and B. Jacrot, J. Mol. Biol., 124,641(1978)
2 V. Luzzati, A. Tardieu and L.P. Aggerbeck, J. Mol. Biol., 131,
   435 (1979)
3 H. Wennerström and B. Lindman, Phys. Rep., 52, 1 (1979)
4 C. Tanford, "The Hydrophobic Effect", 2nd ed., Wiley, New York,
   1980
5 F. Reiss Husson and V. Luzzati, J. Phys. Chem., 68, 3504 (1964)
6 F. Reiss Husson and V. Luzzati, J. Colloid Interface Sci.,
   21, 534 (1966)
7 C. Cabos and P. Delord, J. Physique, Lettres, Paris, 38,
   L-365 (1977)
8 C. Cabos and P. Delord, J. Physique, Paris, 39, 432 (1978)
9 B. Cabane in "Magnetic Resonance in Colloid and Interface Science",
   J.P. Fraissard and H.A. Resing, Editors, p. 321, D. Reidel,
   Dordrecht , 1980
10 B. Cabane,J. Physique, Paris, 42, 847 (1981)
11 T. Zemb and C. Chachaty, Chem. Phys. Letters, 88, 68 (1982)
12 T. Zemb and C. Chachaty, These proceedings
13 D.W.R. Gruen, J. Colloid Interface Sci., 84, 281 and in these
   proceedings
14 K.A. Dill, J. Phys. Chem. , 86, 1498 (1982) and in these
   proceedings
15 A. Guinier, G. Fournet , "Small Angle Scattering of X-Rays",
   Wiley, New York, 1955
16 B. Jacrot, Rep. Progr.Phys., 39, 911 (1976)
17 H.B. Stuhrmann and A. Miller, J. Appl. Cryst., 11, 325 (1978)
18 V. Luzzati and A. Tardieu, Ann. Rev. Biophys.Bioeng. 9, 1 (1980)

19 V. Luzzati, A. Tardieu, L. Mateu and H.B. Stuhrmann, J. Mol. Biol.,
   101, 115 (1976)
20 D.J. Cebula, R.K. Thomas, N.M. Harris, J. Tabony and J.W. White,
   Faraday Discussions of the Chemical Society, 65, 76 (1978)
21 B. Jacrot and G. Zaccai, Biopolymers, 20, 2413 (1981)
22 W. Ruland, J. Appl. Cryst., 4, 70 (1971)
23 B.H. Zimm., J. Chem. Phys., 16, 1093 (1948)
24 M. Corti and V. Degiorgio, J. Phys. Chem., 85, 711 (1981)
25 N.A. Mazer, G.B. Benedek and M.C. Carey, J. Phys. Chem., 80,
   1075 (1976)
26 C.Y. Young , P.J. Missel, N.A. Mazer, G.B. Benedek and
   M.C. Carey, J. Phys. Chem., 82, 1375 (1978)
27 S. Ikeda, S. Hayashi and T. Imae, J. Phys. Chem., 85, 106(1981)
28 B. Cabane and R. Duplessix, J. Physique, Paris, 43, 1529 (1982)
29 D. Taupin and V. Luzzati, to be published
30 J.B. Hayter and J. Penfold, J. Chem. Soc., Faraday Trans. I.,
   77, 1851 (1981)
31 J.B. Hayter and J. Penfold, Mol. Phys., 42, 109 (1981)

32 "SQHP", a Fortran program available on the ILL program Library
33 S.M.F. Tavernier, C.G. Vonk and R. Gijbels, J. Colloid Interface
   Sci., $\underline{81}$, 341 (1981)
34 J.M. Corkill, J.F. Goodman and T. Walker, Trans. Faraday Soc.,
   $\underline{63}$, 768 (1967)
35 J. Clifford and B.A. Pethica, Trans. Faraday Soc., $\underline{60}$, 1483(1964)
36 G.S. Hartley "Aqueous Solutions of Paraffin Chain Salts",
   Hermann, Paris, 1936
37 F.M. Menger, J.M. Jerkunica and J.C. Johnston, J.A.C.S. ,
   $\underline{100}$, 4676 (1978)
38 For a critical analysis of this problem, see :
   H. Wennerström and B. Lindman, J. Phys. Chem., $\underline{83}$, 2931 (1979)

LIGHT SCATTERING AND SMALL-ANGLE NEUTRON SCATTERING INVESTIGATIONS OF DOUBLE-TAILED SURFACTANTS IN AQUEOUS SOLUTIONS

L. J. Magid[*,1,2], R. Triolo[2,3], E. Gulari[4], and B. Bedwell[4]

1. Department of Chemistry, University of Tennessee, Knoxville, Tennessee 37996
2. Chemistry Division, Oak Ridge National Laboratory, Oak Ridge, Tennessee 37830
3. Istituto di Chimica Fisica, Universitá di Palermo, 90123 Palermo, Italy
4. Department of Chemical Engineering University of Michigan, Ann Arbor, Michigan 48109

We have investigated aqueous solutions of sodium p-(1-pentylheptyl)benzenesulfonate (6-PhC$_{12}$SNa) at 45°C using static and dynamic light scattering (LS) and static small-angle neutron scattering (SANS) measurements. The SANS studies in D$_2$O were performed in the concentration range 0.0099-0.0991 M; Guinier radii (R_G) of 13.7-16.6 Å were determined. Scattering curves for the more concentrated solutions show an interaction peak which arises from intermicellar correlations; the effect of added NaCl, at a constant surfactant concentration of 0.0845 M, on these interactions was also studied. The SANS measurements were complemented by static and dynamic LS studies on 0.0845 M 6-PhC$_{12}$SNa in H$_2$O, with and without added salt. Non-single exponential behavior was observed in the intensity autocorrelation functions, and two interpretations of the slow process(es) are briefly discussed.

---

[*] To whom correspondence should be addressed.

INTRODUCTION

Surfactant association in water becomes less cooperative (the c.m.c. region broadens and the optimum aggregation number decreases) as a surfactant's hydrocarbon tail changes from linear to branched.[1] The extent of counterion binding also decreases,[1-3] and the L_1 region (an isotropic solution consisting of water, monomers, and micelles) of the phase diagram becomes less extensive.[4-7] In fact, if the length of the branched alkyl tails is sufficient, the upper limit in surfactant concentration of the L_1 region may fall below the surfactant's (estimated) c.m.c. In such cases the first aggre- gates encountered in the phase diagram are liquid crystals, often lamellar. Surfactants in this class recently have been the subject of extensive investigation[8-11] as synthetic bilayer membrane-formers for use in studies of reaction catalysis, energy storage, etc.

Among the double-tailed surfactants we have been studying[3,12] is a series of sodium alkylbenzenesulfonates which are shown in Figure 1. Sodium p-(1-pentylheptyl)benzenesulfonate (6-PhC$_{12}$SNa)[3,12-14] and sodium p-(1-hexyloctyl)benzenesulfonate (7-PhC$_{14}$SNa)[15] are well-established micelle-formers. There is some doubt over whether sodium p-(1-heptylnonyl)benzenesulfonate (8-PhC$_{16}$SNa), often called[16] Texas #1, displays a c.m.c. in water.[3,17] We have demonstrated[12] recently by SANS measurements

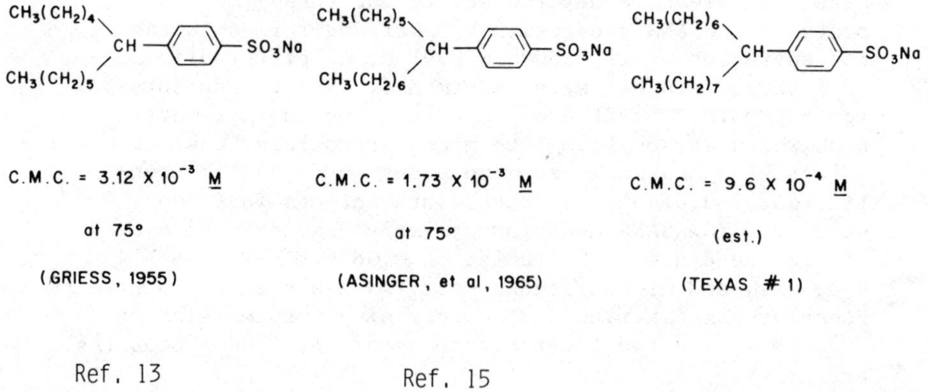

Figure 1. Homologous series of sodium alkylbenzenesulfonates with literature c.m.c. values.

that Texas #1 does form aggregates of micellar size in water at
45°C, but the aggregation may occur by continuous self-association
rather than cooperative micellization.[18]

   This paper describes the results of our SANS and static and
dynamic LS measurements for $6\text{-PhC}_{12}\text{SNa}$ in water and aqueous NaCl
solutions at 45°C. We undertook this investigation seeking answers
to some interesting, and largely unresolved, questions about the
aggregation of double-tailed surfactants. (1) Since the first
liquid-crystalline mesophase one encounters as surfactant concentration increases is lamellar, does a transition in micellar shape
occur from spherical to disklike micelles at the upper limit (in
concentration) of the L_1 region?[19] In the case of dioctanoyl-
phosphatidylcholine, this prediction of disklike micelles antici-
pating the transition to a lamellar mesophase is not borne out:
the LS data have been interpreted by Allgyer and Wells[20] to indicate a sphere-to-rod transition in the L_1 region. (2) Are the [21,22]
geometrical predictions of Israelachvili, Mitchell, and Ninham,
based on the numerical value of $v/a_o l_c$, quantitative? Here v is
the volume per surfactant monomer, a_o is the optimum surface area
per head group, and l_c is the extended length of the surfactant's
hydrocarbon. The value of $v/a_o l_c$ for $6\text{-PhC}_{12}\text{SNa}$ is approximately
0.6, assuming that[3] $v = 500$ $\overset{\circ}{A}^3$, $l_c = 14.5$ $\overset{\circ}{A}$ (vide infra), and
$a_o = 57$ $\overset{\circ}{A}^2$ (Ref. 23). The zero-order theory indicates micelles
are not possible for $v/a_o l_c > 1/2$, with bilayers, vesicles (or
disk-like micelles?) being predicted for $1/2 < v/a_o l_c < 1$.
Cylindrical micelles are predicted for $1/3 < v/a_o l_c < 1/2$. (3)
Do bi- or trimodal distributions of aggregates occur as the upper[24]
limit of the L_1 region is approached? Nagarajan and Ruckenstein
have predicted that with certain double-tailed surfactants the
possibility exists of micelles and vesicles coexisting. Indeed,
Hoffmann[25] has considered that possibility for $6\text{-PhC}_{12}\text{SNa}$.

EXPERIMENTAL SECTION

Surfactant

   The $6\text{-PhC}_{12}\text{SNa}$ used was a generous gift from Henkel, Dusseldorf.
Its c.m.c. at 45°C was 2.7×10^{-3} M (determined conductimetrically[3]).
Elemental analysis: Calcd. C, 62.04; H, 8.39. Found: C, 61.88;
H, 8.59. All samples were prepared by weight in H_2O (for LS) or
D_2O (for SANS) and surfactant concentrations in moles per liter
(M) computed using density data.[3] Samples used in LS were filtered
through 0.2 μm Millipore filters and then centrifuged for several
hours to remove dust.

SANS Measurements

Neutron scattering measurements were performed using the 30m (source-to-detector) SANS instrument of the National Center for Small Angle Scattering Research, Oak Ridge National Laboratory. Scattering intensities were recorded in the Q range [Q = $(4\pi/\lambda)\sin(\theta/2)$, where θ is the scattering angle] $0.012-0.25$ Å^{-1}, which corresponds to sample-to-detector distances of 2.0 and 5.0 meters for neutrons of $\lambda = 4.75$ Å. The samples were contained in cylindrical spectrophotometric cells of 5mm path-length which were thermostatted to ±0.05°C by means of circulation from an external bath. Scattering from sample and solvent was corrected for detector background, detector sensitivity, empty cell scattering, (computed) incoherent scattering and sample transmission; solvent intensity was subtracted from that of sample at each Q. The differences were then converted to radial-average intensity *vs* Q by programs provided by the Center, and absolute cross sections, $d\Sigma/d\Omega$ in cm^{-1}, were computed from calibrations based on the known scattering from pure water or from a vanadium single crystal. The runs at different distances were combined to yield coherent $d\Sigma/d\Omega$ *vs* Q curves. In this Paper, $d\Sigma/d\Omega$ is represented by I(Q).

SANS Data Analysis

For monodisperse, interacting spheres of uniform scattering density I(Q) is given by Equation (1),[26] with N_p the number of

$$I(Q) = \frac{d\Sigma}{d\Omega}(Q) = N_p V_p^2 (\bar{\rho} - \rho_s)^2 [\Phi(Q)]^2 S(Q) \tag{1}$$

particles per cm^3, V_p the sphere volume in Å^3, $\bar{\rho}$ the average scattering length density of the particles and ρ_s the scattering length density of the solvent (6.34×10^{-14} cm/Å^3 for D$_2$O). The single-particle form factor, $\Phi(Q)$, is given for spheres[27] by Equation (2); S(Q) is the interparticle structure function. If the scattering

$$\Phi(Q) = 3 \frac{\sin(QR) - QR\cos(QR)}{(QR)^3} \tag{2}$$

length density is not uniform, but instead can be modeled as arising from a set of N concentric spherical shells, the scattering amplitude density may be expressed as in Equation (3),

$$A(Q) \propto V_1(\rho_1 - \rho_2)\Phi(QR_1) + V_2(\rho_2 - \rho_3)\Phi(QR_2) + \dots$$

$$+ V_N(\rho_N - \rho_s)\Phi(QR_N) \tag{3}$$

where $V_a = (4\pi/3)R_a^3$ and ρ_a is the scattering length density of the ath shell of radius R_a.

SANS curves from solutions of 6-PhC$_{12}$SNa show a prominent interaction peak (due to a peak in S(Q), since [Φ(Q)]^2 is a decreasing function of Q) at concentrations above 0.025 M. In an earlier paper[12] we estimated Guinier radii (R$_G$) for the 6-PhC$_{12}$SNa micelles by analyzing I(Q) for Q > Q$_{peak}$, using Equation (4).

$$I(Q) = I(0)\exp(-Q^2 R_G^2/3) + C \qquad (4)$$

Hayter and Penfold[28,29] have demonstrated that S(Q) for ionic micellar systems can be calculated explicitly by solving the Ornstein-Zernike equation in the mean spherical approximation (MSA), assuming that the structural properties of the solution are dominated by hard-sphere repulsions modified by a finite ion screened Coulomb interparticle pair potential. (At sufficiently high ionic strengths, weakly attractive van der Waals interactions may also contribute, as Corti and Degiorgio have demonstrated for sodium dodecylsulfate.[30]) We[31] have employed MSA to compute S(Q)'s for classical ionic fluids and are presently applying it to the 6-PhC$_{12}$SNa system. Chen and coworkers[32] have used an experimental approach for determining S(Q); [Φ(Q)]^2 is obtained at low enough surfactant and high enough salt concentrations so that S(Q) = 1 over the Q range examined. The result is then used, after multiplication by the appropriate scaling factor, to extract S(Q) from I(Q) for systems where the electrostatic interactions are not completely screened.

It is unlikely that the micellar populations studied in the present work are in fact monodisperse, but the degree of polydispersity is expected to be small and to have little effect on the scattering curves at intermediate Q values. There are a number of particle size distributions which have been employed to evaluate scattering from polydisperse systems.[33,34] If one assumes a log-normal (Gaussian) distribution, Equation (5) holds where $\bar{R}$ is the mean particle radius and σ is the width parameter.

$$f(R,\sigma) = (1/\sigma\sqrt{2\pi}) \exp[-(R-\bar{R})^2/2\sigma^2] \qquad (5)$$

LS Measurements

Static and dynamic light scattering measurements were performed at the University of Michigan. The spectrometer has been described previously;[35] the light source was a 2 watt stabilized argon-ion laser (λ_0 = 514.5 nm), and the intensity autocorrelation functions were measured by a 64 channel Malvern multibit correlator. The modulus of the scattering vector Q is defined as $(4\pi n/\lambda_0)\sin(\theta/2)$,

where n is the refractive index of the sample. In some runs a
dust discrimination routine was employed during data collection.

LS Data Analysis

The angular and concentration dependences of the light scattered from a micellar solution may be expressed[36] according to
Equation (6), where R_{VV} is the Rayleigh ratio (actually

$$\frac{K(C-C_0)}{R_{VV}} = \frac{1}{M_w}[1+Q^2<R_g^2>_z/3] + 2A_2(C-C_0) + 3A_3(C-C_0)^2 + \ldots \quad (6)$$

$R_{VV,soln}-R_{VV,C_0}$), K is $4\pi^2 n_0^2 (dn/dc)^2/N_A\lambda_0^4$, C_0 is the surfactant's
c.m.c., R_g is the radius of gyration, λ_0 is the wavelength of the
incident light and the A's are virial coefficients. Double extrapolation of $K(C-C_0)/R_{VV}$ to zero micellar concentration and zero
angle in principle gives the true micellar molecular weight.

In the dynamic LS measurements the homodyne intensity autocorrelation function, $C(\tau)$, of the scattered light is given by
Equation (7), with $|g^{(1)}(\tau)|$ given by $\exp(-DQ^2\tau)$ for a monodisperse

$$C(\tau) = A(1 + \beta|g^{(1)}(\tau)|^2) \quad (7)$$

system (D is the translational diffusion coefficient of the particles). If a distribution of particle sizes exists, $g^{(1)}(\tau)$ may be
approximated conveniently using the cumulants expansion due to
Koppel,[37] Equation (8), where $\bar{\Gamma} = \bar{D}Q^2$. The variance of the

$$\ln|g^{(1)}(\tau)| = -\bar{\Gamma}\tau + (1/2!)\mu_2\tau^2 - (1/3!)\mu_3\tau^3 + \ldots$$

$$= \sum_{m=1}^{\infty} K_m(\Gamma)(-\tau)^m/m! \quad (8)$$

distribution function is $\mu_2/\bar{\Gamma}^2$, which is a measure of the system's
polydispersity. For a system where particle sizes show a bimodal
distribution, $g^{(1)}(\tau)$ may be given by Equation (9). For spherical

$$g^{(1)}(\tau) = B_1 \exp(-\Gamma_1\tau) + B_2 \exp(-\Gamma_2\tau) \quad (9)$$

particles one may obtain the effective hydrodynamic radius (R_H) from
D using the Stokes-Einstein relationship, $D = k_B T/6\pi\eta R_H$, where η
is the solvent's viscosity, T is the absolute temperature, and k_B
is the Boltzmann constant.

We had hoped to apply the histogram method[38] to analyze combined $C(\tau)$'s obtained by patching together correlation functions
taken at several different delay times (ranging from 1 to 70 μsec).
We present an example of such a patched correlation function below,
but it was evident from our inability to fit the resulting data
that some sort of misnormalization problem occurred in the patching. This is due perhaps to experimental uncertainties in the
computed baselines. Correlation functions for each sample studied
were determined at a minimum of three delay times: 1 or 2 μsec,
13 or 15 μsec and 60 or 70 μsec.

RESULTS AND DISCUSSION

SANS Data in D_2O

Figure 2 shows the effect on the scattering curves of increasing the 6-PhC$_{12}$SNa concentration. The interaction peak moves to
larger Q with increasing concentration. Cabos, et al.,[39] observed
the same behavior for normal micelles of sodium p-n-octylbenzene-
sulfonate in D_2O; the shift to larger Q results from the intermicellar distance decreasing as the surfactant concentration
increases. Scattering from solutions of other polyelectrolytes
shows the same behavior.[40-42]

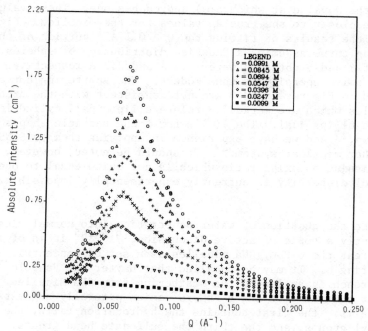

Figure 2.   Scattering from 6-PhC$_{12}$SNa micelles in D_2O at 45°C.

Guinier radii (and from them hydrodynamic radii, assuming spheres) for the micelles were obtained by analyzing[12] $I(Q)$ for $Q > Q_{peak}$, using Equation (4). The results are presented in Table I.

Table I. R_G and R_H Values for $6\text{-}PhC_{12}SNa$ Micelles in D_2O at 45°C

concn, M	R_G, Å	R_H, Å
0.0099	13.7 ± 0.1	17.7 ± 0.1
0.0247	14.5 ± 0.1	18.7 ± 0.1
0.0396	14.8 ± 0.1	19.1 ± 0.1
0.0547	15.6 ± 0.1	20.1 ± 0.1
0.0694	16.6 ± 0.1	21.4 ± 0.1
0.0845	16.6 ± 0.1	21.4 ± 0.1
0.0991	16.4 ± 0.1	21.2 ± 0.1

$$R_G = (0.6)^{1/2} R_H$$

Somewhat smaller radii (by 1 to 2Å) are obtained using Equation (1) for $Q \gtrsim 0.1$ Å^{-1}, which represents a more conservative assessment of the Q value at which $S(Q)$ approaches one. These values should be closer to the true R_H values for the micelles. Figure 3 presents the results of fitting the $Q \gtrsim 0.1$ Å^{-1} portion of the scattering curve assuming a Gaussian distribution of spheres. The values of $\bar{R}$ and σ obtained were 17.4 Å and 2.2 Å respectively; the apparent polydispersity is thus modest. The scattering curves presented in Figures 2 and 3 are not consistent with the presence of micelles which are markedly prolate ellipsoidal, cylindrical, or oblate ellipsoidal. The $\Phi(Q)$'s for these particle shapes are well known,[27] and we have experience in applying them to SANS data for another micellar system.[43] It should be noted, however, that distinguishing an axial ratio which is not quite equal to one from modest polydispersity is currently experimentally impossible.

Since the question of water penetration into normal micelles is currently a topic of active discussion,[19,44,45] it is of interest to treat the $6\text{-}PhC_{12}SNa$ micelles as a set of concentric shells of differing ρ. It may be recalled that Hayter and Penfold[29] took this approach in analyzing the scattering from SDS micelles. The molecular structure of the $6\text{-}PhC_{12}S^-$ anion naturally suggests three shells: the first contains the hydrocarbon tails, the second the phenyl groups, and the third the sulfonate head groups. (This is an idealization of the actual situation, since micelles are

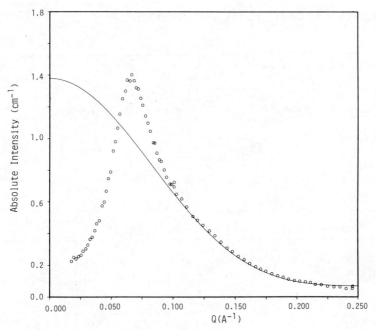

Figure 3. Fit of the high Q scattering for 0.0845 M 6-PhC$_{12}$SNa
assuming a Gaussian distribution of spheres.

known to possess considerable surface roughness.[29,46) If one
assumes an aggregation number of 25 for the micelles,[3,14,47] a dry
C$_{12}$ alkyl tail region will have a radius of 12.9 Å (based on the
molar volume of a dodecyl group), which is slightly longer than
the extended length of the 6-dodecyl group. The phenyl group shell
is estimated, using molecular models, to be 3.5 Å thick; geometrical
considerations indicate the presence of some water in this shell.
The value of 19.3 Å for 0.0845 M 6-PhC$_{12}$SNa (monodisperse sphere
model) is taken for R$_3$, thus defining the head group shells thickness as 2.9 Å. Scattering length densities for the first two shells
were computed using tabulated scattering lengths[26] and molecular
volumes; the results are given in Table II. The value, ρ_3, for
the third shell, was then obtained by using $\bar{\rho}$, obtained from fitting
the data at Q ≥ 0.1 Å^{-1} to Equation (1), and the expression for
A(Q), given by Equation (3) and evaluated at Q = 0. As Table II
indicates, the contrast between D$_2$O and the third shell is very
small. Hayter and Penfold[29] found ρ_3 to be substantially smaller
than ρ_s for SDS micelles, but SDS micelles also have an aggregation
number which is four times that of the 6-PhC$_{12}$SNa micelles.

Table II.   Three-Shell Model for 0.0845 M 6-PhC$_{12}$SNa,* assuming
$\bar{n}$ = 25.

Alkyl region: V_1 = 8992 Å^3 R_1 = 12.9 Å ρ_1 = -3.8 × 10^{-15} cm/Å^3

Benzene rings
 + D$_2$O: V_2 = 18477 Å^3 R_2 = 16.4 Å ρ_2 = 4.63 × 10^{-14} cm/Å^3

V_{bz} = 3525 Å^3 V_{D_2O} = 5920 Å^3

Head groups
 + D$_2$O: V_3 = 30113 Å^3 R_3 = 19.3 Å ρ_3 = 5.96 × 10^{-14} cm/Å^3

$\bar{\rho}$ = 3.64 × 10^{-14} cm/Å^3

---

*Computations based on monodisperse 19.3 Å spheres, I(0) = 1.353 cm^{-1}, ρ_S = 6.34 × 10^{-14} cm/Å^3.

SANS Data in Aqueous NaCl

Figure 4 presents the effect of NaCl on the scattering curves for 0.0845 M 6-PhC$_{12}$SNa. It is not possible with the sodium form of this surfactant to add sufficient NaCl to completely screen the electrostatic interactions between the micelles.[32]

From the scattering curves of Figure 4 at $Q \geq 0.075$ Å^{-1}, the amount of NaCl added does not appear to change significantly the size or shape of the 6-PhC$_{12}$SNa micelles. Indeed, fitting the data for each salt concentration at $Q \geq 0.1$ Å^{-1} to Equation (1) indicates an increase in R from 19.3 Å (no salt) to 20.5 Å (0.0465 M NaCl). However, analysis of the full curves with proper allowance for S(Q) will be necessary to confirm the invariance of $\underline{R}$ with salt. If invariant, one could then in principle extract an experimental S(Q) by computing at $Q \leq 0.1$ Å^{-1} the quantity I(Q)$_{exptl}$/I(I)$_{calc}$, where I(Q)$_{calc}$ comes from the fits described above. The resulting S(Q)'s are not presented here, since Figure 4 can be used to see the resulting trend in S(Q). Other than the shift in peak maximum observed, the most striking effect that adding salt produces is the increase in S(Q), which is due partly to the increase in isothermal compressibility which occurs as the effective range of the micelle-micelle interaction potential decreases.

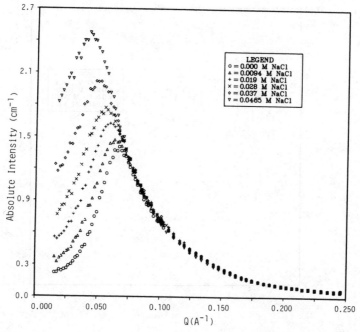

Figure 4.   Effect of NaCl on scattering from 0.0845 M 6-PhC$_{12}$SNa.

Light Scattering: Total Intensity Data

Figure 5 shows the angular dependence of the scattering from 0.085 M 6-PhC$_{12}$SNa at various salt concentrations, and Figure 6 shows the Zimm plot for 6-PhC$_{12}$SNa in water. Note that the Zimm plot is highly distorted which is often characteristic of Zimm plots for polyelectrolytes.[36] Double extrapolation of $K(C-C_0)/R_{vv}$ to $(C-C_0) = 0$ and $\Theta = 0$ yields $M_w^{-1} = 2.2 \times 10^{-5}$. If we make the assumption (reasonable in light of the SANS data) that salt does not substantially affect M_w, we can then use Equation (10) and the data in Figure 5 to estimate the effect of NaCl on the static

$$M^{-1} = M_0^{-1} (1 + k_I' \Phi) \qquad (10)$$

interaction coefficient k_I' (which is proportional to the second virial coefficient). The numerical values are of interest since it has been shown[30,48] that k_I can fact be negative at sufficiently high added salt concentrations. We take $M_0^{-1} = 2.2 \times 10^{-5}$; M^{-1} values from intercepts in Figure 5; Φ, the volume fraction of the 6-PhC$_{12}$SNa micelles in 0.0845 M solution, as 0.0255; and k_I' as $k_I/\bar{v}$, where $\bar{v}$ is the partial specific volume[3] of micellized 6-PhC$_{12}$SNa. The resulting k_I values, in cm^3/g, are 369 (no salt), 250 (0.019 M), 140 (0.028 M) and 33 (0.0465 M), which implies, as did the SANS

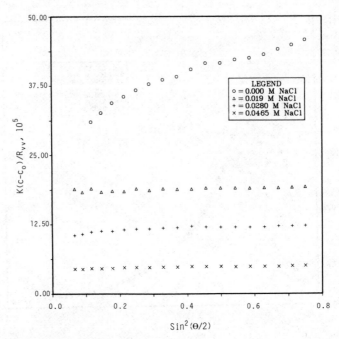

Figure 5. Angular dependence of the scattered light for 0.0845 M
6-PhC$_{12}$SNa at 45°C in water and aqueous NaCl.

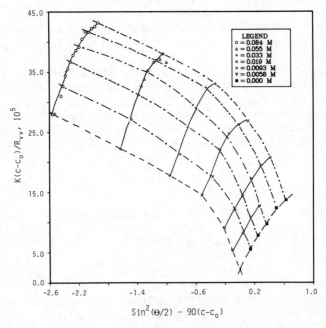

Figure 6. Zimm plot for 6-PhC$_{12}$SNa at 45°C in water.

data, that the electrostatic interactions are not completely
screened for 0.0845 M 6-PhC$_{12}$SNa by 0.0465 M NaCl.

The value for M_w^{-1} obtained from Figure 6 indicates an
aggregation number of 130 for the 6-PhC$_{12}$SNa micelles. This is not
consistent with the SANS data (see earlier discussion on aggregate
radii). A micelle containing 130 monomers would deviate considerably from sphericity. For example, if the micelle were an oblate
ellipsoid, it would have semi-minor and semi-major axes of 17 Å
and 38 Å respectively. The apparent aggregation number obtained
from the total intensity data is undoubtedly anomalously large
because of a small contribution from (a possible) much larger
aggregate (see discussion in the next section).

Dynamic Light Scattering

Figure 7 shows a patched intensity autocorrelation function
for 0.0845 M 6-PhC$_{12}$SNa in the presence of 0.028 M NaCl. This
function was generated from separate autocorrelation functions
accumulated with delay times of 1, 13 and 70 µsec (the full long-
time behavior is not shown). Clearly there are at least two characteristic Γ values (with some associated polydispersity) represented;
we determined them in two different ways: by treating separately
the $C(\tau)$'s obtained at short and long delay times to obtain the
large and small $\bar{\Gamma}$'s respectively (Equation (8)) and by treating
the $C(\tau)$ obtained at intermediate delay times (13 or 15 µsec) to

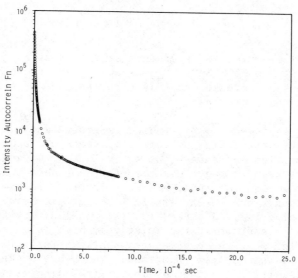

Figure 7. Patched intensity autocorrelation function for 0.0845 M
6-PhC$_{12}$SNa in 0.028 M NaCl at 45°C.

obtain both $\bar{\Gamma}$'s simultaneously (Equations (7) plus (9), hereafter called a double-exponential fit).

Figures 8 and 9 show a cumulants fit for 1 μsec data and a double exponential fit for 13 μsec data respectively. When fitting the long delay time data, the first two data points (which contain a substantial contribution from the large $\bar{\Gamma}$) were omitted. Tables III and IV present the result of fitting the short and long delay time $C(\tau)$'s according to Equation (8). The apparent polydispersities $(\mu_2/\bar{\Gamma}^2)$ are large in both cases, but this is partly a result of including μ_3 in the fit. Inclusion of μ_3 resulted in the best fit in all cases, as indicated by minimization of the sum of the squared errors. The apparent polydispersity in the micellar radii (computed from $\bar{\Gamma}_{large}$) is larger than that obtained from the SANS data. This is undoubtedly due to some contribution from $\bar{\Gamma}_{small}$, even at short delay times. Table V shows the results of the double-exponential fit; the Γ_1's agree reasonably well with the $\bar{\Gamma}$'s from Table IV, but the agreement between the Γ_2's and the $\bar{\Gamma}$'s in Table III is quite poor. The latter is not surprising, since $\exp(-\bar{\Gamma}_{large}\tau)$ ($\bar{\Gamma}_{large}$ from Table III) is only 0.047 to 0.34 at $\tau = 2.6 \times 10^{-5}$ sec (first data point fitted at intermediate delay times). In Table III, the effective hydrodynamic radii of the micelles, which correspond to the $\bar{\Gamma}_{large}$ values, are reported. With no or low amount(s) of added NaCl, $\bar{\Gamma}$ ($= \bar{D}Q^2$) is substantially increased over the true value because of the electrostatic interactions between the micelles. The effective diffusion coefficient is related to the true D_0 by Equation (11), where $k_D{'} = k_D/\bar{v}$ and $k_D{'} = k_I{'} - k_f{'}$, with $k_f{'}$ a hydrodynamic term which takes into

$$D = D_0(1 + k_D{'}\phi) \tag{11}$$

account the volume-fraction dependence of the friction coefficient.[30,48-50] At 0.0465 M NaCl, $\bar{\Gamma}_{large}$ and Γ_2 are the closest to one another, since $k_D{'}$ decreases rapidly (see previous section) as the salt concentration increases.

When salt is added, $\bar{\Gamma}_{small}$ (which is equivalent to Γ_1) increases, and the percent contribution of this component to the overall autocorrelation function decreases (see the values of B_1 and B_2 in Table V). Chu, et al.[40] have observed the same phenomenon in aqueous salt solutions of tRNA, and have interpreted their data as evidence for long range electrostatic interactions between the charged tRNA molecules. However, for the micellar solutions investigated in the present work, the relatively high ionic strengths cast doubt on the occurrence of electrostatic interactions over distances of several hundred Ångstroms. The Debye-Hückel length parameter, $1/\kappa$ where κ is given by $[8\pi I e^2/(\epsilon kT)]^{1/2}$ with the ionic strength I in ions/cm^3, is 20.6, 15.0, 13.5, and 11.5 Å respectively for the four solutions investigated by dynamic LS. Thus the double

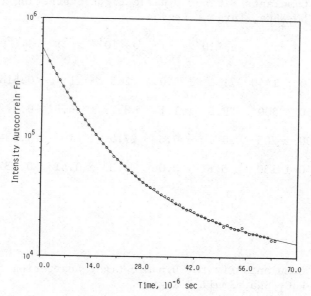

Figure 8.   Cumulants fit for 0.0845 M 6-PhC$_{12}$SNa in 0.028 M NaCl at a delay time of 1.0 μsec.

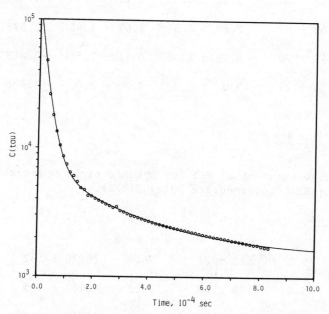

Figure 9.   Double exponential fit for 0.0845 M 6-PhC$_{12}$SNa in 0.028 M at a delay time of 13 μsec.

Table III. Cumulants Fit for Dynamic Light Scattering from
0.0845 M 6-PhC$_{12}$SNa, Large $\bar{\Gamma}$ Region.

[NaCl],M	$\bar{\Gamma}$, s^{-1}	$\mu_2(10^8$ s$^{-2})$	$\mu_3(10^{12}$ s$^{-3})$	$\mu_2/\bar{\Gamma}^2$	$\bar{R}_H$,Å
0.00	117400 ± 1440	71.1 ± 2.0	263 ± 11	0.516	8.8
0.019	89060 ± 900	29.5 ± 1.1	68.7 ± 6.8	0.372	11.6
0.028	69400 ± 205	16.1 ± 0.3	27.5 ± 1.5	0.334	14.9
0.0465	41840 ± 130	5.81 ± 0.06	6.33 ± 0.41	0.332	24.7

Table IV. Cumulants Fit for Dynamic Light Scattering from
0.0845 M 6-PhC$_{12}$SNa, Small $\bar{\Gamma}$ Region.

[NaCl],M	$\bar{\Gamma}$, s^{-1}	$\mu_2(10^5$ s$^{-2})$	$\mu_3(10^8$ s$^{-3})$	$\mu_2/\bar{\Gamma}^2$	$\bar{R}_H$,Å
0.00	858 ± 16	3.22 ± 0.32	2.10 ± 0.30	0.437	1200
0.019	1168 ± 82	7.89 ± 1.51	4.42 ± 1.45	0.578	884
0.028	1342 ± 70	8.79 ± 1.48	5.11 ± 1.38	0.488	770
0.0465	1754 ± 99	15.0 ± 2.2	9.20 ± 2.06	0.488	589

Table V. Double Exponential Fit for Dynamic Light Scattering from
0.084 M 6-PhC$_{12}$SNa (Intermediate Delay Times).

[NaCl],M	B$_1$	Γ_1, s^{-1}	B$_2$	Γ_2, s^{-1}	$R_{H,1}$,Å
0.00	71.1	811 ± 21	76.3	18870 ± 162	1270
0.019	25.4	1123 ± 28	79.2	40460 ± 765	920
0.028	73.6	1841 ± 534	300.9	29380 ± 4700	561
0.0465	120.0	1972 ± 946	644.1	26950 ± 4850	524

layer thickness is much less than even 100 Å in all the solutions studied. (For an aggregation number of 50, the typical intermicellar distance in a 0.084 M solution would be 100 Å.)

If a second aggregate, much larger than the ordinary small micelles, were present, we would have expected the addition of salt to cause some aggregate growth[51] (perhaps slight, because the small micelles themselves do not increase in size in the range of NaCl concentrations used) and hence a decrease in $\bar{\Gamma}_{small}$, rather than the observed increase. It is possible that a small amount of dispersed lamellar liquid crystal might give rise to $\bar{\Gamma}_{small}$, with the size of the dispersion insensitive to the presence of added salt. Because of R^6 intensity weighting, only a very small number concentration (a few parts in 10^9) for a large aggregate need be involved in order to give a sizeable B_1 relative to B_2. No evidence for this large characteristic dimension was obtained from our SANS measurements, because this dimension is too large to be studied using the Q range accessible.

In future work we plan to use the Li form of the surfactant, since it is likely that sufficient LiCl can be added to its solutions to (1) cause $\bar{\Gamma}_{small}$ to disappear, if it is in fact caused by long-range electrostatic interactions, (2) determine whether a micellar size and/or shape change occurs at higher salt concentrations, and (3) enable a better experimental determination of S(Q) to be made.

CONCLUSIONS

Solutions of $6\text{-PhC}_{12}\text{SNa}$ in H_2O or D_2O and in aqueous NaCl solutions contain small approximately spherical micelles which display strong electrostatic interactions. Micelle formation is observed despite the numerical value of $v/a_0 l_c$. SANS measurements produce scattering curves which are dominated by an interaction peak whose maximum moves to larger Q with increasing surfactant concentration and to smaller Q with increasing NaCl concentration. We have not found evidence for a sphere → rod or sphere → disk micellar shape transition. In light of the strongly interacting nature of the micelles, investigations of higher surfactant concentrations (closer to the upper limit of the L_1 region) must await our study of the surfactant's Li form.

We are not able yet to rule out the existence of a bimodal distribution of aggregate sizes in aqueous solutions of $6\text{-PhC}_{12}\text{SNa}$. The intensity autocorrelation functions measured by dynamic light scattering show a slowly decaying component (apparent R_H of 500 to 1300 Å) which may be produced by either a small amount of a large aggregate or by long-range electrostatic interactions. The ionic strengths of the solutions investigated seem too high to make the latter explanation plausible, while the observed dependence of $\bar{\Gamma}_{small}$ on concentration of added NaCl seems inconsistent with the former interpretation. A study of 6-PhC$_{12}$SLi in aqueous LiCl might allow us to decide which interpretation is correct.

ACKNOWLEDGMENTS

Acknowledgment is made to the donors of the Petroleum Research Fund, administered by the American Chemical Society, for support of the work at the University of Tennessee. The Faculty Research/ Development Fund of the University provided assistance with foreign travel.

One of the authors (RT) acknowledges partial support from Italian Ministero della Pubblica Istruzione, Rome.

This research was also supported by the Office of Oil, Gas, and Shale, U. S. Department of Energy, under contract W-7405-eng-26 with Union Carbide Corporation. The SANS measurements were performed at the National Center for Small-Angle Scattering Research (NCSASR). The NCSASR is funded by National Science Foundation Grant No. DMR-77-244-58 through Interagency Agreement No. 40-637-77 with the Department of Energy (DOE) and is operated by the Union Carbide Corporation under contract W-7405-eng-26 with the DOE. We thank H. R. Child and G. D. Wignall of the NCSASR for helpful discussions and assistance in analysis of the SANS results. The holder for liquid samples was constructed by Frank Ward of the Oak Ridge National Laboratory Plant and Engineering Division. The LS measurements were made at the University of Michigan.

REFERENCES

1. H. C. Evans, J. Chem. Soc., 579 (1956).
2. R. Zana, J. Colloid Interface Sci., 78, 330 (1980).
3. L. J. Magid, R. J. Shaver, E. Gulari, B. Bedwell, and S. Alkhafaji, Prepr., Div. Pet. Chem., Am. Chem. Soc., 26, 93 (1981).
4. E. I. Franses, J. E.Puig, Y. Talmon, W. G. Miller, L. E. Scriven, and H. T. Davis, J. Phys. Chem., 84, 1547 (1980).
5. H. Kuneida and K. Shinoda, J. Phys. Chem., 82, 1710 (1978).
6. H. Sagitani, T. Suzuki, M. Nagai, and K. Shinoda, J. Colloid Interface Sci., 87, 11 (1982).
7. R. J. M. Tausk, J. Karmiggelt, C. Oudshoorn, and J. Th. G. Overbeek, Biophys. Chem., 1, 175 (1974).
8. T. Kunitake, J. Macromol. Sci.-Chem., A13, 587 (1979) and references therein.
9. R. A. Moss, T. Taguchi, and G. O. Bizzigotti, Tetrahedron Lett., 23, 1985 (1982).

10. P. Tundo, D. J. Kippenberger, P. L. Klahn, N. E. Prieto,
    T.-C. Jao, and J. H. Gendler, J. Am. Chem. Soc., 104, 456 (1982).
11. Y. Murakami, Y. Aoyama, J. Kikuchi, K. Nishida, and A. Nakano,
    J. Am. Chem. Soc., 104, 2937 (1982).
12. L. J. Magid, R. Triolo, J. S. Johnson, Jr., and W. C. Koehler,
    J. Phys. Chem., 86, 164 (1982).
13. W. Griess, Fette, Seifen, Anstrichm., 57, 24 (1955).
14. D. Bauernschmitt, H. Hoffmann, and G. Platz, Ber. Bunsenges
    Phys. Chem., 85, 203 (1981).
15. F. Asinger, W. Bergen, E. Fanghanel, and K. R. Muller, J. Prakt.
    Chem., 27, 82 (1965).
16. Because of its preparation by W. H. Wade and R. S. Schechter's
    enhanced oil recovery group at the University of Texas.
17. J. E. Puig, E. I. Franses, H. T. Davis, W. G. Miller, and
    L. E. Scriven, Soc. Pet. Eng. J., 19, 71 (1979).
18. W. G. Miller, F. D. Blum, H. T. Davis, E. I. Franses,
    E. W. Kaler, P. K. Kilpatrick, K. J. Nietering, J. E. Puig, and
    L. E. Scriven, these proceedings.
19. B. Lindman and H. Wennerström, Top. Curr. Chem., 87, 1 (1980).
20. T. T. Allgyer and M. A. Wells, Biochemistry, 18, 4354 (1979).
21. J. N. Israelachvili, D. J. Mitchell, and B. W. Ninham, J. Chem
    Soc., Faraday Trans. II, 72, 1525 (1976).
22. D. J. Mitchell and B. W. Ninham, J. Chem. Soc., Faraday Trans.
    II, 77, 601 (1981).
23. F. Van Voorst Vader, Trans, Faraday Soc., 56, 1067 (1960).
24. R. Nagarajan and E. Ruckenstein, J. Colloid Interface Sci., 71,
    580 (1979).
25. H. Hoffmann, Ber Bunsenges, Phys. Chem., 82, 988 (1978).
26. G. Kostorz and S. W. Lovesey, in "Treatise on Materials Science
    and Technology," Vol. 15, G. Kostorz, Editor, pp. 2-64; 230-231;
    Academic Press, New York, 1979.
27. A. Guinier and G. Fournet, "Small-Angle Scattering of X-Rays,"
    pp. 19-28, Wiley, New York, 1955.
28. J. B. Hayter and J. Penfold, Molec. Phys.,42, 109 (1981).
29. J. B. Hayter and J. Penfold, J. Chem. Soc., Faraday Trans. 1,
    77, 1851 (1981).
30. M. Corti and V. Degiorgio, J. Phys. Chem., 85, 711 (1981).
31. R. Triolo and A. M. Floriano, in "Advances in Solution Chemistry,"
    I. Bertini, L. Lunazzi, and A. Dei, Editors, pp. 41-65, Plenum
    Press, New York, 1981.
32. D. Bendedouch, S.-H. Chen, W. C. Koehler, and J. S. Lin, J. Chem.
    Phys., 76, 5022 (1982).
33. W. Burchard, Polymer, 20, 589 (1979).
34. P. van Beurten and A. Vrij, J. Chem. Phys., 74, 2744 (1981).
35. E. Gulari, B. Bedwell, and S. Alkhafaji, J. Colloid Interface
    Sci., 77, 202 (1980).
36. M. B. Huglin, "Light Scattering from Polymer Solutions,"
    Academic Press, New York, 1972.
37. D. E. Koppen, J. Chem. Phys., 57, 4814 (1972).

38. E. Gulari, E. Gulari, Y. Tsunashima, and B. Chu, J. Chem. Phys., 70, 2965 (1979).
39. C. Cabos, P. Delord, and J. C. Martin, J. Phys. Lett., 40, L-407 (1979).
40. A. Patkowski, E. Gulari, and B. Chu, J. Chem. Phys., 73, 4178 (1980).
41. N. Ise and T. Okubo, Accts. Chem. Res., 13, 303 (1980).
42. H. M. Fijnaut, C. Pathmamanoharan, E. A. Nieuwenhuis, and A. Vrij, Chem. Phys. Lett., 59, 351 (1978).
43. R. Triolo, L. J. Magid, J. S. Johnson, Jr., and H. R. Child, J. Phys. Chem., 86, 3689 (1982).
44. F. M. Menger, Accts. Chem. Res., 12, 111 (1979).
45. W. Reed, J. J. Politi, and J. H. Fendler, J. Am. Chem. Soc., 103, 4591 (1981).
46. G. E. A. Aniansson, J. Phys. Chem., 82, 2805 (1978).
47. This number is reasonable since sodium n-octylbenzenesulfonate micelles have an aggregation number of 50, see B. Lindman, M.-C. Puyal, N. Kamenak, B. Brun, and G. Gunnarsson, J. Phys. Chem., 86, 1702 (1982).
48. J. Briggs, R. B. Dorshow, C. A. Bunton, and D. F. Nicoli, J. Chem. Phys., 76, 775 (1982).
49. R. Dorshow and D. F. Nicoli, J. Chem. Phys., 75, 5853 (1981).
50. G. D. J. Phillies, J. Colloid Interface Sci., 86, 226 (1982).
51. S. Hayashi and S. Ikeda, J. Phys. Chem., 84, 744 (1980).

VISCOELASTIC DETERGENT SOLUTIONS

H. Hoffmann, H. Rehage, W. Schorr and H. Thurn

Institut for Physical Chemistry
University of Bayreuth, D-8580 Bayreuth, Postfach 3008, W. Germany

Results of SANS, static and dynamic light scattering, electric birefringence and rheological measurements
of viscoelastic detergent solutions of Cetylpyridinium-
Salicylate (CPySal) are presented and discussed. The
data are consistent with the existence of rodlike
micelles. As long as their length L is smaller than
their distance of separation D, the unsheared solutions
have no elasticity. In this concentration range the L-
values increase linearly with the detergent concentration
and the axis of the rods seem to be randomly oriented.
The SANS-measurements show nearest neighbor order for
the center of masses of the rods. As indicated from
the electric birefringence measurements, the rods have
a rather narrow size distribution. The viscosity and
the elasticity of the solution rise abruptly when the
lengths of the rods approach and surpass their mean
distance. For even higher concentrations the ratio L/D
passes through a maximum. This is evidenced by both
the SANS, electric birefringence and the light
scattering data. In this concentration range some of
the data can be explained on the basis of a theory
which was developed by Doi and Edwards and which
describes the coupling between rotational and
translational motion of overlapping stiff rods.
The elasticity of the solution, however, can not be
explained on this basis alone. The rheological
properties are caused by the existence of a three
dimensional dynamic network which is formed by the

rods as soon as L > D. With increasing concentration
the fraction of the rods which is connected to the
network is increasing and the fraction of free rods is
steadily decreasing. The dynamic nature of the network
is based on by the continous formation and breaking of
contacts. The average number of elastically effective
chains per volume can be determined from the elasticity
modulus from the theory of networks. For high
concentrations this value approaches the number density
of the rods.

INTRODUCTION

Some dilute aqueous detergent solutions are viscoelastic.[1]
The viscoelasticity can be detected easily on the recoil of trapped
bubbles which can be observed when such a solution is sheared and
the shearing is stopped abruptly. The phenomenon of viscoelasticity
has been known for a long time. In spite of many experiments that
have been carried out in such solutions, there still are quite a
few puzzling problems left waiting to be fully explained. The
phenomenon is already mentioned in the classical book on colloid
chemistry by Freundlich "Kapillarchemie I and II."[2] It has been
shown by nmr-[3] and more recently by electric birefringence,[4]
rheological and SANS[5]-measurements that the viscoelastic solutions
contain rodlike micelles. Most of these measurements were carried
out in relatively dilute solutions in which the rotational volumes
of the micelles do not overlap. In a theoretical paper on the
hydrodynamic interaction of rodlike particles, Doi and Edwards[6]
call this range the dilute region. In this region the rotation of
the rods can proceed unhindered. As a consequence the viscosity of
the solution is relatively little affected in this range by the
presence of the colloidal particles. The detailed measurements
that were carried out on CPySal in this concentration range and
the data of which were discussed in detail in four publications[4, 5, 37, 38]
have shown the following:

1. The length of the rods increases more or less linearly
   with the detergent concentration from the concentration
   that marks the transition from globular to rodlike
   micelles.

2. The micellar concentration remains constant in this range.

3. The rods are relatively monodisperse.

4. The unsheared solutions behave as Newtonian liquids.

5. Elasticity in the solutions is induced by shearing.

6. The rods are not randomly distributed in the solution.
   Nearest neighbor order is detected in scattering
   experiments.

7. The elasticity in the unstirred solution begins to appear
    when the rotational volumes of the rods start to overlap.

In order to avoid misunderstanding, the first point needs
some further comment. According to the available theories one
would expect an increasing of the rod length with the square root
of the concentration. The main assumptions in these theories are
however, that there is no intermicellar interaction and that the
ionic strength is constant. None of these basic assumptions is
fulfilled in our case. The deviation of the rod length from the
"square-root-law" seems to be based on the increasing interaction
between the micelles. On the other hand it seems to be indeed
theoretically a very difficult task to consider the unisotropic
interaction energy between charged rodlike aggregates in the
aggregating process. Experiments clearly show that these energies
become high and dominate the scattering behaviour in scattering
experiments. We feel if these interaction terms are correctly
taken into account the theories will be very much different as the
ones which are available now. Intuitively one would expect that
the system tries to avoid situations of strongly overlapping stiff
rods because the rotation of the rods in these situations would be
hindered what would lower the entropy of the system and hence raise
the free Gibbs energy. Conclusions 1 to 3 were mainly based on
electric birefringence measurements which gave single exponential
decay curves with little distortion in the range of nonoverlapping
rods. The dispersity of the rods can be determined from such
curves.

Experimental evidence for a narrowing of polydispersity by
interparticles interactions has been observed in emulsions.[36]
Increasing the electrostatic repulsion between the emulsion droplets
narrowed the size distribution curves considerably. We furthermore
claim that the rodlike micelles in the present systems and in the
absence of large amounts of excess salt are fairly stiff with a
radius of curvature of more than 1000Å. The rigidity of rods will
depend on charge density and ionic strengths, on the interfacial
tension and the chain length of the detergent. There can be other
systems or the same system under different conditions for which the
rods are flexible. For the present situation, we could show however
by the agreements of the L values which were obtained from
rheological measurements with the help of theories made for stiff
rods and the L-values from other measurements that the rods are
fairly stiff.

The overlap of the rods cannot be the only reason for the
elasticity because studies on other systems have shown that the
overlap of rods does not necessarily lead to elasticity.[7] In many
systems this leads only to an increase of the viscosity and the
solutions still behave as Newtonion liquids for not too high shear

rates. In the present paper we will try to find an answer to the
question what happens in solutions when rodlike aggregates start
to overlap. In particular we were interested in the question
whether the lengths of the rods were increasing further with
concentration or whether the rods reached a maximum value and
stopped growing. We also tried to come up with the answer to the
question what is the cause for the elasticity. All the measurements
reported here were carried out on the system CPySal. But similar
results can probably be obtained on other viscoelastic detergent
systems. In this paper we will not go into the theory of the
different methods which were used for the investigations. We will
simply present data and discuss them and we will try to come up
with a structure for the aggregates in the system that can explain
the various phenomena.

EXPERIMENTAL RESULTS

    Results concerning the rheological properties of the system
are given in Figures 1 - 5. The data were obtained with a rotary
(Contraves low shear 30 sinus) and a Zimm-Crothers viscometer with
which very accurate measurements at very low shear rates could be
made.

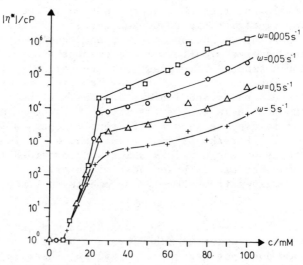

Figure 1. The magnitude of the complex viscosity as a function of
the detergent concentration at different angular frequencies
(CPySal, T = 20°C).

Figure 1 shows the magnitude of the complex viscosity for several
angular frequencies as a function of the concentration. The plots
show an abrupt break at the concentration at which the rods begin
to overlap. (C* = 7,7 mM). The viscous resistance increases several
orders of magnitudes within a rather small concentration range.
There is a second break around 25 mM. Above this concentration η*
increases more slowly. Above the second break, the dynamic
viscosity is frequency dependent. The break itself seems to be
independent of frequency. Up to this concentration the solution
behaves nearly as a Newtonian fluid. Figure 2 gives the storage and
loss modulus G' and G" as a function of the frequency ω for a 25 mM
solution. The magnitude of the complex viscosity is also given in
this figure. It agrees well with the values for the steady state
viscosity η_∞ which was determined under constant shear rate
conditions. This agreement shows that the structures that are
present in the solutions and that are responsible for the
viscoelastic properties are not destroyed by the shearing process.
The situation is quite different for lower temperatures where the
magnitude of the complex viscosity and the steady state viscosity
do not agree. (Figure 3). At temperatures below 20°C the solutions
have a yield value. In rheological measurements a yield value is
assumed to be due to the presence of a network structure[8]. The
network can be destroyed by shearing and as a consequence the

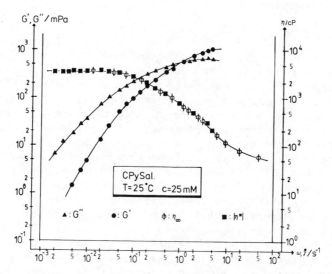

Figure 2.   The storage modulus G', the loss modulus G" and the
magnitude of the complex viscosity η* as a function of the
circular frequency and the steady state viscosity η_∞ as a function
of the shear rate (CPySal, T = 25°C, c = 25mM).

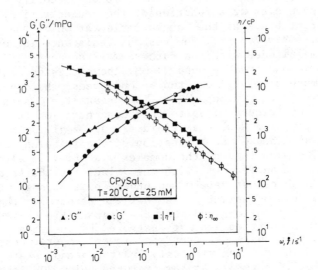

Figure 3.   The storage modulus G', the loss modulus G" and the
magnitude of the complex viscosity |η*| as a function of the
circular frequency and the steady state viscosity η_∞ as a function
of the shear rate (CPySal, T = 20°C, c = 25mM).

magnitude of the complex viscosity and the steady state viscosity
deviate from one another. The storage modulus gives the energy
which is stored in unit volume of the solution being deformed in
harmonic oscillations and the loss modulus gives the energy that
is dissipated. It is interesting to note that the G'-values in
Figure 2 seem to become independent of the circular frequency
above 10 Hz. This means that for angular frequencies which are
higher than 10 Hz the solution behaves as an elastic body. The
frequencies for which G' attain a constant value depend very much
on the concentration of the solution. These characteristic
frequencies are shifted to lower values for higher concentrations
as is shown in Figure 4. The plateau region of the curve is a
consequence of cross-links between the aggregates. If the number
of elastically effective chains between the cross-links increases,
the rubber plateau acquires a higher level. At moderately low
angular frequencies the storage modulus falls rapidly and
eventually vanishes. This is called the terminal zone or flow
region where the molecules are able to move with respect to each
other. At the end of the terminal zone at the left G' becomes
proportional to ω^2. The terminal slope of 2 is characteristic for
solutions in which the viscoelasticity is a relatively minor

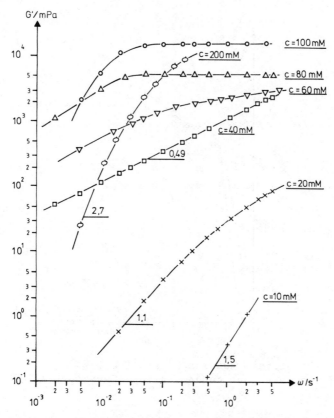

Figure 4.   The storage modulus G' as a function of the angular
frequency at different detergent concentrations (CPySal, T = 20°C).

pertubation of the Newtonian behaviour of the solvent[9]. For
network solutions the slope is considerably smaller. The values
for the storage modulus G' and the loss modulus G" for a circular
frequency of 0,05 Hz are given as a function of the concentration
in a semi log plot in Figure 5.

    The plot indicates that both parameters increase rapidly with
c from the overlap concentration C*. The elasticity of the sheared
system can be measured by a relaxation experiment. When the
solution is subjected to a constant shear rate and this shear rate
is suddenly reduced to zero the shear stress does not vanish
simultaneously but relaxes with one or several characteristic time
constants from which the relaxation shear moduli can be determined.
For most of the solutions at least three time constants could be
evaluated when the shear stress was plotted on a Logscale against
the time. The relaxation experiment after cessation of steady
state flow indicates that at least three different processes
contribute to the decay of the stored energy.

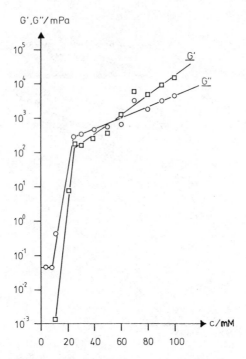

Figure 5.   The storage modulus G' and the loss modulus G" as a
function of the detergent concentration (CPySal, T = 20ºC, ω=0,05s⁻¹).

Typical electric birefringence[10] signals from the dilute and
semidilute region are shown in Figures 6a and 6b. While in the
dilute region the decay of the birefringence falls off with a
single time constant from which the length of the aggregates can
be determined, the fall off in the semidilute range consists of at
least two time constants. The values of the short and the long
relaxation times τ_{n1} and τ_{n2} are given in Figure 7 as a function
of surfactant concentration. It is noteworthy that the decay of the
birefringence contains even much longer relaxation times when E-
pulses are used for the measurements[11]. For long E-pulses only
small electric fields of the order of 100 V/cm are necessary for
the build up of the birefringence. Under these conditions decay
times of the order of seconds and minutes can be observed. The fast
processes which are shown in Figure 6a and Figure 6b are not
observed under these condition. In the following we will only
concern ourselves with the fast processes. The slower effects will
be treated in a separate paper.

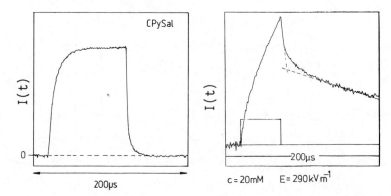

Figure 6.   Typical birefringence relaxation curves in the dilute
(a) and semidilute (b) region for 20mM C_{16}PySal. The amplitudes
give the change of the transmitted light intensity in relative
units. In (a) the decay is single exponential while in (b) the
decay shows two distinct relaxation processes (T = 25°C).

        It is important for the discussion to note that the signal
from Figure 6b is not consistent with a signal as one might expect
it from developing polydispersity. In such a situation one might
expect a single exponential which becomes more and more distorted
as the system becomes polydisperse. This is not the case. The
decay curves could be fitted with two distinct exponentials with
time constants which were up to two orders of magnitude apart. A
signal from a polydisperse system would mainly look like a single
exponential with some faster and slower contribution but never
show a clear break like the curve in 6b.

        The amplitudes of the two processes are given in Figure 8 as
a function of the concentration. The ratio of the amplitudes
showed no electric field dependence. For low concentrations and
small values of the electric fields the electric birefringence
obeyed Kerr's law. Deviations from this law became apparent for
concentrations above 4 mM (Figure 9a). Of course, it was possible
to determine the Kerr constant B, which is given by

$$B = \lim(\frac{\Delta n}{\lambda E^2}) \qquad (1)$$

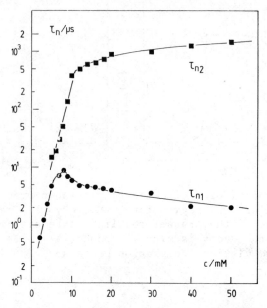

Figure 7.   The short (τ_{n1}) and long (τ_{n2}) birefringence relaxation
times as a function of total surfactant concentration (T = 25°C).

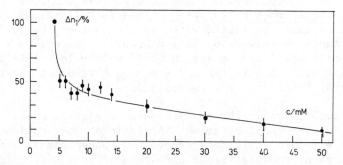

Figure 8.   The amplitude Δn_1 of the fast birefringence relaxation
process is given in percent of the total birefringence amplitude
as a function of surfactant concentration. For concentrations
above 50 mM no fast process was detectable any longer.

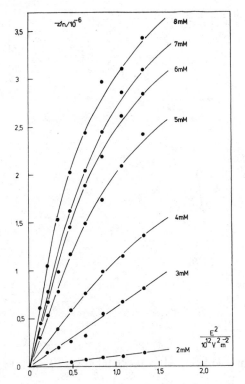

Figure 9a. The total amplitude Δn of the electric birefringence plotted against the square of the field strength for various concentrations of CPySal (T = 25°C).

In Figure 9b, -B is given as a function of concentration, because the birefringence of $C_{16}PySal$ was negative. Interestingly, both the sign and the absolute value of the birefringence for rodlike micelles of the same detergent-ion depend very much on the counter-ion with which the detergent-ion is combined. This is shown in Figure 10 where the birefringence is plotted against the molefraction of the counterions for mixtures of CPySal and $CPyC_7H_{15}SO_3$. The system $CP_yC_7SO_3$ has a positive birefringence and for mixtures of the two systems the birefringence disappears completely even though the rods in both systems are very similar and their size varies very little when the two systems are mixed as is clearly evident from the relatively small change in the orientation times. As will be shown in a separate paper the disappearance of the birefringence at a certain mole ratio of the two concentrations is due to a compensation of the form and the intrinsic birefringence [12]. The intrinsic birefringence is due to

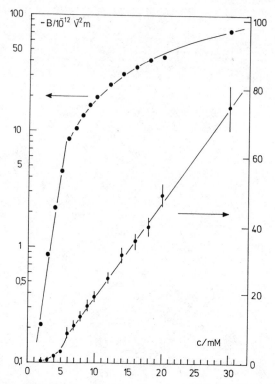

Figure 9b.   The Kerr constant-B of CPySal as a function of total
concentration left hand side logarithmic scale, right hand side
linear scale. The solutions of CPySal showed a negative
birefringence in all cases.

the optical anisotropy of the rods which is based on the different
polarizabilitites of the hydrocarbon-chains along the chain axis
and perpendicular to it. The experimental value of the
birefringence permits therefore to deduce a value for the order
parameter S of the chains inside the rods, a parameter which is
usually only accessible by pulsed nmr-measurements [13]. Electric
birefringence measurements were also carried out to measure the
equilibration time of the solution after the solution had been
heated for a short time period. In this respect viscoelastic
detergent solutions behave quite differently as normal detergent
solutions. As Figure 11 shows it can take hours or even longer
for the solution to reach equilibrium. This is even the case at

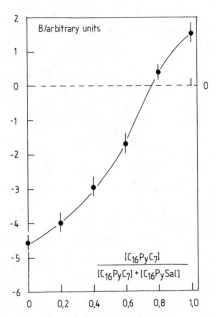

Figure 10. The Kerr constant-B of mixtures of CPySal and CPyC7SO3 as a function of the mole fraction of the counterions.

temperatures considerably above room temperature. At high temperatures the rods are transformed into globular micelles and the birefringence disappears. When the solution is quickly cooled down from this state, the kinetic processes which are necessary for the growth of the globular micelles are not fast enough and the solution is left in a thermodynamic unstable state. The build-up of the rods can thus be followed by the build up of the birefringence and the change of the orientation time with time.

A few small angle neutron scattering (SANS) measurements were carried out on the system.[5] A typical scattering curve is given in Figure 12. The scattering intensity shows a strong maximum as a function of the scattering angle.

Ordinary light scattering measurements were carried out at a wavelength of $\lambda = 6332$ Å. Some results are plotted in fig. 13 where the Rayleigh factor R_Θ is given as a function of total concentration. There is a marked break at a concentration of 7 mM. This break is also apparent in solutions containing excess NaCl. In these solutions the break is shifted to smaller detergent concentrations. In the presence of salt, the light scattering

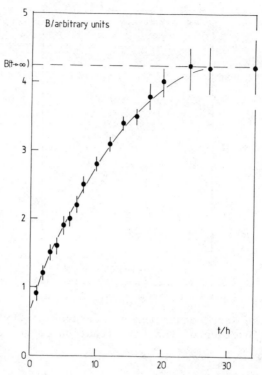

Figure 11.   Plot of the electric birefringence with time for a
$5 \cdot 10^{-3}$M CPySal solution at 25°C. The solution had been treated
to 80°C before it was quenched rapidly in 30 seconds to 25°C again.

shows a marked angular dependence. The ratio of the forward and
backward scattering is also shown in Figure 13. The radius of
gyration can be calculated from this quantity [14] which surprisingly
shows a maximum as a function of the concentration.

     A typical correlation function from dynamic light scattering
measurements [15] is shown in Figure 14. Two time constants had to
be used to fit the correlation function. Both, the fast and the
slow relaxation times, are changing when the concentration is
increased above 7.5 mM. As it is difficult to measure the long tail
of the correlation function with good experimental accuracy, we
restrict ourselves to the short time range, in which the
correlation function is approximately exponential.

     The correlation times obtained in this time region undergo
an abrupt change at 7.5 mM when the concentration is increased
from 6 mM to 8 mM as can be seen in Figure 15. As the correlation
time τ is proportional to the structure factor $S(q)$, this indicates

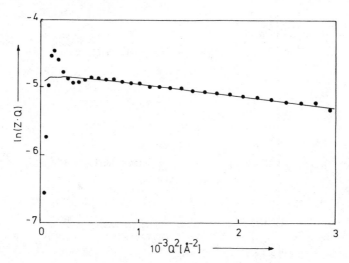

Figure 12.   Logarithmic plot of the product of the scattering
rate Z and the scattering vector Q for SANS measurements for a
10 mM solution of CPySal at 25°C. The drawn line is a fitted curve
with a L-value of 600 Å (for details see ref. 5).

a dramatical change of the repulsive interaction between the
micelles due to the overlap of the rotational volumes. It also
shows that the micellar aggregates can still undergo rapid
translational motions even though the solution is highly elastic
and of high viscosity at concentrations above 7.5 mM.

    On the other hand it is conceivable that the tail of the
correlation function is related to the elasticity of the solution.
It was not observed in other systems which contained rodlike
micelles but were not elastic.[7]

    The change in scattering intensity which results when a
solution is subjected to a temperature jump experiment is shown
in Figure 16. The rise in the scattering intensity as detected
from a scattering angle of 90° is brought about by the orientation
of the rods in the electric field and simultaneously by a decrease
of the nearest neighbor order in the system which is present in
the unperturbed solution as is evidenced from the SANS measurements.
The order can only be restored by translational motion of the rods
after the field has decayed. As a consequence the correlation
time τ_{cl} can also be evaluated from such measurements.

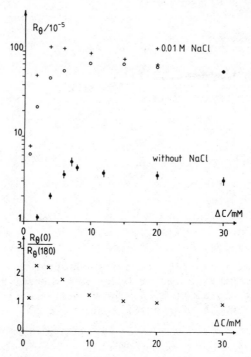

Figure 13.   Logarithmic plot of the Rayleigh factor R_Θ against the CPySal concentration in the micellar state in the absence of NaCl and with 0.01 M NaCl. (+) gives the forward and (O) the backward scattering. The ratio of the forward and backward scattering is given below.

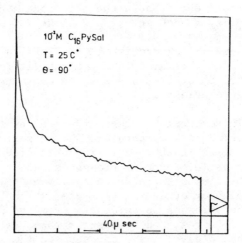

Figure 14.   Typical correlation function from a dynamic light scattering experiment in a 10^{-2} M CPySal solution for a scattering angle of 90°.

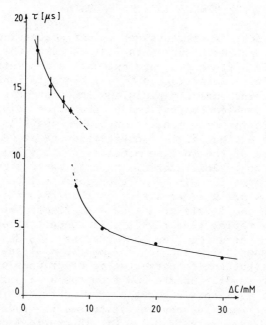

Figure 15.   The fast correlation times as a function of the CPySal concentration for a scattering angle of 90°.

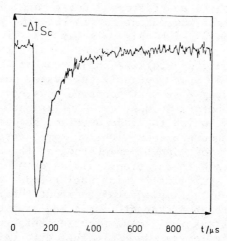

Figure 16.   Change of the scattering intensity of light at an angle of 90° during the discharge of a charged condensor through a CPySal solution (5mM + 5mM NaCl). For detail see ref. (30).

DISCUSSION OF THE DATA

The rheological experiments, the light scattering and the electric birefringence measurements indicate that the solutions above 7 mM behave differently than below this concentration. From this characteristic concentration C* on, the solutions are elastic, the viscosity is increasing rapidly and the scattering intensity is changing little with concentration and the orientation time from the electric birefringence measurements develops a long tail. From the measurements at the dilute concentration range we had to expect that the L-values of the rodlike micelles reach a length that corresponds to about their mean distance of separation at this particular concentration. It seems therefore that the results of the three methods are very sensitive to this transition. There are other data which are little or not at all affected by this point. The conductivity varies linearly in this concentration range when we cross the concentration line of 7 mM. The SANS measurements look about the same below and above 7 mM. We will see that this makes sense and that the changes that do occur when we cross the 7 mM line can be readily understood from the available theories. But let us first make some guesses on how we expect the system to develop if we increase the concentration and the rods would keep on growing with the same rate as below C*. Theoretical considerations predict that a nematic-type phase should form for high rod concentration. Such transitions were first predicted by Onsager[16] and were subsequently somewhat modified.[17] The critical concentration for the nematic phase would be $C = 1/dL^2$ where d is the diameter of the rod and L its length. This would correspond to a state in which the rotational volumes of the rods would overlap even for the case when the rods could rotate only in one plane. If the growth of the rods would continue in the same way as in the dilute region we would reach this point at $2.8 \cdot 10^{-2}$M detergent concentration and the rods would then have a length of about 2 500 Å. But no liquid crystalline phase separates out below a concentration C < 35% w/w. Obviously the system does not behave as expected and something else must occur. Let us have a closer look at the experiments before we make our conclusions. For concentrations above which the rods overlap, the decay of the electric birefringence proceeds with two time constants. According to Doi and Edwards[6] this can be understood on the basis of a coupled rotational and translational diffusion of the rodlike aggregates. In the semidilute region, stiff rods can no longer be completely aligned by a simple rotation about their axes. Each rod can rotate only through a characteristic angle which depends on the overlap of the rods and their concentration before it is stopped by another rod which is in its way. A further orientation is only possible when the other aggregate moves out of the way which is only possible by a translational movement. It can then rotate again until it is stopped by another neighbor and so on. In this model the fast process at the beginning of the decay of electric

birefringence corresponds therefore to the free rotation of the
rods and it is thus still possible to calculate L values from
these time constants. The calculated values are plotted in Figure
17. The data indicate that the rods stop growing after they
overlap and that they actually decrease again in length with an
increase of the concentration. This is a remarkable and somewhat
unexpected result. It can not be reconciled with the available
theories on the concentration dependence of the L-values.[19] These
theories always predict an increase of the L-values with the
square roof of the concentration because they contain only energy
terms for the detergent ion in the monomer and the micellar state
but fail to take into account intermicellar interaction terms for
the aggregation process. These interaction energies are of
considerable size as is clearly expressed in the scattering maximum
in the SANS-data and in the reduced light scattering intensities.

It is this intermicellar interaction energy which stops the
growth of the rods. Experimental evidence for this was recently
found by us in other electric birefringence measurements on a
number of other systems.[20] The significance of the intermicellar
interaction energy for aggregating systems was also recognized by
Nagarajan[21]. It is furhtermore noteworthy that during this meeting
in a number of reports concerning rodlike micelles, the presented
data reflected probably situations where the lengths of rods were
controlled by their interaction energy. This could have been the

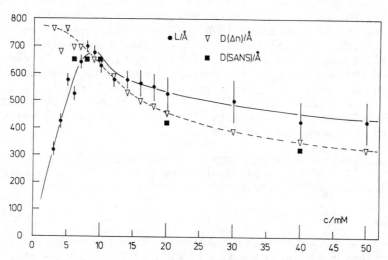

Figure 17.   Plot of the lengths of the micellar rods L and their
mean distance D as determined from the electric birefringence data
and the D-values as determined from the SANS data against the
CPySal concentration.

case for some of the light scattering data which were presented by
D. Nicoli[22] and by P. Stenius[23]. The L-values can now be used to
calculate a mean distance D between the rods when we assume D^3 is
the inverse of the mean number of micelles per unit volume. The D-
values thus calculated are also given in Figure 17. They indicate
that the ratio of L/D remains approximately constant for
concentrations above 30 mM. This will have consequences for the
viscosity as we will discuss. The electric birefringence data thus
point out that the overlap of the rods over their mean distance
remains approximately constant.

The result obtained from the birefringence measurements can
be put to a crucial test by the SANS data. As was shown previously,
such data permit the evaluation of the radius R of the rodlike
micelles, the length L, the mean distance between the rods and the
structure factor S for zero scattering angle.[24] The structure
factor S (q = 0) is the ratio of the measured intensity for q = 0
over the scattering intensity for ideal scattering conditions when
no intermicellar interaction would be present. This intensity can
be extrapolated from the measured intensities at large q values.
The parameter which is the easiest to obtain from the SANS results
is the mean distance between the rods. It can directly be
determined from q_{max} and is given by $2\pi/q_{max}$ if the influence of
the formfactor P(q,L) at large L-values on the scattering maximum
is taken into account. The scattering maximum of the solutions
shifted with increasing concentration of the detergent to larger q
values and hence to smaller D-values. This result tells us right
away that the number of micelles must be increasing with
increasing c. Similar situations have been observed before on a
number of other systems, both forming globular and rodlike
micelles. A particular good set of data showing very clearly the
shift of the maximum with increasing concentration was shown by
Magid in this meeting.[25] The D-values are also given in Figure 17.
They agree remarkably well with the values that were obtained from
the electric birefringence data. The details of the SANS-data will
be given in a separate publication.

The S-values from the SANS-data can be used for the
interpretation of the light scattering results. For interacting
particles the forward scattering intensity is given by

$$I_{sc} \sim (dn/dc)^2 \cdot MG \cdot S \cdot Cg \qquad\qquad (2)$$

where Cg is the concentration in g/cm^3. From the measured
Rayleighfactor R_Θ that is given in Figure 13 it is thus possible
to calculate the molecular weight of the particles when S is known.
From the MG-values it is possible again to calculate L values from
densities for the micelles and the thickness of the rods which are

known from SANS data. The L-values which were determined in this
way agree rather well with the values which were determined from
SANS and electric birefringence data.[5]

So far, we have shown that all the data are consistent with
the proposed model of overlapping stiff rods. The consistency can
also be extended to the dynamic light scattering results. The
correlation function of Figure 14 could be fitted with the two
correlation times τ_{c1} and τ_{c2}. Similar correlation functions have
been reported by Grüner and Lehmann[26] and by Ottewill et al[27] for
latex solutions which had a very small ionic strength. Under such
conditions the electrostatic repulsion between the aggregates
becomes appreciable and this leads to deviation from a single
exponential correlation time. In the extreme case the signal can
be fitted by two time constants. According to Pusey[28] and to Hess
and Klein[29] a very simple equation is valid for the short
correlation time. The reciprocal time constant is given by

$$1/\tau = D_{eff} \cdot q^2 = \frac{D_o}{S} \cdot q^2 \qquad (3)$$

where D_{eff} is an effective diffusion coefficient which is given by
the diffusion coefficient of the aggregates without any interaction
(D_o) and the structure factor. In noninteracting systems the light
fluctuations come about by the decay of concentration waves over
the distance $\frac{2\pi}{q}$. Such waves die out by diffusion of the aggregates.
In the case of systems which contain repulsive interactions the
decay of the concentration wave is controlled by the concentration
gradient and by the repulsive force. The correlation time therefore
is shorter. The long tail of the correlation function contains
according to a theory by Hess and Klein a memory function from
which elastic properties of the solution can be evaluated.
The values for D_{eff} are given in table I. These values can be used
to calculate values for S when we calculate D_o values from the
equation 4 that was given by Doi and Edwards[6] for the translational
diffusion constant.

$$D_o = \frac{kT}{3\pi\eta \cdot L} \ln \left(\frac{L}{d}\right) \qquad (4)$$

These values are also given in the table together with the S-
values. It should be noted that the τ_{c1} values can also be obtained
from relaxation measurements as was demonstrated recently by Platz[30]
In these measurements, the nearest neighbor order in the solution
that is set up by the repulsive forces is perturbed by an electric
field pulse. As a consequence of this perturbation of the order in
the system the scattered light intensity changes what can easily
be monitored by a photodetector. The perturbation of the structure
decays again by a diffusion process and with a time constant τ_{T1}
which is identical with τ_{c1}.

Table I.   Values for D_{eff}, D_0 and S. The values for D_{eff}
were Determined from the Correlation times τ_c using Equation 3.
The values for D_0 were calculated from equation 4 with the use
of the L-values from the Electric Birefringence Measurements.
The S-values are the Ratio of D_0/D_{eff}.

C/mM	$D_{eff}[\frac{cm^2}{s}]$	L/[Å]	$D_0[\frac{cm^2}{s}]$	S(o)
2	$4.72 \cdot 10^{-7}$	220	$3.59 \cdot 10^{-7}$	0.76
4	$5.52 \cdot 10^{-7}$	440	$2.57 \cdot 10^{-7}$	0.45
6	$5.92 \cdot 10^{-7}$	600	$2.13 \cdot 10^{-7}$	0.36
7	$6.3 \cdot 10^{-7}$	670	$1.99 \cdot 10^{-7}$	0.316
8	$10.6 \cdot 10^{-7}$	690	$1.96 \cdot 10^{-7}$	0.18
12	$17.2 \cdot 10^{-7}$	610	$2.11 \cdot 10^{-7}$	0.12
20	$21.7 \cdot 10^{-7}$	550	$2.25 \cdot 10^{-7}$	0.13
30	$29.5 \cdot 10^{-7}$	510	$2.36 \cdot 10^{-7}$	0.08

The viscosity in Equation 4 is the viscosity of the solvent
and not the viscosity of the solution which is measured under
shearing conditions. The translation diffusion along the axis of
a rod is not hindered in the semidilute region by the presence
of the other rods. This is clearly expressed in the values of D_{eff}
which become even larger when er move from the dilute to the
semidilute region even though the viscosity of the solution
increases rapidly. According to Doi and Edwards[31] the viscosity η
of the solution in the semidilute region is given by equation 5.

$$\eta = \eta_s[1+(\hat{c} \cdot L)^3]^3 \tag{5}$$

when $\hat{c}$ is the concentration of the rods. In the semidilute region
the viscosity is much greater than the solvent viscosity η_s. By
expressing c by $\frac{1}{D^3}$ we obtain Equation 6.

$$\eta = \eta_s[1 + (\frac{L}{D})^9] \tag{6}$$

From the L-values and the D-values in Figure 17 we are able to
calculate the viscosity in the semidilute region. The calculated
values thus determined are much lower than the measured values.
This discrepancy shows clearly that the rheological data cannot
be explained alone by the existence of weakly overlapping stiff

rods. Especially the large G'-values cannot be reconciled with
such a model. Doi and Edwards[31] have calculated G'-values for
overlapping rods. For the present situation these values would
be orders of magnitude too low and could not account for the
high G'-values. High elasticity moduli and the existence of a
rubber plateau are usually traced back to the existence of a
three dimensional network.[32] In order to explain fully the
experimental results we have to conclude therefore that some of
the rods form a network when they begin to overlap. A model for
such a network is shown in Figure 18. It seems possible that the
rodlike micelles acquire enough energy to overcome the repulsive
forces between them. They may or may not coalesce. According to
the DLVO-theory energy maxima and minima can exist for charged
particles. Thus two cases appear if two particles have an impact.
First, if the energy of the rods is smaller than the energy maximum,
the micelles have an elastic impact and the time they have
"contact" is neglegible small. By "contact" we mean the distance of
closest approach. Second, if the energy is larger than the energy
maximum, the micelles reach an energy minimum and a minimum of
closest approach. The time of contact, however in this case is
very large compared to the time in the first case and depends on
the depth of the minimum. It should be emphasized that this is
only a qualitative picture and that exact calculations would have
to incorporate anisotropic forces according to the shape of the
particles, the exact Hamaker constant, which is unknown for our

L/a >1

crosslinked dynamic network

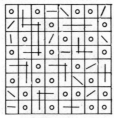

elastic fluid

Figure 18.   Model for a dynamic network which is formed from
individual rods which overlap and make contacts. The circles
represent rods which have their axis perpendicular to the paper
plane. Note: The grid lines should not be mistaken for rods.

system, the exact charge of the micelles, the exact screening of
the charge, and so on. It seems more likely that the rods in the
system keep their identity and the contacts can open up again.
Attractive Van der Waals forces may play a part in the formation
of the contacts by compensating part of the repulsive electrostatic
forces between the rods and thus helping to cross the repulsive
barriers which are between the isolated rods and the rods in
contact. In the present case, this general stabilization energy
may further be enhanced by specific adhesion energies in which
surface energy and specific interaction energies between the
micellar surface may play a key role. It is known for instance
that the elasticity of the system disappears when the salicylate
ion is replaced by the 3-Hydroxybenzoate-ion. For the salicylate-
system it is possible however to control the number of contacts
and hence the elastic properties of the system by the charge
density on the rods. For instance, when CPyCl is added to the
system and the charge density of the rods is thus increased, the
elasticity of the system is lost at a critical charge density even
though the dimensions of the rods do not seem to change very much.
This experiment shows clearly that the elasticity of the system
can not be explained on the basis of entanglements of flexible
rodlike particles as is generally done for more concentrated
polymer solutions. In those systems the volume fraction of the
polymers is usually considerably higher than in the present case.
Furthermore it would seem difficult to understand how the complex
viscosity could have such a marked discontinuity at C*. At C*, the
onset of the elasticity, the lengths of the rods are just long
enough for the contacts to become geometrically possible. On a
statistical basis and without attractive forces between the rods,
only a negligable fraction of the rods would exist accidently as
contacts and entanglements would be practically nonexistent at C*
or slightly above it. Only attractive forces between the rods can
bring the number of contacts above their statistical level and up
to a level which is high enough for the formation of a continuous
dynamic network. We furthermore believe that these attractive
forces in combination with the adhesion energies are the cause for
the rheopectic behaviour of the CPySal solutions at concentrations
below C*. Such solutions show no elasticity in dynamic measurements
when the solutions are at rest. Elasticity in these solutions
however can be induced by stirring or shearing.

Judging from the light scattering and the electric
birefringence data the three-dimensional network would be made up
of only a fraction of all the rods while the majority of the rods
close to the concentration of overlap C* would remain free. With
increasing detergent concentration, the number of contacts is
increasing and the number of free rods is decreasing until
finally all rods are connected to the network. The rheological
data support this hypothesis. From the theory of network it is
possible to calculate the number ν of elastically effective chains

between the crosslinks from the shear modulus of the rubber
plateau by the simple equation[33]

$$G_e = \nu \cdot k \cdot T \qquad\qquad (7)$$

The calculated values of ν as a function of C are given in table
II and plotted in fig. 19. For high concentrations they agree
approximately with the number of rods. It is interesting to note
from Figure 19 that ν increases approximately with the square of
the detergent concentration.

For the simplest model of a viscoelastic fluid which is
usually represented by a dashpot and an elastic spring which are
connected in series this combination is called a Maxwell element.
The storage and loss moduli can be expressed by the shear modulus
Ge as in Equation (8) and (9)

$$G' = Ge \cdot \frac{\omega^2 \cdot \tau^2}{1 + \omega^2 \tau^2} \qquad\qquad (8)$$

$$G'' = Ge \cdot \frac{\omega\tau}{1 + \omega^2 \cdot \tau^2} \qquad\qquad (9)$$

in which τ is the relaxation time for the dynamic network. The
magnitude of the complex viscosity is given by Equation (10)

$$|\eta| = \frac{(G'^2 + G''^2)^{1/2}}{\omega} \qquad\qquad (10)$$

It is obvious from Equation (8) that G' should become
independent of ω for $\omega\tau \gg 1$. For $\omega\tau < 1$ double log plots of G'
against ω should therefore give straight lines with a slope of 2.
This is mostly not the case as is shown in Figure 4 what points out
that viscoelastic properties of the investigated solutions can not
be discribed ba a mechanical model consisting of only one spring
and one dashpot but must be described by a combination of several
maxwell elements.

All the discussed properties can be understood on the basis
of the model that is shown in Figure (18). The dynamic aspect of
the network must be emphasized. The given arrangement is only the
frozen situation for a short period of time. There is a dynamic
aspect to it also. The rods can freely totate through considerable
angles before they are stopped. Furthermore, while the length
distribution of the rods seems to be rather monodisperse, there
will always be some fluctuations. A given rod might therefore be
quite long at one moment and rather short at another moment. Some
rods will dissociate completely and new ones will be formed. That
this dynamic equilibrium is still there in the viscoelastic
solutions can unambiguously be concluded from the slow relaxation

Table II.   The number ν of the Effective Chains Between
Crosslinks and the Concentration $\hat{c}$ of the Rodlike Micelles as a
Function of the Detergent Concnetration for CPySal at 20°C, as
Determined from Electric Birefringence Measurements

C/mM	10	30	50	70	90	100
ν/cm^{-3}	$2,6 \cdot 10^{13}$	$3,0 \cdot 10^{14}$	$9,0 \cdot 10^{14}$	$2,0 \cdot 10^{15}$	$3,4 \cdot 10^{15}$	$4,2 \cdot 10^{15}$
$\hat{c}/cm^{-3}$	$4,5 \cdot 10^{15}$	$1,7 \cdot 10^{16}$	$3,6 \cdot 10^{16}$			

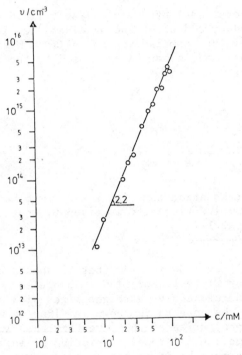

Figure 19.   Double logarithmic plot of the number of elastically
effective chains between the crosslinks against the surfactant
concentration.

this dynamic equilibrium is still there in the viscoelastic
solutions can unambiguously be concluded from the slow relaxation
time in the hour range which can be measured after a perturbation
of the system. It might be tempting to associate the elastic
properties to a distortion of the order in the system under chear
like it is in solids. However this does not seem to be possible as
a comparison between the data of this system and the system $C_{14}Py$-
C_7SO_3 which recently also has been studied in detail[7] clearly

shows. This system also forms rods and behaves in many ways very similar to $C_{16}PySal$. However, there is one noticable difference. These solutions behave nearly as Newtonian liquids even under conditions of overlapping rods. The situation in viscoelastic detergent solutions could be quite similar as in microemulsions when their concentration is increased in these systems. At a critical concentration the conductivity increases very rapidly[34]. This is usually interpreted with the percolation mechanism[35]. The onset of elasticity in the system studied would then correspond to that concentration for which the rods form a network that extends through the whole solution. Only a fraction of the rods might at that point be connected to the network and the majority of them might exist as individual rods.

CONCLUSIONS

Viscoelastic detergent solutions have usually only a narrow concentration region for the existence of globular micelles. While in normal detergent solutions the transition from globular to rodlike micelles is usually induced by the addition of salt, this transition can in viscoelastic detergents already be reached by a modest increase of the detergent concentration. For the studied system CPySal the cmc is $1,5 \cdot 10^{-4}$ M and C_t is $5 \cdot 10^{-4}$ M at $25°$C. For solutions containing no excess salt, the rodlike micelles grow linearly in length with the total detergent concentration in this region. The experimental results indicate that the length L of the rods is approximately given by the simple equation

$$L = ((c^o - cmc)/(C_t - cmc)) \cdot d \cdot 2/3$$

While the linear growth cannot be explained at present with the available theories for aggregating micellar systems, the equation simply reflects the experimental observation that upon crossing C_t all globular micelles begin to grow to rods while the particle concentration remains constant. In this concentration region the rodlike micelles can rotate freely but they show nearest neighbor order in scattering experiments as a consequence of their electrostatic repulsion. The rheological measurements which were carried out with an oscillating viscometer show no elasticity in the detergent solution in this region. Elasticity can however be induced when the solutions are sheared with a constant shear rate over an extended period of time. It is likely that the rods collide by the shearing and form long range structure which can store energy when they are deformed under the continued shear.

The growth of the rods continues with the detergent concentration until the length L' of the rods reaches the same value as their mean distance D. At this particular concentration C*, when L/D = 1, the rheological properties of the solution begin to change dramatically. Even in the unsheared state the solutions are now elastic and both the elasticity and the viscosity increase rapidly with increasing concentration. It is concluded that this dramatic change of the rheological properties is brought about by a network which is formed from the overlapping rods. The network cannot be a permanent one as is clearly evidence from the rheological properties of the solutions. The network is continuously changing by the steady formation and breaking of contacts between individual rods. Only a fraction of the rods is connected to the network as is evidenced from the results of the electric birefringence and dynamic light scattering measurements. When the concentration is increased further the ratio L/D increases and approaches a constant value. Consequently the length of the individual rods passes over a maximum with increasing concentration. This seems to be a general behavior of micellar systems which form rodlike micelles even if the rods do not form networks and the solutions are therefore not elastic. For such systems it was recently also observed that L passes through a maximum. It seems to be that the intermicellar interaction energy between the rods prevents the rods to grow to lengths which are considerably larger than their mean distance.

ACKNOWLEDGEMENTS

    Financial support of this work by the Deutsche Forschungsgemeinschaft and the Fonds der Chemischen Industrie is gratefully acknowledged.

REFERENCES

1. D.Saul, G.J.T.Tiddy, B.A.Wheeler, R.A.Wheeler and E.Willis, J.Chem. Soc.Faraday Trans.I,70, 163 (1974)
    S.Gravsholt, J.Colloid Interface Sci. 57, 575 (1976)
2. H.Freundlich,"Kapillarchemie I und II", Akademische Verlagsgesellschaft, Leipzig (1930)
3. J.Ulmius, H.Wennerström, L.B.A.Johansson, G.Lindblom and S. Gravsholt, J.Phys.Chem. 83, 2232 (1979)
4. H.Hoffmann, G.Platz, H.Rehage, W.Schorr and W.Ulbricht, Ber. Bunsenges. Phys.Chem. 85, 255 (1981)
5. J.Kalus, H.Hoffmann, K.Reizlein, W.Ulbricht and K.Ibel, Ber. Bunsenges. Phys.Chem. 86, 37 (1982)
6. M.Doi and S.F.Edwards, J.Chem.Soc.Faraday Trans.II 74, 560 (1978)
7. H.Hoffmann, H.Rehage, G.Platz, W.Schorr and H.Thurn, Progr. Colloid Polymer Sci., in press
8. J.D.Ferry, "Viscoelastic Properties of Polymers" J.Wiley & Sons, New York 1970

9. R.Darby, in "Viscoelastic Fluids", Marcel Dekker Inc., New York 1976

10. B.R.Jennings, Editor, "Electrooptics and Dielectrics of Macromolecules and Colloids", Plenum Press, New York (1979)

11. H.Hoffmann and W.Schorr, unpublished results

12. J.Rogers and P.A.Winsor, J.Colloid Interface Sci. $\underline{30}$, 247 (1969)

13. A. Johansson, B.Lindman, in "Liquid Crystals and Plastic Crystals", G.W.Gray and P.A.Winsor, Editors, Vol.2, 192, 1974

14. M.Kerker, "The Scattering of Light and Other Electromagnetic Radiation", Academic Press, New York 1969

15. H.Z.Cummins and E.R.Pike, in "Photon Correlation and Light Beating Spectroscopy", Plenum Press, New York (1974)

16. L.Onsager, Anal.N.Y.Acad. Sci. $\underline{51}$, 627 (1949)

17. P.J.Flory and G.Ronca, Mol.Cryst.Liq.Cryst. $\underline{54}$, 289 (1979)

18. M.Doi, J.de Physique $\underline{36}$, 607 (1975)

19. I.N.Israelachvili, D.J.Mitchell and B.N.Ninham, J.Chem.Soc. Faraday Trans.II, $\underline{72}$, 1525 (1976)

20. H.Hoffmann and W.Schorr, J.Phys.Chem. $\underline{85}$, 1418 (1981)

21. R.Nagarajan, J.Colloid Interface Sci., in press

22. D. F. Nicoli, R. B. Dorshow and C. A. Bunton, These proceedings, Vol. 1.

23. P. Stenius, H. Palonen, G. Strom and L. Odberg, These proceedings, Vol. 1.

24. P. Delord and J. C. Martin, J. Phys. (Paris) $\underline{39}$, 432 (1978). J. B. Hayter and J. Penfold, J. Chem. Soc. Faraday Trans I, $\underline{77}$, 1851 (1981). B. Cabane, R. Duplessix and T. Zemb, These proceedings, Vol. 1.

25. L. Magid, R. Triolo, E. Gulari and B. Bedwell, These proceedings, Vol. 1.

26. F.Grüner and W.Lehmann, "Light Scattering in Liquids and Macromolecular Solutions", V.Degiorgio, M.Corti and M. Giglio, Editors, Plenum Publ. Press, New York 1980

27. J.C.Brown, P.N.Pusey, J.W.Goodwin and R.H.Ottewil, J.Phys.A. Gen.Phys. $\underline{8}$, 664 (1975)

28. P.N.Pusey and R.I.A.Tough, in "Dynamic Light Scattering and Velocimetry", R. Pecora, Editor, Plenum Press, New York, 1981

29. W.Hess and R.Klein, Physica A., in press

30. G.Platz, Ber.Bunsenges.physik.Chemie $\underline{85}$, 1418 (1981)

31. M.Doi and S.F.Edwards, J.Chem.Soc.Faraday Trans.II, $\underline{74}$, 918 (1978)

32. W.Oppermann and G.Rehage,Colloid Polymer Sci. $\underline{259}$, 1177 (1981)

33. P.J.Flory, J.Chem.Phys. $\underline{18}$, 108 (1950)

34. H.F.Eicke, in "Microemulsions", I.D.Robb, Editor, Plenum Press, New York, 1982

35. D.G.De Gennes, "Scaling Concepts in Polymer Physics", Cornell University Press, 1979

36.   A.Suzuki, N.F.H.Ho and W.F.Higuchi, J.Colloid Interface Sci.,
      29, No. 3 (1969).
37.   H.Hoffmann, G.Platz, H.Rehage and W.Schorr
      Adv. Colloid Interface Sci., 17, 275 (1982).
38.   H.Hoffmann, G.Platz, H. Rehage and W.Schorr
      Ber. Bunsenges. Phys. Chem. 85, 877 (1982)

THE EFFECT OF INTERMICELLAR INTERACTIONS ON INTERPRETATIONS OF

MICELLAR DIFFUSIVITIES BY DYNAMIC LIGHT SCATTERING

D.F. Nicoli, R.B. Dorshow and C.A. Bunton*

Departments of Physics and Chemistry(*)
University of California
Santa Barbara, California 93106

Using the method of dynamic light scattering, we have determined the diffusion coefficients of micelles in aqueous solution as a function of surfactant and salt concentrations for three related systems: CTAB ($C_{16}H_{33}N(CH_3)_3Br$) + NaBr, MyTAB($C_{14}H_{29}N(CH_3)_3Br$) + NaBr and CTACl($C_{16}H_{33}N(CH_3)_3Cl$) + NaCl. For each system there is a well-defined salt concentration below which the micellar diffusivity increases linearly with surfactant concentration. In this low-salt region, in which the net intermicellar interaction is repulsive, we have obtained good fits to the diffusivity data using linear interaction/DLVO theory. From this procedure we have determined the "minimum-sphere" micellar radius as a function of temperature as well as estimates of the micellar fractional ionization α and the Hamaker constant A for the above systems as well as for SDS($C_{12}H_{25}SO_4Na$) + NaCl. Hence, by this approach one is able to relate variations in these physical parameters to subtle changes in surfactant composition (e.g. alkyl chain length) and counterion properties. Interestingly, significant micellar growth apparently commences at the salt concentration at which the net intermicellar interaction shifts from repulsive to attractive.

455

INTRODUCTION

Two general approaches to the study of micellar diffusivities
by dynamic light scattering have recently had significant impact in
the literature of micelles. In one case, represented by the work
of Mazer et al[1] and Missel et al[2], micellar diffusivities have been
obtained for SDS + NaCl and related systems at relatively high salt
concentrations (e.g. [NaCl] $\gtrsim$ 0.4M). In this "high" salt region it
was assumed that intermicellar interactions are negligible and that
the mean micellar hydrodynamic radius R_h can be obtained directly
from the mean diffusion coefficient D (obtained from the scattered
intensity autocorrelation function) using the Stokes-Einstein expression[3,4] valid at infinite dilution,

$$R_h = \frac{kT}{6\pi\eta D} \tag{1}$$

where k is Boltzmann's constant, T the temperature ($^{\circ}$K) and η the
shear viscosity of the solvent.

Following this approach, Mazer et al and Missel et al obtained
values of R_h for SDS micelles as a function of NaCl concentration
and temperature and observed the micelles to grow dramatically with
increasing [NaCl] and/or decreasing T. The above authors were able
to account for the behavior of R_h (e.g. as a function of T and [SDS])
at relatively high salt concentrations using a "ladder" model of micellar growth. According to this description, the free energy required to add an additional surfactant monomer to a rod-like micellar aggregate, starting with a "minimum-sphere" micelle[5], is independent of the size of the aggregate.

An alternate approach to the analysis of micellar diffusivities
was more recently proposed by Corti and Degiorgio[6]. These authors
showed that over a wide range of NaCl concentrations the SDS micellar diffusivity D (as well as the inverse of the scattering intensity I) is approximately linear in surfactant concentration at a
given temperature. In relatively dilute salt (i.e. [NaCl] $\lesssim$ 0.45M
at 25°C), D increases with [SDS], while at higher values of [NaCl],
D decreases with [SDS]; in the limit of vanishing [SDS] (i.e. at
the cmc), all diffusivities approximately converge to a common
value, D_o, which (using Equation (1)) corresponds to a plausible
estimate of the SDS "minimum-sphere" radius. Corti and Degiorgio
adopted the approach that both the increases and decreases in D
with changing SDS concentration are substantially a consequence of
interactions between micelles. According to this point of view,
the slope of D vs [surfactant] shifts from positive to negative
when [NaCl] exceeds 0.45M (at 25°C) due to a shift in the net intermicellar interaction from repulsive to attractive. Using linear
interaction theory and an intermicellar interaction potential taken
from DLVO theory[7,8], they obtained a set of theoretical fits which

bear a qualitative resemblance to the D vs [surfactant] data. (However, the predicted slopes of D vs [surfactant] deviated significantly from the experimental values at several salt concentrations.) Fits of comparable quality were also obtained for I^{-1} vs [SDS]. Significantly, Corti and Degiorgio assumed that interactions dominate the behavior of micellar diffusivities over a wide range of salt concentrations -- i.e., there is only relatively modest micellar growth at high NaCl concentrations. This conclusion directly contradicts the findings of Mazer et al and Missel et al, who assume that intermicellar interactions are negligible in the high-salt region and that values of D below D_0 correspond to substantial micellar growth for SDS + NaCl and related systems. Others[9-11] have also concluded that for a variety of ionic surfactant/counterion systems there is substantial micellar growth at large [salt].

The approach which we have adopted, described herein and elsewhere[12,13], is intermediate to the two approaches discussed above. Like Corti and Degiorgio, we believe that intermicellar interactions can significantly perturb micellar diffusivities in the concentration ranges of interest. However, unlike these authors, in our theoretical fitting procedure we assume that such interactions dominate the behavior of micellar diffusivities only over the region of dilute salt, where D increases linearly with surfactant concentration, and not at higher ionic strengths, where we have generally observed D to decrease monotonically (but not necessarily linearly) with increasing surfactant concentration. We have found that by confining the interaction analysis to the positive-slope (i.e. net-repulsive) region of D vs [surfactant], we are able to obtain excellent theoretical fits to the experimental data over a wide range of temperatures and salt and surfactant concentrations for a variety of systems. Our findings therefore strongly suggest that an analysis of micellar diffusivities based on the prevalence of intermicellar interactions is valid only in the "dilute" salt region, where there is, at most, only modest micellar growth with increasing salt or surfactant concentration. (However, as discussed below, the actual salt concentration which defines the upper boundary of this "dilute" region varies markedly with the surfactant/counterion system under investigation.) The systems examined thusfar include CTAB($C_{16}H_{33}N(CH_3)_3Br$) + NaBr, CTACl($C_{16}H_{33}N(CH_3)_3Cl$) + NaCl, MyTAB($C_{14}H_{29}N(CH_3)_3Br$) + NaBr and SDS($C_{12}H_{25}SO_4Na$) + NaCl. The first and second systems differ only in the counterion species; the first and third differ only in the length of the n-alkyl group. The last system was analyzed using the published diffusivity data of Corti and Degiorgio[6].

Use of this technique of dynamic light scattering/interaction analysis as outlined below (and in references 6, 12 and 13) has produced a number of useful insights, on both a qualitative and quantitative level. First, we can characterize the interactions which influence micellar diffusivities in dilute salt and determine the range of salt concentration for a particular surfactant/salt system where the linear interaction/DLVO theory is valid. Second, we have established the validity of a simple, non-perturbing technique for the determination of the micellar fractional ionization α (i.e. the extent of charge) when there is negligible micellar growth. As well, we obtain directly the mean hydrodynamic radius a (=R_h, given by Equation (1) with D=D_o) and also an estimate of the Hamaker constant[7,8,14] A, which characterizes the strength of the van der Waals-type attractive interactions between micelles. By comparing these quantities for a series of closely related surfactant/counterion systems, we can thereby relate changes in α, A and a to perturbations in amphiphile structure and counterion physical properties.

THEORY

For the four systems investigated thusfar, the D vs [surfactant] data are consistent with the conclusion that in "dilute" salt the solution consists mostly of "minimum-sphere"[5] micelles, whose mean radius is essentially independent of both surfactant and salt concentrations at a given temperature. That is, there is a well-defined salt concentration below which D vs [surfactant] consists of a nearly ideal "fan" of positive-slope lines[6,12] which, when extrapolated to the cmc, closely converge to a common value of diffusivity, D_o. The hydrodynamic radius of these minimum-sphere micelles is then given by Equation (1), with D=D_o. The slopes of the D vs [surfactant] plots decrease with increasing salt concentration. Briggs et al[15] recently showed for a variety of systems that the slope varies approximately inversely with [salt]; an explanation of this simple functional dependence, based on the theory described herein, was recently provided by Dorshow et al[13].

Following Corti and Degiorgio[6] we assume that the micellar diffusivities in the net-repulsive region can be explained by linear interaction theory,

$$D = D_o \left(1 + (K_t + K_h) \phi \right) \tag{2}$$

where ϕ is the micellar volume fraction and interaction coefficients K_t and K_h are due, respectively, to thermodynamic and hydrodynamic effects. Equation (2) is expected to hold in the limit of small ϕ (e.g. [surfactant] ≲ 0.05M), where higher-order corrections (i.e. ∼ ϕ^2, etc.) can be neglected.

Coefficient K_t is related by a simple numerical factor to the second osmotic virial coefficient[16], well-known in classical light scattering. Coefficient K_h is related to perturbations of the friction factor in the generalized Stokes-Einstein formula[17], due to hydrodynamic (i.e. solvent-streaming) effects and has recently been evaluated by Batchelor[18] and Felderhof[19]. It is convenient to write both K_t and K_h as integral expressions[6,16] involving the pair interaction potential energy (i.e. between two neighboring micelles). It is further convenient to integrate explicitly the contributions to K_t and K_h due to simple hard-sphere repulsion (called K_t^{hs} and K_h^{hs}, respectively), thereby excluding the latter from the overall interaction potential energy $V(x)$. Normalized separation parameter x is defined by $x \equiv (r-2a)/2a$, where r is the distance between the centers of two spherical micelles and a is their radius. To summarize,

$$K_{t,h} = K_{t,h}^{hs} + \int_0^{\infty} dx\, G_{t,h}(x) \left(1 - e^{-V(x)/kT}\right) \qquad (3a)$$

$$\text{where } K_t^{hs} = 8 \qquad (3b)$$

$$K_h^{hs} = -6.44 \text{ (Felderhof)} \qquad (3c)$$

$$G_t(x) = 24(1+x)^2 \qquad (3d)$$

$$G_h(x) = -12(1+x)+\frac{15}{8}(1+x)^{-2}- \frac{27}{64}(1+x)^{-4}- \frac{75}{64}(1+x)^{-5} \qquad (3e)$$
$$\text{(Felderhof)}$$

All that remains to compute the interaction coefficients K_t and K_h is the detailed specification of $V(x)$. Following Corti and Degiorgio, we utilize the expressions appropriate for idealized, charged spherical particles, obtained from DLVO theory[7,8]. The attractive term $V_A(x)$ is given by the London-van der Waals formula[14] for uniform dielectric spheres (Equation (9), reference 12). For spheres of a given radius a, the adjustable parameter which determines the strength of the interaction is the Hamaker constant A. The repulsive term $V_R(x)$ has been calculated by Verwey and Overbeek[7]. The appropriate algebraic solution for $V_R(x)$ depends on the value of κa, where κ is the Debye-Hückel inverse screening length[7,8],

$$\kappa = \left(\frac{8\pi I e^2}{\epsilon kT} \right)^{1/2} \qquad (4)$$

Here, I is the solution ionic strength (ions/cm^3), e the electrostatic charge and ϵ the solvent dielectric constant (T-dependent). For CTAB + NaBr, MyTAB + NaBr and CTACl + NaCl over the ranges of salt concentration of interest, the expression which best describes $V_R(x)$ is that obtained by Verwey and Overbeek for $\kappa a < 1$,

$$V_R(x) = \psi_o^2 \epsilon a\, \frac{e^{-2\kappa ax}}{2(1+x)}\, \gamma \,, \qquad (5)$$

where ψ_o is the surface potential of the micelle, a function of the micellar charge Q and κa; γ is a complicated function of κa and x.

(See Equations (79) and (83), pp 149-150 of the original monograph of Verwey and Overbeek[7] for expressions for ψ_0 and γ. We have calculated $V_R(x)$ exactly using a digital computer; see Figure 1 of reference 12.)

It is important to appreciate that the micellar radius a is not a free parameter in our fitting procedure. Rather, it is obtained with relatively little uncertainty from the extrapolated diffusivity D_0. Indeed, this value emerges from the D vs [surfactant] linear fan independently of the details of linear interaction theory. For a given micellar radius a and fixed κ, the remaining parameter which determines the value of $V_R(x)$ is the micellar charge Q. This can be expressed as $Q = \alpha N$, where N is the micellar aggregation number and α the effective fractional ionization, which is expected to lie in the range $0.1 < \alpha < 0.4$.

As discussed previously[12,13], we have assumed that the dependence of N on [salt] for our systems resembles that found for SDS + NaCl. For the latter system, N appears[20-22] to increase only modestly with increasing [NaCl] as long as one is in the net-repulsive, positive-slope region -- i.e. [NaCl] < 0.45M. The reported values[23] of N for SDS at the boundary of that region, [NaCl]=0.45M, are only about 50% larger than the estimates for zero salt. By contrast, above this "critical" NaCl concentration, N is reported[11,24] to increase greatly with increasing [NaCl], consistent with the dynamic light scattering measurements of Maser et al[1] and Missel et al[2]. Since changes in N over the measured positive-slope region are therefore relatively small compared to those reported outside the region, it is reasonable in the theoretical analysis to assume that N is constant at a given temperature, as discussed previously[12,13]. This is consistent with the observation that the D vs [surfactant] curves for a given T all appear to converge to a single diffusivity value, D_0, at the cmc.

Unfortunately, there remain large uncertainties and discrepancies[25-28] associated with the determination of actual numerical values for N using a variety of techniques, including light scattering (discussed by Kratohvil[29] and others). However, as discussed in references 12 and 13, we settled on the following rough estimates for N in the positive-slope region for each of the systems we examined: CTAB + NaBr, N = 120; MyTAB + NaBr, N = 90; CTAC1 + NaCl, N = 100; SDS + NaCl, N = 95. Fortunately, the success of the theoretical fitting procedure does not greatly depend on the choice of N for a given system. As discussed previously[12], the micellar volume fraction ϕ is inversely proportional to N (i.e. $\phi = 6 \times 10^{20}(4/3)\pi a^3[\text{surfactant}]/N$), while the net interaction coefficient, $K_t + K_h$, is roughly proportional to N over the physical range of interest, as computed from Equations (3a-e). (See Figure 5, reference 12.) Hence, the resulting product $(K_t+K_h)\phi$ (in Equation (2)) is only very weakly dependent on N.

Given a reasonable estimate for N, as outlined above, and a value for radius a, determined from D_0, there are two adjustable parameters which determine the theoretical D vs [surfactant] fits: the fractional ionization α and the Hamaker constant A. The former determines the strength of electrostatic repulsions between charged micelles and has the greatest influence on the slope of D vs [surfactant] at the lowest salt concentrations. Conversely, A determines the strength of van der Waals attractions and has the greatest effect on the fits at the highest salt concentrations, near the boundary of the positive-slope region (i.e. near zero slope). One adjusts both α and A at a given temperature T (typically 40°C) to obtain the best overall agreement of D vs [surfactant] with the data in the positive-slope region. Similar sets of theoretical plots are then computed at each of the remaining temperatures, keeping α, A and N constant (but allowing a to change to reflect the T-dependence of D_0). It should be noted that a slight downward curvature of D vs [surfactant] is obtained at very low salt concentrations (e.g. $\lesssim$ 0.02M) because κ (Equation (4)) is weakly dependent on surfactant concentration through the ionic strength: I = [salt]+ $(1/2)\alpha$[surfactant].

MATERIALS AND METHODS

CTAB, CTACl and MyTAB were purified as previously reported[30]. Sodium bromide and chloride (Mallinkrodt. A.R.) were dried before use. Distilled water was deionized and filtered (0.22μm) using a Milli-Q (Millipore) water purification system. Light scattering measurements (90°) were performed on 0.5-ml samples contained in 6-mm glass culture tubes located at the center of a toluene-filled fluorimeter cuvet, whose temperature was regulated to $\pm 0.05^{\circ}$C by a Peltier thermoelectric element. Other components included a Spectra Physics Model 164 argon-ion laser, EMI type 9789 PMT and Nicomp Model 6864 computing autocorrelator. Intensity-weighted mean diffusion coefficients and associated Stokes radii were obtained from second- and third-order cumulants[31] fits to 48-channel, 4-bit correlation functions with a long-delay measured baseline.

RESULTS AND DISCUSSION

We summarize our experimental and theoretical results for the mean micellar diffusion coefficient D (cm^2/sec) vs [surfactant] and [salt] at 40°C in Figures 1-4 for the systems CTAB + NaBr, MyTAB + NaBr, CTACl + NaCl and SDS + NaCl, respectively. In the latter case, we did not perform dynamic light scattering measurements, but, rather, used the published data of Corti and Degiorgio[6]. The corresponding results for our systems at 15° and 25°C are qualitatively similar to those shown in Figures 1-4 and can be found in refer-

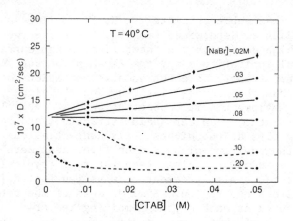

Figure 1. Diffusion coefficient D (cm^2/sec) vs CTAB concentration
(M) at 40°C for a range of NaBr concentrations. The solid curves
are our best theoretical fits (see text); the dashed curves are
meant as guides for the eye only. The diffusivities have been corrected for increases in solvent viscosity due to addition of NaBr.

ences 12 and 13. The solid curves are the best theoretical fits to
the measured data points, obtained following the procedure outlined
earlier. The experimental diffusivities plotted in Figures 1-4
have been corrected for the effects of salt on the solvent viscosity. That is, the observed D values were multiplied by $\eta(20°)/\eta_w(20°)$,
where $\eta(20°)$ is the viscosity of the solvent (salt+water)[32] at
20°C and $\eta_w(20°)$, that of pure water. In this way, variations in
this "water-equivalent" diffusivity with [salt] are the result of
changes in either the micellar hydrodynamic radius a or the intermicellar interactions, and are not simply caused by trivial changes
in the solvent viscosity. Only in the case of CTACl + NaCl were
these corrections significant in the positive-slope region.

As seen in Figures 1-4, the behavior of D vs [surfactant] and
[salt] is qualitatively similar for the four micellar systems examined thusfar. In each case there is a low range of salt concentration where D vs [surfactant] consists of a fan of positive-
slope, approximately linear curves which converge at a common intercept, D_o, at the cmc. At higher salt concentrations, D decreases
with increasing [surfactant], usually nonlinearly and increasingly
precipitously the higher the salt concentration. This behavior is
consistent with the occurrence of substantial micellar growth, as
described by Mazer et al[1], Missel et al[2] and others. (However, it
seems clear that interactions influence the apparent micellar
growth in this region to a limited extent[33]. For example, there is

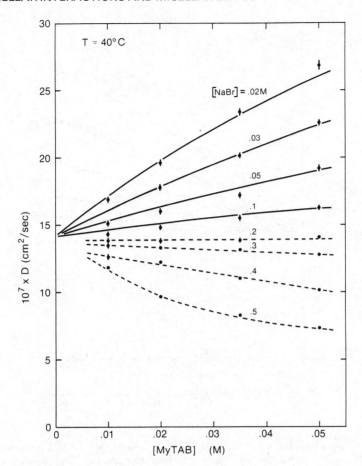

Figure 2. Diffusion coefficient D (cm^2/sec) <u>vs</u> MyTAB concentration
(M) at 40°C for a range of NaBr concentrations. The solid curves
are our best theoretical fits (see text); the dashed curves are
meant as guides for the eye only. The diffusivities have been corrected for increases in solvent viscosity due to addition of NaBr.

a non-monotonic dependence of D on [surfactant] at high salt concentrations in CTAB + NaBr and CTACl + NaCl.)

    There is, however, a striking quantitative difference between
the systems, immediately apparent in Figures 1-4. The salt concentration which marks the boundary of the net-repulsive region
(i.e. where D <u>vs</u> [surfactant] reaches zero slope) varies markedly
with the surfactant/counterion system. This "critical" concentration, $[salt]_{V=0}$, results in zero net interaction (on average)

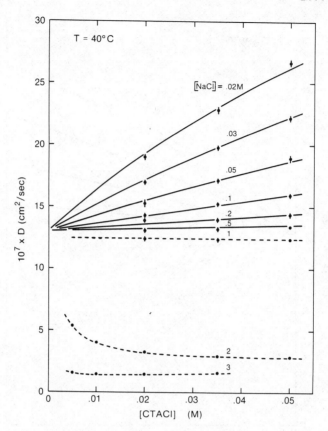

Figure 3. Diffusion coefficient D (cm^2/sec) vs CTACl concentration (M) at 40°C for a range of NaCl concentrations. The solid curves are our best theoretical fits (see text); the dashed curves are meant as guides for the eye only. The diffusivities have been corrected for increases in solvent viscosity due to addition of NaCl.

between micelles; at 40°C, it ranges from $\sim$0.08M for CTAB + NaBr to $\sim$0.5-1M for CTACl + NaCl. The value of $[\text{salt}]_{V=0}$, according to the theoretical model, is intimately related to the values of α and A for a given micellar system.

Before discussing the values found for α and A, we summarize the behavior of the micellar hydrodynamic radius R_h(=a) for the 4 systems examined. Figure 5 shows R_h as a function of T for CTAB + NaBr (closed circles), MyTAB + NaBr (closed triangles), CTACl + NaCl (open circles) and SDS + NaCl (open triangles -- using the published data of Corti and Degiorgio[6]). We observe in all cases that there is modest micellar growth, as measured by the mean hydrodynamic radius, upon cooling. The amount of such growth is, of

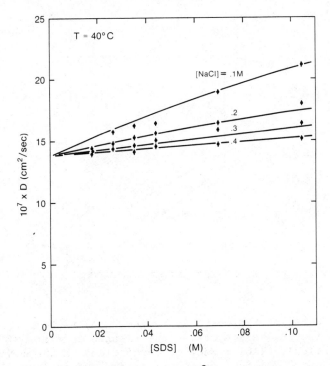

Figure 4. Diffusion coefficient D (cm^2/sec) <u>vs</u> SDS concentration (M) at 40°C for a range of NaCl concentrations, from the data of Corti and Degiorgio[6]. The solid curves are our best theoretical fits (see text). The diffusivities have been corrected for increases in solvent viscosity due to addition of NaCl.

course, much smaller than that observed by dynamic light scattering at high salt concentrations[1,2] (i.e. outside the net-repulsive region). It is instructive to compare the radii for CTAB + NaBr and MyTAB + NaBr, where the only difference is the length of the alkyl chain group. At each temperature the ratio of the respective radii (listed in references 12 and 13) is roughly equal to the ratio of the lengths of the extended alkyl groups, 16/14.

As seen in Figures 1-4, we obtained excellent theoretical fits to the measured micellar diffusivities over the entire net-repulsive region for each system investigated. Furthermore, fits of comparable quality were obtained[12,13] using the same sets of adjustable parameters (but different a values, from Figure 5) at two additional temperatures, 15° and 25°C. Considering the substantial T-dependence of the experimental slopes and the complexity of the T-dependence of $K_t + K_h$ (Equations (3a-e)), this agreement is significant. In the case of SDS + NaCl (Figure 4), we have obtained better theoretical fits to the diffusivity data than those origi-

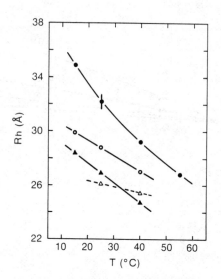

Figure 5. Micellar hydrodynamic radius R_h(Å) <u>vs</u> temperature $T(^\circ C)$
for the surfactant/salt systems investigated thusfar: CTAB + NaBr
(closed circles), CTACl + NaCl (open circles), MyTAB + NaBr (closed
triangles) and SDS + NaCl (open triangles -- from data of Corti and
Degiorgio[6]). The R_h (=a) values were determined from D_o (Equation
(1)).

nally reported by Corti and Degiorgio[6], mainly because we have
confined the interaction analysis to the net-repulsive, positive-
slope region.

    Our best-fit values of fractional ionization α and Hamaker
constant A are listed in Table I. For the three closely related
cationic surfactant systems, the α values lie in a relatively narrow range: 0.22 for both CTAB + NaBr and MyTAB + NaBr and 0.27 for
CTACl + NaCl. The fractional ionization for SDS + NaCl is significantly higher -- 0.36. Values of A (in units of kT at 40°C) range
from 2 for CTACl + NaCl to 15 for CTAB + NaBr. Also listed in Table
I are approximate values of $[salt]_{V=0}$ (at 40°C), which defines the
boundary of the net-repulsive region. These values are obviously
related to α and A; for example, a large value of $[salt]_{V=0}$ can
result from either a large α (i.e. a large electrostatic repulsive
energy, V_R) or a small A (i.e. a small attractive term, V_A), or both.

TABLE I

Micellar Theoretical Interaction Parameters

Surfactant/salt	α	A^* : N	$[salt]_{V=0}$ (M)
CTAB + NaBr	0.22	15 : 120	0.05 - 0.10
MyTAB + NaBr	0.22	7 : 90	0.15 - 0.20
CTACl + NaCl	0.27	2 : 100	0.5 - 1.0
SDS + NaCl	$0.36^{\dagger}$	$4^{\dagger}$: 95	0.40 - 0.45

* Hamaker constant in units of kT (40°C)
† best-fit parameters using the data of Corti and Degiorgio[6]

At our typical lowest salt concentration (0.02M), the magnitude of the electrostatic repulsive energy, $|V_R|$, greatly exceeds that of the attractive term, $|V_A|$, at all but the closest micellar separations. Furthermore, V_R is strongly dependent on α, going as α^2 (i.e. ψ_o in Equation (5) is proportional to α for fixed N). Hence, the slope of D vs [surfactant] is very sensitive to the choice of α and only weakly affected by A. Consequently, this approach of dynamic light scattering/interaction analysis should be an accurate, reliable technique for the determination of the micellar fractional ionization in dilute salt. It should enable one to detect small changes in α due to subtle perturbations in the surfactant/counterion system, of the kind carried out in this study.

From the substitution of MyTAB for CTAB, we conclude that the degree of micellar ionization is relatively insensitive to the alkyl group chain length. However, α varies significantly with the counterion species. When Cl^- is substituted for Br^-, α increases from 0.22 to 0.27. We attribute[13] this change to the increase in the size of the hydrated Cl^- ion relative to hydrated Br^-, which in turn results in less effective penetration of the former into the Stern layer and therefore less effective screening (i.e. neutralization) of the charged micellar head groups. In the case of CTACl + NaCl, a relatively large value of α and a very small value for A results in a large $[NaCl]_{V=0}$ (Table I). That is, over a very large range of NaCl concentration (i.e. $\lesssim$ 1M), micelles of CTACl remain substantially minimum spheres. There is ample confirming evidence[34-37] that micelles of CTACl show much less tendancy to grow with the addition of NaCl than is the case for CTAB plus NaBr.

Finally, we observe that our findings for α are consistent with the range of values reviewed by Anacker[38] for a variety of surfactant systems. Specifically, for CTAB (in zero salt) he lists

α = 0.18-0.22 (determined from emf measurements), while for SDS he quotes α = 0.38 (from conductivity) and 0.27 (by emf). It seems apparent that our light scattering/interaction method yields values for α which are more consistent from system to system than those obtained from the usual classical techniques.

An interpretation of our findings for the Hamaker constant, A, is more difficult, given the relatively poor state of theoretical understanding of the attractive forces between charged, inhomogeneous aggregates, such as micelles, in an electrolyte. All that is available are estimates[39] of A for hydrocarbon droplets in water: A = 0.5-2kT (40°C). The upper end of this range agrees with our smallest value, obtained for CTACl + NaCl (Table I). However, we have recently observed[13] that the counterions in the Stern layer, because of their substantial polarizability, should make a relatively large contribution to the effective attractive interaction. Since the polarizability of Br^- is larger than that of Cl^- (as deduced from their respective molar refractivities[40,41]), we would expect CTAB to possess a larger value of A than CTACl (all other contributions being equal), which is indeed the case.

In conclusion, we draw attention to a finding which we believe is most significant. For all of the systems investigated thusfar, substantial micellar growth (as measured by values of D smaller than D_o) is never observed as long as the net intermicellar interaction remains repulsive -- i.e. as long as [salt] < [salt]$_{V=0}$. That is, all positive-slope D vs [surfactant] curves extrapolate to D_o at the cmc. Hence, the salt concentration which marks the onset of significant micellar growth is the same concentration at which the charged micelles can overcome repulsive forces and begin to collide. This description, based on interactions, must ultimately be compatible with equilibrium models of micellar growth at high salt (e.g. the "ladder" model of Mazer et al[1]), which do not explicitly consider the presence of neighboring micelles.

ACKNOWLEDGEMENTS

We gratefully acknowledge support by the National Science Foundation (Chemical Dynamics Program).

REFERENCES

1. N.A. Mazer, M.C. Carey and G.B. Benedek, in "Micellization, Solubilization and Microemulsions", K.L. Mittal, Editor, Vol. 1, p. 359, Plenum Press, New York, 1977.

2. P.J. Missel, N.A. Mazer, G.B. Benedek and C.Y. Young, J. Phys. Chem. 84, 1044 (1980).
3. B. Chu, "Laser Light Scattering," Academic Press, New York, 1974.
4. B.J. Berne and R. Pecora, "Dynamic Light Scattering," Wiley, New York, 1976.
5. "Minimum sphere" is a convenient term which has been applied to micelles at surfactant concentrations just slightly above the cmc.
6. M. Corti and V. Degiorgio, J. Phys. Chem. 85, 711 (1981).
7. E.J.W. Verwey and J.Th.G. Overbeek, "Theory of the Stability of Lyophobic Colloids," Elsevier, Amsterdam, 1948.
8. P.C. Hiemenz, "Principles of Colloid and Surface Chemistry," Marcel Dekker, New York, 1977.
9. G. Porte and J. Appell, J. Phys. Chem. 85, 2511 (1981); J. Appell and G. Porte, J. Colloid Interface Sci. 81, 85 (1981).
10. H. Hoffmann, G. Platz, H. Rehage and W. Schorr, Ber. Bunsenges Phys. Chem. 85, 877 (1981); also J. Colloid Interface Sci. (in press).
11. S. Ikeda, S. Ozeki and S. Hayashi, Biophys. Chem. 11, 417 (1980).
12. R. Dorshow, J. Briggs, C.A. Bunton and D.F. Nicoli, J. Phys. Chem. 86, 2388 (1982).
13. R.B. Dorshow, C.A. Bunton and D.F. Nicoli, submitted to J. Phys. Chem., 1982.
14. H.C. Hamaker, Physica (Amsterdam) 4, 1058 (1937).
15. J. Briggs, R.B. Dorshow, C.A. Bunton and D.F. Nicoli, J. Chem. Phys. 76, 775 (1982).
16. T.L. Hill, "An Introduction to Statistical Thermodynamics," Addison-Wesley, Reading, Mass., 1960.
17. G.D.J. Phillies, J. Chem. Phys. 60, 976 (1974).
18. G.K. Batchelor, J. Fluid. Mech. 74, 1 (1976).
19. B.U. Felderhof, J. Phys. A: Math. Gen. 11, 929 (1978).
20. S. Hayashi and S. Ikeda, J. Phys. Chem. 84, 744 (1980).
21. P. Lianos and R. Zana, J. Colloid Interface Sci. 84, 100 (1981).
22. K.A. Zachariasse, N.V. Phuc and B. Kozaukiewicz, J. Phys. Chem. 85, 2676 (1981).
23. J.P. Kratohvil, J. Colloid Interface Sci. 75, 271 (1980).
24. P. Lianos and R. Zana, J. Phys. Chem. 84, 3339 (1980).
25. H.J.L. Trap and J.J. Hermans, Proc. K. Ned. Akad. Wet. Ser. B: Phys. Sci. 58, 97 (1955).
26. K.S. Birdi, Kolloid Z.Z. Polym. 250, 731 (1972).
27. D. Attwood, P.H. Elworthy and S.B. Kayne, J. Phys. Chem. 74, 3529 (1970).
28. K.J. Mysels, J. Colloid Sci. 10, 507 (1955).
29. J.P. Kratohvil, Chem. Phys. Lett. 60, 238 (1979).
30. C.A. Bunton, L.S. Romsted and C. Thamavit, J. Am. Chem. Soc. 102, 3900 (1980).
31. D.E. Koppel, J. Chem. Phys. 57, 4814 (1972).
32. "Handbook of Chemistry and Physics," R.C. Weast, Editor, The Chemical Rubber Co., Cleveland, 1971.

33. D.F. Nicoli, R. Ciccolello, J. Briggs, D.T. Dawson, H.W. Offen,
     L.S. Romsted and C.A. Bunton, in "Scattering Techniques Applied
     to Supramolecular and Non-Equilibrium Systems"; S.H. Chen,
     B. Chu and R. Nossal, Editors, p. 363, Plenum Press, New York,
     1981.
34. F. Reiss-Husson and V. Luzzati, J. Phys. Chem. 68, 3504 (1964).
35. G. Lindblom, B. Lindman and L. Mandell, J. Colloid Interface
     Sci. 42, 400 (1973).
36. U. Henriksson, L. Odberg, J.C. Eriksson and L. Westman,
     J. Phys. Chem. 81, 76 (1977).
37. M. Almgren, J.-E. Lofroth and R. Rydholm, Chem. Phys. Lett.
     63, 265 (1979).
38. E.W. Anacker, in "Solution Chemistry of Surfactants," K.L.
     Mittal, Editor, Vol. 1, p. 247, Plenum Press, New York, 1979.
39. J. Visser, Adv. Colloid Interface Sci. 3, 331 (1972).
40. C.K. Ingold, "Structure and Mechanism in Organic Chemistry,"
     P. 142, 146, Cornell University Press, Ithaca, 1969.
41. A. Gordon and R. Ford, "The Chemist's Comparison: A Handbook
     of Practical Data, Techniques and References," p. 154,
     John Wiley, New York, 1972.

LASER-LIGHT SCATTERING STUDY OF NONIONIC MICELLES IN AQUEOUS SOLUTION

Vittorio Degiorgio[1] and Mario Corti[2]

[1]Istituto di Fisica Applicata
Universita di Pavia, Via Bassi 6, 27100 Pavia, Italy
[2]C.I.S.E. S.p.A., 20090 Segrate, Milano, Italy

    n-Dodecyl hexaoxyethylene and n-Octyl tetraoxyethy-
lene glycol monoether ($C_{12}E_8$ and C_8E_4) aqueous solutions
are investigated by static and dynamic light scattering over
a wide temperature range around the minimum in the
cloud curve (critical point). The osmotic isothermal
compressibility (apparent molecular weight of the micelle) does not increase exponentially with temperature T.
but shows a divergence as T approaches the critical
temperature T_c. The divergence of the osmotic compressibility and of the correlation range of concentration
fluctuations follows a simple power-law behavior as a
function of T_c-T, with values of the critical exponents
which depend on the nature of the amphiphile. It is
also found, with the system $C_8E_4-H_2O$, that the low-
concentration phase in the two-phase region above T_c
is still micellar.

INTRODUCTION

In the last two decades there has been carried out a considerable
amount of work dealing with the micellar properties of nonionic
amphiphiles, particularly the alkylpolyoxyethylenes $CH_3(CH_2)_m$
$(OCH_2-CH_2)_n OH$ (called henceafter C_mE_n). Light scattering measurements performed with dilute aqueous solutions of C_mE_n have shown
a considerable increase of the scattered light intensity I_s as
function of the temperature T. Balmbra et al.[1,2] observed an
approximately exponential dependence of the apparent micellar molecular weight on T for C_mE_6, with m = 8, 10, 12, 14, 16. Subsequently,
many papers have reported similar results on various alkylpolyoxyethylenes [3,4] and other nonionic amphiphiles, such as the n-octylsul-
phinylalkanols[5] and the dimethylalkylphosphine oxides[6].

Dilute aqueous solutions of nonionic amphiphiles usually
exhibit a phase separation when the temperature is raised above a
value which depends on the amphiphile concentration c. In the c-T
plane, the single-phase micellar region is separated from the
region in which two immiscible isotropic solutions are present by
a lower consolution curve (cloud curve in the micellar literature).
The minimum in the curve is the critical point, the temperature
and concentration at which the minimum occurs are called the
critical temperature T_c and the critical concentration c_c, respectively. Since it is well known that the approach to the
critical point in binary mixtures is accompanied by long-range
concentration-fluctuations, and hence by a large light-scattering
cross-section, it appears natural to believe that the observed
phenomena are, at least partially, explainable by the existence
of consolute phase boundaries[6,7]. Our laser-light scattering
experiments[7,8] have clearly shown that the main parameter determining solution properties is not the temperature itself but the
temperature distance from the critical point. The results obtained
with $C_{12}E_6$ solutions[8] indicate that the apparent micellar molecular
weight, M_{app}, measured as function of T along the critical concentration line, diverges as the critical point is approached following a power law behavior of the type $M_{app} \sim (T_c-T)^{-\gamma}$, with a
critical exponent $\gamma \simeq 1$. A power-law divergence is also found for
the correlation range of concentration fluctuations measured
by the angular dependence of the scattered intensity or by
the quasielastic light-scattering technique. Apart from the values
obtained for the critical exponents, several features of nonionic
micellar solutions, observed in a temperature region which extends
typically 20°C far away from T_c, are identical to those found in
a large variety of critical systems studied in the last twenty
years. The list of critical systems investigated includes liquid
binary mixtures made of two low molecular weight components[9], like
aniline and cyclohexane or isobutyric acid and water, and macromolecular solutions, like polymer[10] and protein[11] solutions. In
particular, polymer solutions present some similarity with micellar

solutions because the coexistence curve may be strongly asymmetric and may have the minimum at a very low polymer concentration. By making an analogy with critical polymer solutions, it should not be surprising to discover that the strong temperature dependence of the scattered intensity in nonionic micellar solutions may be due to critical concentration fluctuations, that is, to a strongly nonideal behavior. In any case, it is clear that a direct evidence of the fact that the micelle aggregation number changes or not with T cannot be reached by light scattering studies, but requires the use of more local probes, like neutron scattering or nuclear magnetic resonance.

The study of nonionic micellar solutions is important from both fundamental and applied points of view. In order to make a contribution to its full understanding, we have undertaken a systematic light-scattering investigation of aqueous solutions of the amphiphiles C_mE_n. We present in this paper results concerning $C_{12}E_8$ and C_8E_4. The former amphiphile was chosen because of its high critical temperature ($T_c \simeq 76^oC$) which makes easier to study the whole evolution of the system, as T is raised, from the regime of non-interacting micelles to the critical region. The latter amphiphile is interesting because it has a very high critical micelle concentration c_o and a high concentration at the critical point c_c. These features of the C_8E_4-H_2O system enable one to investigate the possible role of monomers in the phase transition and permit also a study of the solution in the two-phase region above T_c.

LIGHT SCATTERING FROM CRITICAL BINARY MIXTURES[12]

The total intensity of scattered light from a binary mixture is I_s+I_r, where I_s is the contribution of concentration fluctuations and I_r the contribution of density fluctuations. In our case, I_r practically coincides with the contribution of the solvent. The intensity I_s depends on the scattering angle θ according to the Ornstein-Zernike relation

$$I_s = \frac{I_{so}}{1+k^2 \xi^2} \tag{1}$$

where $k = (4 \pi n/\lambda)\sin \theta /2$ is the modulus of the scattering vector, λ being the wavelength of incident light and n the index of refraction of the medium, and ξ is the correlation length of concentration fluctuations. The extrapolated scattered intensity at zero scattering angle I_{so} is related to the derivative of the osmotic pressure with respect to the concentration, $(\partial \pi/ \partial c)_{T,p}$, as follows:

$$I_{so} = Ac \left(\frac{dn}{dc}\right)^2 k_B T (\partial \pi / \partial c)^{-1}_{T,p} \tag{2}$$

where A is an instrumental constant, dn/dc the refractive index increment, k_B the Boltzmann constant and T the absolute temperature. The quantity $(\partial \pi / \partial c)_{T}$ is, in general, a function of T and c which takes very simple forms in two important limiting cases. When the system is far from the critical point, and the solution is sufficiently dilute.

$$k_B T (\partial \pi / \partial c)^{-1}_{T,p} = \frac{M}{1 + 2Bc} \tag{3}$$

where B is the second virial coefficient and M is the molecular weight of the solute particles. When the system is close to the critical point both $(\partial \pi / \partial c)^{-1}_{T,p}$ and ξ are found to diverge as $T \rightarrow T_c$, and the following relation holds

$$(\partial \pi / \partial c)^{-1}_{T,p} \propto \xi^{2-\eta} \tag{4}$$

where η is much smaller than one. If the system approaches the critical point in the single-phase region at the critical concentration c_c, the two quantities diverge according to a simple power-law behavior

$$(\partial \pi / \partial c)^{-1}_{T,p} \propto \varepsilon^{-\gamma} \quad , \quad \xi = \xi_o \varepsilon^{-\nu} \tag{5}$$

where $\varepsilon = (T_c - T)/T_c$ and the two critical exponents γ and ν are related through Eq (4), that is, $\gamma = \nu (2-\eta)$.

Turning now to dynamic light scattering, we recall that the time-dependent part of the intensity-correlation function measured at a scattering angle θ , corresponding to a scattering vector $\vec{k}$, is simply the square of the correlation function of the concentration-fluctuations component with wave vector $\vec{k}$. Therefore the mass-diffusion coefficient D_m is derived from the decay time τ_c of the intensity-correlation function as $D_m = (2k^2 \tau_c)^{-1}$. When T is sufficiently close to T_c, D_m is found to depend on k. The hydrodynamic mass-diffusion coefficient D is derived as the limit of $\overline{D_m}$ as k goes to zero. For small values of the product $k\xi$, the first correction is, according to the mode-mode coupling theory,

$$D_m = D(1 + \frac{3}{5} k^2 \xi^2) \tag{6}$$

The diffusion coefficient D is related to ξ by the Stokes-Einstein formula

$$D = h \frac{k_B T}{6\pi\eta\xi} \qquad (7)$$

where $h \simeq 1$ and η is the macroscopic shear viscosity of the solution. When the system is far from the critical point and the solution is sufficiently dilute,

$$D = D_o (1 + k_D c) \qquad (8)$$

where D_o is the translational diffusion coefficient of the solute particles and k_D reflects the effect of two-particle interactions.

MATERIALS AND METHODS

The main characteristics of the light scattering setup are given in our previous paper [13]. A very detailed description can be found in Ref. 14. Measurements of viscosity and of refractive index increments are made with standard apparatus (Ubbelohde viscometer and differential refractrometer).

High-purity $C_{12}E_8$ was obtained from Nikko Chemicals, Tokyo. The compound C_8E_4 was synthetized and purified by Grabo et al.[15]. The sample preparation procedure is the same as described in Ref.8.

The phase diagram of C_8E_4 in H_2O, as taken from Ref.16, is shown in Fig. 1. The critical micelle concentration (cmc, the concentration above which micelles are formed; this should not be confused with the concentration c_c at the critical point) curve which separates the monomeric phase I from the ordinary micellar phase II was obtained from surface tension measurements. The cmc decreases from 3.8 mg/cm^3 at 6°C to 1.1 mg/cm^3 at 60°C. The phase boundary which separates phase II from phase III where two isotropic micellar solutions coexist was determined by observation of the critical opalescence. It shows a rather flat minimum at $T_c = (40.3 \pm 0.1)$°C and $c_c = 70$ mg/cm^3. The phase diagram of $C_{12}E_8$ is similar to that shown in Fig. 1, but $T_c = 76.1$°C and $c_c = 15$ mg/cm^3, and the cmc curve is shifted toward much lower concentration.

EXPERIMENTAL RESULTS

We have reported in Fig. 2 the logarithm of the scattered intensity measured at $\theta = 90°$, as function of the temperature, for the $C_{12}E_8$-H_2O solution at the amphiphile concentration $c_c = 15$ mg/cm^3.

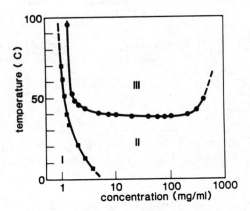

Figure 1.   Phase diagram of C_8E_4 in H_2O at atmospheric pressure.
Region I: monomeric amphiphile in water; region II: micellar solution; region III: two coexisting micellar solutions. (from Ref. 16).

Below 40°C, I_s changes very little with T. Above 40°C, I_s increases
considerably, and seems to diverge as the temperature approaches
the critical temperature T_c.

The measured temperature dependence of the refractive index
increment dn/dc of the $C_{12}E_8$ solution is well described, in the
range 15-75°C, by the linear law

$$dn/dc = 0.135 - 0.00024 \ (T-293) \qquad (9)$$

where T is the absolute temperature in K and dn/dc is expressed
in cm^3/g.

When the temperature distance from T_c is less than a few
degrees, I_s becomes k-dependent. By measuring I_s at two distinct
scattering angles and using eq. (1), we derive the extrapolated
scattered intensity I_{so}.

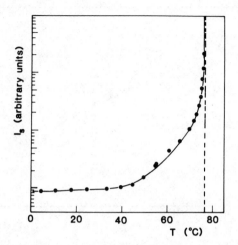

Figure 2 - Scattered light intensity measured as function of the temperature for a $C_{12}E_8-H_2O$ solution at the amphiphile concentration of 15mg/cm^3.

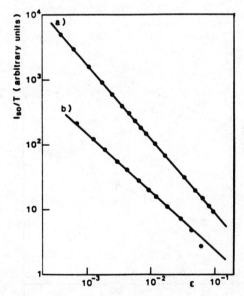

Figure 3 - The quantity I_s/T proportional to osmotic isothermal compressibility of the solution $C_8E_4-H_2O$ (curve a) and of the solution $C_{12}E_8-H_2O$ (curve b) as function of the reduced temperature $\varepsilon=(T_c-T)/T_c$ along the critical isoconcentration line $c_c = 15$ mg/cm^3.

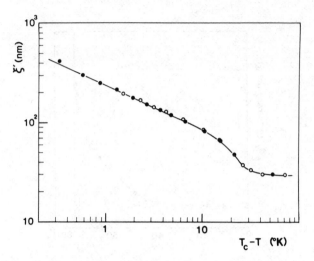

Figure 4 - The correlation range ζ', obtained from dynamic light scattering data, plotted as function of T_c-T. The measurement is performed on the $C_{12}E_8-H_2O$ system at the concentration 15 mg/cm^3.

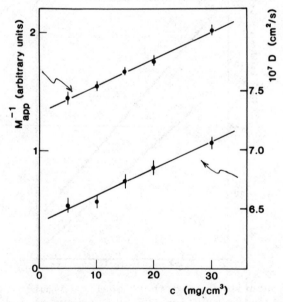

Figure 5 - The apparent micellar weight and the mass diffusion coefficient of $C_{12}E_8-H_2O$ solutions at 20°C as function of the amphiphile concentration.

Once dn/dc and I_{so} are known experimentally, it is possible
to use Eq. (2) to derive the osmotic compressibility of the solution.
The behavior of I_{so}/T as function of the normalized temperature
distance from the critical point, $\varepsilon = (T_C-T)/T_C$, is shown in Fig.3.
We have reported in the same figure also the results relative to
C_8E_4 solutions, taken at the concentration of 70 mg/cm^3. We have
obtained an absolute calibration for the osmotic compressibility
by measuring the turbidity near the critical point[16]. The results
of Fig. 3 are well described by the power law $(\partial\pi/\partial c)_{T,p}^{-1} = B\varepsilon^{-\gamma}$
with $T_C = 40.285°C$, $B = (1.17 \pm 0.1) \times 10^{-3}$ cm^{-2}s^2 and $\gamma = 1.15 \pm 0.02$
for the C_8E_4 solution, and $T_C = 76.075°C$, $B = (2.60 \pm 0.3)$ cm^{-2}s^2
and $\gamma = 0.90 \pm 0.03$ for the $C_{12}E_8$ solution.

Since the $C_{12}E_8$ data reported in Fig. 3 were taken with a
solution prepared few days before the light-scattering run, we repeated subsequently the measurements by using a freshly prepared
solution. We found a small aging effect in the critical temperature
($T_C = 76.500$ for the freshly prepared solution), but the ε-depen-
dence of static and dynamic parameters was not appreciably modified.

We have derived from the dynamic light scattering data, by
using eqs. (6) and (7), the correlation range $\xi' = \xi/h$. The results
relative to $C_{12}E_8$ solution at the critical concentration are shown
in Fig. 4. In the region $T_C-T < 20°C$, ξ' follows a simple power-law
behavior as a function of T_C-T, with a critical exponent $\nu = 0.45$
± 0.02. In the region $T_C-T > 40°C$, ξ' is practically constant at
3.5 nm. The C_8E_4 data are of similar quality, and show that ξ'
follows a power-law behavior in the range 10-40°C with a critical
exponent $\nu = 0.57 \pm 0.03$. The constant h is found in Ref. 16 to be
1.1, which means that ξ' is 10% smaller than the correlation range
ξ one would measure from the angular dependence of the scattered
intensity.

The properties of the $C_{12}E_8$ micelles below 40°C have been de-
rived by measuring the concentration dependence of $M_{app} = k_BT(\partial\pi/\partial c)_{T,p}^{-1}$. The data relative to 20°C are shown in Fig. 5. We have
verified that the temperature dependence of the results is very
weak in the range 20-37°C.

In order to study the behavior of C_8E_4 solutions at low concentration and in the two-phase region we have performed static
and dynamic light scattering measurements along the isotherm T =
44.84°C in the concentration range 0.5-4 mg/cm^3, and along the
isoconcentration line c = 4 mg/cm^3 in the temperature range 15-
46.2°C. Note that the two paths cross on the coexistence curve
(at c = 4 mg/cm^3) which means that the data taken above 44.84°C
refer to the two-phase region. Because of the geometry of the
experimental arrangement, only the low-concentration phase was
accessible to light scattering measurements (the high-concentration
phase constitutes a thin layer floating upon the low-concentration

phase). The interpretation of light-scattering data obtained along the two paths is hampered by the effect of a small amount of water-insoluble impurities which give a small peak in the scattered light intensity near the cmc. We have reported about similar effect in Triton X-100 solutions (see also Ref. 17 for a quantitative study of this effect in SDS solutions). We summarize the main results obtained with the low-concentration C_8E_4 solutions: i) the C_8E_4 micelle has a molecular weight M $\simeq$ 25000 (which corresponds to an aggregation number around 80) and a hydrodynamic radius 2.5 nm; ii) the low-concentration phase above T = 44.84°C is still micellar and shows an apparent molecular weight which decreases with increasing temperature. the few available points are not accurate enough to try a power-law fit. However, we can say, that the data are consistent with a behavior, $(\partial c/\partial\pi)_{T,p} = B' \epsilon^{-\gamma}$, along the coexistence curve where γ has the same value measured along the critical concentration line and B' $\simeq$ B/3.8.

DATA INTERPRETATION AND DISCUSSION

Micelle Size

The light-scattering data of Figs. 2 and 4 suggest that the $C_{12}E_8$ solution below 40°C behaves as a solution of weakly interacting micelles having an aggregation number independent of the temperature. The concentration dependence of M_{app} and D shown in Fig. 5 cannot be attributed to an actual change in the aggregation number m because it is unreasonable to find a decrease of m with concentration. If we assume that m does not depend on concentration, we can fit the data of Fig. 5 with eqs. (3) and (8). We find M=65000, $k_I = 2B = 16.3$, $D_0 = 6.4 \times 10^{-3}$ cm^2/s, and $k_D = 3.3$.

The values of M and D_0 obtained are in good agreement with literature data[18,19]. The hydrodynamic radius of the $C_{12}E_8$ micelle, derived from D_0 by the Stokes-Einsten relation, is 3.4 nm. The positive values of the slopes k_I and k_D are indicative of repulsive intermicellar interactions. Since the micelles are not charged, the observed slopes should be only due to excluded volume effects. The theoretically predicted slopes for a dilute system of hard spheres are, however, half the experimentally observed values. It is not clear whether this discrepancy can be explained by taking into account the strong hydration of the polyoxyethylene chains.

The slopes k_I and k_D are found to be independent of the temperature in the investigated range 20-37°C. Above 40°C the slopes become negative, that is, M_{app} increases with the concentration. Clearly the existence of the phase transition is an indication of the fact that, above a certain temperature, attractive interactions must exist among the micelles.

With regard to the C_8E_4 micelle, it is difficult to find a region in the c-T plane in which the light scattering results may simply be interpreted in terms of individual particle properties. In any case, the value of the hydrodynamic radius obtained coincides with the neutron scattering result of Zulauf and Rosenbusch[20], whereas a discrepancy exists for the micellar molecular weight between our result and the result of Ref.20.

Apparent micellar growth and critical concentration fluctuations

The power-law divergences shown in Figs. 3 and 4 can be explained by the increase in the correlation range ξ of concentration fluctuations, in analogy with critical consolution points in a large variety of binary mixtures, without invoking a growth of the micellar aggregation number with T. However, it is not possible to exclude, on the basis only of light-scattering data, that the micelle size somewhat changes with T. Information on micelle size can be obtained by using techniques which are not sensitive (or weakly sensitive) to the collective effects probed by light scattering. Although some of the reported experiments may not have a straightforward interpretation, there is now some evidence in favour of the hypothesis that in some systems micelle size does not change dramatically with T, as shown by the following list of experiments:

a) NMR linewidth measurements on $C_{12}E_6$-D_2O in the temperature range 30-50°C[21].
b) Electric birefringence on C_8E_4-H_2O. The fact that no birefringence is observed indicates that micelles remain small and globular even if M_{app} increases[22].
c) Kinetic measurements on Triton X-100. Relaxation after a T-jump does not show a change of micelle size with T[23].
d) Neutron scattering experiments on C_8E_4, C_8E_5 and $C_{12}E_6$ solutions. Measurements performed at large k show a micelle size which is not influenced by temperature[20,24].
   Indirect evidence is given by the following results:
e) Usually rapid changes of size are accompanied by a considerable polydispersity. No large polydispersity is observed in our data, or by the comparison between light-scattering and osmometry data (see Fig. 8.3 in Ref. 25).
f) The effects of additives on M_{app} can be fully explained by considering the shift in the position of the critical point induced by the additive. Besides the data mentioned in Ref. 8, one could consider the recent results on solutions of the commercial surfactants Brij 92 and 96[26].

It should, however, be added that at the Lund Symposium we became aware of other results which appear to favour micellar growth, such as NMR self-diffusion measurements on $C_{12}E_5$ by Wennerström and Lindman (unpublished), fluorescent probe measurements on

$C_{12}E_6$ by Almgren and Lofroth (see these Proceedings), and neutron scattering data by Ravey (unpublished).

The temperature dependence of the $C_{12}E_8$ data is qualitatively similar to that observed with $C_{12}E_6$ solutions[8]. There is a low temperature region, $T < 40°C$, in which the micelle size is constant, and a critical region, $64°C < T < T_C$ in which the solution behaves as a critical binary mixture and the micelle size is probably constant (but it may not coincide with the low-temperature value). In the intermediate region, $40°C < T < 64°C$, micelle interactions are certainly present, but it is not possible to state with certainty whether the change of M_{app} (and of ξ') is totally due to nonideality effects. With regard to C_8E_4 solutions, all the available data at c_C belong to the critical region. Since T_C is low for this system, it is not possible to extend the measurements down to the temperature region characterized by constant-size non-interacting micelles.

Two-phase region

Many authors believe that in the region above the consolution curve (cloud curve) there are no micelles because surfactant molecules are segregated from the aqueous phase. According to this view, which does not seem to have been tested by any experiment, one phase contains singly dispersed surfactant monomers and the other is a surfactant phase containing dissolved water. C_8E_4 solutions are well suited for a study of the two-phase region because T_C is low and both c_C and the cmc are relatively large.

By simple inspection of the phase diagram in Fig. 1 it is clear that the isotropic micellar solution separates above the consolution curve into two isotropic micellar solutions. Of course, for temperatures much larger than T_C, the low concentration phase becomes more and more dilute, and presumably, merges into the cmc curve; whereas the high concentration phase may become anisotropic, depending on the nature of the nonionic amphiphile. The light scattering measurements described in the previous section confirm, at least for the low-concentration branch of the consolution curve, that the solution is still micellar. The observation that M_{app} decreases by increasing T above T_C represents a further confirmation of the correction of the description in terms of critical binary mixtures. It would be difficult to explain, within the frame of a theory based only on changes of micelle structure, why the micelle size grows to an infinite aggregation number at T_C and decreases progressively, as T become larger than T_C, in the two-phase region.

Analogy with polymer solutions

It may be interesting to see whether some understanding of

the behavior of nonionic micellar solutions can be gained by making an analogy with polymer solutions. The problem has been, up to now, completely open. We report in the following some qualitative considerations which could be helpful in developing a quantitative approach.

A typical polymer solution (say, for instance, polystyrene in cyclohexane[10]) may show the following behavior: at high temperatures the solvent behaves as a good solvent and one obtains a single phase. By reducing the temperature the solvent becomes less "good", and below the so-called θ-temperature polymer-solvent interactions are less favourite than polymer-polymer and solvent-solvent interactions. The solution shows an upper consolution point, below which it separates into two phases. Several features of the separation process are qualitatively explained by the Flory-Huggins theory which consists of a lattice model solved by a mean-field approximation[27]. The Flory-Huggins theory predicts that the critical volume fraction $\emptyset_c$ is given by $(1+m^{1/2})^{-1}$, where m is the number of monomers in the polymer molecule. When m=1, the coexistence curve is symmetric around its maximum at a polymer fraction of 50%. By increasing m, $\emptyset_c$ is shifted toward lower concentrations and the coexistence curve becomes less and less symmetric. The critical temperature is predicted to be little dependent on m for large m. The critical exponents calculated from the Flory-Huggins theory are the mean-field exponents. The available data indicate that this theory represents only a qualitative description of the situation.

More relevant for the analogy with the nonionic amphiphile solutions are the polymer solutions presenting a lower consolution point, that is, showing phase separation when the temperature is increased. This is, for instance, the behavior of polyoxyethylene polymers in water[28]. The fact that such solutions show a lower consolution point may be physically explained by the observation that the solubility of polyoxyethylene chains in water is due to hydrogen-bond formation between the ether oxygens of polyoxyethylene and water[29]. By increasing T, the hydrogen bonds may break, and water becomes a less "good" solvent for the polymer. Formally, it is possible to describe lower consolution points by using decorated Ising models, as shown by Wheeler and coworkers[30].

The consolution curves obtained with the $C_mE_n-H_2O$ system are qualitatively similar to those of high molecular weight polyoxyethylene polymers in water because they have an upward curvature and they are strongly asymmetric. Starting from this observation, there have been some attempts to apply the Flory-Huggins theory to nonionic amphiphile solutions[31], but without attempting to describe critical properties.

The simplest possible model could assume that the micelle behaves as a polymer having a molecular weight coincident with the weight of the hydrophilic part of the micelle. For instance a $C_{12}E_8$

micelle with an aggregation number 120 would be equivalent to a polyoxyethylene polymer composed of 8x120 = 960 ethylene oxide monomers. The larger asymmetry and the smaller critical concentration of the $C_{12}E_8$-H_2O system with respect to C_8E_4-H_2O would reflect, therefore, the fact that the $C_{12}E_8$ micelle is larger than the C_8E_4. However, the trend of critical temperatures is in contrast with that expected from the Flory-Huggins theory because T_c is larger for $C_{12}E_8$. Furthermore the observed ratio between the c_c of C_8E_4 and that of $C_{12}E_8$ should be, according to the same theory, the reciprocal of the square root of the ratio between numbers of ethylene oxide monomers in the micelle. If we take the micelle aggregation numbers measured far from T_c this latter ratio is about 2, implying a ratio 1.4 between critical concentration, whereas the observed value is 7/1.5 = 4.7.

CONCLUSIONS

We have presented a static and dynamic light-scattering study of nonionic polyoxyethylene micelles in aqueous solution. The most important experimental findings are:

a) the apparent molecular weight of nonionic micelles does not increase exponentially with temperature but shows a divergence when T approaches the critical temperature T_c;

b) the divergence of M_{app} (and of the apparent hydrodynamic radius) follows a simple power-law behavior as function of T_c-T;

c) the critical exponents describing the power-law divergence depend on the nature of the amphiphile;

d) the consolution curve does not cross the cmc curve (for the system C_8E_4-H_2O) which means that the low-concentration phase in the two-phase region is micellar.

The results obtained show that critical concentration fluctuations play an important role for the description of solution properties in a large temperature region around T_c. The critical behavior of micellar solutions presents however, some peculiarities not observed in usual critical binary mixtures. This indicates that the usual models of critical phenomena may not be appropriate for describing critical micelle solutions.

A very important step toward the understanding of the problem could be represented by a direct measurement of the temperature-dependence of micelle size. Although there are experiments indicating that such a dependence is small for some systems, there appears to be no general agreement about this point, and it is still unclear whether the results obtained with a specific system can be extrapolated to all the homologs.

ACKNOWLEDGEMENTS

Some of the data presented in this work have been obtained in collaboration with M. Zulauf. Some measurements have been performed with the help of C. Minero.

This work was partially supported by CNR/CISE Contract N.81.00938.02

REFERENCES

1) R.R. Balmbra, J.S. Clunie, J.M. Corkill and J.F. Goodman, Trans. Faraday Soc. 58, 1661 (1962).
2) R.R. Balmbra, J.S. Clunie, J.M. Corkill and J.F. Goodman, Trans. Faraday Soc. 60, 979 (1964).
3) R.H. Ottewill, C.C. Storer and T. Walker, Trans. Faraday Soc. 63, 2796 (1967).
4) D. Attwood, J. Phys. Chem. 72, 339 (1968).
5) J.M. Corkill, J.F. Goodman and T. Walker, Trans Faraday Soc. 63, 759 (1967).
6) K.W. Herrmann, J.G. Brushmiller and W.L. Courchene, J. Phys. Chem. 70, 2909 (1966).
7) M. Corti and V. Degiorgio, Opt. Commun. 14, 358 (1975).
8) M. Corti and V. Degiorgio, Phys. Rev. Letters 45, 1045 (1980); J. Phys. Chem. 85, 1442 (1981).
9) D. Beysens, A. Bourgou and P.Calmettes, Phys.Rev.A 26, 3589 (1982).
10) N. Kuwahara, D.V. Fenby, M. Tamsky and B. Chu, J. Chem. Phys. 55, 1140 (1971); S.P. Lee, W. Tscharnuter, B. Chu and N.Kuwahara, J. Chem. Phys. 57, 4240 (1972).
11) C. Ishimoto and T. Tanaka, Phys. Rev. Letters 39, 474 (1977).
12) H.L. Swinney in "Photon Correlation and Light Beating Spectroscopy", H.Z. Cummins and E.R. Pike, Editors, p. 331, Plenum Press, New York, 1974.
13) M.Corti and V. Degiorgio, Ann. Physique (Paris) 3, 303 (1978).
14) V. Degiorgio, M. Corti and C. Minero, Nuovo Cimento D(to appear).
15) M. Grabo, J.P. Rosenbusch, and M. Zulauf, J. Mol. Biol (to be published).
16) M. Corti, V. Degiorgio and M. Zulauf, Phys. Rev. Letters 48, 1617 (1982).
17) M. Corti and V. Degiorgio, Chem. Phys. Lett. 49, 141 (1977).
18) C. Tanford, Y. Nozaki and M.F. Rohde, J. Phys. Chem. 81, 1555 (1977).
19) M. Corti, V. Degiorgio, R. Ghidoni and S. Sonnino, J. Phys. Chem. 86, 2533 (1982).
20) M. Zulauf and J.P. Rosenbusch, J. Phys. Chem. (to be published).

21) E.J. Staples and G.J. Tiddy, J. Chem. Soc. Faraday Trans. 1,
     74 2530 (1978).
22) H. Hoffmann, H.S. Kielman, D. Pavlovic, G. Plats and W. Ulbricht,
     J. Colloid Interface Sci. 80, 237 (1981).
23) G. Plats, Ber. Bunsenges, Phys. Chem. 85, 1155 (1981).
24) R. Triolo, L.J. Magid, J.S. Johnson and H.R. Child,
     J. Phys. Chem., 86, 3689 (1982).
25) C. Tanford, "The Hydrophobic Effect", p. 82, Wiley, New York,
     p. 401, Addison-Wesley, Reading, 1960.
26) A.A. Al-Saden, A.T. Florence, T.L. Whateley, F. Puisieux and
     C. Vaution, J. Colloid Interface Sci. 86, 51 (1982).
27) T.L. Hill, "An Introduction to Statistical Thermodynamics",
     p. 401, Addison-Wesley, Reading, 1960.
28) S. Saeki, N. Kiwahara, M. Nakata and M. Kaneko, Polymer 17,
     685 (1976).
29) R. Kjellander and E. Florin, J. Chem. Soc. Faraday Trans. I,
     77, 2053 (1981).
30) G.R. Andersen and J.C. Wheeler, J. Chem. Phys. 69, 3403 (1978).
31) J. Goldfarb and L. Sepulveda, J. Colloid Interface Sci. 31,
     454 (1969).

LIGHT SCATTERING FROM CONCENTRATED SOLUTIONS OF SODIUM OCTANOATE

MICELLES

M. Drifford, T. Zemb, M. Hayoun, and A. Jehanno

Département de Physico-Chimie, CEN.Saclay
91191 Gif-sur-Yvette Cedex, France

Sodium octanoate micelles have been studied by
light scattering at surfactant concentrations up to 1.8 M.
The micellar weight as determined by intensity measurements
shows an increase in the aggregation number from 11 at the
c.m.c. to 26 at 1.8 M. Dynamic light scattering indicates a
"reactive" diffusion which is a purely translational effect
due to the chemical exchange between the micelles and free
monomers.

INTRODUCTION

The micelles formed in aqueous solutions of sodium octanoate
at concentrations above the critical micellar concentration
(c.m.c. $\simeq$ 0.4 M) have been extensively studied[1,2]. The determination
of aggregation numbers by different techniques[3,4] and numerical
evaluations[5] is difficult because this is a limiting case of micelle
formation with a large c.m.c. and a very low aggregation number.

In this system at large volume fractions ($\phi \leqslant 0.4$), the intermicellar correlations resulting from repulsive interactions between
the charged particles are important. A recent neutron scattering
study[6] shows a strong contribution from the intermicellar structure
factor $S(q)$. The application of modern liquid theory to colloïdal
systems[7,8,9] now allows an analysis of the scattering data from
solutions of charged micelles at any concentration and gives some
fundamental information on the structure of the micelles[10] or
microemulsion[11]. The calculation of the analytical structure factor
$S(q)$ using the mean spherical approximation with a repulsive

electrostatic potential between interacting charged micelles[8] has been used to analyse neutron scattering data.

In this paper we report light scattering measurements made at various concentrations. Results have been extrapolated to the c.m.c. (0.4 M) and we have determined the aggregation number at different concentrations using the values of $S(q \to 0)$ extracted from neutron scattering data. Dynamic light scattering measurements have also been made to determine the hydrodynamic radius of the micelles. At low volume fractions, the diffusion coefficient, D_{eff}, is perturbed by strong exchange between monomers and micelles. At 0.6 M, there are 0.4 M monomer units (c.m.c.) and $\cong 0.02$ M micelles. The data have been analysed using the crude model proposed by Phillies [12].

EXPERIMENTAL

Sodium octanoate ($CH_3-(CH_2)_7COO$ Na) was purified as described in Reference 13 and twice recrystallised from ethyl-alcohol.

The samples were dissolved in pure water (double distilled, with a resistivity of 18 MΩ.cm). The solutions were filtered through a 0.22 μm millipore membrane into the light scattering cell. The samples were centrifuged at 5000 rpm for an hour before measurement. For dynamic measurements the 500 mW beam of an argon-ion laser (wavelength 5145 Å) Spectra Physics SP 165 model is focused at the center of the 8 mm diameter cylindrical sample cell. For static studies, a square sample cell was used, intensities being measured at 90° to the incident beam. The temperature was kept at 28°C $\pm$ 0.1. The intensity and the autocorrelation function were recorded with a photomultiplier EMI 9863 KB 100 followed by a 64 channels K 7025 Malvern digital correlator. The autocorrelation function observed by homodyne detection was computed in the multibit mode an analysed on a Commodore computer. Data were analysed using the cumulative method which was used for the study of polyelectrolytes [14]. The effective translational diffusion coefficient D_{eff} is deduced from the average decay rate $\bar{\Gamma}$. The first cumulant is obtained from the extrapolation of $\bar{\Gamma}_i$ versus $\bar{\Gamma}_i \tau_{max}$ ($\bar{\Gamma}_i$ being the linear factor of the ith order polynomial fit and τ_{max} is 64 times the sample times $\Delta\tau$). Good agreement is found between $\bar{\Gamma}1$ ($\bar{\Gamma}1 \times \tau_{max} \to 0$) and ($\bar{\Gamma}2 \times \tau_{max} \to 0$) for most of the samples investigated. For each sample the intensity was measured during 20 seconds and obtained from the time averaged rate of photocounts entering the correlator . Background scattering due to monomers at the c.m.c. was subtracted from the observed data before calculating the relative scattered intensity. The measured intensities were converted into absolute intensities using as a reference value the intensity, I_B, scattered by triply distilled and millipore filtered benzene under the same conditions 15.

THEORETICAL BACKGROUND

The intensity of scattered radiation (X-ray-Neutron-Light) from a solution of charged droplets has the form :

$$I(q) = P(q) \, S(q) \qquad\qquad (1)$$

where q is the wave number, $S(q)$ is the structure factor which expresses the influence of micelle-micelle interactions on the scattering, and $P(q)$ is the intraparticle structure factor.

The light is essentially scattered by concentration fluctuations of the micelles in the aqueous phase. The excess scattering may be written[16] :

$$R(q) = K \, c \, M \, P(q) \, S(q) \qquad\qquad (2)$$

Here $q = \dfrac{4\pi n}{\lambda_o} \sin \theta/2$ is the scattering wave vector, θ is the scattering angle, n is the refractive index of the dispersion and λ_o the wavelength of the incident light in vacuo.

M is the micellar weight, c is the weight concentration, and K is the optical constant . For polarized incident light at $\theta = 90°$, K is given by :

$$K = 4\pi^2 n^2 \left(\frac{dn}{dc}\right)^2 \lambda_o^{-4} \, N_A^{-1} \qquad\qquad (3)$$

where $\left(\dfrac{dn}{dc}\right)$ is the refractive index increment with respect to the concentration.

At $\theta = 90°$ for $\lambda = 5145 \; \overset{\circ}{A}$, $q \cong 10^{-3} \; \overset{\circ}{A}{}^{-1}$. For this q range, we have $P(q) \cong 1$. Since there are no long range interactions in this system, we have $S(q) = S(o)$ from Equation (2), one obtains :

$$R_{q\to 0} = R_{90} = KcMS(\theta) \qquad\qquad (4)$$

or

$$R_{90} = \frac{Kc \quad RT}{\left(\frac{\partial \Pi}{\partial c}\right)} \qquad\qquad (5)$$

where Π is the osmotic pressure, $R = kN_A$ is a gas constant and T is the absolute temperature.

For low concentration $c \to c.m.c.$ or $\phi \to 0$

$$\frac{Kc}{R_{90}} = \frac{1}{RT}\left(\frac{\partial \Pi}{\partial c}\right) = \frac{1}{M}\,(1 + 2A_2 c) \qquad\qquad (6)$$

Extrapolation to $\phi \to 0$ therefore allows one to deduce M and the second virial coefficient A_2. For each ϕ, the quantities R_{90}, c, K are known, and in order to obtain the aggregation number, it is necessary to have the value of S(0) at any ϕ.

For neutron scattering, the parameters which appear in S(q) are : the volume fraction of micelles Φ, the screening length κ, the apparent charge Z and the radius of the micelles. S(q) is calculated in closed analytical form, by solving the Ornstein-Zernike equation in the mean spherical approximation assuming that the repulsive electrostatic potential determines the intermicellar correlation.[8]

There are two ways to determine S(0) as a function of the volume fraction. Φ.
- From theoretical calculation knowing only the volume fraction occupied by the hard sphere micelles (i.e. with hydration layer) and the apparent charge of the micelles which enables κ to be obtained. These data can be extracted from a neutron scattering study[6]. Typical values are $|Z| \simeq 10$ and $\kappa = 5\text{Å}$.

- From experimental data S(0) can be obtained directly from small angle neutron scattering at very low $q(\sim 10^{-2} \text{ Å}^{-1})$. In this case, the absolute scattered intensity corresponds to S(0). This is only true if there is no long range attractive term in the potential interaction between micelles.

In our study, these two methods gave the same results suggesting that the attractive interaction is weak for these small micelles. For SDS and cationic micelles [16, 17], the theoretically predicted value of S(0), i.e. the osmotic compressibility differs from the measured value since one needs a complementary attractive potential to fit the data [16].

The autocorrelation function of the scattered light is given by:

$$S(q,t) = \exp \left\{ - K_1(q,t) + K_2(q) \frac{t^2}{2!} + \ldots \right\} \tag{8}$$

For interacting systems such as anionic micellar solutions :

$$K_1(q,t) = D_{eff} \, q^2 \tag{9}$$

Where D_{eff} is the effective translational diffusion coefficient.

For micellar solution at very low concentration :

$$D_{eff} = D_o \left[1 + K \, \phi \right] \tag{10}$$

K is the thermodynamic and hydrodynamic correction coefficient.

D_0 was obtained by extrapolating the hydrodynamic radius, given by the Stokes-Einstein relation, to the c.m.c.:

$$R_H^\circ = \frac{kT}{6\pi\eta D_o} \qquad (11)$$

where η is the viscosity of the solvent.

RESULTS AND DISCUSSION

Typical plots of the absolute scattered intensity R_{90} against the volume fraction ϕ_{HS} are shown in Figure 1.

The volume fraction ϕ_{HS} of the micelle was taken as the sum of the partial molar volume of the Na octanoate molecule[18] and a water hydration shell of about 8-9 water molecules per monomer unit[6,19,20]. In this study, the volume fraction of micelle in solution is considered as independent of the aggregation number of the micelle.

Assuming that there are no long range attractive terms in the intermicellar potential, S(0) was deduced from neutron scattering spectra at the lowest q value measured. Using Equation (4), the absolute value of R_{90} can be calculated for different aggregation numbers since S(0) is independent of micellar radius (dashed line in Figure 1).

The maximum in the scattered intensity, R_{90}, is obtained for a volume fraction of 12 % : this is a well-known result for a system of concentrated hard spheres or microemulsion droplets in oil[21,22,23].

The experimental curve of R_{90} has a maximum for $\phi_{HS} \simeq 0.2$ and it is no longer possible to consider that the aggregation number is independent of the surfactant concentration. Using Equation (4) and the value of S(0) deduced from neutron scattering data or MSA calculations, $|Z| = 10$ and $\kappa = 5$ Å, the aggregation number is extracted as a function of surfactant concentration. The results are summarized in Table I and Figure 2, where they are compared with the data obtained by fitting the observed scattering intensity in the whole ϕ range[6]. Although the numerical values obtained for the aggregation numbers differ slightly, there seems to be a similar linear relation between aggregation number N and surfactant concentration.

A similar result was obtained with SDS micelles in water by neutron scattering[24] and fluorescence quenching[25]. This problem is still a subject of discussion and some contoversial results exist in the literature[26]. For SDS, all methods give the same result

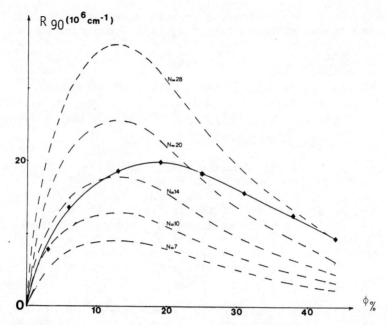

Figure 1. Reduced light scattering R90 of Na Octanoate as a function of volume fraction ϕ_{HS}.

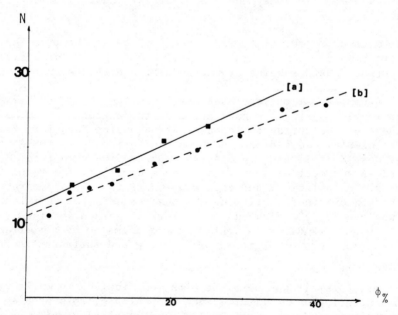

Figure 2. Comparison between aggregation number N for Na octanoate micelle deduced (a) from neutron scattering (b) from light scattering as a function of ϕ_{HS}.

at the c.m.c. ($N \simeq 65$) but the determination of aggregation number without extrapolation to $\phi \rightarrow 0$ necessitates using models whose effects are too difficult to ascertain.

This interpretation is in disagreement with thermodynamic models for micelle formation where the aggregation number may be independent of surfactant concentration. In our case when monomers are added above the c.m.c., some are absorbed into existing micelles of constant shape.

Model thermodynamic calculations did not predict this behaviour when assuming a constant charge[2], but more refined electrostatic calculations[5] predict an increase in the aggregation number.

In this study, the concentration of free monomers has been considered equal to the c.m.c. Kamenka and Lindman [27] have recently shown that this is not the case and that the monomer concentration decreases with increasing surfactant concentration. This effect could give a small systematic error which could explain the differences in aggregation number found by neutron and light scattering ; the values of ϕ_{HS} could be in error by up to 10%.

Table I shows the radii of these micelles with and without the water of hydration. The "hydrated" radius varies from 10 Å to 14 Å in agreement with neutron data[6] and with dynamic light scattering near the c.m.c.

Typical plots of the measured effective translational diffusion coefficient as a function of ϕ_{HS} are shown in Figure 3.

D_{eff} (ϕ) shows a minimum at the volume fraction which corresponds approximately to the maximum of $R_q(\phi)$. This behaviour has already been noted for diffusion coefficients of microemulsions in oil [22, 28] and the extrapolation to $\phi \rightarrow 0$ gives the hydrodynamic radius of the droplet.

In our case, the extrapolation gives a very low value for the apparent hydrodynamic radius $R_H^{app} \simeq 7$ Å, taking the viscosity of micellar solution at the c.m.c. as $\eta \simeq 1.12$ cp.

From Table I at the c.m.c., the hydrodynamic radius is approximately 10 Å corresponding to $D_o \simeq 2.10^{-6}$ cm^2/s. Between volume fraction of 0.15 and 0.4, D_{eff} is approximately linear with ϕ_{HS}. The slope is positive indicating a predominently repulsive potential. A crude linear extrapolation to $\phi \rightarrow 0$ gives 2.10^{-6} cm^2/s and $R_H^o \simeq 10$ Å.

Table I - Concentration of micellized amphiphile C-c.m.c., hard-sphere volume fraction ϕ_{HS}, structure factor at low angle S(0), mass M and aggregation number N of sodium octanoate micelle at room temperature. R_{HY} is the hard sphere radius and R_I the core radius.

C-c.m.c. M/1	ϕ_{HS} % (a)	S(o)	$10^6 R_{90}$ cm^{-1} (b)	M	N	R_I Å(c)	R_{Hy} Å(d)
0.1	2.94	0.68	7.7	1690	11	8.3	10.8
0.2	5.87	0.55	15.9	2160	14	9.0	11.7
0.3	8.81	0.39	17.7	2260	15	9.1	11.9
0.4	11.75	0.30	18.9	2350	15	9.2	12.1
0.6	17.62	0.18	20.0	2760	18	9.8	12.8
0.8	23.49	0.115	18.9	3060	20	10.1	13.2
1.0	29.36	0.07	15.8	3370	22	10.4	13.6
1.2	35.24	0.04	12.6	3920	25	10.9	14.3
1.4	41.11	0.025	9.4	4010	26	11.0	14.4

(a) $\phi_{HS} = (\bar{v} \times 154.7 + 9 \times 18).10^{-3}(C\text{-c.m.c.})$; $\bar{v} = 0.851$ cm^3/g

(b) R_{90} is measured with $\frac{dn}{dc} = 0.156$ cm^3/g (ref.18)

$R_{90} = 4.029$ 10^{-7} $(cm/g)^2$ $mole^{-1}$

$R_{90\ v,v+H}^{b,28°C}$ (5145 Å) $= 3.2.10^{-5}$ cm^{-1}

(c) R_I is calculated from M without hydration water
(d) R_{Hy} is calculated from M with 9 water molecules[20].

Table II - Effective diffusion coefficient and apparent radius R* for Na-octanoate micelle at room temperature.

C-c.m.c. (mole/ℓ)	$10^6 D_{eff}$ (cm^2/s)	$10^6 \Delta D_{eff}$ (cm^2/s)	R_{app}^{*} (Å)	ΔR_{app} (Å)
0.034	1.8	0.1	10.5	0.6
0.054	1.54	0.06	12.3	0.5
0.074	1.43	0.06	13.2	0.6
0.094	1.40	0.04	13.5	0.4
0.114	1.18	0.02	16.0	0.3

* $\eta \simeq 1.17$ cp with aqueous solution 0.04 Na Octanoate + 3 M NaCl.

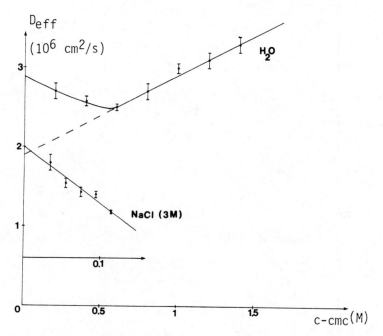

Figure 3. Measured diffusion coefficient as a function of surfactant concentration Na Octanoate.

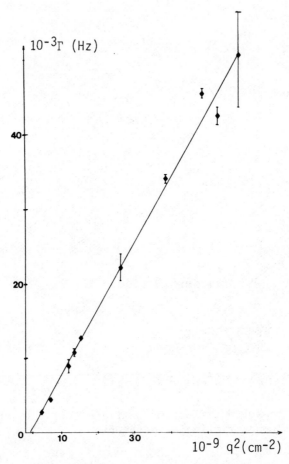

Figure 4. Linewidth of autocorrelation function $(D_{eff}\, q^2)^{-1}$ as a function of q^2 at 0.6 M. Error bars were evaluated from reproducibility of the measurement.

The difference between the experimental data and the linear evolution of D_{eff} has been considered as a "reactive" contribution to the effective translational diffusion coefficient.

Phillies[12] proposes that monomer-micelle exchange can contribute to the micellar translational diffusion.

This effect will be important for small charged micelles with a large c.m.c. and a low aggregation number. He estimates that the position of a micelle is perturbed when it gains or loses a monomer unit. With sodium octanoate at c.m.c.($\simeq$ 0.4 M) there are 20 times more monomer units than small micelles ($N \simeq 10$) and each surfactant monomer in an octanoate micelle is therefore, on the average, replaced $\sim$ 500 times during the relaxation time $(2 \ D^T_{eff} \ q^2)^{-1}$.

This "reactive" diffusion is essentially a translational effect and not a conventional chemical reaction which is independent of q^2 [29]. Figure 4 shows that the linewidth Γ is linear in q^2, with $\Gamma \to 0$ as $q^2 \to 0$ for c = c.m.c. = 0.2 M in surfactant concentration. If Brownian motion and reactive diffusion are independent, then their contribution to Γ will be additive giving[12] :

$$\Gamma = (D^T_{eff} + D^R) \ q^2 = D_{eff} \ q^2 \qquad (12)$$

D^T_{eff} is the effective translational diffusion coefficient.

D^R is the "reactive" diffusion coefficient due to micelle-monomer exchange.

At $\phi_{HS} \to 0$, $D_{eff} \simeq 2.9.10^{-6}$ cm^2/s and the linear extrapolation gives $D_o \simeq 1.9 \ 10^{-6}$ cm^2/s from Equation (10). The value of D_R is 1.10^{-6} cm^2/s and $D_R/D_o \cong 0.5$.

Applying the model of Phillies to sodium hexylsulfate (SHS)[12] gives D_R/D_o as 0.35.

To justify the importance of the large c.m.c. in the reactive diffusion at low concentration, some scattering data on Na octanoate in 3M NaCl were collected. In this case the c.m.c. is 0.04 M[20] and the volume fraction is small ($\phi < 3$ %). In Figure 3, the variation of the effective diffusion coefficient is shown and the extrapolation to $\phi \to 0$ gives 2.10^{-6} cm^2/s. The negative slope indicates an attractive potential with a modification of the apparent micellar size from 10 Å to 16 Å. No evidence of reactive diffusion was found in this case since the c.m.c. is too low.

Phillies[12] gives a simplified estimation for D_R by a random motion due to micelle-monomer exchange :

$$6 \, D_R = R \, \langle \ell^2 \rangle \tag{13}$$

R being the stepping rate diffusion and $\langle \ell^2 \rangle$ an average jump length due to micelle-monomer-solvent forces. In the first approximation R can be equal to $2R^+$ at equilibrium. R^+ is the rate constant of the exchange reaction between monomer and micelle.

At $\phi \to 0$ $D_R \sim 10^{-6}$ cm^2/s and $k^+ \simeq 10^{+9}$s^{-1} [26].

$\langle \ell^2 \rangle \sim 30$ A^2 $\qquad$ $\langle \ell \rangle \simeq 5.5$ Å

The model of Phillies was used to estimate ℓ from the hydrodynamic radius and aggregation number

$$\langle \ell \rangle = \frac{r_o}{1 + n^{-1/3}} \tag{14}$$

if $r_o \simeq 10$ Å and $n = 12$ at $\phi \to 0$, $\langle \ell \rangle$ is equal to 7 Å. There is a qualitative agreement between the two values of $\langle \ell \rangle$, which correspond approximately to the mean end-to-end vector as determined by NMR methods[13].

In another paper [30] we considered the effect of monomer-micelle exchange in terms of reaction rates rather than dynamics. It affects the diffusion coefficient in the following way : the micelle can be considered as diffusing on its own or can be decomposed into fast diffusing monomers which reform into another observable micelle.

The net effect of the two states diffusion will be an increase in the effective micellar diffusion coefficient. A quantitative treatment of this effect is obtained using a normal modes analysis[30]. A good agreement between the calculated curves and the experimental data obtained from Sodium Octanoate and Sodium Hexyl Sulfate clearly shows that the observed minimum results from the micelle-monomer exchange. At large volume fractions the micelle-micelle interaction (thermodynamic and hydrodynamic contribution) dominates the measured values of D_{eff}.

The polydispersity could also affect the concentration dependence of the diffusion coefficient. Near the c.m.c., the size distribution could be large and as ϕ increases, the distribution could be narrower. Both $D_{eff}(\phi)$ and $I(\phi)$ curves would be modified in agreement with the experimental observation. But in the limit of rapid monomer micelle exchange, the polydispersity is undetectable by light scattering [31].

No direct measurement of polydispersity as reflected in a

variance can be exhibited. The experimental "polydispersity factor" $\frac{\mu_2}{\bar{\Gamma}^2}$ is very low and near the c.m.c., $\bar{\Gamma}_1$ and $\bar{\Gamma}_2$ are equivalent : The autocorrelation function has a form of single exponential.

It is only in the slow reaction limit (SDS, CTAB) which will yield a sum of exponentials that polydispersity can be detected [32], [33,34].

CONCLUSION

The neutron and light scattering experiments used to determine the aggregation number in micellar systems are independent of each other except that $S(q \to 0)$ is transferred from one experiment to the other. The only assumption used here is that long range interactions are negligible.

The average aggregation number for Na octanoate micelles increases linearly with surfactant concentration from 11 ± 1 at the c.m.c. to 25 ± 2 at 1.8 M. This increase can complicate the interpretation of light scattering data if there is no complementary information on the electrostatic interactions, such as provided by neutron scattering experiments. It may be possible, using on appropriate thermodynamic model, to show the correlation between this increase in aggregation number and the decrease of free monomer in concentrated systems observed by Kamenka and Lindman[27].

From dynamic light scattering measurements at low micellar concentration a perturbation in the Brownian motion is observed.

This "reactive"diffusion is purely translational and arises from the rapid chemical exchange between micelles and monomers. This increase in the effective diffusion constant disappears at high volume fraction ($\phi \cong 0.15$). An average step length of 7 Å is deduced from experimental data. Some further experiments to confirm this effect are in progress on sodium hexyl sulfate micelles and cationic micelles with a short alkyl chain.

REFERENCES

1. B. Wennerström and B. Lindman, Phys. Reports 52, 1 (1979).
2. F. Eriksson, J.C. Eriksson and P. Stenius in "Solution Chemistry of Surfactants", K.L. Mittal, Editor, Vol. I, p. 297, Plenum Press, New York, 1979.
3. B. Lindman and B. Brun, J. Colloid. Interface Sci. 42, 388 (1973).

4. P.O. Persson, T. Drakenberg and B. Lindman, J. Phys. Chem. 83, 3011 (1979).
5. G. Gunarsson and H. Wennerström, to be published.
6. J.B. Hayter and T. Zemb, Chem. Phys. Letters, 93, 91 (1982).
7. W. Agterof, J. Van Zomeren and A. Vrij, Chem. Phys. Lett. 43, 363 (1976).
8. J.B. Hayter and J. Penfold, Molec. Phys. 42, 109 (1981).
9. L. Belloni, Thèse 3ème Cycle Paris 1982, Note CEA N 2292.
10. J.B. Hayter and J. Penfold, J. Chem. Soc. Faraday Trans. I, 77, 1851 (1981).
11. D.J. Cebula, D.Y. Myers and R.H. Ottewill, Colloid Polymer Science, 107, 96 (1982).
12. G.D. Phillies, J. Phys. Chem. 85, 3541 (1981).
13. T. Zemb and C. Chachaty, Chem. Phys. Letters 88, 68 (1982).
14. P. Tivant, P. Turq, M. Drifford, A. Magdalenat, and R. Menez, Biopolymers 22, 2 643 (1983).
15. S. Hayashi and S. Ikeda, J. Phys. Chem. 84, 744 (1980).
16. M. Corti and V. Degiorgio, J. Phys. Chem. 85, 711 (1981).
17. R. Dorshow, J. Briggs, G.A. Bunton and D.F. Nicoli, J. Phys. Chem. 86, 2395 (1982).
18. P. Ekwall, H. Eikrem and L. Mandell, Acta Chem. Scand. 17, 8 (1963).
19. B. Lindman and B. Brun, J. Colloid Interface Sci. 42, 388 (1973).
20. R. Friman, K. Pettersson and P. Stenius, J. Colloid Interface Sci. 53, 90 (1975).
21. A.K. Van Helden and A. Vrij, J. Colloid Interface Sci. 78, 312 (1980).
22. A.M. Cazabat, D. Langevin and A. Pouchelon, J. Colloid. Interface Sci. 73, 1 (1980).
23. A.A. Calje, W. Agterof and A. Vrij, in "Micellization Solubilization and Microelumsions", K.L. Mittal, vol. 2, pp. 779-790, Plenum Press, New York, 1979.
24. J.B. Hayter and Penfold, to be published.
25. F.C. De Schryver, Y. Croonen, E. Geladé, M. Vander Auweraer, J.C. Dederen, E. Roelants and N. Boens, These proceedings.
26. P. Lianos and R. Zana, J. Colloid Interface Sci. 84, 100 (1981).
27. N. Kamenka and B. Lindman, personal communication.
28. A. Graciaa, J. Lachaise, P. Chabrat, L. Letamendia, J. Rouch, C. Vaucamps, M. Bourrel and C. Chambu, J. de Phys. Lett. 38, 253 (1977).
29. G.B. Phillies, J. Colloid. Interface Sci. 83, 226 (1982).
30. P. Turq, M. Drifford, M. Hayoun, A. Perara and J. Tabony, in press in J. de Phys. Lett. (1983).
31. D.F. Evans, D.J. Mittchele and B.W. Ninham, to be published.
32. M. Weissman, J. Chem. Phys. 72, 231 (1980).
33. P.N. Pusey, J. Phys. A11, 119 (1978).
34. M.M. Kops-Werkhoven, C. Pathmamanoharan, A. Vrij and H.M. Fijnaut, J. Chem. Phys. 77 (12), 5913 (1982).

NMR AND ESR STUDIES OF DIBUTYLPHOSPHATE MICELLAR AGGREGATES

S. Belaïd and C. Chachaty

Département de Physico-Chimie, CEN. de Saclay
91191 Gif sur Yvette Cedex, France

Sodium dibutylphosphate micellar solutions were investigated by ^{31}P, ^{13}C and ^{23}Na NMR as well as by ESR, using VO^{2+} as paramagnetic probe, the CMC being determined by conductometry. The aggregation number of DBP$^-$ was estimated from the reorientation correlation time of the vanadyl probe bound to the surface of micelles. The value of N = 11 $\pm$ 2 at 300 K was confirmed by the concentration dependence of the ^{31}P longitudinal relaxation at 36.5 and 111.7 MHz. The temperature dependence of the CMC and of the aggregation number were interpreted in terms of size distribution and area per polar head assuming a spherical or cylindrical shape of the micelles. The conformations and dynamics of DBP$^-$ molecules in the aggregates were investigated by ^{31}P and ^{13}C relaxation in diamagnetic solutions or in the presence of paramagnetic Ni^{2+} ions bound to the micellar surface. This procedure allows the determination of the trans and gauche populations of the oxyalkyl chains as well as the reorientation correlation time of DBP$^-$ molecules and the rates of segmental motions in the non polar core of the micelles.

INTRODUCTION

Recent experiments showed us that short chain dialkylphosphates like sodium dibutyl[1] and dihexylphosphates are subject to micellar aggregation in aqueous solutions, giving rise to liquid crystalline phases at high solute concentrations (≥ 3M). This unexpected behaviour prompted us to investigate by magnetic resonance spectroscopy and conductometry some of the physical properties of these solutions. These methods allow in particular a determination of the thermodynamic parameters of micellization and consequently an estimate of the size distribution and of the area per amphiphile at the micelle surface. The ^{31}P and ^{13}C NMR provides also subsidiary information on the conformations and dynamical behaviour of dibutylphosphate (DBP⁻) in the monomeric and aggregated states, which are mainly obtained from relaxation experiments.

EXPERIMENTAL

The dibutylphosphoric acid from OSI, France, contains appreciable amounts of tributylphosphate and monobutylphosphoric acid. The former was removed by CCl_4 extraction from sodium hydroxyde solution. The dibutylphosphoric acid was released by acidification of this solution, extracted with CCl_4, and recovered by pumping off the solvent under vacuum. Most of the NMR experiments were performed with Varian XL 100 or Bruker WH 90 spectrometers. ^{31}P relaxation measurements at 111.7 MHz were made in the laboratory of Dr. Gueron at the Ecole Polytechnique, Palaiseau, France. The ESR spectra of spin probes were recorded on a Varian E9 X band spectrometer.

In experiments using paramagnetic divalent ions (Ni^{2+} and VO^{2+}) as nuclear relaxation reagents or ESR spin probes, the metal ion to DBP⁻ molecular ratio was varied over a range of 10^{-3} to 5×10^{-2}. The linear dependence of the paramagnetic relaxation upon this ratio shows that the properties of the micellar solution are not appreciably perturbed by addition of small amounts of divalent ions. This point was confirmed by the ESR of the DBP-VO^{2+} complex where the correlation times given by the line widths are independent of the VO^{2+} concentration.

CRITICAL MICELLE CONCENTRATION AND AGGREGATION NUMBERS

The critical micelle concentration (CMC) was obtained by conductometry at variable temperatures. It was found that the plot of the equivalent conductivity vs. the DBP concentration is linear and shows a sudden change of the slope at the CMC (Figure 1). A first estimate of the aggregation number was provided by ESR experiments on VO^{2+} complexes of DBP.

A detailed study of the complexation of VO^{2+} has shown that above the CMC, this ion binds to two DBP molecules of the micelles but not to the monomer. The reorientation correlation time $(\tau_R)_{ag}$

of the micelle-VO^{2+} complex was obtained from the width of individual 8 lines (S = 1/2, I = $\frac{7}{2}$) or from the total width of the spectrum below and above τ_R = 2 x 10^{-10}s, respectively, as previously reported[1]. The aggregation number N is given by the ratio of correlation times $(\tau_R)_{ag}/(\tau_R)_{mono}$ assuming that τ_R is proportional to the molecular volume and to the microviscosity of the solution. ESR line widths measurements on a nitroxide spin probe (tempoamine) showed us that the microviscosity increases from 0.01 cP to 0.02 cP at 300 K for DBP$^-$ varying in the 1M-3M range, whereas the bulk viscosity increases from 2.5 to 25 cP under the same conditions, yielding unlikely values of the aggregation number (N < 3) which are in disagreement with other observations. $(\tau_R)_{mono}$ was taken as one half that of the dibutylphosphoric acid dimer measured by ^{31}P nuclear relaxation and by the ESR of its VO^{2+} complex in chloroform solution. After correction for viscosity, these two determinations gave $(\tau_R)_{mono}$ = 5 x 10^{-11} s for DBP$^-$ in aqueous solutions. This procedure yields 9 < N < 12 at 300 K, the temperature dependence of the aggregation number being expressed by the empirical relation :

$$N(T) = N(300) \exp \left\{ \frac{\Delta E_N}{R} \left(\frac{1}{T} - \frac{1}{300} \right) \right\} \tag{1}$$

with ΔE_N = 1.5 kcal.mole^{-1}.

A more accurate determination of N at 300 K was made by ^{31}P relaxation at 36.5 and 111.7 MHz (Figure 2). The relaxation rate $(T_1^{-1})_{31_P}$ is constant up to [DBP$^-$] $\simeq$ 0.8 M and is inversely proportional to [DBP$^-$] from 1.5 M to 3 M. The nucleation of the lamellar phase occurs for [DBP$^-$] $\simeq$ 3 M and gives rise to a steep increase of the relaxation rate of ^{31}P in both the isotropic and liquid crystalline phases, the spectra of which are partially superimposed. Assuming that the mass action law model holds one has[2] :

$$\frac{[DBP_N^-]}{[DBP_1^-]^N} = k_N \text{ and CMC} \simeq k_N^{1/1-N} \tag{2}$$

where the subscripts 1 and N refer to the monomer and micelles, respectively. Under fast exchange conditions of the monomer between the aqueous and micellar phases, the observed relaxation rate for monodisperse micelles is :

$$\left(T_1^{-1} \right)_{31_P} = \left\{ \left(T_1^{-1} \right)_1 [DBP_1^-] + N[DBP_N^-]\left(T_1^{-1} \right)_N \right\} / [DBP^-] \tag{3}$$

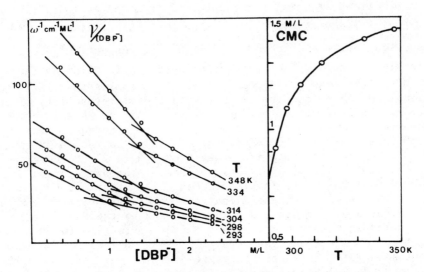

Figure 1. Determination of the CMC by conductivity measurements.

This equation is well fitted to experimental results by taking

$$\left(T_1^{-1}\right)_1 = 0.048 \text{ s}^{-1}, \left(T_1^{-1}\right)_N = 0.23 \text{ s}^{-1} \text{ ; and } \left(T_1^{-1}\right)_1 = 0.15 \text{ s}^{-1},$$

$\left(T_1^{-1}\right)_N = 0.71 \text{ s}^{-1}$ at 36.5 and 111.7 MHz, respectively, with N = 10
and $k_N = 1.25 \text{ (mole/1)}^{1-N}$. The limiting relaxation rates at 36.5MHz
are determined by the vertical axis intercepts of the linear
portions of the T_1^{-1} vs. $\left[\text{DBP}^-\right]^{-1}$ plots. The relevant values of T_1^{-1}
at 111 MHz are calculated from the ^{31}P chemical shift anisotropy
and dipolar couplings with nearby protons as shown below (see
Table II). Equation 2 yields CMC = 0.97 mole 1^{-1} at 300 K, in
excellent agreement with conductivity measurements (Figure 1). A
similar procedure was applied to the concentration dependence of
^{23}Na chemical shifts δ and linewidths $\Delta\delta_{1/2}$ at 23.8 MHz. By taking
the values of k_N and N given above, the Equation 3 may be fitted
to experimental data, taking $\delta_N = 15$ Hz, $\delta_1 = 1$ Hz ; and
$(\Delta\nu_{1/2})_N = 40$ Hz, $(\Delta\nu_{1/2})_1 = 4$ Hz (Figure 3). It should be pointed
out that as in the case of ^{31}P, Equation 3 does not hold for DBP-
concentrations above 2.5 M.

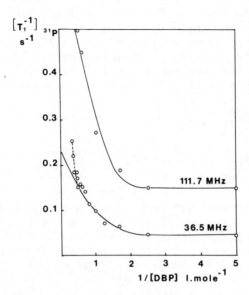

Figure 2. Concentration dependence of ^{31}P longitudinal relaxation
rate. The solid lines are calculated from Equation 3.

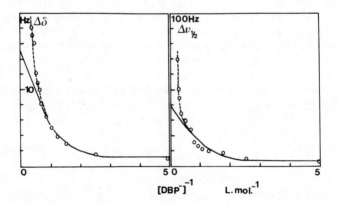

Figure 3. Concentration dependence of ^{23}Na chemical shifts ($\Delta\delta$)
and line widths ($\Delta\nu_{1/2}$). The solid lines correspond to Equation 3.

THERMODYNAMIC PARAMETERS OF MICELLIZATION

These parameters were derived from the temperature dependence of the CMC and aggregation numbers given by conductivity and magnetic resonance experiments. Assuming the multistep equilibrium model of micellization[3,4], we consider the micelles of aggregation number N as distinct species whose free energy is ΔG_N°. The mole fraction of amphiphilic molecules in these aggregates is then :

$$x_N = N \, x_1^N \, \exp\left[- N \, \frac{\Delta G_N^{\,\circ}}{RT}\right] \tag{4}$$

where x_1 denotes the mole fraction of free amphiphile molecules, the total mole fraction of DBP$^-$ molecules being $x_T = x_N + x_1$ for monodisperse micelles.

ΔG_N° is the sum of an attractive and a repulsive component denoted as ΔU°_N and W_N which are given in calories mole^{-1} by Tanford's empirical expression[4] :

$$\Delta U_N^\circ = 1.6 \left[\frac{-2000 - 700(n_c - 1)}{RT}\right] + \frac{25(A_H - 21)}{RT} \tag{5}$$

$$W_N = \frac{\alpha}{A_R RT}$$

n_c is the number of carbons per alkyl chain, the factor 1.6 accounts for an increase of 60 % of the free energy change associated with the transfer of the amphiphile from the aqueous phase to the micelle when we are dealing with a double alkyl chain instead of a single one[4]. α is an adjustable parameter, A_H is the area per amphiphile in Å^2 at the interface between the nonpolar core and the hydration shell and A_R the area per polar head.

For spherical micelles of size N, the areas A_H and A_R are given by [5] :

$$A_{H,R} = \frac{4\pi}{N}\left[\left(\frac{3NV_o}{4}\right)^{1/3} + \delta_{H,R}\right]^2 \tag{6}$$

where :

$$V_o = 2(27.4 + 26.9(n_c - 1)) \tag{7}$$

with $n_c = 4$. V_o is the volume of the non polar core, δ_H and δ_R its effective distances to the hydration shell and to the surface of localization of electrical charges, respectively. At concentrations significantly larger than the CMC ($1.5 M \leqslant DBP^- \leqslant 3M$) we have assumed for simplicity that the concentration of monomeric amphiphile is equal to the CMC so that:

$$x_T = x_{CMC} + N \; x_{CMC}^N \; \exp(- \frac{N \Delta G_N^\circ}{RT}) \qquad (8)$$

The parameters $\alpha, \delta_H, \delta_R$ were adjusted to verify Equation 8, yielding $\alpha = 2.2 \times 10^5$ cal $\mathring{A}^2$ mole^{-1}, $\delta_H \simeq 1.4 \; \mathring{A}$ and $\delta_R \simeq 4 \; \mathring{A}$. It should be pointed out that δ_H and δ_R are respectively very close to the $C_1 - O_3$ and $C_1 - O_1$ distances (1.4 and 3.9 A) O_1 and O_3 being phosphoryl and oxybutyl oxygens (see Figure 4). Thus the electric charges are located on the surface containing the phosphoryl oxygens and the interface of the hydration shell with the non polar core including the oxybutyl oxygens.

We have also determined the free enthalpy ΔH_N° as well as the standard entropy of micellization ΔS_N° from the expressions [6] :

$$\Delta H_N^\circ = -(1-\frac{1}{N}) RT^2 \frac{d}{dT} \ln(CMC) - \frac{RT^2}{N^2} \frac{dN}{dT} \qquad (9)$$

$$\Delta S_N^\circ = \frac{\Delta H_N^\circ - \Delta G_N^\circ}{T} \qquad (10)$$

They are given in Table I together with the relevant values of A_H and A_R.

The area A_R per polar head at 300 K corresponds to a radius of ca. 12 A for an aggregation number of 10, as compared with $\sim 9\mathring{A}$ using molecular models without hydration of the phosphate group. This difference results likely from the assumption of monodispersity as well as from the discrepancy between the spherical model adopted here and the actual shape of the micelle. We have also considered the model of micellar aggregates of cylindrical shape with hemispheral ends[5]. At 300 K this model gave us $A_H = 94 \; \mathring{A}^2$, $A_R = 156 \; \mathring{A}$ the length of the cylindrical section being $L = 10.7 \; \mathring{A}$. This length is comparable to the thickness of the nonpolar domains given by X rays diffraction of the NaDBP/D$_2$O lamellar phase at the same temperature ($L = 10 \pm 0.5 \; \mathring{A}$). Above 320 K the values calculated for L become however quite unlikely.

A distribution of micellar size has been calculated from our experimental data by means of the model of Eriksson et al[7] using the expression :

$$\Delta G_N^\circ = k_o + k_\gamma N^{-1/3} + k_e N^{2/3} \qquad (11)$$

k_o and k_γ were taken from ref. 7 with n_c = 7. k_e was adjusted to verify the equation :

$$x_T = x_1 + \sum_2^{N_{max}} x_N \qquad (12)$$

with N_{max} = 20, extrapolating $x_T - x_1$ to zero at variable DBP$^-$ concentrations to obtain the CMC value given by conductivity. A more elaborated treatment should take into account the decrease of the monomer concentration above the CMC[8]. Figure 5 shows however that the procedure outlined above yields a mean aggregation number of 13 at 300 K, in reasonable agreement with ESR and NMR determinations, the width of the distribution at half height being of the order of 6.

Table I. Thermodynamic Parameters of the Aggregation of DBP$^-$ (spherical shape)

T	N	$(-\Delta G_N^\circ)^{exp}$	$(-\Delta G_N^\circ)^{calc}$	$-N\Delta S_N^\circ$	A_H	A_R
	(x)			(xx)		
(K)		(kcal mole^{-1})		(caL deg^{-1} mole^{-1})	($\overset{\circ}{A}{}^2$)	
270	12.9	2.0	2.07	15.6	114	173
280	11.7	2.01	2.01	16.1	119	183
290	10.6	1.95	1.95	16.3	124	192
300	9.75	1.8	1.88	16.4	129	201
310	8.99	1.82	1.82	16.4	133	210
320	8.34	1.75	1.75	16.3	138	219
330	7.76	1.68	1.68	16.1	142	228
340	7.26	1.61	1.61	15.9	146	237
350	6.82	1.5	1.53	15.6	150	245
360	6.42	1.46	1.47	15.3	154	254

x Given by the empirical relation $N = 0.8 \exp\left(\frac{1500}{RT}\right)$

xx Calculated with $\Delta H_N^\circ = -2.22$ kcal.mole^{-1}

CONFORMATIONS OF DBP$^-$ FROM NUCLEAR SPIN RELAXATION

The longitudinal relaxation rates of ^{31}P in the monomer and micellar states (Figure 2) result from the chemical shift anisotropy (CSA) and from the dipolar interaction with nearby protons. For an isotropic overall motion, the former contribution is given by :[9]

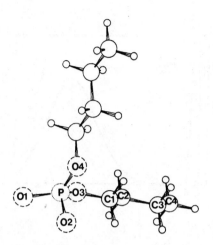

Figure 4. DBP$^-$ molecule in its most probable conformation : the O_1-P-O_3-butyl (or O_2-P-O_4-butyl) residue is all trans and C_1-O_3-P-O_4 is gauche.

$$T_1^{-1} \text{ CSA} = \frac{3}{10} \left(H_o \gamma_P \right)^2 \sigma_1^2 \left(1 + \frac{\eta^2}{3} \right) \frac{\tau_R}{1 + \omega_P^2 \tau_R^2} \qquad (13)$$

where H_o is the magnetic field strength, ω_P and γ_P the ^{31}P Larmor frequency and magnetogyric ratio. σ_1 is one of the principal values of the chemical shift tensor and $\eta = (\sigma_3 - \sigma_2)/\sigma_1$, the asymetry parameter. Here we have taken $\sigma_1 = 110$, $\sigma_2 = -17.5$ and $\sigma_3 = -76$ ppm as in barium diethylphosphate[10]. The dipolar contribution results mainly from the time dependent dipolar coupling of ^{31}P with the proton of $(CH_2)_1$ groups (Figure 4). The ^{31}P – H distances are modulated by the trans $\rightleftharpoons$ gauche isomerization of the P-O-C_1-C_2 fragment. The populations of the trans (T) and gauche (G^+, G^-) rotamers are : [11,12]

$$P_T = \frac{W_2}{2W_1 + W_2} \quad , \quad P_{G^+} = P_{G^-} = \frac{W_1}{2W_1 + W_2} \qquad (14)$$

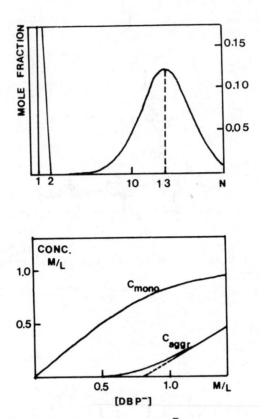

Figure 5. Size distribution of DBP⁻ micelles at 300 K.

W_1 and W_2 being the rates of $T \rightarrow G^+$, G^- and G^+, $G^- \rightarrow T$ transitions. It may be expected that the rate W_3 of the $G^+ \rightleftarrows G^-$ transition is negligibly small compared to W_1 and W_2 since the corresponding potential barrier is of the order of 10 kcal.mole^{-1}. Previous works[12] showed us that the time scale for internal motions of an aliphatic chain at room temperature is of the order of 10^{-10} s so that W_2 was fixed to 10^{10} s^{-1}. In the calculation of internuclear distances, the bond angles and lengths were taken from ref. 13. The adjustable parameters for the calculations of the ^{31}P relaxation were the correlation time τ_R and the ratio W_1/W_2 (or the probability P_T). The expressions for the dipolar contribution may be found in ref. 12. The experimental and computed ^{31}P relaxation rates are given in Table II.

Table II. ^{31}P Longitudinal Relaxation Rates in DBP$^-$ Monomer and Aggregates at 300 K.

		36.5 MHz		111.7 MHz	
		monomer	micelle	monomer	micelle
(T_1^{-1})exp. s^{-1}		0.048	0.23	0.15	0.71
(T_1^{-1})calc.	CSA	0.0125	0.067	0.1167	0.594
	dipolar	0.035	0.162	0.0347	0.1245
	CSA + dip.	0.0475	0.229	0.1514	0.7185

Parameters : $P_T = 0,625$, $W_1 = 3 \times 10^9 \ s^{-1}$, $W_2 = 10^{10} \ s^{-1}$, $W_3 = 10^8 \ s^{-1}$, $(\tau_R)_{mono} = 6.5 \times 10^{-11} \ s$, $(\tau_R)_{mic.} = 3.5 \times 10^{-10} \ s$.

The population of the trans rotamers of the POCC fragment given by relaxation is in agreement with the one obtained from the $^3J_{PC} = 7.1 \pm 0.1$ Hz coupling using the Equation[14] :

$$^3J_{PC} = 9.5 \cos^2 \phi - 0.6 \cos \phi \tag{15}$$

where ϕ is the POCC dihedral angle which yields $J_t = 10.1$ Hz, $J_g = 2.075$ Hz and therefore $P_T = (^3J_{PC} - J_g)/(J_T - J_g) = 0.62 \pm 0.02$.

The conformations of the POCCCC fragments were derived from the ^{13}C relaxation induced by Ni^{2+} which forms a bidentate complex with monomeric or aggregated DBP$^-$. A study of the complexation of $M^{2+} = Ni^{2+}$, VO^{2+} and Co^{2+} ions by DBP showed that M^{2+} is located on the bissector of $O_1 P O_2$ with $d(M^{2+} - P) \simeq 2.7 \pm 0.1$ Å. The relaxation rate of a carbon i of the butyl chain is given by [15] :

$$\left(T_{1M}^{-1}\right)_i = \frac{2}{15} (\gamma_C \gamma_S \hbar)^2 S(S+1) \left\{ \frac{3\tau_{c_1}}{1+(\omega_C \tau_{c_1})^2} + \frac{7\tau_{c_2}}{1+(\omega_S \tau_{c_2})^2} \right\} <r_i^{-6}> \tag{16}$$

where S denotes the electron spin, $\tau_{c_1} = (\tau_i^{-1} + T_{1e}^{-1})^{-1}$, $\tau_{c_2} = (\tau_i^{-1} + T_{2e}^{-1})^{-1}$, τ_i being the effective correlation time of the motion of C_i and $T_{1,2e}$ the electron spin relaxation times. For $H_o \simeq 20$ kgauss, $T \simeq 300$ K, we have $T_{2e} < T_{1e} \sim 2 \ 10^{-11} s \ <<\tau_i$ and therefore

$\tau_{c_1} \sim T_{1e}$, $\tau_{c_2} \sim T_{2e}$ [16]. $\langle r_i^{-6} \rangle$ is a statistical average over the 3^4 possible conformers of O_1PO_3CCCC :

$$\langle r_i^{-6} \rangle = \frac{\sum_u (r_i^{-6})_u \exp \left(-\frac{E_u}{RT}\right)}{\sum_u \exp \left(-\frac{E_u}{RT}\right)} \qquad (17)$$

with :

$$E_u = \sum_{n=1}^{4} \delta_j (\Delta E_G)_j + n' \Delta E_{G^+G^-} \quad (\delta_{Tj} = 0, \ \delta_{G_j} = 1) \qquad (18)$$

$(\Delta E_G)_j = RT \ln(\frac{W_2}{W_1})_j$ is an adjustable parameter which corresponds to the energy difference $E_G - E_T$ about the jth bond and $E_{G^+G^-} \simeq 2kcal$ $mole^{-1}$ an energy increment associated with the sterically disallowed G^+G^- local form. n' is the number of occurrence of this local form in the molecular conformation u and n is the number of bonds.

Table III shows the values of $(\Delta E_G)_j$ and the relevant probabilities P_T for a $O_1-P-O_3-C-C-C-C$ fragment of DBP⁻ which have been taken to fit the calculated relaxation rates of ^{13}C induced by Ni^{2+} to the observed ones, after correction for intermolecular contributions and proton induced relaxation. The methods of calculation are reported in references 12 and 17.

The data of Table III allow calculations of the probabilities of all the conformers of DBP⁻ molecule assuming for simplicity that the two oxybutyl fragments behave similarly (Figure 6). The validity of this approach was confirmed by introducing these data in the calculations of the $^{13}C-^1H$ dipolar relaxation[18] taking $\tau_R = 6 \times 10^{-11}$ s and $\tau_R = 3 \times 10^{-10}$ s for the monomer and micellar states, respectively (Figure 7). The value of τ_R given by nuclear relaxation for DBP⁻ aggregates is less than half that measured by ESR, using the VO^{2+} probe. This difference is indicative of the reoriental freedom of the DBP⁻ molecules in the aggregates resulting in part from the fast conformational changes of the alkyl chains. It explain also why the polydispersity of the micelle size (Figure 5) has virtually no influence on the quasi isotropic reorientation correlation time τ_R governing the nuclear spin relaxation of ^{31}P and ^{13}C in diamagnetic solutions. It should be pointed out also that the paramagnetic relaxation of ^{13}C, confirmed by experiments

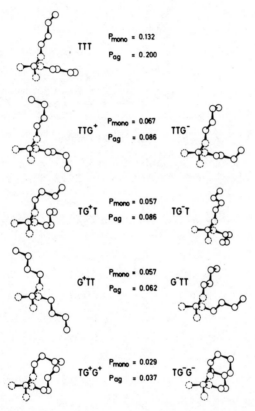

Figure 6. Main conformers of DBP⁻. The probabilities P_{mono} and P_{ag} are calcalated from the data of table III, assuming a gauche conformation for the $C_1-O_3-P-O_4$ residue (see Figure 4).

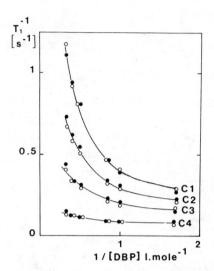

Figure 7. Experimental (●) and computed (o) relaxation rates of ^{13}C induced by ^{1} adjacent proton ν = 25.2 MHz, T = 300 K).

Table III. Experimental and Calculated ^{13}C Relaxation Rates Induced by Ni^{2+}.

Monomer		C_1	C_2	C_3	C_4
$\left(T_{1M}^{-1}\right)_i / \left(T_{1M}^{-1}\right)_1$ Exp.		1	0.5	0.23	0.12
Calc.		1	0.5	0.22	0.13
		$P-O_3$	O_3-C_1	C_1-C_2	C_2-C_3
$\Delta E_j (cal.mole^{-1})$		900	400	500	500
P_T		0.7	0.50	0.54	0.54
Micelle		C_1	C_2	C_3	C_4
$\left(T_{1M}^{-1}\right)_i / \left(T_{1M}^{-1}\right)_1$ Exp.		1	0.43	0.20	0.10
Calc.		1	0.43	0.19	0.10
		$P-O_3$	O_3-C_1	C_1-C_2	C_2-C_3
$\Delta E_j (cal.mole^{-1})$		1800	700	500	500
P_T		0.91	0.62	0.54	0.54

on diamagnetic solutions, indicates an increase of the population of the all-trans conformer of the O_1-P-O_3-butyl (or O_2-P-O_4-butyl) residue upon micellization. A similar effect observed in the case of sodium octanoate[16] appears as a consequence of hydrophobic interactions which confine the alkyl chains in the non polar core of the micelle. Thus, the magnetic resonance experiments reported here lead to a rough but quite complete description of the micellization of one of the smallest amphiphilic molecule investigated up to now.

ACKNOWLEDGEMENTS

We are greatly indebted to Drs. M. Gueron and J.L. Leroy (Ecole Polytechnique, Palaiseau, France) for ^{31}P relaxation experiments performed in their laboratory.

REFERENCES

1. S.Belaïd and C.Chachaty, J.Colloid Interface Sci. 86, 277 (1982).
2. B.Lindman and H. Wennerström, Topics in Current Chemistry, 87, 1 (1980).
3. C. Tanford, J. Phys. Chem., 78, 2469 (1974).
4. C. Tanford in "Micellization, Solubilization and Microemulsions" Vol. I pp. 119-131, K.L. Mittal Editor, Plenum Press, New York, 1977.
5. E. Ruckenstein and R. Nagarajan, ibid., Vol. I, p. 133-149.
6. N. Muller, in "Solution Chemistry of Surfactants" Vol. I, p. 229-240, K.L. Mittal Editor, Plenum Press, New York (1979).
7. F. Eriksson, J.C. Eriksson and P. Stenius, ibid. p. 297-310.
8. B. Lindman, C. Puyal, N. Kamenka, B. Brun and G. Gunnarson, J. Phys. Chem., 86, 1702 (1982).
9. A. Abragam, "The Principles of Nuclear Magnetism," Chap.VIII, Clarendon Press, Oxford, 1961.
10.R.G. Griffin, L. Powers, J. Herzfeld, R. Haberkorn and P.S. Pershan, in "Proc. XIXth congress AMPERE", p. 257-260, Heidelberg, 1976.
11.A.Tsutsumi, Molec. Phys. 37, 111 (1979).
12.A.Tsutsumi, J.P. Quaegebeur and C. Chachaty, Molec. Phys. 38, 1717 (1979).
13.B.Pullman, N. Gresh and H. Berthod, Theoret. Chim. Acta 40, 71 (1975).
14.G.Govil and I.C.P. Smith, Biopolymers 12, 2589 (1973).
15.J. Reuben, G.H. Reed and M. .Cohn, J. Chem. Phys. 52, 161 (1970).
16.T.Zemb and C. Chachaty, Chem. Phys. Letters 88, 68 (1982).
17.C.Chachaty, B. Perly and G. Langlet, J. Magn. Res. 50, 125 (1982).
18.A.Tsutsumi and C. Chachaty, Macromolecules 12, 429 (1979).

CONFORMATIONAL CHANGE OF SURFACTANTS DUE TO ASSOCIATION;

RAMAN SCATTERING AND CARBON-13 NMR STUDIES

H. Okabayashi and K. Matsushita*

Department of Industrial Chemistry, Nagoya Institute of
Technology, Gokiso, Showa-ku, Nagoya 466, Japan
*NMR Application Laboratory, JEOL, Akishima, Tokyo, Japan

Trans-3-hexenoate ion has four possible rotational isomers
(ske-skew, skew-skew', cis-skew and cis-cis). From Raman
studies of trans-3-hexenoate ion, we find that the skew-
skew and cis-skew forms are stable in aqueous solution
and the skew-skew form is more stable in the micelle
state. Carbon-13 NMR chemical shifts are discussed in
relation to conformational change on micellization with
emphasis on the C-C single bond adjacent to the trans
C=C double bond. The concentration dependence of carbon
-13 chemical shifts was investigated for trans-3-hexeno-
ate ion. The shifts of all carbon-13 atoms are almost
constant below the CMC (1.22M). However, above this
concentration, the signals of C-1 and C-3 are shifted
up-field, whereas those of C-2, C-4, C-5 and C-6 are
shifted downfield. This observation is ascribed to the
conformational change of this ion accompanying micelle
formation. The atomic charge densities of trans-3-hexe-
noate ion skeleton were calculated by the INDO method
for rotational isomers, and were used to explain the
carbon-13 chemical shift change described above.

517

INTRODUCTION

The conformational changes of the hydrocarbon moiety of
surfactant molecules, upon micellization, have been observed for
some simple surfactants. The all-trans form is stable in the micelle for normal pentanoate and hexanoate ions. This applies to
surfactant molecules having longer hydrocarbon chains.[1] For trans-
2-hexenoate ion having five possible rotational isomers,[2] the skew-
gauche form is preferentially stabilized in the micelle. A conformational change of the hydrophilic part has also been observed
in the case of normal acylsarcosinate ions.[3,4]

The surfactant-protein interaction has been extensively
studied, and the secondary structural change of a protein caused
by the surfactant binding has been observed.[5-7] The α-helical
content of bovine serum albumin decreases with an increase in the
amount of sodium dodecylsulfate and the conformational changes
about the C-S-S-C bonds of the protein molecule also occur.[7]
The study of surfactant-protein interaction has important biological implications in relation to the lipid-membrane protein interaction in the biomembranes. The interaction between biological
macromolecules could possibly result in a conformational change
of the molecules. The interaction of DNA with histones
brings about a conformational change in the DNA or histones,[8] and a
conformational change due to a higher degree of association of the
histone octamer has also been observed.[9,10]

The conformational change of a molecule in the dissociation-
association system seems to have fundamental significance, since
this could be important in the function or activity of the biological substances.

In the present work, the concentration dependence of the
carbon-13 chemical shifts of the trans-3-hexenoate ion is reported,
and is discussed in relation to conformational change on micelle
formation with emphasis on the C-C single bond adjacent to the
trans C=C double bond. The atomic charge densities of this ion
skeleton were calculated by the INDO method for possible rotational isomers and were used to explain a trend in the carbon-13
chemical shift change on micellization.

EXPERIMENTAL SECTION

Raman Spectrum Measurements. Raman spectra were recorded on a
JASCO R-800 Raman spectrometer with the 514.5 nm line of an Ar ion
laser. The spectrum of the crystalline trans-3-hexene was obtained
at liquid nitrogen temperature, after repeated annealing until no
further changes were observed in the spectra.

Carbon-13 NMR Measurements. Carbon-13 chemical shifts (TMS: external reference) and carbon-13 spin-lattice relaxation times (T_1)

were obtained using a JEOL-FX-100 Fourier transform spectrometer, at 32 °C. Measurement of a time-domain of 16384 points was used for each carbon-13 chemical shift. The T_1 values were measured by the Inversion Recovery Fourier Transform method (π -τ-π/2 sequence) with a spectral width of 5000 Hz and 8192 points in the time domain. The pulse recycling time (t_{pr}) was varied to satisfy the relation $t_{pr} \gtrsim 5T_1$.

Materials. Potassium trans-3-hexenoate (PT3H) was prepared from potassium hydroxide and trans-3-hexenoic acid. Trans-3-hexene was purchased from Tokyo Kasei Co..

RESULTS AND DISCUSSION

(I) Molecular Conformations of PT3H in Aqueous Solution.
     The Raman studies of trans-3-hexene give useful information about the conformations of PT3H in aqueous solution. For the crystalline and liquid states of trans-3-hexene, the Raman spectra were obtained and are shown in Figure 1 (a).

     Rotational isomerism in trans-3-hexene has been studied by Shimanouchi et al.,[11] using the infrared spectra in the 600-300 cm^{-1} region, and it has been concluded that the skew-skew', skew-skew and cis-skew forms coexist in the liquid state but only the skew-skew' form is stable in the crystalline state. The Raman bands of the crystalline trans-3-hexene at 483 and 323 cm^{-1} correspond to the infrared bands at 483 and 323 cm^{-1}, respectively, and are assigned to the vibrational mode of the skew-skew' form. Other Raman bands of the liquid state which disappeared in the crystal are assigned to the vibrations of the skew-skew and cis-skew forms, as is shown in Figure 2.

     The Raman spectra of PT3H were recorded for the crystalline state and aqueous solutions, and were compared with those of trans-3-hexene (Figure 1 (a) and (b)), and the normal vibration results calculated by Shimanouchi et. al.[11]. According to the normal vibrations of trans-3-hexene, it was assumed that the skew-skew and cis-skew forms coexist in the crystal PT3H, but the former is predominant. The Raman bands of PT3H in aqueous solution closely correspond to those of the crystal. This observation indicates that the skew-skew and cis-skew forms also coexist in the aqueous solution. The concentration dependence of the Raman bands of PT3H in aqueous solution was also investigated. The intensities of the Raman bands of the skew-skew form increase relative to those of the cis-skew form with an increase in concentration. This observation implies that the skew-skew form of the trans-3-hexenoate ion is more stable in the micelle. A more detailed study of the molecular vibrations of the PT3H ion is going on and will be reported separately.

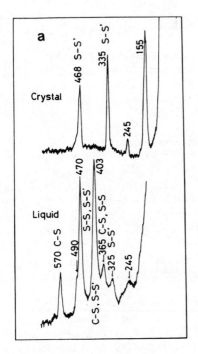
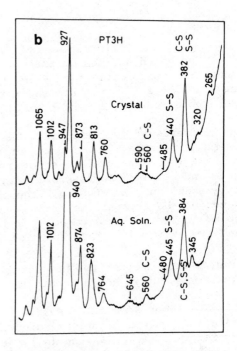

Figure 1. Raman spectra of trans-3-hexene (a) and PT3H (b).

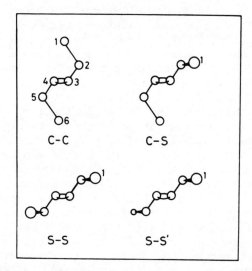

Figure 2. Possible rotational isomers of carbon-skeleton of PT3H
and numbering of carbon atoms. C: Cis, S: Skew, S': Skew'.

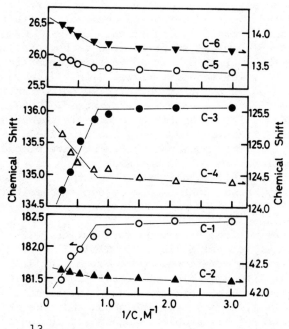

Figure 3. ^{13}C chemical shift changes of PT3H at various
concentrations (32 °C).

(II) Concentration Dependence of Carbon-13 NMR Chemical Shifts.

For surfactants in aqueous solution, carbon-13 chemical shift changes of hydrocarbon moiety on micellization have been investigated.[18-22]   The observed down-field shift has been explained as caused by an increased proportion of the trans segments of the alkyl groups in the micellar state.

For the α-carbon atom, the carbon-13 chemical shift change on micellization depends on the nature of the polar groups.[12]   For example, micellization results in an up-field shift of α-carbon for the alkylsulfate ion and a down-field shift for alkylsulphonate.[13] These observations are probably due to a change in the electronic state of the polar group and the near-by methylene groups on micellization. In the monomer state, the α-carbon-13 atom of the alkyl-sulfate ion resonates at much lower magnetic field than that in the sulphonate ion, and this is due to the substituent effect which can be exerted on the δ-carbon atom.[13]   For the trans-2-hexenoate ion having unsaturated hydrocarbon chain, it has been demonstrated by UV spectra that conformational changes markedly affect the electronic state of this ion and the carbon-13 chemical shift.[14] Especially, in simple surfactants, the effect of polar group on the electronic state of a whole molecule should not be negligible in relation to carbon-13 chemical shift.

In the present study, carbon-13 chemical shifts of PT3H in D_2O solution were also measured at various concentrations. Figure 3 shows the observed carbon-13 chemical shifts for PT3H as a function of the inverse concentration. For C-2, C-4, C-5 and C-6, the carbon-13 chemical shift of each carbon atom remains almost constant below 1.22 M, but above this concentration it shifts rapidly to lower magnetic fields (Table I). This concentration is regarded as the critical micelle concentration (CMC) obtained by the C-13 NMR chemical shift method. Thus, the observed down-field shift is ascribed to micelle formation. For C-1 and C-3, micellization leads to a rapid up-field shift of the signals (Table I). This observation can be interpreted as caused by the conformational change of the PT3H ion on micellization. That is, the preferential stabilization of the skew-skew form on micellization may bring about such a carbon-13 chemical shift change.

Similar observations have been made for the PT2H (potassium trans-2-hexenoate) ion (also listed in Table I),[14] and have been attributed to the conformational change upon micelle formation. The PT2H ion has five possible rotational isomers (skew-trans, skew-gauche, skew-gauche', cis-trans and cis-gauche). From the Raman study of PT2H, it has been indicated that the skew-trans and skew-gauche forms are stable in aqueous solution and the latter form is preferentially stabilized in the micelle.

Table I. ^{13}C Chemical Shifts (ppm) of PT3H and PT2H in the Monomer and Micellar States.

| | PT3H | | | | | |
	C-1	C-2	C-3	C-4	C-5	C-6
Monomers	182.17	42.20	136.08	124.33	25.88	13.70
Micelles	181.13	42.43	134.10	125.35	26.03	14.23
$\delta_m^{a)}$	-1.04	+0.23	-1.98	+1.02	+0.15	+0.53

| | PT2H | | | | | |
	C-1	C-2	C-3	C-4	C-5	C-6
Monomers	177.51	147.21	127.49	34.62	22.24	14.16
Micelles	175.80	143.95	128.92	35.12	22.85	14.92
$\delta_m^{a)}$	-1.71	-3.26	+1.43	+0.50	+0.61	+0.76

a) $\delta_m = [\delta]_{micelle} - [\delta]_{monomer}$

Table II. Atomic Charge Densities of Stable Rotational Isomers for PT3H and PT2H, and Estimated ^{13}C Chemical Shift Change on Micellization.

| | PT3H | | | | | |
	C-1	C-2	C-3	C-4	C-5	C-6
Cis-Skew	3.53128	4.05866	3.92615	4.05644	3.92267	3.94214
Skew-Skew	3.53152	4.05754	3.93099	4.05376	3.92416	3.94104
Estimated CS Change $^{a)}$	-$^{b)}$	+$^{b)}$	-	+	-	+
Observed CS Change $^{a)}$	-	+	-	+	+	+

| | PT2H | | | | | |
	C-1	C-2	C-3	C-4	C-5	C-6
Skew-Trans	3.53828	4.08518	4.00398	3.93581	3.94072	3.94532
Skew-Gauche	3.53807	4.08648	4.00079	3.93560	3.93327	3.94470
Estimated CS Change	+	-	+	+	+	+
Observed CS Change	-	-	+	+	+	+

a) CS Change: Chemical shift change on micellization

b) Plus and minus signs mean down-field and up-field shifts, respectively.

(III)   Atomic Charge Density Change of Molecular Skeleton on
Micellization.   Theoretical studies on carbon-13 chemical
shifts have been carried out for various compounds having saturated hydrocarbon chains.[15-17]   From these results, knowledge of
atomic charge densities of a surfactant skeleton seems to be useful to our understanding of the carbon-13 chemical shift changes
on micellization.

We calculated the atomic charge densities of skeletal atoms
for the possible rotational isomers of the PT2H and PT3H ions.
Table II shows the atomic charge densities of the skeletal atoms
for the skew-trans and skew-gauche forms of PT2H. We can omit
those of other isomers, which are less stable than skew-trans
and skew-gauche in aqueous solution and have very small populations.

The preferential stabilization of the skew-gauche form leads
to a decrease of the atomic charge densities of C-1, C-3, C-4, C-5
and C-6, and an increase of that of the C-2 atom. The relative
change in these numbers rather than the absolute values should be
emphasized here.

We assume here that only the atomic charge density is
responsible for the carbon-13 chemical shift. When the skew-
gauche form of PT2H is preferentially stabilized in the micelle
state, the contribution of the skew-gauche form to the carbon-
13 chemical shift becomes larger. For PT2H, a trend of the down-
field shift upon micellization can be estimated for C-1, C-3, C-4,
C-5 and C-6 and a trend of upfield shift for C-2 (Table II).
This estimation agrees with the observed trend except C-1.

For PT3H, the atomic charge densities of the cis-cis and
skew-skew' forms could be omitted, because their population are
very small in the micelle. Thus, the same estimation is also
possible for the PT3H ion. The population of the skew-skew form
increases in the micelle state, compared with that of the cis-
skew form. The trend of carbon-13 chemical shifts on micellization could be estimated as shown in Table II. For C-1, C-2, C-3,
C-4 and C-6, the estimated trend agrees with the observed one
except for C-5.

For both PT3H and PT2H, we conclude that such an estimation
is in fairly good agreement with the observed one. That is, one
can explain qualitatively the observed carbon-13 chemical shift
change on micellization by the preferential stabilization of a
specific isomer and the change in the atomic charge density.

(IV) Concentration dependence of ^{13}C spin-lattice relaxation time.
    For the PT3H-D_2O solutions, the carbon-13 spin-lattice relaxation time was also measured at various concentrations, and these values in the monomer and micellar states are listed in Table III. The T_1 values of all carbon-13 nuclei rapidly decrease above the CMC, while these remain almost constant below this concentration. The change of the T_1 value on micellization is due both to slow molecular motions in the micelles and to the restricted internal motions of the hydrocarbon chain on conformational changes. However, a further detailed interpretation of these results is difficult.

Table III. Spin-lattice relaxation $(NT_1, \text{sec})^a$ of the ^{13}C nuclei in PT3H (32 °C).

	$C-1^b$	C-2	C-3	C-4	C-5	C-6
Micelles^c	6.67	1.28	0.78	0.83	2.56	6.99
Monomers $(3.29 \times 10^{-4} M)$	46.86	8.00	5.63	6.56	8.98	13.13

^{a}N: The number of protons attached to carbon.

^{b}The T_1 value measured for the non-protonated carbon.

^{c}The T_1 value was obtained from the intercepts at the infinite concentration.

REFERENCES

1. H. Okabayashi, M. Okuyama and T. Kitagawa, Bull. Chem. Soc. Jpn. 48, 2264 (1975).
2. H. Okabayashi and M. Abe, J. Phys. Chem., 84, 999 (1980).
3. H. Takahashi, K. Kihara, Y. Nakayama, H. Hori, H. Okabayashi and M. Okuyama, J. Colloid & Interface Sci., 54, 102 (1976).
4. H. Okabayashi, K. Kihara and M. Okuyama, in " Colloid and Interface Science," M. Kerker, R. L. Rowell and A. C. Zettlemoyer, Editors, Vol. II, p. 357, Academic Press, New York, (1976) .
5. M. C. Chen, R. C. Lord and R. Mendelsohn, J. Am. Chem. Soc., 96, 3038 (1974).
6. J. A. Reynolds and C. Tanford, Proc. Natl. Acad. Sci. U. S., 66, 1002 (1970).
7. K. Aoki, H. Okabayashi, S. Maezawa, T. Mizuno, M. Murata and K. Hiramatsu, Biochim. Biophys. Acta, 703, 11 (1982).

8. A. J. Adler, D. G. Ross, K. Chen, P. A. Stafford,
    M. J. Woiszwillo and G. D. Fasman, Biochemistry 13, 616 (1974).
9. A. Stein and D. Page, J. Biol. Chem., 255, 3629 (1980).
10. H. Okabayashi, N. V. Beaudette and G. D. Fasman,
    unpublished results.
11. T. Shimanouchi, Y. Abe and Y. Alaki, Polymer Journal, 2, 199
    (1971).
12. U. Kragh-Hansen and T. Riisom, J. Colloid Interface Sci.,
    66, 428 (1978).
13. H. Okabayashi, T. Yoshida, K. Matsushita and Y. Terada,
    Chemica Scripta, 20, 117 (1982).
14. K. Matsushita and H. Okabayashi, Chemica Scripta, 15, 69(1980)
15. T. Yonezawa, I. Morishima and H. Kato, Bull. Chem. Soc. Jpn.,
    39, 1398 (1966).
16. J. A. Pople and M. Gordon, J. Am. Chem. Soc., 89, 4254 (1967).
17. V. B. Cheney and D. M. Grant, J. Am. Chem. Soc., 89, 5319
    (1967).
18. T. Drakenberg and B. Lindman, J. Colloid Interface Sci. 44,
    184 (1973).
19. B. -O. Persson, T. Drakenberg and B. Lindman, J. Phys. Chem.
    80, 2124 (1976).
20. U. Kragh-Hansen and T. Riisom, J. Colloid Interface Sci. 66,
    428 (1978).
21. H. Maeda, S. Ozeki, S. Ikeda, H. Okabayashi and K. Matsushita,
    J. Colloid Interface Sci. 76, 532 (1980).
22. K. Matsushita and H. Okabayashi, Chemica Scripta 15, 69 (1980).

AN NMR STUDY OF PARAMAGNETIC RELAXATION INDUCED IN OCTANOATE

MICELLES BY DIVALENT IONS

T. Zemb and C. Chachaty

Centre d'Etudes Nucléaires de Saclay
Département de Physico-Chimie
91191 Gif-sur-Yvette Cedex, France

Paramagnetic relaxation induced in the octanoate
chain by manganese and nickel ions has been measured at
2 observation frequencies. The concentration dependence
allows the free monomer contribution to be separated
from that due to the micelles.
    The observed relaxation induced in the chain
decreases from the head to the tail. This is consistent
with a model of chains with a preferred trans conformation, provided a surface hydration condition due to the
spherical interface is taken into account. The results
allow us to evaluate the probability of occurrence of
the conformers in the micellar state.

INTRODUCTION

The knowledge of the conformations of alkyl chains inside a
micelle is essential to the understanding of the problem of their
internal structure, up to now a subject of considerable controversy.
Models currently proposed are :

- Menger 's model, emphasizing water penetration in the core[1].

- The surfactant-block model, where the interface is not spherical[2].

- The model of Gruen which takes into account the constant density
  in the core[3].

- The model proposed by Dill and Flory, which predicts an order
  parameter increasing at the end of the chain, due to spherical
  packing[4].

While X ray diffusion experiments show that the micellar core behaves as a quasiliquid[5], Raman spectroscopy indicates that the population of the local trans conformers of the chains is larger than in hydrocarbons, but quantitative data on this point are still lacking. ^{13}C NMR studies[6] of the dipolar interaction between the protons and the carbons of the chains at different field strength could be interpreted in terms of a slow overall motion superimposed on a fast motion due to conformation interchange. The order parameter of the carbons with respect to the interface were obtained by this method.

To compare the predictions of the different packing models, we need information in terms of distance from the paramagnetic ion. To this end we used the nuclear spin relaxation induced by a paramagnetic ion complexed at the micellar surface.

This study has been performed using sodium octanoate micelles whose physical properties are fairly well known (see for instance ref. 7), in particular the aggregation number at high concentration which has been obtained from neutron scattering[8] using the method of Hayter and Penfold[9,10]. The use of paramagnetic ions as relaxation reagents in the investigation of the structure of sodium dodecyl sulfate (SDS) has been recently reported[11], the experimental data being interpreted by means of a simplified model of chain packing where the methylene and methyl group are likened to spherical beads stacked in a circular box. In the present work, the experimental data were compared to those calculated from the geometry and probabilities of all the possible chain conformers for different models of segmental motions, according to a procedure developed in our laboratory[12,13,14].

EXPERIMENTAL

Prolabo or BDH octanoic acid was neutralized by sodium hydroxide in ethanol solution, dried under vacuum at 100 °C and recrystallized twice from this solvent. The purity of the sodium octanoate was checked by perchloric acid titration with crystal violet as indicator[15] and then by quasielastic light scattering measurement of the effective diffusion coefficient, until disappearance of long distance index fluctuations[16] which induce abnormally high values of light scattering. The longitudinal relaxation time T_1 of ^{1}H and ^{13}C were measured by inversion recovery at several frequencies by means of BRUKER WH 90, VARIAN CFT20 and CAMECA TSN 250 spectrometers. In ^{13}C experiments the pulse delays were optimized by the FIRFT method[17] and corrections were made for pulse imperfections and radiofrequency field inhomogeneity according to ref. 18.

RESULTS AND DISCUSSION

Mn^{2+} and Ni^{2+} ions were chosen as relaxation reagents because of their very different electron spin T_{1e} relaxation times, of the order of 10^{-9}s and 10^{-11}s respectively. The determination of the paramagnetic nuclear relaxation rates was made under the following assumptions :

- the formation constants of the M^{2+}-octanoate complexes are independent of the aggregation number.

- the pseudophase model of micelles[19] holds, as previously shown[18]. M and C_t denoting the total concentration of the metal and octanoate ions, the molar fraction f of octanoate bound to M^{2+} in micelles and in the monomeric state of premicellar aggregates given by :

$$f_1 = \frac{M_1}{C_t} q_1 \quad , \quad f_2 = \frac{M_2}{C_t} q_2 \qquad (f_1 \ll 1) \tag{1}$$

q_1 and q_2 being the coordination numbers, with $M_1 + M_2 \simeq M$. Under fast exchange conditions, the observed relaxation rate is :

$$W^{\cdot\cdot}_{obs} = 1-f_1-f_2)W_o + f_1 W_{1M} + f_2 W'_{1M}$$

$$W_{obs} = W^{\cdot\cdot}_{obs} - W_o \simeq f_1 W_{1M} + f_2 W'_{1M} \tag{2}$$

W_o is the relaxation rate in octanoate in the absence of paramagnetic ions, W_{1M} and W'_{1M} being the M^{2+} induced relaxation rates for M_1/C_t or M_2/C_t = 1. The pseudophase model yields :

$$f_2 = p \frac{C_c q_2}{C_t} \quad , \quad f_1 = p \frac{C_t - C_c}{C_t} q_1 \tag{3}$$

C_c being the critical micelle concentration (CMC = 0.4 M at 28 °C D_2O) and $p = M/C_t \simeq 10^{-3} - 10^{-4}$. After correction for the proton induced relaxation of ^{13}C measured on diamagnetic solutions, one has :

$$W_{obs} = W^{\cdot\cdot}_{obs} - W_o = \frac{pC_c}{C_t} (q_2 W'_{1M} - q_1 W_{1M}) + pq_1 W_{1M} \tag{4}$$

Kamenka et al.have established recently by autodiffusion measurements that the concentration of monomers decreases below the

cmc at high concentration of surfactants[20]. Since the induced relaxation rate W_{obs} is extremely small in dilute monomeric solutions,
we can assume the pseudo phase model is sufficiently valid to extrapolate the observed relaxation in the micellized state.

The vertical axis intercept of the linear portion of W_j^{obs} vs
$1/C_t$ yields $(W_{1M})_j$, q_1 for a nucleus j in the micellar state
(Figure 1). The linearity of $W_j^{obs} = f(1/C_t)$ shows that in the
concentration range investigated here, the formation constants of
the M^{2+} octanoate complexes are nearly independent of the aggregation number. The deviation from linearity observed in Figure 1
for $C_t < 0.5$ M and $C_t > 1.8$ M are respectively due to the non
validity of the pseudophase model below C_c and to the increase of the
micelle size at high surfactant concentrations[21].

The observed relaxation rate is the sum of intramolecular and
intermolecular relaxation (outer sphere relaxation denoted as OSPR),
the latter being measured separately as shown later.

In the case of micelles the intermolecular relaxation rate can
be measured as shown below. The distance i_j of any carbon j to M^{2+}
is derived from the intramolecular relaxation $(W_{1M})_j$ by the
equation[22].

$$(W_{1M})_j = \frac{2}{15} S (S+1)(\gamma_I \gamma_s \hbar)^2 \left[\frac{3\tau_{c1}}{1+(\omega_I \tau_{c_1})^2} + \frac{7\tau_{c2}}{1+(\omega_s \tau_{c_2})^2} \right] <r^{-6}> \quad (5)$$

Where S is the total electron spin, γ_I and γ_S the nuclear and
electron magnetogyric ratios and $\omega_{I,S}$ the relevant Larmor ·
frequencies. The correlation times of Equation 5 are

$\tau_{c1,2} = (\tau_R^{-1} + T_{1,2e}^{-1})^{-1}$, τ_R being the reorientation correlation time of
the micelle and $T_{1,2e}$ the electron spin relaxation times. $<r^{-6}>$ is
a weighted average on the 729 (3^6) possible conformers of the alkyl
chain. The use of Equation 5 needs a preliminary determination of the
geometry of the binding site of the metal ion.

Averaging over $<r^{-6}>$ rather than $<r^{-3}>$[22] implies that the
paramagnetic relaxation does not depend appreciably on the absolute
rates of internal motions (see for instance ref. 23). Experiments
and calculations performed on small molecules show indeed that the
influence of fast motions on the paramagnetic relaxation of nuclei
remote from the metal ion is rather small[14].

In the present case the relative values of the ^{13}C relaxation
rates induced by Mn^{2+} are nearly independent of the spectrometer
frequency with $(W_{1M})_j^{63MHz} / (W_{1M})_j^{20MHz} = 0.76 \pm 0.04$. Likewise
the ratio of the Ni^{2+} and Mn^{2+} induced relaxation rates is

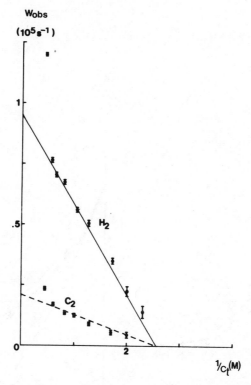

Figure 1. Determination of the absolute relaxation rates W_{1M} x q
in the octanoate - Mn^{2+} complex. Spectrometer frequencies :
250 MHz (^{1}H), 20 MHz (^{13}C).

independent of the carbon number from C_3 through C_8 with
$(W_{1M})^{Ni}_j$ / $(W_{1M})^{Mn}_j$ = 1.6 x 10^{-2} $\pm$ 0.002 (Table I).

    Moreover, there are no temperature effects on the relative
relaxation rates between 25 and 50 °C. These results indicate also
that the fast exchange condition τ^{-1}_h >> $(W_{1M})_j$ is valid for all the
nuclei j under our experimental conditions, τ_h being the lifetime
of the M^{2+} octanoate complex.

    In the calculations of relaxation rates the intermolecular
and outer sphere contribution (OSPR) micelles has to be taken
into account as pointed out above. This contribution has been
estimated from the relaxation induced by the Gd^{3+} -EDTA complex
(EDTA = ethylene diamine tetracetate), the electron spin relaxation
time of which is comparable to that of Mn^{2+} (under the same experimental conditions, the EDTA-Mn^{2+} complex is diamagnetic). The
Gd^{3+}-EDTA complex does not bind to the carboxylate ion and it induces
only an intermolecular relaxation inside the micelle. The outer
sphere relaxation rate of a nucleus in a shell of radius ρ is
expressed as[11] :

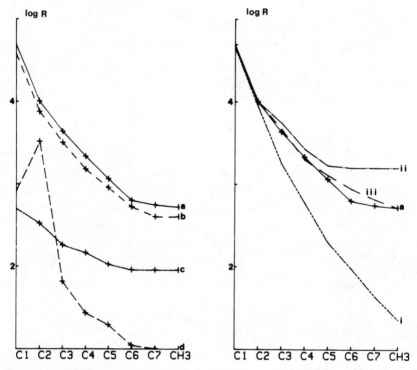

Figure 2. A – ^{13}C relaxation rates R : (a) induced by Mn^{++} at
20 MHz ; (b) induced by Mn^{++} at 63 MHz; (d) induced by Ni^{++} and
(c) intermolecular relaxation induced by Gd^{3+} – EDTA complex.

B – ^{13}C computed relaxation rates at 20 MHz : model i, model ii,
model iii. Experimental values (a) (see text).

Table I. Paramagnetic Longitudinal Relaxation Rates $W_{1M}(s^{-1})$ in Sodium Octanoate Micelles in D_2O at 28 °C given for Coordination Number 2 and $M_1/C_t = 1$.

	ion	Spectrometer frequency	C1	C2	C3	C4	C5	C6	C7	C8
$(W_{1M})^{13}C$	Mn^{2+}	20 MHz	49000	10000	4300	2150	1150	630	555	525
		63 MHz	38000	7500	3150	1500	900	525	400	400
$(W_{1M})^{13}C$	Ni^{2+}	20 MHz	825	3250	65	27	19.5	11	10	9
$(W_{1M})^{13}C$	Gd^{3+}	20 MHz	-	330	180	150	105	90	90	90

	ion	Spectrometer frequency	H2	H3	H4	H5	H6	H7	H8
$(W_{1M})^{H}$	Mn^{2+}	250 MHz	-	46000	24000	-	-	-	2750
		90 MHz		135000	-	-	-	-	7500

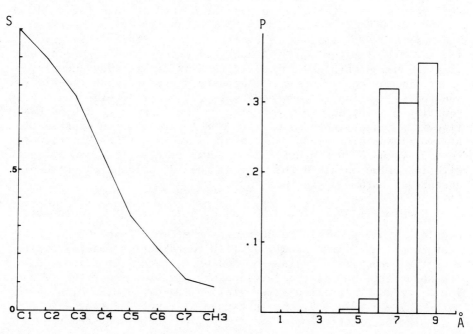

Figure 3. a) – Order parameter S_j computed for model iii (see text) b) – Normalized radial distribution of CH_3 from the carbon of the headgroup.

$$(W_{1M})_{OSPR} = K \, \frac{R^2 + \rho^2}{(R+\rho)^4 (R-\rho)^4} \tag{6}$$

Where R is the radius of a sphere of paramagnetic ions surrounding the micelle and K a constant. The calculations of $(W_{1M})_{OSPR}$ were done with R = 18 Å and R = 15 Å for the Gd-EDTA complex and Mn^{2+}, respectively. We found that the ratio $(W_{1M})_{OSPR}^{Mn2+} / (W_{1M})_{OSPR}^{Gd3+}$ is nearly constant and equal to ~ 4 for $\rho < 10$ Å, providing thus a simple correction of the observed relaxation rates. In the case of the Ni^{2+} -octanoate complex, it was assumed that the relative contribution of the OSPR to the overall relaxation rates is the same as for the relevant Mn^{2+} complex.

Figure 2A shows the experimental longitudinal relaxation rates induced by Mn^{2+} and Ni^{2+}. In the case of Ni^{2+} an abnormally high relaxation rate is observed for C_2. A similar effect has been pointed out in longchain amines, two bonds away from the nitrogen directly attached to Ni^{2+} and attributed to scalar and dipolar relaxation rates induced by comparatively high local s and p electron spin densities[24].

After correcting for outer sphere relaxation the ratio of the Mn^{2+} relaxation rates of carbon 13 at 63 and 20 MHz yields $\tau_{C_1} \simeq 1.3 \; 10^{-9}$s in a field of $\simeq 20$ k gauss. Likewise the ratio of the Mn^{2+} and Ni^{2+} induced relaxation rates in this field gives $\tau_{C_1} \simeq 2 \times 10^{-11}$s for the $Ni^{2+}-$ Octanoate complex, in good agreement with the value obtained for pyridine-N-oxide-Ni^{2+} aquo complexes under similar conditions[25]. It appears therefore that the Ni^{2+} induced relaxation is governed by T_{1e}, the term containing τ_{C_2} in Equation 5 being negligibly small. On the other hand X band ESR linewidth measurements as well as NMR ^{13}C relaxation experiments at two frequencies give $5 \; 10^{-9} < T_{1e} < 10^{-8}$s for the Mn^{2+} complex in a field $H_o \simeq 20$ k gauss and $\tau_R = 1.5 \times 10^{-9}$s, in good agreement with the value obtained for τ_R from the ^{13}C relaxation in the absence of paramagnetic ions[26].

Knowing $\tau_{C_1} \sim \tau_R$ for Mn^{2+} and $W_{1M}q_1$ for the Mn^{2+} octanoate complex, it is possible to determine the C_1-Mn^{2+} and C_2-Mn^{2+} distances. Mn^{2+} forms likely a first coordination sphere complexe with the carboxylate residue of sodium octanoate as suggested by its low solubility and by the large observed relaxation rates (Table I). The C_1 and C_2 relaxation rates yield $d_1(Mn-C_1) = 2.93 \; q_1^{1/6}$ and $d_2(Mn-C_2) = 3.82 \; q_1^{1/6}$, the Mn-oxygen distance d_3 being 2.23 Å [27]. A single intersection point of the C_1, C_2 and O centered sphere of radii d_1, d_2 and d_3 is obtained for $q_1 = 2$ only assuming that q_1 is an integer. The relevant geometry corresponds to a monodentate complex involving two adjacent octanoate molecules.[12]

It is now possible to compare the values of $<r^{-6}>_j$ computed for the different carbons with those derived from experimental data provided that a convenient model is found for the alkyl chain conformations in the micellar state. This average may be expressed as :

$$<r^{-6}>_j = \frac{\sum_u (r^{-6})_{j,u} \exp(-\frac{E_u}{RT})}{\sum_u \exp(-\frac{E_u}{RT})} \qquad (7)$$

E_u being the total conformational energy of one of the conformers of the alkyl chain. These conformations correspond to the trans $\rightleftarrows$ gauche isomerization of the alkyl chain, the energy difference between these local conformers being ΔE_g. Potential energy calculations show that the occurrence of the g^+g^- local form, giving rise to a folding back of the chain is sterically disallowed and involves an additional energy $\Delta E_{g^+g^-}$. The event of an alkyl chain segment entering into the metal coordination sphere estimated to $R_C \simeq 3$ Å may be inhibited by an energy increment ΔE_{sp} such that $\exp(-\frac{E_u}{RT}) \simeq 0$ for any conformation u where $r_{j,u} < R_C$. In some conformations, it happens also that a segment of the alkyl chain comes out of the micellar hydrocarbon core and becomes hydrated. This process involves an additional energy $\alpha\Delta E_h$ which takes into account the hydration energy of a methylene ($\alpha\Delta E'_h$ for methyl groups). α is set to 1 if the hydration is complete, i.e. the group stays in the solvent. α is set to 0.5 if the group is on the interface.

Neutron scattering experiments performed on octanoate micelles under indentical conditions[8] show that we are dealing with aggregates of 25–30 molecules with 9 water molecules/polar head. The radius D_1 of the polar core is ~ 10 Å, the hydration shell spreading between D_1 and $D_2 = 15$ Å. The total energy of an octanoate conformer in the micellized state is :

$$E_u = n_{1,u} \Delta E_{gt} + n_{2,u} \Delta E_{g^+g^-} + \Delta E_{sp} + \Delta E_H (n_{3,u} + 0.5\, n_{4,u}) \qquad (8)$$

$$+ \Delta E'_H (n'_{3,u} + 0.5\, n'_{4,u})$$

$n_{1,u}$ and $n_{2,u}$ denote the number of $g^{\overset{+}{-}}$ and $g^{+}g^{-}$ local forms respectively; $n_{3,u}$ and $n_{4,u}$ are the number of methylene and methyl groups entering into the micelle hydration shell or the aqueous phase respectively. For methyl group $n'_{3,u}$ and $n'_{4,u} = 1$ or 0.

All the possible conformations of octanoate were generated by a computer program, the relevant energies E_u being calculated according to three models :

i) the alkyl chain is rigid and in the all trans conformation ($E_g \approx 10^4$ cal/mole)

ii) the alkyl chain is flexible, the gauche and trans conformations being equiprobable ($\Delta E_g = 0$). The hydration is not taken into account, but the sterical constraints are considered by taking $\Delta E_{sp} \geq 10^4$ cal/mole and $\Delta E_{g^+g^-} = 2000$ cal/mole.

iii) the trans local conformers are more probable than the gauche ones with $\Delta E_g = 550$ cal/mole as in hydrocarbons[28], ΔE_{sp} and $E_{g^+g^-}$ are the same as above and the hydration energy of the chain segments is $\Delta E_H = 800$ cal/mole and $E'_H = 1600$ cal for CH_2 and CH_3.

Figure 2B shows clearly that only the relaxation rates computed for model (iii) are in reasonable agreement with the experimental values. Thus the chain conformations differ from those of a hydrocarbon essentially on account of their confinement in the non polar core of the micelle which gives rise to an increase in the trans population from 60 % up to more than 80 % at the chain end. This result is not very dependent on the α parameter for the chain segments entering in the micelle hydration sphere.

As the probability of all tbe different conformers is known from the model, one can calculate an order parameter $S_j = \frac{1}{2}$ $(3 < \cos^2 \theta_j > - 1)$ where θ_j is the angle between the bisector of the carboxylate residue and the normal to the plane of the methylene groups or the symmetry axis of the methyl (Figure 3a). The comparison of S_j with the order parameters of octanoate in the lamellar phases[29] shows that the flexibility of alkyl chains is larger in micelles than in liquid crystalline phases. An other criterion of flexibility of alkyl chains is the distribution of Cl-CH_3 distances. The probability of folding back of the chain is not negligible although more than 90 % of the methyl group are located between 6 and 9 Å from the polar head. (Figure 3b).

Figure 4 shows the cylindrical projection of the distribution or carbon for models i, ii, iii quoted above. This representation shows clearly the effect of the confinement of the chain, due to hydrophobic effects.

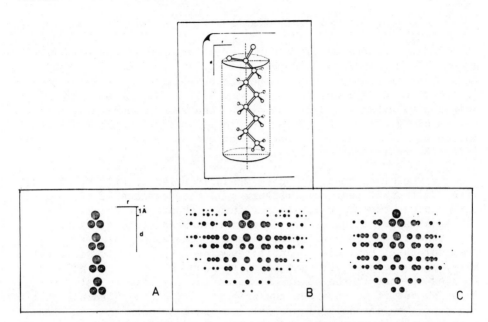

Figure 4. Cylindrical projection of the probability of finding a methylene group on a given point. Number of circles is proportional to the logarithm of the probability of finding a group on a given point. (A for model i, B for model ii, C for model iii).

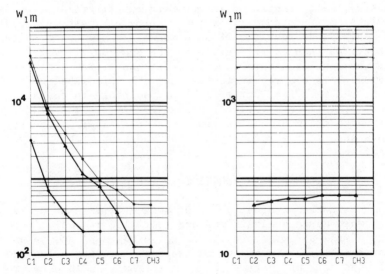

Figure 5. a) Observed relaxations induced by Mn^{2+} (20 MHz) (o) octanoate at 20 % weight (▲) octanoate in the system octanoate 18 % w, pentanol 10 % w, water 72 % w. (●) pentanol chain in the same system. b) relaxation induced by a spin label in octanoate 20 % weight in D_2O.

APPLICATIONS

Adding benzene or cyclohexane at saturation to the octanoate micelle (1.2 M) had no measurable effect on the induced relaxation W_{obs}. This is possibly due to the low solubility of these compounds in the micelle ($\simeq$ 1 molecule per micelle). Adding 1M salt to the solution results in a decrease of the charge of the micelle without noticeable growth[30]. Here also, there is no effect on the conformation of the chains ; we conclude here that there is little charge effect on chain conformations.

On the other hand, we have measured relaxation in the micelle with addition of pentanol. (Figure 5a). The decrease of the relaxation along the chain is steeper than without pentanol : we conclude that the chains are more ordered when pentanol is added. This observation is consistent with the conclusion inferred from chemical shift analysis[31].Quantitative interpretation is difficult here, since we need information on the concentrations of all the species present in the solution and on the competition between the ligands. But if we consider the relaxation induced on the aliphatic chain of the pentanol (Figure 5a), we can conclude from the parallelism between the pentanol and the octanoate chain that conformation of the pentanol is disordered, likewise the octanoate polar head is on the interface and 70 % on the bounds on the average are in the trans configuration.

Lastly, we attempted to induce paramagnetic relaxation using a spin label : (N-octanoyl-4-amino-2-2-6-6-tetramethylpiperidine N oxy). This probe consists of an aliphatic chain and a nitroxide headgroup. The induced relaxation is the same on all carbons of the chain : the dipolar interaction is independent from the depth of the carbon. This clearly indicates that the spin label does not remain at the interface but that it moves freely in the apolar core (Figure 5b). This observation casts some doubt on the validity of studies which claim a localisation of a spin label on the basis of arguments such as the polarity of the headgroup without any direct verification.

CONCLUSION RELEVANT TO THE METHOD

A quantitative evaluation of paramagnetic relaxation is only possible when certain conditions are fulfilled : namely fast exchange of the paramagnetic ion on the headgroups and an unique correlation time governing the relaxation. Interpretation becomes simpler when relaxation time (T_{1e}) is faster than internal motions, so that relaxation is proportional to $\langle r^{-6} \rangle$

Two cases can arise :

- There is no specific complexation (sulfate headgroups). One measures only outersphere relaxation which has to be spherically averaged. The origin of distances is the interface and therefore the aggregate has to be spherical[11].

  - There is specific complexation. One has to subtract outer sphere relaxation but the measurement is independent of the actual shape of the aggregate. The origin for distances is the ion and not the interface.

CONCLUSION RELEVANT TO THE OCTANOATE MICELLE

We found approximately 70 % trans conformations in each position of the chain. But the possibility that a chain segment comes out of the inner core is not ruled out. The work of Cabane[11] and Van Bocksteale et al[32] provides further experimental evidence of it.

Local order parameters are consistent with those deduced by Ahlnäs et al[6] : we confirm that order parameters decrease from the head to the tail and this point is inconsistent with Dill and Flory model[4]. Chains in the micellar core have a larger motional freedom than in the corresponding lamellar phase.

The model (iii) presented above is very similar to the model of Grüen[3], except that we did not impose any polar head radial distribution, nor "lateral pressure." All measurements performed are consistent with a micelle formed of very flexible chains ($\simeq$ 30 % gauche conformers) which leads to a rapidly decreasing order parameter and no accumulation of methylene groups near the geometrical center. Compared to chains in a bulk liquid, chains in a micelle are mainly constrained by hydration on the surface.

Lastly, it would certainly be interesting to compare our measurement of the mean distances of an ion to the predictions given by current models of micelle structure.

ACKNOWLEDGEMENTS

We thank J.B. Hayter for neutron scattering experiments and data reduction, M. Drifford for light scattering control measurements and B. Cabane for helpful discussions.

REFERENCES

1. F.M. Menger and J.M. Jerkunica, J. Am. Chem. Soc. 100, 688 (1978).
2. P. Fromherz, Chem. Phys. Lett. 77, 460 (1981).
3. D.W.R. Gruen, J. Colloïd Interface Sci. 84, 281 (1981).
4. K.A. Dill and P.J. Flory, Proc. Natl. Acad. Sci. USA 78, 676 (1981).
5. H. Wennerström and B. Lindman, Phys. Reports 52, 1 (1979).
6. T. Ahlnäs O. Söderman, H. Walderhaug and B. Lindman, these proceedings, Volume 1.
7. B.O. Persson, J. Drakenberg and B. Lindman, J. Phys. Chem. 83, 3011 (1979).
8. J.B. Hayter and T. Zemb (submitted to Chem. Phys. Lett.)
9. J.B. Hayter and J. Penfold, J. Chem. Soc. Faraday Trans. I, 77, 1851 (1981).
10. J.B. Hayter and J. Penfold, Molec. Phys. 42, 109 (1981).
11. B. Cabane, J. Physique 42, 847 (1981).
12. T. Zemb and C. Chachaty, Chem. Phys. Letters 88, 68 (1982).
13. T.Yasukawa and C. Chachaty, Chem. Phys. Letters 43, 565 (1976).
14. A. Tsutsumi, J.P. Quaegebeur and C. Chachaty, Molec. Phys. 38 1717 (1979).
15. R.Friman, K. Pettersson and P. Stenius, J. Colloïd Interface Sci. 53, 90 (1975).
16. M. Corti and V. Degiorgio, Chem. Phys. Letters 49, 212 (1977).
17. D. Canet, G.C. Levy, I.R. Peat, J. Magn. Res. 18 199 (1975).
18. J.J. Led, Molec. Phys. 40, 1296 (1980).
19. B. Lindman and H. Wennerström, Topics in Current Chemistry 87, 1 (1980).
20. N. Kamenka, personal communication.
21. A. Boussaha and H.J. Ache, J. Phys. Chem. 85, 2444 (1981).
22. J. Reuben, G.H. Reed, M. Cohn, J. Chem. Phys. 52, 1617 (1970).
23. R.F. Lenkinski and J. Reuben, J. Am. Chem. Soc. 98, 4065 (1976).
24. T. Yasukawa and C. Chachaty, Chem. Phys. Lett. 51, 311 (1977).
25. J.C. Ronfard-Maret and C. Chachaty, J. Phys. Chem. 82, 1541, (1978).
26. H. Wennerström, Björn Lindman, O. Söderman, T. Drakenberg et J. B. Rosenholm, J. Am. Chem. Soc. 101, 6860 (1979).
27. Molecular Structure and Dimensions (Vol. A1) publ. by International Union of Crystallography (Utrecht) (1972).
28. P. Flory in "Statistical Mechanics of chain molecules" pp. 133-140, Interscience publishers, New York, (1969).
29. U. Henriksson, L. Odberg and J.C. Erikson, Mol. Cryst. Liq. Cryst. 30, 73 (1975).
30. R. Friman, K. Pettersson and P. Stenius, J. Colloïd Interface Sci. 53, 90 (1975).
31. J.B. Rosenholm, T. Drakenberg and B. Lindman, J. Colloïd Interface, Sci. 63, 538 (1978).
32. M. Van Bockstaele, J. Jelan, H. Martens, J. Pul et S.C. de Schryver et J.C. Dederen, Chem. Phys. Lett. 70, 605 (1980).

ESR STUDY OF SPIN LABELS IN SURFACTANT MICELLES

P. Baglioni, M. F. Ottaviani, G. Martini and E. Ferroni

Institute of Physical Chemistry
University of Florence
Via G. Capponi 9, 50121 Firenze, Italy

In this paper the results obtained by using TEMPOL, TM-TEMPAMINE[+], Octanoyl-, Dodecanoyl-, and Hexadecanoyl -TEMPO as spin labels for micellar solutions of C_8- -SO_4Na, C_{12}-SO_4Na and C_{16}-SO_4Na are reported. The analysis of ESR lineshape, of reorientational correlation time, τ, and hyperfine coupling constant, $<A_N>$, have been carried out to obtain details on these systems. TEMPOL, TM-TEMPAMINE[+] and Octanoyl-TEMPO did not give micellar solutions at the concentrations ($\simeq 10^{-4}$M) used. The ESR spectra obtained with these probes were easily interpretable over a wide range of surfactant concentrations. On the contrary, Dodecanoyl and Hexa-decanoyl-TEMPO give micellar solutions at concentration $<10^{-4}$M. The ESR spectra of these probes in C_{12}-SO_4Na and C_{16}-SO_4Na respectively are therefore complicated by adsorptions due to species in different environments i.e., in spin probe micelles and in surfactant micelles. Kinetic parameters have been obtained for the chemical exchange among different environments.

INTRODUCTION

The electron spin resonance (ESR) of spin probes and spin labels has proved to be very useful in different research fields such as polymers, proteins, lipids, membranes, nucleotides, detergents and micellar systems.[1]

In particular, from ESR studies of spin probes and spin labels in micellar systems information can be obtained on the dynamics of

541

the micellization process, the stabilitity and motion of the micellar aggregates and the critical micelle concentration.[2,3] Thermodynamic parameters have also been obtained in some cases.[4] The ESR
parameters, that usually are analyzed to obtain such information,
are the nitrogen hyperfine coupling constant,$< A_N >$, and the correlation time for the motion, τ.

In this paper we summarize some results obtained by using several nitroxide radicals in aqueous solution of surfactants.

EXPERIMENTAL

The surfactants used, sodium octyl-sulfate (SOS), sodium dode-
cyl-sulfate (SDS), and sodium hexadecyl-sulfate (SHS), were obtained
from K & K, recrystallized two times from mixtures of water-alcohol,
and subsequently dried at 50 °C under moderate vacuum. The nitroxide
radicals used in this work are shown in Table I. Stock solutions of

Table I. Nitroxide radicals used in this work.

TEMPOL: 4-hydroxy-2,2,6,6-tetramethyl-
piperidine-1-oxyl.

TM-TEMPAMINE[+]: 4-trimethylammonium-2,2,6,6-
tetramethyl-piperidine-1-oxyl.

C_8-TEMPO: 4-octanoyloxy-2,2,6,6-tetra-
methyl-piperidine-1-oxyl.

C_{12}-TEMPO: 4-dodecanoyloxy-2,2,6,6-tetra-
methyl-piperidine-1-oxyl.

C_{16}-TEMPO: 4-hexadecanoyloxy-2,2,6,6-
tetramethyl-piperidine-1-oxyl.

SOS, SDS, and SHS 1.0 M, 3.5×10^{-1} M, and 3.0×10^{-2} M in deoxygenated and twice distilled water were prepared. This solutions were properly diluted with water and spin probe solution were added to obtain the required concentrations. The spin probe concentration was 10^{-4} M for TEMPOL, TM-TEMPAMINE$^+$, and C_8-TEMPO to avoid dipolar broadening of ESR lines, and $\simeq 10^{-5}$ M for the other probes.

The spectra were recorded using a model 200tt Bruker ESR spectrometer operating in the X-band. Modulation amplitude was kept 0.16 G to avoid signal overmodulation; power > 10 dB were used to prevent saturation effects. The cavity temperature was maintained constant, within ± 1 °C, at the values given in the text (Result Section) by using the Bruker ST 100/700 variable temperature assembly. The g and $\langle A_N \rangle$ values were measured by exact field calibration using a Hewlett-Packard model X532B frequencymeter and by comparison with potassium peroxylaminedisulfonate (g = 2.0054, $\langle A_N \rangle$ = 13.0 G).

ESR LINESHAPE OF NITROXIDE RADICALS

The predominant relaxation mechanism for nitroxide radicals ($S = \frac{1}{2}$) is given by the time modulation of $\underset{\sim}{A}$ and $\underset{\sim}{g}$ anisotropic tensors. Depending on the value of the correlation time for the motion, τ, different times of motion can be observed. If an isotropic motion is assumed, values of τ less than 10^{-9} s are typical of fast motion conditions: isotropic $\underset{\sim}{g}$ and $\underset{\sim}{A}$ are obtained that are the averaged values of the principal components of the corresponding tensors. When τ is in the range 3×10^{-7} s to 10^{-9} s, the modulation rate of $\underset{\sim}{A}$ and $\underset{\sim}{g}$ anisotropic tensors is comparable with their anisotropy. A complex lineshape is obtained, that is called "slow motion spectrum". A rough evaluation of the correlation time is given by:[5]

$$\tau = \alpha(1 - T'_{zz}/T_{zz})^{\beta} \tag{1}$$

where T'_{zz} and T_{zz} are the separations of the two extreme peaks in the observed spectrum and the spectrum in frozen solution, and α and β are coefficients that depend on the nitroxide. More accurate values of τ are usually obtained from spectral computation. Finally with $\tau > 3 \times 10^{-7}$ s solid like spectra are obtained from which components of anisotropic $\underset{\sim}{g}$ and $\underset{\sim}{A}$ tensors can be measured.

All the systems investigated in this work gave ESR spectra in fust tumbling conditions. In this case the linewidth of each hyperfine component is given by the expression:[6]

$$\Delta H(m_N, m_H) = A + Bm_N + Cm_N^2 + B'm_H + C'm_H^2 + Dm_N m_H \qquad (2)$$

in which both the m_N and m_H dependence is included. However, since the ratio A_N / A_H is about 40, the m_H dependence can be neglected[7] and the linewidth can be expressed by this simplified expression:

$$\Delta H(m_N, 0) = A + Bm_N + Cm_N^2 \qquad (3)$$

where:

$$B = 0.103\omega_e \left[\Delta g \Delta A + 3 \delta g \delta A \right] \times \tau_B \times \left[1 + \frac{3}{4(1 + \omega_e^2 \tau_B^2)} \right] \quad (3a)$$

$$C = 1.81 \times 10^6 \left[(\Delta A)^2 + 3(\delta A)^2 \right] \times \tau_c \times \left[1 - \frac{3}{8(1 + \omega_N^2 \tau_C^2)} \right.$$
$$\left. \frac{1}{8(1 + \omega_e^2 \tau_C^2)} \right] \qquad (3b)$$

$$\Delta A = T_{zz} - \tfrac{1}{2}\left[T_{xx} + T_{yy} \right] \quad ; \quad \delta A = \tfrac{1}{2}\left[T_{xx} - T_{yy} \right]$$
$$\Delta g = g_{zz} - \tfrac{1}{2}\left[g_{xx} + g_{yy} \right] \quad ; \quad \delta g = \tfrac{1}{2}\left[g_{xx} - g_{yy} \right]$$
$$\omega_N = 8.8 \times 10^6 < A > \quad ; \quad \omega_e = 5.97 \times 10^{10} \ \text{Hz}$$

and A contains terms arising from g and A anisotropies together with other terms not due to the motion (i.e. spin rotation, additional unresolved superhyperfine structure, etc.).

If the conditions of the isotropic motion hold, i.e.

$$\tau_B = \tau_C = \tau \qquad (4)$$

Assuming the Debye diffusional model (the rotating molecule is suggested to be a sphere of radius R):

$$\tau = \tau_R = (6D)^{-1} \qquad (5)$$

where D is the coefficient for the rotational diffusion:

$$D = kT/8\pi\eta R^3 \qquad (6)$$

(η being the viscosity of the environment). Thus:

$$\tau_R = 4\pi \eta R^3 / 3kT \qquad (7)$$

For the evaluation of τ, the B and C parameters of equations

(3a) and (3b) can be used:

$$B = [\Delta H(1) - \Delta H(-1)]/2 \qquad\qquad (8)$$

$$C = [\Delta H(1) + \Delta H(-1)]/2 - \Delta H(0) \qquad (9)$$

In the absence of hydrogen superhyperfine structure, the following approximate equations have been used in several cases[8-10]:

$$\tau_B = K\Delta H(0)\{[h(0)/h(-1)]^{\frac{1}{2}} - [h(0)/h(1)]^{\frac{1}{2}}\} \qquad (10)$$

$$\tau_C = K\Delta H(0)\{[h(0)/h(-1)]^{\frac{1}{2}} + [h(0)/h(1)]^{\frac{1}{2}} - 2\} \qquad (11)$$

where $h(i)$ is the height of the peak i and K is a constant whose value depends on the ESR parameters of the nitroxide used. In most of the spectra obtained in the systems investigated in this work, partial resolution of the hydrogen superhyperfine structure was observed so that equations (10) and (11) cannot be used. The true linewidth was obtained by spectral computation using a double precision Fortran program by summing derivatives of Lorentian lines with variable widths[11].

Several complications may induce appreciable errors in the evaluation of τ by the above procedure: for instance, i) non Brownian motion, that leads to different values for τ_B and τ_C; ii) conformational interconversion, particularly in the six membered ring; iii) dipolar broadening, and iv) spin exchange. Only point iv) has some relevance in this study because of micelle formation of the spin labels used (see below). In these cases, i.e. for high nitroxide concentration, the Heisenberg spin exchange term must be added in the spin Hamiltonian describing the system i.e.,

$$\mathcal{H}_{ss} = J(+)\underset{\sim}{S}_1 \cdot \underset{\sim}{S}_2 = \tfrac{1}{2}J^2(\underset{\sim}{S}^2 - \underset{\sim}{S}_1^2 - \underset{\sim}{S}_2^2) = \tfrac{1}{2}J(\underset{\sim}{S}^2 - \tfrac{3}{2}) \qquad (12)$$

where $S^2 = (\underset{\sim}{S}_1^2 + \underset{\sim}{S}_2^2)$ and $J = 2J'$, J' being the exchange integral. Depending on the bimolecular collision rate between radicals as compared with energy separation between mono- and biradical, either line broadening of each transition with concomitant decrease of the coupling constant or single exchange narrowed signal may be observed.

NITROGEN HYPERFINE COUPLING CONSTANT

The value of $<A_N>$ gives information on the polarity of the medium: thus it is of particular importance for our purposes. It is known that the unpaired electron in a nitroxide radical is largely delocalized in the $2p\pi^*$ nitrogen orbital perpendicular to the xy plane, that contains the structure[12]:

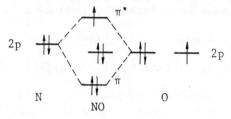

The approximate axial symmetry ($T_{xx} \simeq T_{yy}$) that is usually assumed is well in agreement with this molecular orbital scheme:[12]

Two resonance structures can be written for the nitroxide group:[13]

$$\overset{\backslash}{\underset{/}{N}}-\bar{\underset{}{O}} \quad \longrightarrow \quad \overset{\backslash}{\underset{/}{N}}\overset{(+)}{-}\bar{\underset{}{O}}|^{-}$$

I   II

Clearly, both the solvent polarity and its capacity in hydrogen bond forming favour the pseudo-ionic structure II. Several relation-ships have been proposed to relate $\langle A_N \rangle$ with polarity, Kosower fac-tor and dielectric constant of the environment. For instance, Engberts and Stout[14] suggested a linear dependence of $\langle A_N \rangle$ on ε only, while Yoshioka showed that the hydrogen bond also must be taken into account.[15]

RESULTS AND DISCUSSION

Figure 1 shows the ESR spectra of the nitroxide radicals at concentration $\leq 10^{-4}$ M without added surfactants. The two spin probes, TEMPOL and TM-TEMPAMINE^{+}, and the spin label C_8-TEMPO show the usual three-line spectra with partially resolved superhyperfine structure, while C_{12}-TEMPO and C_{16}-TEMPO show more complex absor-ptions.

In the case of C_{12}-TEMPO two distinct signals are clearly iden-tified: a) a three-line absorption with $\langle A_N \rangle = 17.05$ G, which is due to spin label monomer dissolved in the water phase in isotropic motion with $\tau_c = 5.0 \times 10^{-11}$ s; and b) single exchanged narrowed signal with $\Delta H \simeq 9$ G, due to spin label micelle in slow chemical exchange with the bulk solution. From the linewidth of each compo-nent of the triplet a relaxation time $T_2^{-1} \simeq 6 \times 10^{-10}$ s^{-1} can be eva-luated: thus the residence time of the monomer in the micelle should be $\simeq 6.0 \times 10^{-10}$s.

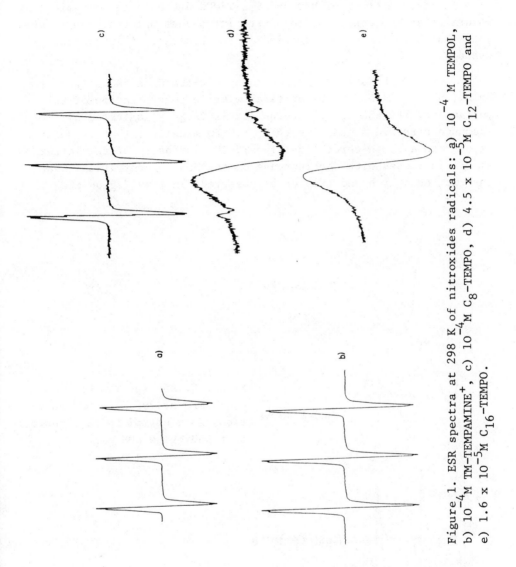

Figure 1. ESR spectra at 298 K of nitroxides radicals: a) 10^{-4} M TEMPOL, b) 10^{-4} M TM-TEMPAMINE$^+$, c) 10^{-4} M C_8-TEMPO, d) 4.5×10^{-5} M C_{12}-TEMPO and e) 1.6×10^{-5} M C_{16}-TEMPO.

For C_{16}-TEMPO only the exchange narrowed signal is observed without resolution of separate three lines due to free nitroxide radical. No residence time can therefore be calculated for the monomer in this case.

The fact that C_{12}-TEMPO and C_{16}-TEMPO are able to form micellar solutions at the concentrations used introduces spectral complications that have to be kept in mind in the analysis of surfactant systems.

Figure 2 shows the ESR spectra of C_{12}-TEMPO in SDS solutions below, near, and above the critical micelle concentration of the surfactant. The ESR spectra below cmc is largely dominated by exchange narrowed signal due to C_{12}-TEMPO micelles, that is superimposed to a broad three-line signal. The intensity of the latter signal increases with the increase of the SDS concentration. Near the cmc the only broad triplet is observed. In this region the

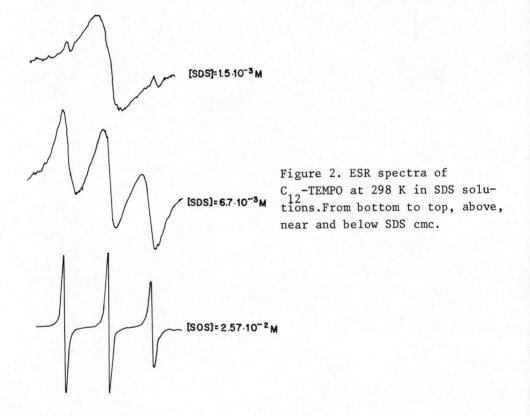

[SDS]=1.5·10⁻³M

[SDS]=6.7·10⁻³M

[SOS]=2.57·10⁻²M

Figure 2. ESR spectra of C_{12}-TEMPO at 298 K in SDS solutions. From bottom to top, above, near and below SDS cmc.

linewidth of each component is at its maximum, then progressively
decreases and stabilizes well above SDS cmc. The hyperfine
coupling constant near cmc shows a sharp decrease with respect to
bulk water ($<A_N>$ = 17.05 G) and increases again above cmc until it
is stabilized at 16.85 G. The correlation time of C_{12}-TEMPO can
accurately be evaluated only above the cmc and is equal to 2.4×10^{-10}
s. The above data can be interpreted assuming that an increasing
number of monomers of SDS surfactant enter in the C_{12}-TEMPO micelle,
thus diluting the paramagnetic label; the lineshape progressively
turns from an exchanged-narrowed spectrum (below cmc) to the usual
three line spectrum (above cmc) by passing through an exchange broad-
ned spectrum with very low $<A_N>$ near the cmc. The SDS molecule C_{12}-
-TEMPO ratio accounts quite well for this assumption. Near the cmc
this ratio is $\simeq$ 100. Due to the fact that the aggregation number
of SDS micelle is 86, SDS micelle containing more than one spin

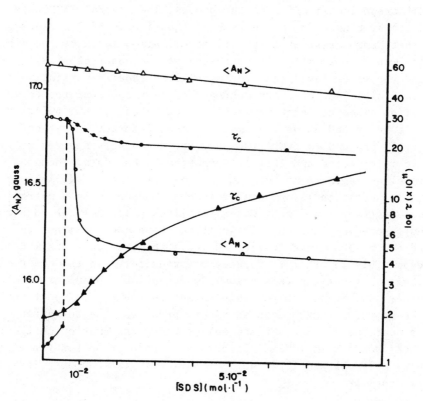

Figure 3. Nitrogen hyperfine coupling constant,$< A_N>$, and correlation time for the motion, τ_c,of TEMPOL (open and closed triangle)
and TM-TEMPAMINE$^+$ (open and closed circle) as a function of SDS
concentration, T = 298 K.

label molecule are therefore expected and this fact leads to
line broadening. From the spectral lineshape and $\langle A_N \rangle$ values,
it is quite reasonable to assume that C_{12}-TEMPO acts essentially
as a surfactant molecule; in fact it is able to form micelles and
to enter in SDS micelles in which it seems to behave almost in
the same manner as SDS does.

Figure 3 shows the variation of the nitrogen coupling constant
and of the correlation time of TEMPOL and TM-TEMPAMINE$^+$ as a
function of the SDS concentration. The sudden decrease of $\langle A_N \rangle$
for TM-TEMPAMINE$^+$ and the increase of the τ_c for both spin probes
indicate a marked change of polarity of the environment and of the
motion of the probes. The decrease in $\langle A_N \rangle$ and the increase in τ_c
start at 8.3×10^{-3} M, i.e. at a value very close to the cmc reported for this surfactant.[16,17] The $\langle A_N \rangle$ of TEMPOL decreases almost
linearly with increasing SDS concentration. This findings agree
for different localization of the probes. The neutral TEMPOL is
not directly bonded to the surface of the micelle of SDS. We suggest that TEMPOL resides in the first solvation layer of the micelle. Its mobility is strongly decreased at the onset of micellization
because the water layer near the micelle surface is partially immobilized. The lack of sharp decrease of $\langle A_N \rangle$ confirms the above
suggestion since no large variation is expected for the polarity
of water below and above the cmc. A direct interaction between the
positively charged TM-TEMPAMINE$^+$ and the negatively charged micelle
surface is expected and this is supported by ESR data. When the
SDS micelles begin to be formed, $\langle A_N \rangle$ decreases by $\simeq 4\%$. This considerable variation is comparable to what happens when a nitroxide
radical is transferred from water to a less polar solvent such as
ethanol. Moreover, the correlation time above cmc seems not to
depend on the SDS concentration, which means that only the micelle
is responsible for the τ_c values. Unfortunately in the cmc region
spectral complications (see below) do not allow a precise calculation of τ_c and it is not possible to obtain the τ_c value at the
cmc. In summary, TEMPOL acts as a spin probe for the solvation
layers near the micelle surface and provides information on their
dynamics; whereas TM-TEMPAMINE$^+$ acts as a spin probe for the micelle surface itself and it is able to follow variations of polarity
and of mobility.

Results in line with those given by C_{12}-TEMPO are obtained
with C_{16}-TEMPO (see Figure 4). In this case also, the ESR spectrum
consists of a single exchange-narrowed signal below SHS cmc, that
converts in a broad three-line spectrum near the cmc and in a

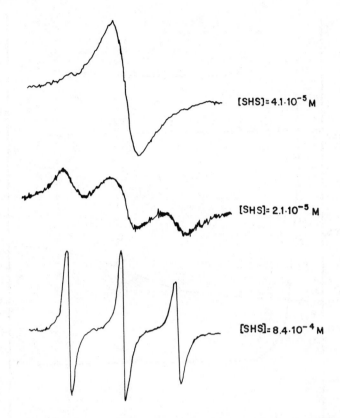

Figure 4. ESR spectra of C_{16}-TEMPO in SHS solutions below, near and above SHS cmc, T = 323 K.

narrow three-line spectrum above the cmc. Considerations similar to those made for C_{12}-TEMPO hold in this case too. Some differences are observed in the value of the correlation time for the motion (τ_c = 2.8 x 10^{-10} s) and in the variation of < A_N > , the latter does not pass through a minimum near cmc, because no appreciable fraction of free C_{16}-TEMPO in bulk water is observed at the spin label concentration used (≈1.6 x 10^{-5} M). The < A_N > shows an increase from low SHS concentration and stabilizes at 16.4 G above the cmc. The < A_N > and τ_c variations for TEMPOL and TM-TEMPAMINE[+] as a function of SHS concentration (Figure 5) show almost the same trend as observed in SDS solution; the largest differences are the higher decrease of < A_N > of TM-TEMPAMINE[+] and the higher increase of τ_c of the same spin probe in SHS solution than in SDS solution.

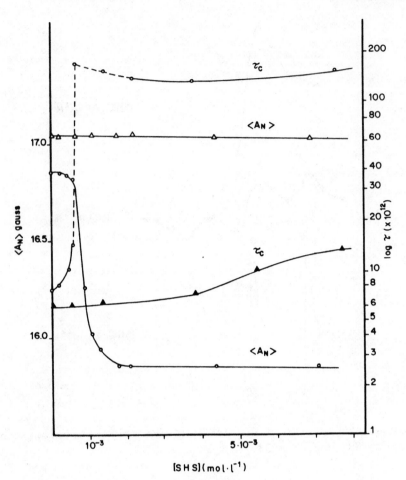

Figure 5. Nitrogen hyperfine coupling constant, $<A_N>$, and correlation time for the motion, τ_c, of TEMPOL (open and closed triangle) and TM-TEMPAMINE[+] (open and closed circle) as a function of SHS concentration (T = 323 K).

In this case also, the point of τ_c increase and $<A_N>$ decrease allows a quite accurate determination of cmc of SHS and this value (5.5×10^{-4} M) is very near to those obtained using other methods (5.8×10^{-4} M and 4.5×10^{-4} M).[16,18]

No spectral complications arise neither from the analysis of C_8-TEMPO ESR spectra nor from the analysis of TEMPOL and TM-TEMP-AMINE[+] ESR spectra in SOS solutions. As it was outlined before, C_8-TEMPO does not give micellar solution at a concentration of

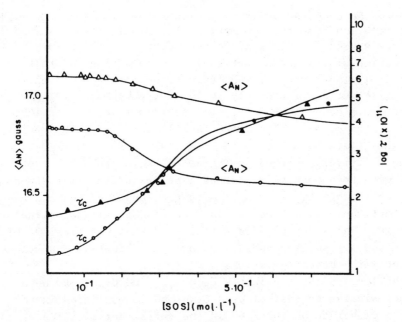

Figure 6. Nitrogen hyperfine coupling constant, $<A_N>$, and correlation time for the motion, τ_c, of TEMPOL (open and closed triangle) and TM-TEMPAMINE$^+$ (open and closed circle) as a function of SOS concentration (T = 298 K).

10^{-4} M. Thus the ESR spectra of C_8-TEMPO in SOS solution show progressive changes from that in bulk solution to that in SOS micelle: i) the coupling constant $<A_N>$ starts to decrease from 17.05 G at SOS cmc to 16.65 G above cmc; and ii) the correlation time τ_c, also markedly increases from 5×10^{-11} s near cmc to 1.2×10^{-10} s well above cmc. In addition, the linewidth does not pass through a maximum as it was observed with C_{12}-TEMPO in SDS but only increases near the cmc to a constant value above this concentration. Analogous behavior is shown by TM-TEMPAMINE$^+$ and TEMPOL (see Figure 6). The decrease of TM-TEMPAMINE$^+$ $<A_N>$ below and above the cmc is about 0.38 G, well below that observed with SDS and SHS (0.62 G and 1.0 G respectively). The fact that $<A_N>$ depends on the hydrocarbon chain length in the order SHS > SDS > SOS may reflect the relative ability of the SHS, SDS, and SOS micelles to solubilize TM-TEMPAM-INE$^+$. The degree of hydration[19] of the micelle surface must also be taken into account. No similar dependence is observed for TEMPOL nitrogen hyperfine coupling constant, which is not too sensitive to the micelle forming surfactant. However the data in Figures 3, 5, and 6 indicate that environments with slightly different polarity are sensed by TEMPOL in SOS, SDS, and SHS solutions.

From the combined analysis of lineshape, coupling constant
and correlation time for the motion of the spin probes and of the
spin labels used in this work we can understand the nature of the
localization of the paramagnets, and the following information is
obtained: i) The neutral TEMPOL spin probe should be localized near
the micelle surface (probably in the Gouy-Chapman layer) with its
hydrophilic side (the -OH group) inserted in the water layers
hydrating the micelle. The TEMPOL mobility reflects the mobility of
these water layers.

ii) The positively charged TM-TEMPAMINE[+] spin probe interacts directly with the negatively charged micelle surface. The almost direct proportionality of $< A_N >$ with the length of the chain suggests that the $>N\dot{-}0$ bond is inserted in the palisade structure
immediately below the micelle surface and that the micelles of SOS,
SDS and SHS are swollen and/or hydrated differently. The values of
τ_c agree with these suggestions.

iii) The long chain spin labels C_8-, C_{12}- and C_{16}-TEMPO behave in
a manner similar to that of C_8-, C_{12}- and C_{16}-sulfates in that they
are able either to be inserted in the surfactant micelles or to
produce micelles themselves (as in the case of C_{12}- and C_{16}-TEMPO).
Information "inside" the micelle and on the micellization process
can be inferred using these nitroxide.

Details on the kinetics of association and dissociation processes can also be deduced by an analysis of the linewidth and of
the lineshape of TM-TEMPAMINE[+] near the cmc of each surfactant (see

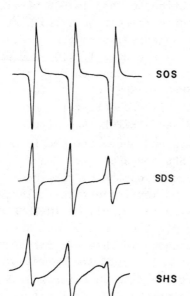

Figure 7. ESR spectra of TM-
-TEMPAMINE[+] near cmc of SOS, SDS
and SHS.

Figure 7). Following the treatment proposed by several authors
to micellar equilibria in solution of ionic surfactants we may
distinguish two relaxation processes in the presence of spin probe
R:

$$(n - r)S + rR \xrightleftharpoons[k'_d]{k'_f} S_{(n-r)}R_r$$

where $r = 1, 2, \ldots\ldots$ with residence time τ_{R1}; the relaxation
time τ_{R1} should be long enough to render lineshape independent of
this process. The second process is:

$$S_{(n-r)}R_r \xrightleftharpoons[k'_-]{k'_+} S_{(n-r)}R_{(r-1)} + R$$

where the residence time of the radical R, τ_{R2}, falls in the ESR
time scale: thus this process can be followed from the analysis of
the ESR lineshape.

The ESR lineshape is determined by the relative rate of chemical exchange P_{exch} between the bulk and the micelle with respect
to the separation between the ESR lines of the probe in both environment, $\delta\omega$ (in rad. s^{-1})[21]. Under the condition,

$$P_{exch} \ll \delta\omega$$

two separate absorption are observed. This is the case of TM-
-TEMPAMINE$^+$ in SHS solutions. However, for the condition

$$P_{exch} \gg \delta\omega$$

the observed spectrum is a weighted average of the signals of the
probe in water and in the micelle. This is the case of TM-TEMPAMI-
NE$^+$ in SOS, where the linewidth increases near the cmc, without
passing through a maximum. In the intermediate region, the linewidth, $1/T_2$, depends on both relaxation time of the probe in water,
$1/T_{2o}$, and on the frequency exchange; the linewidth passes, in
this case, through a maximum.[21] This is what happens near the cmc
in the case of TM-TEMPAMINE$^+$ in SDS solutions.

The evaluation of the residence times using this procedure
gives $\tau_{R2} \gg 5.7 \times 10^{-7}$ s, $\tau_{R2} = 0.93 \times 10^{-7}$ s and $\tau_{R2} < 1.8 \times 10^{-7}$
s for TM-TEMPAMINE$^+$ in SHS, SDS and SOS respectively; to be compa-
red with the frequencies of surfactant monomer exchange 1.7×10^{-5}
s, 1.0×10^{-7} s and 1.0×10^{-8} s respectively as obtained by using
other techniques.[18] Thus the residence time of the spin probe in
the three surfactant systems have the same trend as those of
surfactant monomers.

These results, together with those discussed above, prove that the addition of nitroxide radicals (unable to form micelles) in surfactant systems introduces only small perturbations on the properties of the systems. When micelles of spin labels are present, a careful analysis of the ESR lineshape allows deduction of useful information on the system under study.

ACKNOWLEDGEMENTS

Thanks are due to the Consiglio Nazionale delle Ricerche (CNR) "Piani Finalizzati Chimica Fine e Secondaria" for financial support.

REFERENCES

1. L. J. Berliner, Editor "Spin Labeling: Theory and Applications", Academic Press, New York, vol. I (1976), vol. II (1979).
2. A. S. Waggoner, O. H. Griffith and C. R. Christensen, Proc. Natl. Acad. Sci., 57, 1198 (1967).
3. C. Jolicoeur and H. L. Friedman, J. Solution Chem., 7, 813 (1978).
4. H. Yoshioka, J. Colloid Interface Sci., 63, 378 (1978).
5. S. A. Goldman, G.V. Bruno and J. H. Freed, J. Phys. Chem., 76, 1858 (1972).
6. D. Kivelson, J. Chem. Phys., 33, 1094 (1960).
7. C. Jolicoeur and H. L. Friedman, Ber. Bunsenges Phys. Chem., 75, 248 (1971).
8. J. Martinie, J. Michon and A. Rassat, J. Am. Chem. Soc., 97, 1818 (1975).
9. H. Yoshioka, J. Phys. Chem., 82, 103 (1978).
10.E. D. Sprague, D. C. Duecker and C. E. Larrabee Jr., J. Am. Chem. Soc., 103, 6797 (1981).
11.M. F. Ottaviani, P. Baglioni and G. Martini, (1982), J. Phys. Chem., in press.
12.R. Brière, H. Lemaire and A. Rassat, Bull. Soc. Chim. France, 3273 (1966).
13.B. N. Knauer and J. J. Napier, J. Am. Chem. Soc., 98, 4395 (1976).
14.G. Stout and J. B. F. N. Engberts, J. Org. Chem., 39, 3800 (1976).
15.H. Yoshioka, Chemistry Lett., 1477 (1977).
16.H. C. Evans, J. Chem. Soc., 579 (1956).
17.B. D. Flockhart, J. Colloid Interface Sci., 16, 484 (1962).
18.E. A. G. Aniansson, S.N. Wall, M. Almgren, H. Hofmann, I. Kielmann, W. Ulbricht, R. Zana, J. Lang and C. Tondre, J. Phys. Chem., 80, 905 (1976).

19. H. Wennerström and B. Lindman, J. Phys. Chem., $\underline{83}$, 2931 (1979).
20. See reference 18 and references therein.
21. A. Carrington and A. D. McLachlan, "Introduction to Magnetic Resonance", Harper and Row, New York, 1967.

SPIN LABEL STUDY OF MOLECULAR AGGREGATES

M. Schara and M. Nemec

Jozef Stefan Institute, E. Kardelj University of
Ljubljana
61000 Ljubljana, Yugoslavia

The distribution of amphiphilic molecules between
membrane and buffer solutions, containing the amphiphile
aggregates, is described by a simple model. The phenomenological relations describing the partitioning were derived and compared with the electron paramagnetic resonance measurements of the bovine erythrocyte-spin labeled
methyl ester of the palmitic acid system. The apparent
variation of the partition coefficient with the amphiphile concentration is explained.

INTRODUCTION

Biological membranes do have the capacity to accumulate materials from the neighbouring solutions. The concentration in the
membrane can rise by several orders of magnitude above the water
solution concentration of these molecules[1]. A sufficient concentration of the molecules accumulated might be a prerequisite for
the proper functioning of the membrane. Some scarcely soluble molecules can enter the membrane via the monomeric solution in water
as it was proven for cholesterol[2]. The nonspecific binding of molecules to the membrane is usually described by the partition coefficient of the solute between the membrane and water solution.
Large variation in partition coefficient are obtained depending on
the method used. It was suggested that molecular aggregation could
have influenced these results[3]. Introduction of larger quantities
of amphiphilic spin probes into membranes is often exposed to the
nonlinear responses between the amounts of the spin probe added,
the sample membrane volume fraction and the membrane acceptance of
the spin probe. These responses can influence further paramagnetic

559

resonance measurements. The equilibrium distribution of amphiphilic molecules between the membrane and water solution containing aggregates of these molecules will be described.

EXPERIMENTAL

The erythrocytes were prepared from freshly collected citrated bovine blood. They were stored in Tris-HCl buffer at pH 7.4. The erythrocyte samples obtained after centrifugation contained 1.7 x x 10^{10} cells/ml.

The spin labeled palmitic acid methyl ester MeFASL(10,3) was used as the amphiphile and object for the EPR measurement. The spin probe was injected as ethanol solution into the Tris buffer to obtain the final amounts of the spin probe 2.9, 21.7 and 43.4 x 10^{15} MeFASL (10,3) molecules/ml. The amount of ethanol in the samples was below 0.5 volume percent. To these solutions different numbers of cells were added and after 15 minutes the samples were centrifuged and the cells were measured in 1 mm inner diamter glass capillaries. The measurements were performed on the Varian E-9 EPR spectrometer. The intensities were normalized in a double cavity with respect to the strong pitch standard. The numbers of molecules obtained have only a relative meaning in this experiment.

It should be mentioned that the electron paramagnetic resonance spectra of this spin probe in buffer solutions are at the limit of intensity detectability; namely, the aggregated MeFASL(10,3) molecules exhibit strongly broadened EPR spectra so that only molecules in the monomeric state can be measured. Our estimate is about 7 x 10^{-7} mol/l in the buffer solution. To determine the amount of aggregated molecules, the buffer solutions were extracted by chloroform. The solubility of these molecules in chloroform eliminates the exchange interactions typical for aggregates and therefore the total amount of MeFASL(10,3) can be detected. The same situation is for membranes. The shape of the spectra is different, but the intensity is proportional to the concentration of the dissolved molecules.

RESULTS AND DISCUSSION

For low concentrations of the amphiphile the partition coefficient defined as the ratio between the concentration of these molecules in the membrane and the neighbouring buffer can be used. The relative numbers of molecules for a given volume ratio can be written as

$$\frac{N}{N_m} = 1 + \frac{1}{K} \frac{V_o}{V_m} \qquad (1)$$

Here N is the total and N_m is the membrane bound number of amphiphile molecules in the sample, and V_o/V_m is the volume ratio between the buffer solution and the membrane.

At higher amounts of the spin probes in the cell suspension samples these molecules do aggregate. Using the formalism of the free energy minima described previously[4] for the equilibrium between the molecular aggregates and the monomerically dissolved molecules we write down the relation for all three: the membrane, the monomer and the average aggregate size, g. It is obvious that this approximation would be valuable only for monodisperse systems with a narrow size distribution of aggregates.

The free energy ϕ_t for the membrane and buffer solution dissolved molecules can be written as

$$\phi_t = \phi_s + \phi_m \qquad (2)$$

where

$$\phi_m = N_{om}\mu^0_{om} + N_m\mu^0_m + kT(N_{om}\ln(N_{om}/N_{mt}) + N_m\ln(N_m/N_{mt}))$$

and

$$\phi_s = N_o\mu^0_o + N_1\mu^0_1 + N_g\mu^0_g + kT(N_o\ln(N_o/N_{st}) + N_1\ln(N_1/N_{st}) +$$

$$+ \frac{N_g}{g}\ln(N_g/gN_{st}))$$

N_i is the number of molecules, μ^0_i the standard free energy per molecule, g the aggregation number, k the Boltzmann constant, T the absolute temperature. The subscripts mean: om membrane molecules - - the membrane solvent molecules, m membrane dissolved spin probe molecules, o water molecules, 1 the monomeric and g the polymeric- - aggregated spin probe molecules dissolved in the buffer solution, st and mt designate the total amount the molecules in the buffer solution and in the mebrane respectively, N is the total number of the spin probe in the sample. Using the relation

$$N = N_1 + N_m + N_g = \text{const.} \qquad (3)$$

as a constraint in the free energy minimum determination for

$$\phi = \phi_t + \lambda(N - N_1 - N_m - N_g) \qquad (4)$$

where λ is the Lagrange multiplier. Solving the system of equations $\partial\phi/\partial N_i = 0$ where $i = 1,m,g$ and the assumption that $K = N_m V_o/N_1 V_m$ is a constant, the relative amount of the total to the membrane bound spin probe in the sample can be written

$$\frac{N}{N_m} = 1 + \frac{V_o}{V_m} \frac{1}{K} \{1 + g(\frac{N_1}{N_{st}})^{g-1} \exp |g(\mu_1^o - \mu_g^o)/kT|\} \tag{5}$$

Equation (5) can be expressed as

$$\frac{N}{N_m} \sim B + \frac{A}{K} \frac{V_o}{V_m} \tag{6}$$

where $V_o/V_m \sim N_{st}/N_{mt}$ is the volume ratio between the buffer and membranes. These coefficients can be estimated from the linear plot of the experimental data.

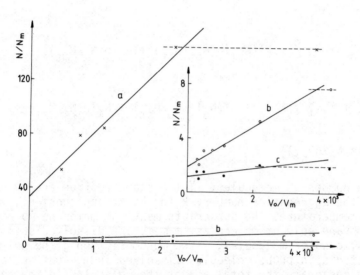

Figure 1. The ratio between the total and membrane bound number of Me FASL (10,3) molecules in Tris buffer dispersions plotted against the volume ratio of the buffer and membranes of this sample. The total amount of the spin probe N are: a) 42×10^{15}, b) 21.7×10^{15} and c) 2.9×10^{15} molecules per ml. The insert enlarges the low N plots.

The results obtained for the total to membrane bound spin probe ratio with respect to the variation of the ratio between the buffer solution and membrane volumes are plotted and compared with the linear relation given in Equations (5) and (6). Larger inclinations A/K of the lines are observed for larger total amounts of the spin probe in the sample N. At lower N the inclination increase is proportional to N in accord with Equation (5), while for larger N the inclination rise can be explained by the free energy exponent α increase. Probably μ_g decreases due to the apperance of larger aggregates[4] in the buffer solution.

At high values of V_0/V_m the limiting value N_m designated by a dashed horizontal line in Figure 1 a further decrease of the membrane volume at a constant value of the sample volume does not alter the concentration of the spin probe in the membrane. The relative amount N_m/N of the membrane bound molecules decreases linearly with increasing N in samples with large V_m/V_0. On the other hand the total number N_m of the membrane bound molecules increases initially with N and decreases afterwards at higher N. This observation is in accord with the observed decrease of the binding capacity with N as shown in Equation (5), since the apparent partition coefficient K/A decreases. The values for K/A obtained from curves a), b) and c) are 19, 658 and 2941. On the other hand corresponding values of the partition coefficient evaluated from the limiting values given in Figure 1 are 1.600, 2.900 and 33.700. We have shown the sources of large differences in the partition coefficient determinations.

These results on erythrocytes will be tested also on other membrane and amphiphile systems to confirm the proposed description.

CONCLUSION

An example is presented how the molecular aggregates can influence the uptake of substances into the membrane and to limit the acceptance of larger quantities of amphiphiles. These findings may have important implication in living systems.

REFERENCES

1. S. Pečar, M. Šentjurc, M. Schara, M. Nemec, B. Sorg and E. Hecker, Cancerogenesis 7, 541 (1982).
2. E. Bojesen, Nature 299, 276 (1982).
3. M.J. Conrad and S.J. Singer, Biochemistry 20, 808 (1981).
4. E. Ruckenstein and R. Nagarajan, in "Micellization, Solubilization and Microemulsions", K.L. Mittal, Editor, Vol. 1, p. 133, Plenum Press, New York, 1977.

MICELLAR STRUCTURE AND WATER PENETRATION STUDIED BY NMR AND OPTICAL SPECTROSCOPY*

Klaas A. Zachariasse, Boleslaw Kozankiewicz[1], and
Wolfgang Kühnle
Max-Planck-Institut für biophysikalische Chemie
Am Fassberg, Postfach 968, D-3400 Göttingen
West Germany

The molecular structure of aqueous micellar solutions and microemulsions is studied by spectroscopic
methods, employing molecules that respond to polarity.
The spectroscopic methods used are: fluorescence
(spectra, lifetimes and quenching studies) and NMR.
In the fluorescence experiments, the probe molecules
are present in extremely low concentration (below
10^{-6} M), thereby minimizing the possible perturbation
of the probe environment. The probes are predominantly
solubilized in the micellar phase. Their location can
be investigated utilizing NMR spectroscopy (ring
current effect).

By employing as probes amphiphilic molecules such
as $Py(CH_2)_nCOOH$ and $Py(CH_2)_nN(CH_3)_3Br$ with Py=pyrene,
it could be shown that the pyrenyl end group penetrates more and more into the interior of sodium
dodecyl sulfate (SDS) micelles when n increases from
3 to 15. The polarity of the direct probe environment
decreases from a value somewhat larger than that of
methanol, for $Py(CH_2)_3COOH$ solubilized in the surface
region of the micelle, to a value approaching that of
hexadecane for the central region of the micelle
(for n=15). It is concluded that the water content in
the SDS micelles decreases when going from the surface
region to the micellar interior. The latter was found
to be free of water molecules.

---

* Dedicated to Professor Friedrich Boberg on the occasion of
his 60th birthday.

1. INTRODUCTION

In view of the rather simple overall composition of aqueous micellar solutions, amphiphilic molecules present in groups of roughly 100 units[2] in water, it seems at first sight surprising that the exact molecular structure of these micelles is still an open question. This can be seen from the continuing debate[3,4,5] concerning a number of, partly conflicting, molecular micelle models: a) the alkane-sphere model (Hartley, 1939[6]); b) the double-layer model (McBain, 1942; Philippoff, 1950[7]); c) the porous cluster model (Menger, 1978[3]) and d) the surfactant-block model (Fromherz, 1981[5]). This last model is a synthesis between the alkane droplet (a) and the bilayer (b) concepts.

In a discussion of micellar structure and stability, the role of the water molecules, more precisely the extent of water penetration into the micelles,[3] obviously is of prime importance. In the present paper, this question of the distribution of water molecules between the surface region and the centre of micelles is studied employing fluorescent probes. These probe molecules, derivatives of pyrene (see below), are predominantly solubilized in the micellar phase, due to their low solubility in water (6.7 x 10^{-7} M for pyrene)[8]. As the presence of water molecules in an environment made up out of alkyl chains will reveal itself as an increase in the overall polarity, those properties of the probe molecules have to be employed that are a function of the polarity of their direct surroundings. Such properties are:

a) the fluorescence decay time, which decreases in the case of pyrene with increasing solvent polarity;[9]

b) the intensity distribution of the vibrational peaks in the fluorescence spectrum, the Ham effect.[10-12]

An important factor in the interpretation of measurements with probes, apart from an assessment of possible perturbations exerted by these molecules,[13] is the consideration of the exact probe location in the medium under investigation. It is now generally assumed that aromatic molecules such as naphthalene[14] or pyrene[15] are located in the outer-sphere of the micelles, close to the bulk aqueous phase. This has been deduced from NMR measurements[14] and from studies of the Ham-effect.[15] Further evidence comes from experiments utilizing the change in the relative peak intensities in the fluorescence spectrum of pyrene, as a function of variations in surfactant chain length, counterion, salt addition, and temperature.[16] The pyrene molecule appears to be solubilized in a region of the micelle that is only slightly less polar than that observed with the pyridinium N-phenolbetaine $E_T(30)$, a polarity probe for aqueous interfaces.[17,18] As this latter molecule is solubilized in the aqueous surface region of the micelles,[17] it is concluded that the pyrene molecule is located only slightly further away from the bulk aqueous phase than is the $E_T(30)$ molecule. Therefore, although pyrene can probe the outer-sphere of the micelles, it cannot reveal anything about the properties of the inner region of the micelles. More specifically, most probe molecules used to investigate water penetration into aqueous micellar aggregates,[19-22] only report on the presence of water molecules in the micellar surface region, where these probe molecules are predominantly solubilized.

It is therefore essential in these investigation of micelles, to be able to locate the same probe molecule selectively in specific locations of the micelle, i.e. at different distances from the micellar surface region. A well-known method to obtain such a selective location, is to chemically attach the probes to different positions of the alkyl chains of the surfactant molecules. Examples are the carbonyl probe 8-ketohexadecyltrimethylammonium bromide used in studies of water penetration in micelles[23] and the anthroate esters employed in investigations of phosphatidylcholine bilayers[24]. In the present paper, pyrenyl probes[25,26] have been employed that are chemically linked to the end of the alkyl chains of amphiphilic molecules: the (1-pyrenyl)-substituted alkylcarboxylic acids $Py(CH_2)_nCOOH$ (n=3,5,8,9,11 and 15) and the compound (12-(1-pyrenyl)dodecyl)trimethylammonium bromide $Py(CH_2)_{12}N(CH_3)_3Br$.

However, this method does not a priori provide absolute certainty about the probe location in micelles, as will be shown in section 3a. Hence, an independent method to determine the probe location is needed.

2. EXPERIMENTAL

The n-(1-pyrenyl)alkylcarboxylic acids $Py(CH_2)_nCOOH$ were synthesized as follows. For n=3: reduction with hydrazine hydrate of 3-(1-pyrenoyl)propionic acid, obtained from a reaction between pyrene and succinic anhydride. For n=5,8,9: The monoethylester of the appropriate alkyldicarboxylic acid is synthesized from the diester and the dicarboxylic acid. The monoester is transformed into the acid chloride, which is reacted with pyrene to form the pyrenoylalkylcarboxylic ester. This ester is reduced with hydrazine hydrate to the pyrenylalkylcarboxylic acid. For n=11 and 15, a procedure similar to that employed with n=5,8,9 is used, utilizing the monoethylester (obtained from the diester with $Ba(OH)_2$). N-(12-(1-pyrenyl)-dodecyl)-N,N,N-trimethylammonium bromide $Py(CH_2)_{12}N(CH_3)_3Br$, was synthesized starting from the ethylester of the corresponding (1-pyrenyl)alkylcarboxylic acid, which was reduced to the alcohol with $LiAlH_4$. This alcohol is transformed into the bromide, and then reacted with gaseous trimethylamine in

a benzene/methanol solvent mixture. The purity of the compounds
was checked with NMR and elemental analysis. The source of the
surfactant molecules, sodium dodecyl sulfate (SDS), hexadecyltrimethylammonium bromide (CTAB) and hexadecyltrimethylammonium chloride (CTAC), has been described in Reference 17. The oil in water
(o/w) microemulsion SDS(M) was prepared by mixing the surfactant
(1.25g), 3.4 ml of 1-pentanol (Merck), 0.6 ml of hexadecane (Merck),
and 13.5 ml of water.

The ^{1}H NMR spectra were recorded on a Bruker WH-270. 360 free
induction decays were accumulated and then transformed to the
spectra. The fluorescence decay times were determined, at 376 nm,
with a N_2-Laser system (Lambda Physik). The fluorescence spectra
were run on a Hitachi-Perkin Elmer MPF-2A spectrofluorimeter. All
solutions in the fluorescence experiments were degassed using the
freeze-pump-thaw method (5 cycles). The probe/surfactant ratio
was 1:20 for the NMR spectra and lower than 1:1000 in the fluorescence experiments.

3. RESULTS AND DISCUSSION

a. Quenching of Fluorescent Probes by Micellar Counterions (Br^-)

The fluorescence of pyrene[27] and of 1-methylpyrene (MePy)
solubilized in deoxygenated hexadecyltrimethylammonium bromide
(CTAB) micelles (0.05M in H_2O), is strongly quenched by the Br^-
counterions. This leads to a shortening of the fluorescence decay
time of MePy (125ns at 30°C) as compared to a lifetime of 210ns
in hexadecyltrimethylammonium chloride (CTAC) micelles, where Cl^-
does not quench the pyrenyl fluorescence (Table I). This quenching
action of Br^- on the fluorescence of MePy is a consequence of the
fact that Br^- and MePy are both located in the aqueous micellar
surface region (see above).

It was expected, as discussed in the Introduction, that the
pyrenyl end group of a probe molecule such as $Py(CH_2)_{12}N(CH_3)_3Br$,
when cosolubilized with CTAB, would be located in the micellar
interior, at any rate further away from the micellar surface than
in the case of MePy. This would then result in a considerably
longer fluorescence lifetime of the pyrenyl end group than that
observed with MePy, as the Br^- ions are supposed not to be able
to enter into the micellar interior.

Surprisingly, however, the - single exponential - fluorescence
decay time of $Py(CH_2)_{12}N(CH_3)_3Br$ in CTAB has practically the same
value (128ns at 30°C) as that observed for MePy, and is considerably shorter than the lifetime in CTAC (203ns), see Table I.

Table I.   Fluorescence Decay Times τ of 1-Methylpyrene and
           $Py(CH_2)_{12}N(CH_3)_3Br$ in CTAB and in CTAC Micelles
           (0.05M in H_2O) at 30°C.

Probe Molecule	τ(ns) in CTAB	τ(ns) in CTAC
1-methylpyrene	125	210
$Py(CH_2)_{12}N(CH_3)_3Br$	128	203

This must mean that the location of the pyrenyl end group of
$Py(CH_2)_{12}N(CH_3)_3Br$ in the CTAB micelles is similar to that of MePy,
i.e. close to the micellar surface, in spite of the chemical linkage
to the end of the aliphatic chain. This leads to the conclusion
that the pyrene moiety of $Py(CH_2)_{12}N(CH_3)_3Br$ has turned back to a
location in the aqueous surface region of the CTAB micelles.

It is therefore important first to ascertain the location of
the probe molecules (section 3b) before the lifetime measurements
(section 3d) and the quenching experiments (section 3e) are discussed.

b. NMR studies of probe location in anionic and cationic micelles

The only method presently available to determine the location
of aromatic probe molecules, such as naphthalene and pyrene, in
micelles is based on the ring current effect these molecules exert
on the different H atoms of the surfactant molecules, resulting in
changes in the chemical shifts in the NMR spectra.[14] The relative
magnitude of the chemical shift difference ΔΔσ is a measure of the
relative mean distance of the probe molecule to the particular
H atom.[14,17] Using this NMR technique, it was deduced that naphthalene is located near the polar head groups of the CTAB micelle,[14]
i.e. close to the micellar surface. Similar results were obtained
for pyrene[16] and for $E_T(30)$.[17] These conclusions were supported
by other methods, such as the Ham effect, as mentioned above.
(see section 3c).

NMR spectra were determined for CTAB micelles with and without
the probe molecule $Py(CH_2)_{12}N(CH_3)_3Br$ (section 3a), see Figure 1.

The differences in chemical shift, ΔΔσ, between the NMR spectra
with and without probe have been plotted for the various H atoms
of the CTAB surfactant molecule (Figure 2), for the probe/surfactant

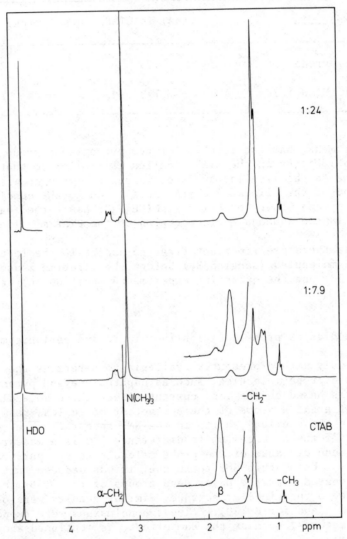

Figure 1. ^{1}H NMR spectra of the probe molecule $Py(CH_2)_{12}N(CH_3)_3Br$ in aqueous micellar solutions of hexadecyltrimethylammonium bromide (CTAB, 0.059 M) at 35°C, without probe and with a probe/surfactant ratio 1:24 and 1:7.9.

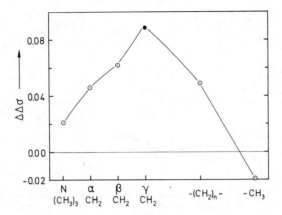

Figure 2.   The difference in chemical shift, $\Delta\Delta\sigma$, obtained from the
^{1}H NMR spectra of an aqueous micellar solution of hexadecyltrimethylammonium bromide (CTAB, 0.059 M) at 35°C, with and without
the probe molecule $Py(CH_2)_{12}N(CH_3)_3Br$ (probe/surfactant ratio 1:24).

ratio 1:24.[28] It is seen from Figure 2 that the effect of the ring
current of the probe molecule has its largest value for the polar
trimethylammonium group $N(CH_3)_3$ and for the α-, β-, and γ-CH_2
groups. The change in the chemical shift of the terminal CH_3 group
is of opposite sign than that for the other H atoms. This is caused
by the fact that the CH_3 groups are predominantly not located over
the aromatic ring. This reverses the sign of the change in the
chemical shift.[29] This is even the case for the fairly large probe/
surfactant ratio 1:7.9. From the NMR spectra for this probe/sur-
factant ratio, it is seen that the probe has an effect only on
part of the $-CH_2$-groups. It is concluded from the NMR spectra that
the pyrenyl moiety of $Py(CH_2)_{12}N(CH_3)_3Br$ is indeed located near
the polar head group and near the CH_2 groups that are adjacent to
the micellar surface. This completely supports the conclusions
derived in the previous section on the basis of the fluorescence
quenching data.

In a further attempt to introduce pyrenyl probe molecules
into the micellar interior, a series of molecules $Py(CH_2)_nCOOH$
with different n was utilized. These molecules were introduced
into micellar solutions of SDS (0.1M) in a probe to surfactant
ratio of 1:24. The difference $\Delta\Delta\sigma$ in the chemical shifts of the
different proton resonances, (α-CH_2, β-CH_2, γ-CH_2, the remaining
methylene groups ($-CH_2$-), and the terminal CH_3 groups) with and
without probe were then determined. The results so obtained with
$Py(CH_2)_nCOOH$, n=3,9,11 and 15, with 1-methylpyrene (MePy) and with
$Py(CH_2)_{12}N(CH_3)_3Br$, have been depicted in Figure 3. It is seen
that in the case of MePy the ring current effect on the terminal
CH_3 groups of the SDS surfactant molecules is smaller than on the
other H atom resonances, the strongest effect being observed for

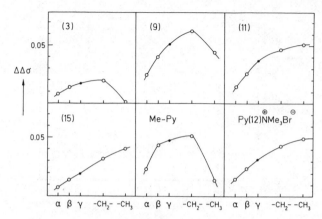

Figure 3. The difference in chemical shift, $\Delta\Delta\sigma$, obtained from
the ^{1}H NMR spectra of an aqueous micellar solution of sodium
dodecyl sulfate (SDS, 0.1 M) at 35°C, with and without a probe
molecule from the series of $Py(CH_2)_nCOOH$ (with n=3,9,11 and 15),
1-methylpyrene (MePy) and $Py(CH_2)_{12}N(CH_3)_3Br$.

the β-CH_2 and, especially, for the middle methylene groups (-CH_2-).
With $Py(CH_2)_3COOH$, a pattern closely similar to that found for
MePy is observed. For the $Py(CH_2)_nCOOH$ molecules with longer alkane
chains the ring current effect of the probe on the terminal CH_3
group increases with n, reaching its highest value for n=15. The
cationic probe molecule $Py(CH_2)_{12}N(CH_3)_3Br$ solubilized in the
anionic SDS micelles presents a pattern similar to that observed
with the $Py(CH_2)_nCOOH$ probe with n=11. These observations indicate
that it is possible to introduce the pyrenyl end group of the probe
molecules $Py(CH_2)_{12}N(CH_3)_3Br$ and $Py(CH_2)_nCOOH$ with n=11 and 15,
into the central region of the SDS micelles.[30] This could not be
achieved, with the first probe molecule, in CTAB micelles (section
3a).

c. Influence of the polarity of the medium on the vibrational
   structure of the fluorescence spectra of pyrene and its
   derivatives (the Ham effect)

As stated in the Introduction, pyrene can be used as a probe
of the polarity of its direct environment, as the vibrational
structure in the fluorescence spectrum changes as a function of the
polarity of the medium, the Ham effect. This is illustrated, as an
example, by the fluorescence spectra of pyrene in a number of
solvents and micellar solutions depicted in Figure 4. The intensity
ratio of the vibrational peaks at 374nm and 395nm, I_1/I_4, increases
with solvent polarity, e.g. from liquid paraffin ($\varepsilon = 2.2$) to
methanol ($\varepsilon = 32.7$). The intensity ratio I_1/I_4 for a series of
solvents at 50°C, has been plotted in Figure 5, against the polarity

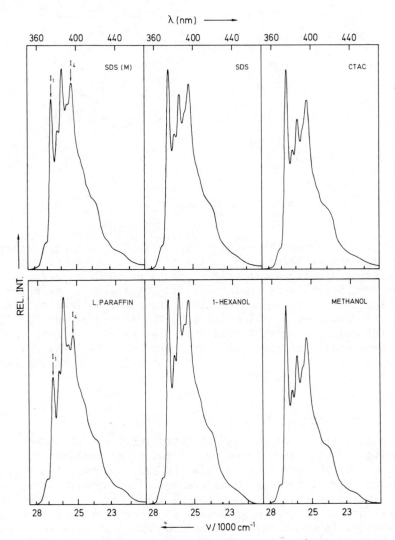

Figure 4.   Fluorescence spectra of pyrene $(5 \times 10^{-6}$ M), at around 30°C, solubilized in a number of solvents: liquid paraffin $(\varepsilon = 2.2)$, 1-hexanol $(\varepsilon = 13.3)$ and methanol $(\varepsilon = 32.7)$, lower row; and, upper row, in two aqueous micellar solutions: sodium dodecyl sulfate (SDS, 0.1 M) and hexadecyltrimethylammonium chloride (CTAC, 0.05 M) and in the o/w microemulsion SDS(M), see Experimental Section.

parameter $E_T(30)$. It is seen that the solvents appear in this plot in two classes, the aprotic and the protic solvents, similar to what has been reported elsewhere.[17]

The I_1/I_4 ratio observed (see Figure 4), for the micellar solutions of SDS and CTAC, the micelles that are discussed in this paper, therefore correspond to two polarity values: either to that

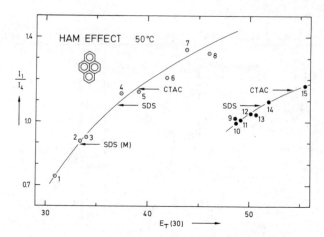

Figure 5. The intensity ratio I_1/I_4 of the vibrational peaks at
374nm (I_1) and at 395nm (I_4) in the fluorescence spectrum of pyrene
in different aprotic and protic solvents at 50°C, as a function of
the solvent polarity parameter $E_T(30)$. The solvents are numbered
as follows: 1: n-hexane; 2: n-dibutylether; 3: diisopropylether;
4: tetrahydrofuran; 5: chloroform; 6: 1,2-dichlorethane; 7: di-
methylformamide; 8: acetonitrile; 9: 2-propanol; 10: 1-hexanol;
11: 1-pentanol; 12: 1-butanol; 13: 1-propanol; 14: ethanol;
15: methanol.

of ethanol/methanol or to that of tetrahydrofuran/chloroform,
depending on which one of the two correlation lines in Figure 5
is taken. For the o/w microemulsion SDS(M), however, the I_1/I_4
value can only be correlated to the series of aprotic solvents,
being similar to that observed for dibutylether. This shows that
the environment of pyrene in SDS(M) does not contain as many water
molecules as in the case of the aqueous micellar solutions, being
approximately similar to a mixture of hexadecane (the oil) and
1-pentanol (the cosurfactant). The pyrene probe is thus solubilized
away from the surface of the microemulsion, which can be considered
to be a "swollen micelle".[17]

As the Ham effect is much less pronounced for the alkyl-sub-
stituted pyrene derivatives utilized here,[13], than for the pyrene
itself, the discussion in what follows will concentrate on the
polarity dependence of the decay times, the Ham effect being
employed as additional support.

d. Fluorescence decay times and Ham effect of $Py(CH_2)_nCOOH$
   solubilized in anionic and cationic micelles. The influence
   of polarity (water penetration).

The fluorescence lifetime of pyrene (see above) and pyrene
derivatives, such as 1-methylpyrene (MePy) and $Py(CH_2)_nCOOH$, decreases with increasing solvent polarity, especially when the
polarity is increased by the presence of water molecules, see
Table II.

As was shown above, the molecules $Py(CH_2)_nCOOH$ can be selectively solubilized at various distances from the surface region of
SDS micelles. The polarity dependence of the fluorescence decay
times of the pyrenyl end group in the different $Py(CH_2)_nCOOH$
molecules can, therefore, be employed to investigate the polarity,
i.e. the presence of water molecules, at different distances from
the surface of the SDS micelles.

The fluorescence decay times of the series of probe molecules
$Py(CH_2)_nCOOH$ solubilized in SDS micelles (0.1M), are plotted as a
function of n in Figure 6a. The decay times of $Py(CH_2)_nCOOH$ in
methanol have also been depicted for comparison (see Table II).
It is seen that for small values of n (n=3, 5 and 8) the single
exponential[31] decay time in SDS is smaller than that in methanol,
indicating a polarity higher than that of methanol for the probe
environment. For larger values of n, the decay time of $Py(CH_2)_nCOOH$
in SDS becomes progressively larger, having a value equal to that
in methanol for n=9, and reaching finally, for the longest methylene
chain (n=15), a value equal to that of hexadecane. The environment
of the $Py(CH_2)_{15}COOH$ probe, as in the case of all other probe
molecules, is completely isotropic. This can be derived from the
fact that the decay times are single exponential. Similar results
are obtained with $Py(CH_2)_nCOOH$ solubilized in micellar solutions
of sodium tetradecyl sulfate (STS).[16]

It is concluded from these results that the polarity of the
direct probe environment decreases with increasing n. This
signifies that the pyrenyl end group of $Py(CH_2)_nCOOH$ encounters a
progressively smaller number of water molecules in its surroundings,
when n becomes larger, finding an environment practically free of
water for n=15. This environment, as deduced from the NMR results
described in the preceding section, is the region of the micelle
where the probe molecule is located in the vicinity of the terminal
CH_3 group and the methylene groups ($-CH_2-$), further down than the
$\gamma-CH_2$ group of the alkane chain of the SDS molecule. In a spherical
micelle model, this is the central micellar region. We therefore

Table II. Fluorescence Decay Times τ of 1-Methylpyrene (MePy) and $Py(CH_2)_nCOOH$ with n=3,5,15 as a Function of Solvent Polarity at 30°C.

solvent (ε)[a]	$f-\frac{1}{2}f'$[b]	$E_T(30)$[c]	τ(MePy)	$\tau(Py(CH_2)_nCOOH)$		
		(kcal/mol)	(ns)	(ns)		
				n=3	n=5	n=15
Hexadecane (2.06)	0.104	(30.9)	240	(271)	250	237
C_2H_5OH (24.73)	0.379	51.9	227	--	--	--
CH_3OH (33.14)	0.393	55.5	225	245	237	229
CH_3OH/H_2O = 9:1 (39.94)	0.396	56.1	216	232	225	220
CH_3OH/H_2O = 3:1 (47.58)	0.397	60.0	206	217	210	205
H_2O (78.54)[d]	0.455	63.1	--	137	130	--

a) dielectric constant at 25°C.

b) solvent polarity scale: $f-\frac{1}{2}f' = (\varepsilon-1)/(2\varepsilon+1)-\frac{1}{2}(n^2-1)/(2n^2+1)$ at 25°C.

c) solvent polarity parameter, see Reference 18.

d) at pH 8.8.

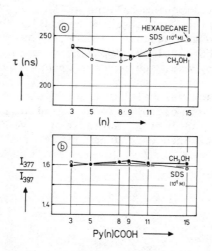

Figure 6. a) The fluorescence decay times (⊙) at 30°C, of the
series of probe molecules Py(CH$_2$)$_n$COOH solubilized in aqueous
micellar solutions of sodium dodecyl sulfate (SDS, 0.1 M), as a
function of n, the number of the methylene groups in Py(CH$_2$)$_n$COOH.
The decay times in methanol (●) and in hexadecane (coincident with
the value for n=15 in SDS) have also been indicated.
b) plot of the peak intensity ratio I (377)/I (397) in the
fluorescence spectrum of Py(CH$_2$)$_n$COOH in SDS micelles (⊙) and in
methanol (●), as a function of n, the number of methylene groups
in Py(CH$_2$)$_n$COOH (see text).

conclude that indeed the interior of the SDS micelles does not
contain water molecules. The extent of water penetration from the
micellar surface, or better the question which carbon atoms of the
detergent molecules are part of the surface region in contact with
water molecules, is difficult to ascertain with precision. This is
not surprising, as a micelle is a dynamic entity, in constant
motion. From an inspection of the NMR spectra, we are led to the
conclusion that water does not penetrate much further into the
micelles than to the γ-CH$_2$ group. The concentration of water
molecules radially diminishes when going from the aqueous surface
region to the micellar interior. This is the result expected, from
a purely geometric reasoning, for a spherical micelle in which the
surfactant molecules stretch out from the center, leaving more and
more space for water molecules as the micellar surface is approached.

From the fluorescence quenching experiments (Br^-) and the NMR
spectra discussed above, it was concluded that in cationic micelles
of hexadecyltrimethylammonium halides (CTAB and CTAC), the
$Py(CH_2)_{12}N(CH_3)_3Br$ probe and also the $Py(CH_2)_nCOOH$ molecules are
solubilized close to the aqueous surface, in contrast to what was
observed with the SDS micelles. Apparently, as was also reported
to be the case for pyrene, the pyrenyl groups have a strong affinity
for the tetraalkylammonium head groups, possibly due to complex
formation.[15,32] An analysis of the fluorescence decay times of the
series of $Py(CH_2)_nCOOH$ molecules solubilized in CTAC micelles
fully supports this conclusion. As is seen from Figure 7a, the
single exponential fluorescence decay times of the $Py(CH_2)_nCOOH$

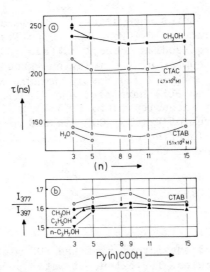

Figure 7.   a) The fluorescence decay times (◉), at 30°C, of the
series of probe molecules $Py(CH_2)_nCOOH$ solubilized in aqueous
micellar solutions (0.05 M) of hexadecyltrimethylammonium chloride
(CTAC) and hexadecyltrimethylammonium bromide (CTAB) as a function
of n, the number of the methylene groups in $Py(CH_2)_nCOOH$. The decay
times in methanol (●), in ethanol (▼) and in 1-propanol (▲) have
also been indicated.
b) Plot of the peak intensity ratio I (377)/I (397) in the
fluorescence spectrum of $Py(CH_2)_nCOOH$ in CTAB micelles (◉) and
in methanol (●), ethanol (▼) and 1-propanol (▲) as a function of
n, the number of the methylene groups in $Py(CH_2)_nCOOH$ (see text).

probes in CTAC have, for all n, values that are smaller than those observed in CH$_3$OH. They are clearly larger than those observed in H$_2$O (for n=3 and 5), however.

In micellar solutions of CTAB, the decay times of the Py(CH$_2$)$_n$COOH molecules have considerably smaller values than in CTAC (Figure 7a), due to the quenching action of the Br$^-$ counterions. The dependence of the lifetimes on the number of the methylene groups of the Py(CH$_2$)$_n$COOH probes has the same pattern as that observed for the CTAC micelles (Figure 7a). This shows that, as in the case of CTAB, the pyrenyl end group of all Py(CH$_2$)$_n$COOH probes is located in the aqueous micellar surface region, where they can readily interact with the Br$^-$ counterions present in their direct surroundings.

The fluorescence intensity ratio I(377)/I(397) of the Py(CH$_2$)$_n$COOH molecules, depicted in Figure 7b as a function of n, reports a higher polarity for the probe surroundings in CTAB micelles than that observed in methanol. For CTAC similar results are observed.[16] This fully supports the conclusions reached above on the basis of the decay time measurements. See also Figure 6b.

e. Fluorescence quenching with Ag$^+$ ions

It has been concluded in the preceding sections that the probe molecules Py(CH$_2$)$_n$COOH, and also Py(CH$_2$)$_{12}$N(CH$_3$)$_3$Br, are solubilized at increasing mean distances from the aqueous surface region of SDS micelles, when n increases from 3 to 15. Simultaneously, the polarity, i.e., the percentage of water molecules in the probe environment, was observed to decrease with increasing n. These conclusions can be tested by experiments with quenchers such as the silver ion Ag^{+33-35} that are present in the aqueous phase. Such an ion then quenches the fluorescence of the pyrenyl end groups of the probe molecules, when these are present at the aqueous micellar interface.

$$^1\!\overset{*}{Py}(CH_2)_n COOH + (Ag^+)_{H_2O} \xrightarrow{k_1} {}^3\!\overset{*}{Py}(CH_2)_n COOH + (Ag^+)_{H_2O} \qquad (2)$$

It is then expected that the quenching efficiency of the silver ions on the fluorescence of the Py(CH$_2$)$_n$COOH probes (i.e. the value of the rate constant k_1) should decrease with increasing n. This rate constant k_1 can be determined by measuring the fluorescence decay time of the probe molecules in the presence $(1/\tau)$ and absence $(1/\tau_o)$ of the quencher at a given concentration [Ag$^+$]:

$$k_1[Ag^+] = 1/\tau - 1/\tau_o \qquad (3)$$

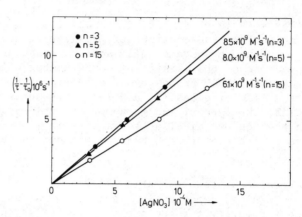

Figure 8. Flourescence quenching by AgNO₃ of the series of probe molecules Py(CH₂)ₙCOOH solubilized in aqueous micellar solutions of sodium dodecyl sulfate (SDS, 0.1 M) at 30°C. The difference in the reciprocal lifetime of Py(CH₂)ₙCOOH in SDS solutions with $(1/\tau)$ and without $(1/\tau_0)$ AgNO₃, has been plotted as a function of the quencher concentration.

The decay time difference $(1/\tau - 1/\tau_0)$ is plotted, in Figure 8, against the quencher concentration [AgNO₃], for three probe molecules Py(CH₂)ₙCOOH (n=3,5 and 15) in a micellar solution of SDS at 30°C. In all cases the linear plots pass through the origin. The slopes of the lines give the value of the rate constants k_1 (Equation 3) for the different probe molecules.

The values of k_1, obtained in this manner, for the series of Py(CH₂)ₙCOOH probes with different n (and for Py(CH₂)₁₂N(CH₃)₃Br), solubilized in sodium dodecyl sulfate micelles (0.1 M SDS in water, at 30°C) are collected in Table III. Also values for $k_1/k_1(3)$, the ratio of the quenching rate constants of the particular probe molecules and that of Py(CH₂)₃COOH, are presented.

It is seen (Table III) that Ag$^+$ quenches the pyrenyl fluorescence irrespective of the number of methylene groups n, from n=3 to 15. However, the value of k_1 decreases with increasing n, having the largest value $(8.5 \times 10^9 M^{-1}s^{-1})$ for Py(CH₂)₃COOH and for MePy, a value which is not much smaller than that measured in H₂0 $(11.2 \times 10^9 M^{-1}s^{-1})$ at the same temperature.[16] This is under-

Table III. Fluorescence Quenching of Pyrenyl Probe Molecules with AgNO$_3$ in Sodium Dodecyl Sulfate Micelles (SDS, 0.1M in H$_2$O) at 30°C.

Probe Molecule	k_1[a] $(10^9 M^{-1} s^{-1})$	$k_1/k_1(3)$[b]
Py$(CH_2)_3 COOH$[c]	8.5	1.00
Py$(CH_2)_5 COOH$[c]	8.0	0.95
Py$(CH_2)_9 COOH$	7.5	0.88
Py$(CH_2)_{11} COOH$	6.5	0.76
Py$(CH_2)_{15} COOH$[c]	6.1	0.72
Py$(CH_2)_{12} N(CH_3)_3 Br$	6.8	0.80
1-methylpyrene	8.5	1.00

a) $k_1 = (1/\tau - 1/\tau_0)/[Ag^+]$: see text, Equation 3 and Figure 8.
b) the ratio of the quenching rate constants of the particular probe molecule and that of Py$(CH_2)_3 COOH$.
c) data refer to micellar solutions at pH 9.5.

standable, as Py$(CH_2)_3 COOH$ appears (see above) to be solubilized in the micellar surface region, which contains a large number of water molecules. The pyrenyl moiety is then directly accessible to the Ag$^+$ ions. The decrease of the quenching efficiency for the pyrenyl probes with increasing alkane chainlength, as shown by the decrease in the ratio $k_1/k_1(3)$ (Table III), signifies that the mean location of the probe molecules goes further away from the micellar surface as n increases. This supports the conclusions regarding the probe location derived in the preceding sections.

The fact that even the fluorescence of Py$(CH_2)_{15} COOH$ is quenched, although the direct environment of its pyrenyl moiety is practically free of water molecules and hence of Ag$^+$ ions (section 3d), means that the pyrenyl end group present in the micellar interior (int) can very well diffuse towards the micellar

surface (surf) within its singlet excited state lifetime (235ns).
It follows immediately from the formula[27] $\bar{x} = \sqrt{2Dt}$ that the pyrenyl
end group can diffuse over a mean distance x of 2 nm (the approximate radius of a SDS micelle[36]) in 25 ns, assuming a diffusion coefficient D of $8 \times 10^{-7} cm^2 sec^{-1}$ at $30°C^{[37,38]}$. This leads to the
following mechanism for the quenching reaction:

$$(^1\overset{*}{Py}(CH_2)_n COOH)_{int} \underset{k_{-d}}{\overset{k_d}{\rightleftharpoons}} (^1\overset{*}{Py}(CH_2)_n COOH)_{surf} \xrightarrow{k_q(n)[Ag^+]} \quad (4)$$

$$(^3\overset{*}{Py}(CH_2)_n COOH)_{surf}$$

The overall quenching rate constant k_1 (mechanism 2) then depends
on the rate constants for diffusion of the probe molecule from the
micellar interior towards the surface (k_d) and vice verse (k_{-d}),
apart from the quenching rate constant per se $(k_q(n))$:

$$k_1 = \frac{k_d\, k_q(n)[Ag^+]}{(k_{-d} + k_q(n)[Ag^+])} \quad (5)$$

As k_d/k_{-d} equals $[Py]_{surf}/[Py]_{int}$, the concentration ratio of the
pyrenyl moiety at the surface and in the micellar interior, the
overall quenching efficiency (described by k_1) will decrease with
decreasing values for the ratio $[Py]_{surf}/[Py]_{int}$, i.e. with
increasing n for the series of probe molecules $Py(CH_2)_n COOH$. This
decrease in $[Py]_{surf}/[Py]_{int}$, when n becomes larger, follows from
the NMR experiments described above (section 3b).

4. CONCLUSION

Aromatic probe molecules such as pyrene or naphthalene probe
only the outer micellar surface region, in a manner not very
different from probe molecules such as $E_T(30)$ that are solubilized
close to the bulk aqueous phase. The fact that such molecules
report a high value for the local polarity (roughly comparable to
methanol) signifies that the surface region (including the α-,
β- and γ-CH_2 groups) contains a considerable amount of water
molecules. These probe molecules, however, do not present information on the nature of the micellar interior.

When the fluorescent probe molecule (i.e. pyrene) is attached
to an amphiphilic molecule with a long alkane chain such as with
$Py(CH_2)_n COOH$ for n=11 and 15, this probe can penetrate into the
interior of SDS micelles revealing the complete absence of water
molecules. The probe location can be ascertained using NMR experiments based on the ring current effect. The pyrenyl moiety of
$Py(CH_2)_n N(CH_3)_3 Br$, however, remains at the surface of CTAB micelles,
in contrast to what is observed for this probe in SDS micelles.

Even probe molecules whose direct environment is completely free of water, as in the case of $Py(CH_2)_{15}COOH$, can be quenched by Ag^+ ions, exclusively present in the aqueous phase. This is understandable, as the pyrene moiety can very well diffuse over distances of 2 nm during its singlet excited state lifetime.

5. REFERENCES

* The results discussed in this paper have been presented (in part) at the VIII. IUPAC-Symposium on Photochemistry, Seefeld (Austria), July 1980 (Book of Abstracts, pp. 144-146).

1. Present address: Institute of Physics, Polish Academy of Sciences, Al. Lotnikow 32-46, 02-668 Warszawa, Poland.
2. C. Tanford, "The Hydrophobic Effect: Formation of Micelles and Biological Membranes", Wiley-Interscience, New York, 1980.
3. F.M. Menger, Acc.Chem.Res., 12, 111 (1979).
4. H. Wennerström and B. Lindman, J.Phys.Chem., 83, 2931 (1979).
5. P. Fromherz, Chem.Phys.Letters, 77, 460 (1981).
6. G.S. Hartley, Kolloid-Z., 88, 22 (1939).
7. W. Philippoff, J.Colloid Sci., 5, 169 (1950).
8. D. Mackay and W.Y. Shiu, J.Chem.Eng.Data, 22, 399 (1977).
9. A. Nakajima, Bull.Chem.Soc. Japan, 46, 2602 (1973).
10. A. Nakajima, Bull.Chem.Soc. Japan, 44, 3272 (1971).
11. K. Kalyanasundaram and J.K. Thomas, J.Am.Chem.Soc., 99, 2039 (1977).
12. P. Lianos and S. Georghiou, Photochem.Photobiol., 30, 355 (1979).
13. K.A. Zachariasse, in "Fluorescent Techniques and Membrane Markers in Cancer and Immunology: Membrane Dynamics, Cellular Characterization and Cell Sorters", P. Viallet, Editor, Elsevier, Amsterdam, in press (1983).
14. J. Ulmius, B. Lindman, G. Lindblom, and T. Drakenberg, J.Colloid Interface Sci., 65, 88 (1978).
15. M. Almgren, F. Grieser, and J.K. Thomas, J.Am.Chem.Soc., 101, 279 (1979).
16. K.A. Zachariasse and B. Kozankiewicz, 1983, in preparation.
17. K.A. Zachariasse, Nguyen Van P., and B. Kozankiewicz, J.Phys.Chem., 85, 2676 (1981).
18. Ch. Reichardt and K. Dimroth, Fortschr.Chem.Forsch., 11, 1 (1968).
19. K. Kalyanasundaram and J.K. Thomas, J.Phys.Chem., 81, 2176 (1977).
20. J.C. Dederen, L. Coosemans, F.C. De Schryver, and A. Van Dormael, Photochem. Photobiol., 30, 443 (1979).
21. F.M. Martens and J.W. Verhoeven, J.Phys.Chem., 85, 1773 (1981).
22. J.C. Russell, D.G. Whitten, and A.M. Braun, J.Am.Chem.Soc., 103, 3129 (1981).
23. F.M. Menger, J.M. Jerkunica, and J.C. Johnston, J.Am.Chem.Soc., 100, 4676 (1978).

24. K.R. Thulborn, in "Fluorescent Probes", G.S. Beddard and
    M.A. West, Editors, pp. 113-141, Academic Press, London, 1981.
25. E. Sackmann, Z. Physik. Chem. (Frankfurt am Main), 101, 391
    (1976).
26. S.S. Atik, M. Nam, and L.A. Singer, Chem.Phys.Letters, 67, 75
    (1979).
27. M.A.J. Rodgers and M.F. Da Silva E Wheeler, Chem.Phys.Letters,
    43, 587 (1976).
28. Probe molecules can in principle lead to a distortion of the
    structure of the micelles.[13] However, for a probe/surfactant
    ratio 1:40 the plot of $\Delta\Delta\sigma$ for the various H atoms gives a
    pattern similar to the one depicted in Figure 2. This supports
    our assumption that the probes do not essentially perturb the
    micellar structure.
29. F.A. Bovey, "Nuclear Magnetic Resonance Spectroscopy",
    pp. 64-71, Academic Press, New York, 1969.
30. The patterns of $\Delta\Delta\sigma$ (Figure 3) represent the product of two
    distributions, the distribution of the probe molecule with
    respect to the various H atoms in the surfactant molecule and
    the distribution of these H atoms between the micellar surface
    and center.[17] The central micellar region is considered to be
    a volume, comprising a large part of the total volume of the
    (dynamic) micelle, down from the $\gamma-CH_2$ group to the center
    (see text).
31. The fact that the decays are single exponential indicates that
    the shortening of the lifetime is not due to the presence of
    probe molecules in the bulk aqueous phase (see Table II).
32. Investigations[16] of this interaction in the system $Py(CH_2)_3COOH$/
    tetraethylammonium chloride (1.4M) in H_2O (pH 8.8, at 30 °C)
    show that the fluorescence decay time of $Py(CH_2)_3COOH$ (146 ns,
    cf. Table II) is not decreased by the presence of the tetra-
    alkylammonium compound per se.
33. T. Saito, S. Yososhima, H. Masuhara, and N. Mataga, Chem.Phys.
    Letters, 59, 193 (1978).
34. M.A.J. Rodgers and M.F. Da Silva E Wheeler, Chem.Phys.Letters,
    53, 165 (1978).
35. J.C. Dederen, M. Van der Auweraer, and F.C. De Schryver,
    J.Phys.Chem., 85, 1198 (1981).
36. M.F. Emerson and A. Holtzer, J.Phys.Chem., 71, 1898 (1967).
37. This value for the diffusion coefficient D of pyrene in a SDS
    micelle (0.1M) at 30°C, is based on the value for D in hexane
    (2.93 x 10^{-5} cm^2sec^{-1}, taken from Reference 38) and the
    viscosity (19cP at 30°C) of the environment of 1,3-di(1-pyrenyl)
    propane in SDS, reported in Reference 13. The diffusion co-
    efficient of the pyrenyl end group in $Py(CH_2)_nCOOH$ will be
    smaller than that of pyrene.
38. E.G. Meyer and B. Nickel, Z.Naturforsch., 35A, 503 (1980).

SOLUBILIZATION AND WATER PENETRATION INTO MICELLES AND OTHER ORGANIZED ASSEMBLIES AS INDICATED BY PHOTOCHEMICAL STUDIES[1]

David G. Whitten, João B. S. Bonilha, Kirk S. Schanze,
and Joseph R. Winkle
Department of Chemistry
University of North Carolina
Chapel Hill, North Carolina 27514

This paper presents a summary of studies using three kinds of photochemical and photophysical probes to monitor solubilization sites and related phenomena in micelles and vesicles. The three probes include a functionalized surfactant ketone whose photoreactivity is strongly dependent upon medium viscosity and polarity, a series of p-donor-p'-acceptor substituted azobenzene dyes whose absorption spectra and thermal $cis \rightarrow trans$ isomerization rates are strongly influenced by solvent polarity and protic-aprotic nature, and a luminescent extramicellar metal complex whose quenching can be used to monitor the binding of certain cations to anionic surfactant assemblies. All of these probes lead to a consistent picture of micelle structure and the nature of the solubilization of polar organic substances by surfactants in aqueous organized assemblies. The studies indicate that there is considerable water penetration into supposedly hydrophobic regions of the assemblies and suggest that elimination of hydrocarbon-water contacts is a major driving force in the solubilization of many substances of moderate polarity.

INTRODUCTION

Much recent work has focused on the use of photochemical reactivity to explore the structure and properties of surfactant assemblies.[2-12] Probably the property of these assemblies of greatest importance is their ability to solubilize a wide variety of disparate reagents, often to relatively large effective concentrations. Some of the major questions under active investigation concern the nature of the solubilization process and the binding sites for different substrates. In our studies we have made use of a variety of photochemical probes; in particular we have synthesized and investigated a number of functionalized surfactant molecules containing reactive groups in the hydrophobic portion of amphiphilic molecules.[7,9,13-15] We have attempted to use the reactivity of these molecules to assess the properties of the "hydrophobic" portions of various surfactant assemblies including micelles, "swollen" micelles, microemulsions, and vesicles.[16] In the present paper we review some recent work from our laboratories employing three different types of photochemical probes. The first part discusses some studies using as an "intrinsic" probe the surfactant ketone 1. The second and third parts of this paper focus on nonsurfactant reagents 2-4. While some of these investigations have been described independently in other papers,[7,17,20] the current paper reviews the results of these studies together in such a way that a rather clear picture of micellar structure and the solubilization process for hydrophobic molecules emerges.

CH_3-⟨benzene⟩-$\overset{O}{\overset{\|}{C}}$-$(CH_2)_n$-$\overset{O}{\overset{\|}{C}}$-OH

1   a: n = 14
     b: n = 10

$$\left[\begin{array}{c} \end{array} \right]^{4-}_3$$

2

CH_3-$\overset{+}{N}$⟨pyridine⟩-⟨pyridine⟩$\overset{+}{N}$-CH_3

3

$\underset{C_2H_5}{\overset{C_2H_5}{N}}$-⟨benzene⟩-N=N-⟨benzene⟩-NO_2

4

Norrish Type II Reactivity of a Surfactant Ketone in Micelles and Vesicles

Among the most widely studied of all photoreactions is the Norrish Type II photoelimination of ketones possessing a γ-hydrogen atom as shown in Equation (1). The quantum yield of this reac-

$$\qquad (1)$$

tion is strongly dependent on medium viscosity and polarity.[21] Thus in highly rigid media such as polymers below the glass transition temperature, crystals, or supported multilayers, many carbonyl containing segments are unable to attain the required configuration for reaction and hence the diradical intermediate generated by intramolecular hydrogen atom abstraction cannot be formed.[7,22] In these cases the quantum yield is low and only a small fraction of the total molecules reacts. In less rigid media where the requisite geometry for intramolecular hydrogen atom abstraction can be attained the quantum yield increases and in most cases the efficiency of forming the biradical is believed to be near unity.[21]

However, even under conditions where the efficiency of formation of the biradical is unity, the overall quantum yield is much lower due to the reverse hydrogen abstraction which regenerates the ground state of the starting material. In the presence of polar solvents which can serve as hydrogen bond acceptors the biradical is stabilized and the quantum yield for the cleavage products approaches unity.[21,23] Thus for a typical aryl ketone, n-butyrophenone, the quantum yield in benzene is 0.3 while in t-butyl alcohol it climbs to 1.0.[21]

Our initial studies with the surfactant ketones 1a and 1b showed relatively low reactivity (ϕ^{II}(1a) in benzene = 0.2) in fluid hydrocarbons similar to their non-surfactant counterparts.[7] In t-butyl alcohol or water the quantum yields for 1a and 1b approach unity. In studies with 1a in sodium dodecyl sulfate (SDS) micelles it was found that the quantum yield was 0.80, only slightly lower than in t-butyl alcohol; similar results were obtained with octanophenone and 1b.[18] The results with 1a were regarded as especially significant since this ketone is completely water-insoluble in the absence of surfactant. More recent work with 1a has included a study of its reactivity over a wide temperature range in several different surfactant solutions.[18] Table I compares its reactivity in aqueous solutions of SDS, cetyl trimethylammonium chloride (CTAC), the vesicle forming dioctadecyl-dimethylammonium chloride (DODAC), and dipalmitoyllecithin (DPL). These results show striking contrasts between the behavior of the ketone in homogeneous solution,

Table I. Comparison of Type II Photoelimination Quantum Yields
for Surfactant Ketone $\underline{1a}$ in Different Aqueous Surfactant Solutions.[a]

| T, °C | ϕ_{II} | | | |
	SDS[b]	CTAC[c]	DODAC[d]	DPL[d]
20–28			0.22±0.02	
30	0.81±0.01	0.72±0.01	0.27±0.02	0.28±0.03
32			0.31±0.03	
36			0.33±0.03	
40	0.81±0.01	0.72±0.01	0.34±0.03	
45			0.40±0.03	0.38±0.03
50	0.79±0.01	0.73±0.01		
60	0.79±0.01	0.74±0.02		

[a]Irradiations at 254 nm. [b]3.22×10^{-4} M ketone in 0.02 M surfactant.
[c]1.25×10^{-4} M ketone in 6×10^{-3} M surfactant. [d]1:20 ketone/sur-
factant.

the micellar media and the vesicles. In the latter case there are
pronounced differences above and below the phase transition temperatures. In Table II are listed values of the quantum yield for
the type II photoelimination for $\underline{1a}$, $\underline{1b}$, and the nonsurfactant ketone octanophenone in different media. Taken together these results
are interpretable as follows. The solution phase behavior of the
surfactant ketones $\underline{1a}$ and $\underline{1b}$ is similar to that of "normal" alkyl-
aryl ketones and shows high sensitivity to the medium environment
with high quantum efficiences in polar protic solvents.

Table II. Quantum Yields for Type II Photoelimination for Ketones
$\underline{1a}$, $\underline{1b}$, and Octanophenone in Different Media.

| | ϕ_{II} | | |
	Benzene	SDS	t-Butyl Alcohol
$\underline{1a}$	0.27	0.81	1.00
$\underline{1b}$	–	0.84	1.00
octanophenone	0.29[a]	0.76	1.00[a]

[a]Value taken from reference 4.

The results for micellar solutions of all the ketones studied are most consistent with location of the ketone chromophore in a highly polar environment capable of forming hydrogen bonds with the O-H group of the biradical intermediate. This could be accounted for either by location of the ketone chromophore near the micelle-water interface or to substantial penetration of water into the vicinity of the carbonyl group if it resides in the micelle interior. This point will be returned to later. In contrast to the micelle solutions the vesicles give results consistent with residence of the ketone in a much more hydrophobic site. In general, the behavior in the vesicles seems fairly straight-forward. Above the phase transition temperature (36° for both DODAC and DPL)[24] both vesicle systems show quantum yields rather higher than in benzene solution but well below micelles and alcohol solutions. No other products than the cleavage products have been detected and there is no bleaching of the carbonyl chromophore transition upon irradiation as might be anticipated if intermolecular hydrogen abstraction leading to photoreduction were occurring. The behavior here is thus consistent with efficient formation of the biradical in a non-polar, non-hydrogen bonding environment. Hence the ketone reports an environment in which there is considerable freedom of motion of the hydrocarbon chain but relatively little water content, at least in the interior where the ketone would be expected to reside. In contrast the significantly lower quantum yields for photoelimination below the phase transition temperature are probably most consistent with a hydrophobic environment but one which is somewhat more rigid. Since even at the low temperatures we find no evidence for other photoproducts or reactions which might be anticipated if the excited carbonyl were prevented from abstracting the γ-hydrogen intramolecularly, it is not clear that the lower quantum yields are due to a reduction in the efficiency of forming the biradical. While this could be the case, the lower quantum yields could also be accounted for by efficient formation of the biradical in a more restrictive environment which provides enhanced return to the starting material in what would be analogous to a solvent "cage" effect.

As mentioned above, the micelle behavior of the different ketones is consistent with location of the carbonyl in each case in a relatively hydrophilic environment. The big question here centers around the carbonyl location within the micelle and whether or not it is a valid "probe" of the micelle interior. Since previous studies with similar functionalized surfactants containing the stilbene chromophore in a similar position to the aryl ketone in 1a and 1b also yield a picture of the micelle interior as being relatively accessible to hydrophilic reagents,[14] it is tempting to suggest the ketone results are not inconsistent with it residing "within" the micelle. Nonetheless, it is certainly true that the aryl ketone chromophore might be sufficiently polar to seek out a surface site. To test this, we carried out studies with the three ketones

1a, 1b, and octanophenone in SDS micelles in the presence of
Eu^{3+}.[4,18] As described below and elsewhere, inorganic ions such
as Cu^{2+} and Eu^{3+} bind strongly to anionic assemblies; however their
binding occurs via an ion exchange process and thus they are associated largely with the anionic head groups at the micelle-water interface.[19,20] Eu^{3+} quenches the type II elimination of aryl-alkyl ketones with a rate constant, $k_q \sim 8 \times 10^9 M^{-1}s^{-1}$, near diffusion-controlled.[25] Adding Eu^{3+} to micellar solutions of the ketones resulted in very little quenching of the reaction from 1a and 1b,
and no quenching of the reaction of octanophenone as shown in Table
III. These results are striking and suggest rather strongly that
neither in the case of the surfactant ketones 1a and 1b nor in that
of the non-surfactant octanophenone does the carbonyl chromophore
spend much time in the region of the head groups. Thus even for
the case of 1a where a maximum of 20% quenching is observed with ca
one Eu^{3+} per micelle for an effective local concentration of ca.
0.07 M, the quenching is much less than the value of 90% quenching
which would be obtained in a homogeneous aqueous solution. The
differences in the quenching observed for the three different ketones are probably due to the mismatch in chain lengths for 1a and
SDS; this also may be reflected in the slightly higher quantum
yields for 1a in SDS compared to octanophenone in the same medium
and 1a in CTAC.

The two results for the ketones, taken together, suggest then
that in the micelles the ketone resides in the interior, relatively
remote from the charged head groups but nonetheless in a site which
is relatively polar and thus probably easily accessible at any instant, to water. As pointed out previously similar behavior has been
observed with other "intrinsic" probes such as the surfactant stilbenes[14] so that a consistent picture emerges from these studies of
a micelle structure relatively open with extensive hydrocarbon-water
contact.

Table III. Quenching of the Type II Photoelimination of SDS Solutions of Ketones 1a, 1b, and Octanophenone by Eu^{3+}.

Ketone[a]	ϕ_{II}^o	$\phi_{II} [Eu^{3+}]=2\times10^{-4}M$	% Quenching
1a	0.81±0.01	0.66±0.04	18
1b	0.84±0.02	0.76±0.03	10
Octaphenone	0.76±0.03	0.75±0.02	~0

[a]Ketone concentration $3.22 \times 10^{-4}M$; [SDS]= 0.02 M in each case.

Ion-Binding to Micelles and Vesicles

Electron transfer quenching of luminescent excited states has become a very well established and widely investigated phenomenon. A particularly well investigated system is that involving the luminescent cation tris(2,2'-bipyridine)ruthenium(II)$^{2+}$ and the organic quencher-oxidant N,N'-dimethyl-4,4'-bipyridine^{2+} (methyl viologen) (3).[26-28] Quenching of the excited state of the ruthenium complex occurs at nearly diffusion-controlled rates and is accompanied by conversion of 3 to its one-electron reduced product. In a number of investigations it has been established that this is a general phenomenon which can occur under a variety of conditions including aqueous surfactant solutions.[27,28] The work relevant to the present discussion involves studies using the anionic ruthenium complex 2 and anionic assemblies with 3 and other cations as quenchers for the luminescent excited state.[20] Since 2 is anionic, and relatively hydrophilic, it does not associate with anionic assemblies. Its excited state lifetime and luminescence spectra are unaffected by the addition of anionic surfactants. In the absence of surfactants the luminescence of 2 is strongly quenched by 3 and inorganic cations such as Cu^{2+} or Eu^{3+}. The quenching is partially static and partially diffusional; by addition of moderate amounts of Na^+ the quenching can be rendered totally diffusional.[19,20] Addition of anionic surfactants attenuates the luminescence quenching of the various cations by binding the cations to the surfactant assemblies.[20] Since the luminescence intensity provides a sensitive measurement of the concentration of quencher-cation in the aqueous phase, the concentration of bound quencher can be readily determined and the equilibrium constant for the reaction given by Equation 2 evaluated, where SA refers to the surfactant assembly. This

$$Q_{(H_2O)} \quad \underset{\longleftarrow}{\overset{K_A}{\longrightarrow}} \quad Q_{(SA)} \qquad (2)$$

has been done for several anionic micelles and for dicetylphosphate vesicles. Table IV lists some values determined for K_A for 3 and Cu^{2+} at different temperatures together with values of $\Delta H°$ and $\Delta S°$ estimated from the temperature dependence.

Several interesting results emerge from these studies. Discussing first the micellar systems, it is found that K_A increases with surfactant chain length for Q = 3 but is more or less constant for Cu^{2+}. For the micellar systems binding of 3 is characterized by a small value of $\Delta H°$ and a relatively larger positive $\Delta S°$; clearly the binding of 3 is favored mainly by the entropic term. The reverse is true for Cu^{2+}; here the binding is favored by $\Delta H°$ and the thermodynamic parameters are typical of that observed for a "classic" association equilibrium governed by coulombic attraction or donor-acceptor interactions. The thermodynamic parameters observed for binding of 3 to the micellar assemblies are similar to those observed for the micellization process of several surfactants

Table IV. Equilibrium Constants and Thermodynamic Quantities for
Binding of Cations to Anionic Surfactants in Water.

Surfactant	T, °C	K	$\Delta H°$ kca/mol	$\Delta S°$ eu
Binding of 3				
Sodium Decyl Sulfate	25°	2100±100	4.9	32±4
	35°	2750± 50		
Sodium Dodecyl Sulfate	25°	4150±300	3.6	29±1
	35°	4910±300		
	45°	6120±100		
Sodium Tetradecyl Sulfate	35°	6540±700	-0.7	15±1
	45°	6580±1000		
Sodium Dicetyl Phosphate	20°	5890±250	5.8	37
	25°	6980±300		
	30°	8190±350		
Binding of Cu^{2+}				
Sodium Dodecyl Sulfate	25°	1980	-5.6	-3.6
	35°	1640		
	45°	1380		
Sodium Tetradecyl Sulfate	35°	1870	-6.8	-7.0
	45°	1650		

and suggest that here a "hydrophobic effect" may also be operating.[29]
The results suggest strongly that organic cations such as 3 and inorganic cations such as Cu^{2+} or Eu^{3+} bind at different sites and that
the binding is governed by rather different forces.

A reasonable interpretation of these results is that Cu^{2+} and
Eu^{3+} bind by a conventional attraction to the anionic head groups
in an ion-exchange type process. In contrast 3 (and by inference
many other organic ions)[30] binds by what could be defined as a hydrophobic interaction in which the organic ion penetrates more
deeply into the assembly and releases water from regions where hydrocarbon-water contact exists. The release or expulsion of water
on binding of the organic cation would account for the positive entropy and the driving force for the binding.

The results obtained thus far concerning binding of 3 to anionic vesicles suggest that similar factors may operate here as
well. The only system investigated thus far is anionic dicetylphosphate in the low temperature phase (the phase transition temperature for DCP vesicles is 46 °C).[31] In the case of the vesicles
it was not possible to study binding of Cu^{2+} since addition of even

small amounts of Cu^{2+} caused precipitation of the surfactant. With 3, perhaps surprisingly, the equilibrium constants for binding on the outside only (the vesicles were not sonicated on addition of 3) are in the same range as those for sodium hexadecyl sulfate. The thermodynamic parameters calculated for the three temperatures stud- ied--20°, 25°, and 30°--suggest once again an entropically control- led binding which could be attributed to hydrophobic binding con- current with water expulsion. Since some of our studies with sur- factant stilbenes and methyl viologen suggest relatively little in- teraction between the two in the low temperature phase of DCP,[32] it is possible that the hydrophobic bonding observed here is occur- ring only at the outer portion of the surfactant chains where cur- vature of the vesicle may generate substantial water-hydrocarbon contacts.

Azobenzene Dye Isomerization as a Probe of Solubilization Phenomena in Surfactant Assemblies

The final "photochemical probe" discussed in this paper con- cerns a thermal isomerization of some substituted azobenzene dyes which is initiated and studied by a flash photolysis technique.[17] The dyes studied are p-donor-p'-acceptor azobenzenes of which 4 has been most widely investigated. The cis or syn isomer of 4 can be produced photochemically and its relatively rapid isomerization (Equation 3) can be monitored spectroscopically. This reaction has

$$(3)$$

been studied in homogeneous solution for a wide variety of solvents and it has been found that the reaction is strongly solvent-depen- dent.[17,33] The key features of the reaction which make it especial- ly useful as an indication of solvent properties are its indepen- dence of solvent viscosity and its strong dependence on solvent pol- arity and protic-aprotic character. The independence of viscosity has been verified in both nonpolar and polar solvents.[34] The depen- dence on solvent polarity and protic-aprotic nature can be demon- strated in a number of ways and explained in terms of the isomeriza- tion mechanism.

In initial studies it was found that the absorption spectrum of the trans isomer of 4 is strongly red shifted with increase in solvent polarity.[33] The rate constant for *cis→trans* isomerization, $k_{c \to t}$, also shows an increase with increasing solvent polarity. Both of these phenomena can be accounted for in terms of increasing contributions of the dipolar form (5) shown in Equation 4. While most

$$R_2N\text{-}\langle\text{phenyl}\rangle\text{-}N=N\text{-}\langle\text{phenyl}\rangle\text{-}NO_2 \longleftrightarrow R_2\overset{+}{N}=\langle\text{phenyl}\rangle=N\text{-}N=\langle\text{phenyl}\rangle=\overset{+}{N}\overset{O^-}{\underset{O^-}{}} \tag{4}$$

5

simple azobenzenes isomerize by a mechanism involving inversion through one of the azo nitrogens,[35] the p-donor-p'-acceptor azobenzenes evidently isomerize by rotation around the nitrogen-nitrogen bond and the increase in importance of contributing structure 5 in polar solvents reduces the bond order and hence the activation energy $\Delta G^{\ddagger}_{c \to t}$. Attempts to correlate the spectra with $\Delta G^{\ddagger}_{c \to t}$ with quantitative measures of solvent polarity lead to interesting and instructive results. While several correlations can be made,[36] the clearest results obtained thus far involve the use of the Kosower "Z" values[37] and $\Delta G^{\ddagger}_{c \to t}$. When $\Delta G^{\ddagger}_{c \to t}$ for 4 (and the several other dyes having similar donor-acceptor substitution) is plotted vs Z (Figure 1) two linear relationships are clearly

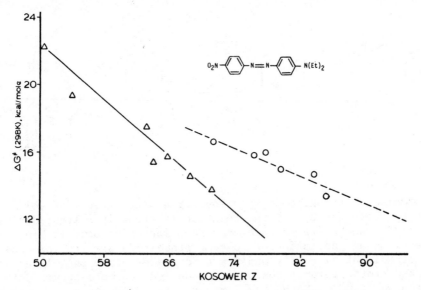

Figure 1. Plot of $\Delta G^{\ddagger}_{c \to t}$ for dye 4 vs Kosower "Z" values. The line at the upper right is for protic solvents, the line at the left is for aprotic solvents.

discernible.   Thus there is one good linear correlation for aprotic
solvents and another for protic solvents.[17]   The two plots show
rather clearly that for two solvents at comparable Z, the isomerization reaction is much faster for the aprotic solvent.   Although a
number of reasons could be advanced for this behavior, a simple explanation is that protic solvents hydrogen bond to the amino nitrogen lone pair (6) and thus reduce the importance of the dipolar
structure 5.

$$S-H$$

$$R_2\overset{\cdots}{N}-\!\!\langle\bigcirc\rangle\!\!-N\!\!=\!\!N\!\!-\!\!\langle\bigcirc\rangle\!\!-NO_2$$

6

     Since dye 4 is water insoluble but can be solubilized in a
variety of aqueous surfactant solutions, its isomerization in these
media offers an interesting probe which in many ways complements
the previously discussed results.   Table V lists $\Delta G^{\ddagger}_{c\to t}$ values for
4 in four different aqueous surfactant solutions together with the
Z values extrapolated from the plot in Figure 1.   Since there are
two lines, there are two extrapolated values of Z for each solution, one for a protic medium and the other for an aprotic microenvironment.   The "protic" values range near 94, the value for H_2O,
while the "aprotic" values are in the range 70-75.   At first inspection it might seem that the "protic" value should apply, particularly in view of the fact that the solutions are primarily water and
much evidence has already been presented that considerable water
penetration into simple micelles occurs.   However studies of charged
amphiphilic alkylpyrdinium iodides (containing the "Kosower" chromophore) indicate that these ions experience a microenvironment
having a polarity corresponding to Z=70-80 in micelles and 65-75 in
phospholipid vesicles[38-40].   These charge molecules would scarcely be
expected to reside in sites less polar than the neutral dye 4.
Further, extrapolation of the surfactant solution $G_{c\to t}$ values for
4 using the aprotic line yields Z values of 70-75; these are in
agreement with the Z values determined using the alkyl pyridinium
iodides.   Thus, we deduce that 4 may reside in a relatively aprotic
environment in the surfactant solutions.

Table V. Free Energies of Activation and Extrapolated "Z" Values
for Dye 4 in Aqueous Surfactant Solutions.

Surfactant	$\Delta G^{\ddagger}_{c\to t}$	"Z"/protic	"Z" aprotic
SDS	12.2	94	75
CTAB	12.4	92	74
Brij-35	14.3	83	70
DODAB	13.0	90	72

SUMMARY

The results of these three independent studies present a picture of micellar structure and solubilization phenomena which is quite different from the classical "liquid hydrocarbon solvent" model. Thus the intrinsic probes 1a and 1b (as well as other probes studied elsewhere) show evidence for simple micelles having a relatively "open" structure in which there are extensive hydrocarbon-water contacts or interface regions. The solubilization of reagents such as the organic cation 3 or the polar, but neutral, dye 4 occurs at these "interfacial sites" concurrent with expulsion of water and elimination of unfavorable hydrocarbon water contacts. The solubilization of these reagents which might be regarded as "hydrophobic-hydrophilic" essentially dehydrates a portion of the surfactant assembly. While our results show this phenomenon is especially clear and important in simple micelles, the limited data gathered thus far suggest it may also be important for more ordered bilayer vesicles and perhaps in other surfactant assemblies.

ACKNOWLEDGMENT

We are grateful to the National Science Foundation (Grant No. CHE-8121140) for support of this work. J. B. S. Bonilha (Visiting Scholar from Department of Chemistry, Faculdade de Filosofia Ciencias e Letras de Ribeirão Preto da Universidade de Sao Paulo, 14100-Ribeirão Preto S. P.-Brasil) acknowledges research fellowship support from the Fundacao de Amparo à Pesquisa do Estado de Sao Paulo (FAPESP 16-quimica 80/0292) and research fellowship from the Conselho Nacional de Desenvolvimento Cientifico e Technologico (CNPq 200.555-81-QU).

REFERENCES

1.   "Photochemical Reactivity in Organized Assemblies" 31. Paper 30: G. S. Cox, M. Krieg, D. G. Whitten, submitted for publication.
2.   N. J. Turro, M. Gratzel, and A. M. Braun, Angew. Chem., Int. Ed., Engl., 19, 675 (1980).
3.   M. Almgren, F. Grieser, and J. K. Thomas, J. Am. Chem. Soc., 102, 3188 (1980).
4.   N. J. Turro, K.-C. Liu, M. F. Chow, Photochem. Photobiol., 413, 26 (1978).
5.   R. Breslow, S. Kitabatake, and J. Rothbard, J. Am. Chem. Soc., 100, 8156 (1978).
6.   M. F. Czarniecki and R. Breslow, J. Am. Chem. Soc., 101, 367 (1979).
7.   P. R. Worsham, D. W. Eaker, and D. G. Whitten, J. Am. Chem. Soc., 100, 7091 (1978).
8.   M. W. Geiger and N. J. Turro, Photochem. Photobiol., 26, 221 (1977).

9. J. C. Russell, S. B. Costa, R. P. Seiders, and D. G. Whitten, J. Am. Chem. Soc. 102, 5678 (1980).

10. K. A. Zachariasse, Chem. Phys. Lett., 57, 429 (1978).

11. K. A. Zachariasse, Chem. Phys. Lett., 73, 6 (1980).

12. N. J. Turro, M. Aikawa, and A. Yekta, J. Am. Chem. Soc., 101, 771 (1979).

13. D. G. Whitten, D. W. Eaker, B. E. Horsey, R. H. Schmehl, and P. R. Worsham, Ber. Bunsenges Phys. Chem., 82, 858 (1978).

14. J. C. Russell, D. G. Whitten, and A. M. Braun, J. Am. Chem. Soc., 103, 3219 (1981).

15. F. H. Quina, and D. G. Whitten, J. Am. Chem. Soc., 97, 1602 (1975).

16. J. C. Russell and D. G. Whitten, J. Am. Chem. Soc., in press.

17. K. S. Schanze, T. F. Mattox, and D. G. Whitten, J. Am. Chem. Soc., 104, 1733 (1982).

18. J. R. Winkle and D. G. Whitten, manuscript in preparation.

19. J. B. S. Bonilha, T. K. Foreman, and D. G. Whitten, J. Am. Chem. Soc., in press.

20. T. K. Foreman, W. M. Sobol, and D. G. Whitten, J. Am. Chem. Soc., 103, 5333 (1981).

21. P. J. Wagner, Accts. Chem. Res., 4, 168 (1971).

22. G. H. Hartley, and J. E. Guillet, Macromolecules 1, 165 (1968).

23. N. J. Turro, K.-C. Liu, M. F. Chow, Photochem. Photobiol., 413, 26 (1978).

24. J. H. Fendler, Accts. Chem. Res., 13, 7 (1980).

25. R. D. Samll, Jr., and J. C. Scaiano, J. Phys. Chem., 81, 2126 (1977)

26. C. R. Bock, T. J. Meyer, and D. G. Whitten, J. Am. Chem. Soc., 96, 4710 (1974).

27. R. H. Schmehl and D. G. Whitten, J. Am. Chem. Soc., 102, 1938 (1980).

28. D. G. Whitten, Accts. Chem. Res., 13, 83 (1980).

29. C. Tanford, "The Hydrophobic Effect: Formation of Micelles and Biological Membranes," 2nd Ed., Wiley Interscience, New York, 1980.

30. D. G. Whitten, R. H. Schmehl, T. K. Foreman, J. Bonilha, and W. M. Sobol, ACS Symposium Ser., 177, 37 (1982).

31. J. R. Escabi-Perez, A. Romero, S. Lukac, and J. H. Fendler, J. Am. Chem. Soc., 101, 2231 (1979).

32. D. G. Whitten, J. C. Russell, T. K. Foreman, R. H. Schmehl, J. Bonilha, A. M. Braun, and W. Sobol, in "Chemical Approaches to Understanding Enzyme Catalysis: Biomimetic Chemistry and Transition State Analogs," Eds. B. S. Green, Y. Ashani, and D. Chipman, Editors, p. 66, Elsevier Pub.Co., Amsterdam, 1982.

33. P. D. Wildes, J. G. Pacifici, G. Irick, and D. G. Whitten, J. Am. Chem. Soc., 93, 2004 (1971).

34. K. S. Schanze, unpublished results.

35. H. Rau and E. Lüddecki, J. Am. Chem. Soc., 104, 1616 (1982), and references therein.

36. D. G. Dong and M. A. Winnik, Photochem. Photobiol., 35, 17 (1982).

598 D. G. WHITTEN ET AL.

37. E. M. Kosower, J. Am. Chem. Soc., $\underline{80}$, 3253 (1958).
38. A. Ray and P. Mukerjee, J. Phys. Chem., $\underline{70}$, 2138 (1966).
39. P. Mukerjee and A. Ray, J. Phys. Chem., $\underline{70}$, 2144 (1966).
40. E. Südholter, J. Engberts, and D. Hoestra, J. Am. Chem. Soc., $\underline{102}$, 2457 (1980).

CRITIQUE OF WATER PENETRATION STUDIES IN MICELLES USING EXTRINSIC PROBES

K.N.Ganesh,P. Mitra and D.Balasubramanian*

Centre for Cellular and Molecular Biology

Hyderabad 500 009, India

The sites of solubilization of some classes of 'water penetration probes' in micelles have been studied. These probes, which invariably contain polar or aromatic groups in them, are shown to be inherently interfacially active. The effect of such an activity is seen to be enhanced when the probes are solubilized in systems of large surface area-to-volume ratios such as amphiphile aggregates. NMR results reveal that these 'water penetration probes' are located largely in the headgroup region of micelles. They would thus be expected to monitor the polarity and water content of this region rather than those of the micellar interior. Thus, studies using such probes do not necessarily prove the existence of water inside micelles and related assemblies.

INTRODUCTION

The current debate on the issue of micelle structure and the extent of water penetration into the body of these aggregates is relevant in a larger perspective that encompasses the structure and properties of other supramolecular assemblies of amphiphiles and lipids. These include reverse micelles, lamellae, bilayers and liposomes, and biological membranes - since the general principles of organization of all these are thought to be similar in nature.[1-3] The presence of significant amounts of water inside micelles would imply similar situations in bilayers, and in artificial and natural membranes as well. Yet one finds the rates of self-diffusion of water in micelles, and of water diffusion among reverse micelles and across lipid bilayers to be rather low.[4] The permeability of water across artificial lipid membranes, in the absence of pores, channels or carriers, is also quite low.[5] These results argue for the presence of very little water in the main body of these aggregates. It is thus of relevance to know whether a micelle is structurally "an oil drop with an ionic or polar coat"[4,6-8], or "a porous cluster with substantial water in the inside"[9-15], or a "surfactant block assembly"[16], or any other.

Structural investigations on amphiphile aggregates often make use of external additives as solubilized probes that monitor the polarity, microviscosity and similar microenvironmental features. Since such molecules report on the properties of their "cybotactic" regions[17], it is important to establish the actual site of occupation of the added probe in the host aggregate. A probe that is largely located in the core of a micelle would be expected to report on the properties of the core, while one that is positioned in the headgroup region would largely monitor the properties of this region. It is our belief that much of the conflicting interpretations about the structures of micelles, based on the results using external probes, is due to a lack of appreciation of this point.

Table I lists several extrinsic probes that have been recently used in studies on micelle structure. It is worthwhile to note that all these molecules contain polar groups and/or conjugated π-electron systems. There is evidence in literature that benzene and higher arenes are located largely in the headgroup region of ionic micelles.[18-24]. Mukerjee and coworkers[18,19] have shown that benzene is mildly interfacially active, and have suggested that the mild surface activity of the aromatic or polar solubilizates tends to be greatly amplified when placed in micelles and related assemblies whose surface area per cm^3 would be as large as 10^6-10^7 cm^2, leading to a large fraction of the solubilized molecules accumulating at interfacial sites.

Table I. Some typical "water penetration probes" of micelles, compiled from recent literature.

Probe	Method	Ref.
1. Aromatic probes		
acridine	fluorescence quantum yield	25
1-methyl indole	fluorescence quantum yield and lifetime	26
pyrene and derivatives, related arenes (emitters); aromatic quenchers	fluorescence hetero excimer emission	27
methyl viologen (acceptor); 3-methyl indole, pyrene, BDNH (donors)	charge transfer spectra	28
2. Keto probes		
octanal, 1-naphthaldehyde, dihexyl ketone	^{13}C NMR	10
8-keto HTAB	"	10
ω-keto dodecanoates	"	29
(+) trans-2-chloro-5-methyl cyclohexanone; 17α-hydroxy-5α-pregnan 20-one	ORD	13
steroid enones	reduction by aq. BH_4^-	14

In light of these observations, it becomes important to
establish where the water penetration probes are located when they
are solubilized in micelles. Are these probes inherently surface
active? Do they position themselves near the headgroup region of
micelles, or are they located in the micellar core? Does the
polarity or the water content that they sense reflect that of the
micellar surface, core, or an average of the two environments? How
far do its properties reflect the dynamic features - such as probe
translocation between sites, and differential segmental protrusion
as suggested by Aniansson[30]- of surfactant monomers in the
micelles? We describe some experiments that attempt to provide
answers to some of these questions. First, we have looked at the
issue of the sites of solubilization of some representative aromatic optical water penetration probes by the NMR method described
earlier[20-24]. The results suggest that all these probes are
largely located in the headgroup region of the host micelles.
Next, we have measured the interfacial activity of some ketone
probes by tensiometry and by NMR methods, and find that these
polar molecules are also surface-active and are positioned in the
headgroup regions of micelles. Based on all these results, we
conclude that most of the water penetration probes used to date
in the study of micelle structure are located largely in the
headgroup region of the micelles in which they are solubilized,
and thus report on the polarity and the water content of this
region and not necessarily of the micellar interior.

RESULTS AND DISCUSSION

I. Aromatic optical probes

The method used to establish the sites of solubilization were
basically the same adopted earlier[20-24], wherein the ring-current-
induced shielding and deshielding effects produced by the added
aromatic probe on the NMR signals of the protons in various segments of the surfactant molecules are followed in the micellar
phase. Acridine, which has been used as a water penetration probe
by Wolff[25], and 1-methyl indole which has been used by Turro
et al.[26] for a similar purpose were used as representative solubilizates in cationic (CTAB, DTAB), anionic (SDS) and nonionic
(Brij 58, Octyl glucoside (OG)) micelles. Benzophenone, which is
thought to be located near the headgroups in micelles[31], was also
studied for comparison. The results of the study, presented in
detail elsewhere[32], are illustrated in Figures 1 and 2 and may be
summarized as follows.

1.   All the three solubilizates are largely located near the
      headgroups of the various ionic micelles. While benzophenone
      and acridine display maximum shielding effects on the γ-CH$_2$
      groups of the ionic micelles, 1-methyl indole elicits the

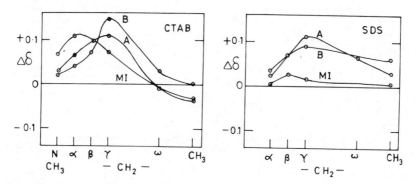

Figure 1.   Shifts in the NMR signals of various surfactant protons
caused by the ring current effects of the added probe. $\Delta\delta$, in ppm,
is the difference between the chemical shift value of the same set
of protons with and without the added probe. + refers to an up-
field shift and - to a downfield shift. The x axis shows the
assignment of the signals to the various protons of the surfactant
chain. The probes are A = acridine, B = benzophenone, and MI = 1-
methyl indole. CTAB and SDS are the usual abbreviations.

maximum response from the β-CH$_2$ segment. It is not clear
whether 1-methyl indole is positioned closer to the headgroup
than the other two, or the topography of the ring current
shielding domains of the three probes are such that differential effects are observed if all the three probes are positioned largely in the "Stern layer". The alternate possibility of water penetrating upto the β-or even γ-CH$_2$ groups of
these micelles cannot be assessed by our results at this
stage. In the case of the nonionic OG micelles, similar
results are obtained but a detailed analysis will have to
await unequivocal assignment of the signals to the various
sets of protons in the molecule. With the other nonionic
micelles of Brij 58, the results suggest that all the three
probes are largely located in the junction region of the
polyoxethylene segment and the cetyl chain.

2.   All the protons, including the ω-CH$_2$ and ω-CH$_3$ of the surfactant chain are affected by the solubilizate, albeit to diffe-
rent extents. It would thus appear that the probe is able to

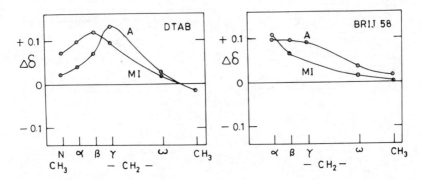

Figure 2. Shifts in the NMR signals of the various protons of
DTAB and Brij 58, caused by the added probe. See Figure 1 legends
for further details.

move (translocate) between the headgroup region and the interior at a rate faster than what is resolvable by such NMR
experiments, yet with an average residence time heavily
weighted towards the former site. It is worth noting here
that the added probe does not generate resolvable multiple
signals, but only a peak shift of each set of protons. The
additional or alternate possibility exists of the surfactant
monomers in the micelles executing differential protrusions
into the bulk phase[30], resulting in all the protons experiencing the probe-induced ring current effects, to varying extents.
Our experiments, per se, are not in a position to distinguish
between the various dynamic processes that are possible.

3. The interior protons of SDS and Brij 58 experience a greater
    extent of the probe-induced effects than those of CTAB or
    DTAB. This might suggest a greater mobility of the probes
    within SDS and Brij 58, and/or more facile protrusions of the
    monomers in these micelles. It is also interesting that
    acridine and 1-methyl indole induce downfield (deshielding)
    shifts of the ω-CH_3 signals in CTAB and DTAB, but not in the
    others. (A similar deshielding of the ω-CH_3 signals of CTAB
    was observed earlier upon pyrene solubilization[23]). This
    fact, combined with the observation that the probes induced
    peak-splitting of the interior ω-CH_2 manifold, is indicative

of changes in the interactions and micro dynamics of the amphiphile, and perhaps some orientational features of the probe.

4.   A study of the variation of the probe-induced NMR effects as a function of the probe/surfactant mole ratio did not suggest any probe translocation in the micelle at higher concentrations (the mole ratio was varied from 1:10 to 1:3). The results in Figures 1 and 2 are with a probe/surfactant ratio of 1:8.

Based on these results, we suggest that all the extrinsic aromatic optical probes used in micellar structural studies (e.g. those listed in Table I) are largely positioned in the headgroup region of the host micelles. These probes will thus be expected to monitor the polarity and water content of this region, and not of the micellar interior.

II. Ketone probes

The other class of molecules that have been used as water penetration probes in micelles are carbonyl compounds, several of which are also listed in Table I. The solvent polarity-dependent variation displayed by the keto carbon CMR chemical shift has been used as a method to investigate the water content and porosity of micelles.[10,29] A similar use has been made of the polarity-sensitive conformational changes that occur in some optically active alicyclic ketones.[13] Likewise, the ease of reduction of micelle-bound steroid enones by bulk phase $NaBH_4$ has also been used as a method to monitor micellar water content.[14]

The basic assumption made in these studies has been that the keto probes, being lipophilic and essentially water-insoluble, are solubilized in the micellar interior and monitor the polarity of this region. However, in the light of our above discussion on lipophilic aromatic probes, it becomes important to establish the actual sites of location of the ketone probes (and their functional groups) in micelles before interpreting the results obtained by using them. We have studied this problem in some detail[33], and summarize the results below.

Figure 3 reveals that a cyclic ketone, an aromatic ketone and an ester, and a steroid enone are surface active, with interfacial activities higher than that of benzene[18]. Table II shows that the measurement of the carbonyl carbon chemical shifts of, e.g., testosterone in various situations provides a CMR method of monitoring interfacial activity. It also shows that an amplification of the effects of this activity can be achieved by solubilizing such a mildly polar molecule in systems of increasing surface area-to-

Table II. ^{13}C NMR chemical shifts of ketones in anhydrous organic solvents, water: oil interfaces and microdroplets, and in micelles.

	δ, ppm*	$\Delta\delta$**	$\Delta\delta$***	$\Delta\delta_{micelle}$
Benzophenone	126.65[a]	+ 0.21 ppm	-	-
4-cholestene-3-one	131.85[b]	+ 0.21 ppm	-	-
Testosterone	129.05[c]	+ 0.25 ppm	+ 0.40 ppm	+ 5.9 ppm in SDS. Same value in CTAB also.

a : in heptane
b : in benzene
c : in carbon tetrachloride

* w.r.t. external standard dioxane 67.4 ppm

** downfield shift obtained when water was added to the solution in the NMR tube, producing a bulk interface

*** downfield shift obtained when water was added to the solution in CCl_4: hexane (density 1) and agitated, to produce isodense buoyant microdroplets (much larger interfacial area than in case **)

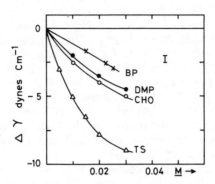

Figure 3. Reduction of the interfacial tension of CCl_4:H_2O at
24 ± 1°C by benzophenone (BP), dimethyl phthalate (DMP), cycloheptanone (CHO), and testosterone (TS).

volume ratios - bulk interface, water:oil microdroplets, and
micelles. The large downfield shift of over 5 ppm observed for the
keto carbon signal of the probe in micelles (see Table II) would
suggest that the probe is predominantly at surface sites of the
host micelle.

A direct way of ascertaining the location of a nonaromatic
probe such as testosterone is to use a variant of the ring current
induced shielding experiments discussed earlier. In this case, we
reverse the roles by choosing surfactants that contain aromatic
headgroups and let the solubilizate experience the ring current
effect.[33] Such a comparative experiment revealed that the NMR
signal of the A-ring H-4 proton of testosterone is upfield shifted
by 0.22 ppm in sodium dodecyl benzene sulfonate (SDBS) micelles,
when compared to its value in SDS micelles. (Likewise, the signal
was upfield shifted in cetyl pyridinium chloride (CPC) micelles by
0.07 ppm compared to the value in CTAB. The effects on the other
protons of the probe were significantly smaller). With the assumption that the environment of the probe in SDBS and in SDS (cf. CPC
and CTAB) is essentially the same, the upfield shift in SDBS or in
CPC would be consistent with the interpretation that the probe is
largely located in the headgroup region of the micelle and experiences the shielding effect of the aromatic headgroup of the surfactant in the assembly.

The results of this section indicate that ketone probes are inherently interfacially active, and such activity is amplified when the probe is solubilized in micelles, leading to a large fraction of the probe molecules in the headgroup region of these aggregates. That benzophenone[31] and dimethyl phthalate[34] are so located in micelles has been shown earlier. The present results on testosterone would indicate, by analogy, that other sterones listed in Table II might also be surface-located in micelles. This would make the rapid chemical reduction of micelle-solubilized steroid enones by bulk phase borohydride understandable.[33] Furthermore, our results on the interfacial activity of cycloheptanone suggest that the ORD probe (+)-trans-2-chloro-5-methylcyclohexanone might be located in the "Stern layer" of CTAB and SDS micelles. In this connection, it is not surprising to note that the polarity sensed by this ORD probe when dissolved in heptane and when added to reverse micelles of Aerosol OT (with a water:surfactant mole ratio = 40) in heptane is the same.[13] A reverse micelle, for obvious structural reasons, does not possess as high an interfacial area as a normal micelle does. This, added to the high partition coefficient of the lipophilic probe favouring heptane and the large bulk phase volume of heptane, would drive the probe to be in the bulk oil phase and minimize the fraction of molecules inserting into the reverse micelle-water interior interface.

III. Intrinsic probes

While the above discussion has been concerned with extrinsic probes, there are studies reported using ω-functionalized detergent molecules as intrinsic probes, either as pure micelles by themselves or as additives in other micelles. Some examples are: dodecanoates functionalized with nitrito group at the 5,6,7,8 or 10 positions, or with keto group at 6,7 or 10 positions[29]; 11-(3-hexyl-1-indolyl) undecyltrimethylammonium bromide and sodium 11-(3-hexyl-1-indolyl) undecyl sulfate[26], and stearic acid labelled with nitroxide spin moiety at the 5,12 and 16 positions.[35] The methods used have been optical absorption, NMR, fluorescence and ESR spectroscopy. In these cases of intrinsic probe studies, it is assumed that the reporting functional group is perforce imbedded in the interior of the amphiphile aggregate, for structural reasons. Yet the results on micelle structure obtained using these are not very different from those using similar extrinsic probes. It is not clear whether this represents a situation where: (i) the polar functional group has dragged water molecules into its vicinity, or (ii) the analysis of the results suffers from our lack of knowledge of the molecular basis of spectral changes, as with CMR[4,36], or (iii) the results are indicative of differential protrusions into the bulk phase by the surfactant monomers.[30] The additional and distinct possibility of conformational alterations in such intrinsic probe molecules in a fashion that leads the polar ω-functional group to seek the water interface cannot be dismissed at the moment. Monolayer studies on nitroxide derivatives of stearic acid suggest that both the carboxylate group and the nitroxide moiety are attached to water in the interface.[37] Ramachandran et al.[38], in a recent paper, have suggested this possibility for two stearic acid probes functionalized with the nitroxide moiety at the 5 and 12 positions. Hamilton and Small have shown that the triglyceride triolein inserts into lecithin bilayers with a conformational orientation that is different from the one in its neat liquid phase[39]; and also that cholesteryl oleate adopts a bent conformation in lecithin vesicles, which enables the ester group to be positioned at the water interface.[40] (The lattice model representation of micelles[3] indeed allows chain bending of the detergent molecules in the assembly and finite probabilities for interior and even terminal segments to reside near the water interface. The surfactant block model[16] also suggests slight bending of the monomers, and 'fatty patches' in the surface.)

MATERIALS AND METHODS

The details about the materials, the experimental methods and conditions, and analysis of data have been presented elsewhere.[32,33]

ACKNOWLEDGEMENTS

We thank the University of Hyderabad, BHEL R&D Hyderabad, Indian Institute of Science, Bangalore and JEOL, Japan for access to their NMR facilities, and Dr. Harigopal of RRL, Hyderabad for the White tensiometer. P. Mitra thanks the CSIR India for a Senior Research Fellowship.

REFERENCES

1. C. Tanford, "The Hydrophobic Effect", 2nd ed., Wiley-Interscience, New York, N.Y., 1980.
2. J.N. Israelachvili, S. Marcelja, and R.J. Horn, Quart. Rev. Biophys., 13, 121 (1980).
3. K.A. Dill and P.J. Flory, Proc. Natl. Acad. Sci. U.S.A., 78, 676 (1981).
4. H. Wennerstrom and B. Lindman, J. Phys. Chem., 83, 2931 (1979).
5. R.I. Sha'afi, in "Membrane Transport", S.L. Bonting, Editor, Elsevier, Amsterdam, 1981.
6. G.S. Hartley, Quart. Rev. Chem. Soc., 2, 152 (1948).
7. P. Mukerjee, in "Solution Chemistry of Surfactants", K.L. Mittal, Editor, Vol.1, p.153, Plenum Press, New York, N.Y., 1979.

8.   B. Lindman and H. Wennerstrom, Topics Current Chem., 87, 1
     (1980).
9.   B. Svens and B. Rosenholm, J. Colloid Interface Sci., 44, 495
     (1973).
10. F.M. Menger, J.M. Jerkunica, and J.C. Johnston, J. Am. Chem.
     Soc., 100, 4676 (1978).
11. F.M. Menger, Acc. Chem. Res., 12, 111 (1979).
12. F.M. Menger, J. Phys. Chem., 83, 893 (1979).
13. F.M. Menger and B.J. Boyer, J. Am. Chem. Soc., 102, 5936
     (1980).
14. F.M. Menger and J.M. Bonnicamp, J. Am. Chem. Soc., 103, 2140
     (1981).
15. F.M. Menger, H. Yoshinaga, K.S. Venkatasubban, and A.R. Das,
     J. Org. Chem., 46, 415 (1981).
16. P. Fromherz, Ber. Bunsenges. Phys. Chem., 85, 891 (1981).
17. E.M. Kosower, "Introduction to Physical Organic Chemistry",
     John Wiley, New York, N.Y., 1968.
18. P. Mukerjee and J.R. Cardinal, J. Phys. Chem., 82, 1620 (1978).
19. P. Mukerjee, J.R. Cardinal, and N.R. Desai, in "Micellization,
     solubilization and Microemulsions", K.L. Mittal, Editor, Vol.1,
     p.24, Plenum Press, New York, N.Y., 1977.
20. J.C. Eriksson and G. Gillberg, Acta Chem. Scand., 20, 2019
     (1966).
21. J.H. Fendler and L.K. Patterson, J.Phys.Chem., 75, 3907 (1971).
22. M. Gratzel, K. Kalyanasundaram, and J.K. Thomas, J. Am. Chem.
     Soc., 96, 7869 (1974).
23. J. Ulmius, B. Lindman, G. Lindblom, and T. Drakenberg,
     J. Colloid Interface Sci., 65, 88 (1978).
24. M. Almgren, G. Grieser, and J.K. Thomas, J. Am. Chem. Soc.,
     101. 279 (1979).
25. T. Wolff. Ber. Bunsenges. Phys. Chem., 85, 145 (1981).
26. N.J. Turro, Y. Tanimoto, and G. Gabor, Photochem. Photobiol.,
     31, 527 (1980).
27. Y. Waka, K. Hamamoto, and N. Metaga, Photochem. Photobiol. 32,
     27 (1980).
28. F.M. Martens and J.W. Verhoeven, J. Phys. Chem., 85, 1773
     (1981).
29. P.de Mayo, J. Miranda, and J.B. Stothers, Tetrahedron Lett.,
     22, 509 (1981).
30. G.E.A. Aniansson, J. Phys. Chem., 82, 2805 (1978).
31. J.H. Fendler, E.J. Fendler, G.A. Infante, P.S. Shih, and
     L.K. Patterson, J.Am. Chem. Soc., 97, 89 (1975).
32. K.N. Ganesh, P. Mitra, and D. Balasubramanian, J. Phys. Chem.,
     86, 4291 (1982).
33. P. Mitra, K.N. Ganesh, and D. Balasubramanian, J. Phys. Chem.,
     submitted.
34. J.W. McBain and H. McHan, J. Am. Chem. Soc., 70, 3838 (1948).
35. H. Yoshioka, J. Am. Chem. Soc., 101, 28 (1979), and references
     cited therein.

36. H. Wennerstrom and B. Lindman, J. Phys. Chem., 83, 3011 (1979).
37. D.A. Cadenhead and F. Muller-Landau, Adv. Chem. Ser., 144, 294 (1975).
38. C. Ramachandran, R.A. Pyter, and P. Mukerjee, J. Phys. Chem., 86, 3198 (1982), and accompanying papers by them in the same issue.
39. J.A. Hamilton and D.M. Small, Proc. Natl. Acad. Sci., U.S.A., 78, 6878 (1981).
40. J.A. Hamilton and D.M. Small, J. Biol. Chem., 257, 7318 (1982).

THE SIZE OF SODIUM DODECYL SULFATE MICELLES WITH VARIOUS ADDITIVES:

A FLUORESCENCE QUENCHING STUDY

Mats Almgren[*] and Shanti Swarup[*]

Physical Chemistry 1, Chemical Center
University of Lund
P.O.B. 740, S-220 07 Lund, Sweden

Fluorescence quenching methods for the determination of micelle aggregation numbers have been critically evaluated recently. It was concluded that the simple, static fluorescence quenching method can be used with confidence for small micelles if it is complemented by some time-resolved measurements. The probe/quencher pair $Ru(bpy)_3^{2+}$ / 9-methylanthracene has been studied extensively in SDS-micelles. Studies at varying salt concentration suggest this pair to be useful for micelles up to an aggregation number of about 130.

This probe-quencher pair was employed in the study of Sodium Dodecyl Sulfate (SDS) micelles with additives of various kinds. It was found that the surface density of sulfate headgroups, N_{SDS}/A, which is also a surface charge density, σ/e, can be used as a size determining variable. For a number of alcohols (n-alkanols from C_4 to C_{10}, benzyl alcohol, t-butanol) as additives approximately the same dependence of N_{SDS}/A on X_A (molefraction additive in the micelle) was obtained. The relationship was approximately linear over a broad range of compositions.

On addition of saturated hydrocarbons the micelles responded by growing until the initial surface area per head group was restored. The same respons was also obtained, surprisingly, on addition of cationic surfactants. Bulky hydrophobic counterions of tetraalkylammonium and tetraphenylphosphonium salts give a decreasing SDS headgroup surface density, which also results on addition of nonionic surfactants.

613

INTRODUCTION

Two important methodological advances make it possible to determine, easily and reliably, the aggregation number and composition of micelles under various conditions of temperature, solvent composition, additives, concentrations, etc. The first is the Fourier Transform NMR Pulsed-Gradient Spin-Echo Self-Diffusion method of Stilbs[1-3], which is very powerful for the determination of how surfactant and polar and nonpolar additives are distributed between micelles and water. Partitioning data from Stilbs have been used in this study. Secondly, a number of fluorescence quenching methods[4-6] or excimer fluorescence methods[7-9] have evolved for the determination of micelle sizes. The first method of this kind, introduced by Turro and Yekta[4], used a long-lived fluorescence probe, $Ru(bpy)_3^{2+}$, which is bound to anionic micelles. The luminescence from this agent is very effectively quenched by 9- methyl anthracene(MeA), a hydrocarbon which is almost insoluble in water and very strongly prefers the micelle medium. Turro and Yekta assumed that fluorescence was observed only from the fraction of excited $Ru(bpy)_3^{2+}$ that was localized on micelles without quenchers. They also assumed that the quenchers were randomly distributed over the micelles without interactions among themselved or with $Ru(bpy)_3^{2+}$. For micelles of equal size, this means a poissonian distribution. From the fluorescence quenching results, the concentrations of micelles could be calculated and thereby the the number average micelle aggregation number.

The assumption that the luminescence from micelles with quenchers was totally negligible was later shown to be an oversimplification[10], permissible only for small micelles[5]. The range of usefulness of the method was greatly extended and its reliability greatly enhanced by the introduction of time resolved fluorescence quenching methods, using $Ru(bpy)_3^{2+}$/MeA or other probe/quencher pairs. The pyrene excimer method is a special case. It has the merit of being useful for all kinds of micelles irrespective of charge type. The exploitation of possible probe/quencher pairs in this field has just started. With promising new pairs, it will become possible to study other types of aggregation phenomena as well.

Figure 1 presents a comparison between values of N_{SDS}, the aggregation number of sodium dodecylsulfate micelles, at 25^0C with various amounts of NaCl added. The quasielastic light scattering results by Mazer et al.[11] are lower than values obtained by other methods at low salt concentration but higher at high salt concentration. The deviations at high salt concentrations are partly due to the fact that the micelles in this region are very polydisperse and that the various methods measure different types of averages[5,6].

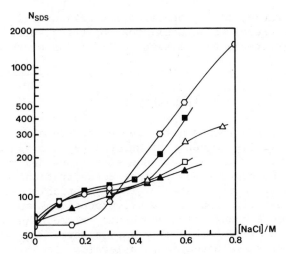

Figure 1. Variation of aggregation number of SDS with concentration of NaCl determined by various methods.
    : Quasielastic light-scattering, Reference 11;
▲: Static fluorescence quenching, Reference 4;
■: Pyrene-excimer method, Reference 9;
□: Time-resolved fluorescence quenching, Reference 16;
○: Classical light-scattering and sedimentation equilibrium,
     References 13-15;
●: Static fluorescence quenching, Reference 12;
△: Time-resolved quenching, Reference 5.

     A comparison between the results from the different fluore-
scence quenching methods shows that below a salt concentration of
about 0.4 M -- or an aggregation number less than about 110-120--
all the methods give about the same results (with the exception of
Turro and Yektas' value at [NaCl] = 0.15 M, which probably is too
low, since a redetermination with this method at 0.1 M gave a
higher aggregation number [12]). These values also compare favour-
ably with those obtained by the 'classical' method[13,14] in very
careful experiments, as discussed by Kratohvil [15].

     The conclusion from this comparison is that the fluorescence
quenching methods are reliable for small micelles, which in the
case of SDS means an aggregation number less than about 120,
corresponding to a spherical hydrophobic volume of radius ≈ 21 Å.
Furthermore, for these small micelles, the simple, static Turro
and Yekta method is as reliable as the more cumbersome time-resolved methods. It has therefore been used to determine micelle

sizes in this range, making occasional checks on critical compositions with time resolved methods. Most of the results have been
presented in greater detail elsewhere.[12,17] An overview will be
given here.

EXPERIMENTAL

Details of the method have been presented elsewhere.[5,12,17]
From measured micelle concentrations, the aggregation numbers were
calculated using the known total concentrations and either
estimated free concentrations or free concentrations calculated
from Stilbs' distribution data[2-3]. The hydrophobic radius R was
calculated by assuming that the alkyl groups from surfactant and
additives formed a spherical hydrophobic volume. The surface area
$4\pi R^2$ is thus the area of this hydrophobic core. $N_{SDS}/4\pi R^2 = \sigma/e$ is
the charge density evaluated at this level.

RESULTS AND DISCUSSION

Alcohols. The behaviour of SDS micelles on addition of alcohols
is at first sight very complex. Figure 2 illustrates part of this
complexity. The size of SDS-micelles is plotted vs. the concentration of free alcohol in the water pseudophase, calculated from
Stilbs' partitioning data[2]. For the lowest alcohols: ethanol,
n-propanol, and tert-butanol, the behaviour is initially very
similar, with all showing a slow decrease in size. At higher alcohol concentration, the micelle size seems to increase again (the
results at these high concentrations of alcohols are uncertain,
and must be tested by other measurements). n-Butanol and n-pentanol show more pronounced minima in micelle size. These minima
occur at the same alcohol concentrations where minima in the cmc
values have been reported previously.[18] Changing solvent properties may be the cause for this pattern only for the lowest alcohols. As shown in Figure 3, the longer alcohols give more and more
shallow minima, which eventually disappear altogether and are
replaced by a steady growth (from n-octanol) of micelle size up
to the solubility limit. Results obtained by Lianos and Zana [19]
at high concentrations of SDS and alcohols (in the microemulsion
region) show an even more complex behavior.

In Figure 4, the surface charge density of SDS-alcohol micelles is shown as a function of alcohol content, for all alcohols
used from C_4 and up (mainly normal alcohols but also tert.-butyl-
alcohol and benzylalcohol). The plot shows a remarkably similar
behaviour for all alcohols. The surface charge density is thus of
overriding importance for the size of the micelles. Alcohols
induce a decrease in surface charge density which is linear over
a wide composition range.

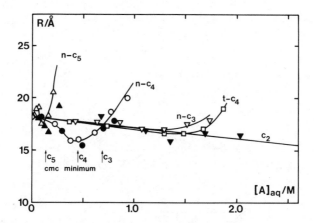

Figure 2. Radius of SDS micelles with added alcohols. Two concentrations of SDS were used on additions of n-pentanol (n-C$_5$) and n-butanol (n-C$_4$), 0.037 M (filled symbols) and 0.130 M (open symbols). The concentration of SDS was otherwise 0.13 M. The concentration of alcohol in the aqueous pseudophase was calculated using Stilbs' distribution data.[3]

   As discussed earlier,[17] the reason for this simple empirical relationship between charge density and composition can be understood qualitatively from the recent theoretical work by Jönsson, Wennerström and others in Lund.[20] The micellar size is determined by the balance between the strive to reduce the interfacial contact between hydrocarbon and water (the hydrophobic effect) and the electrostatic repulsive energy of the headgroups. Jönsson et al.[20] have shown that the remarkably simple equation

$$2E_{el}/A = \gamma \qquad (1)$$

holds for lamellar systems; for micelles of finite size a further entropic term has to be added. In this equation γ is an interfacial tension and A the interfacial area at the level of water-hydrocarbon contact. Jönsson[20a] has also shown from the Poisson-Boltzmann equation that E_{el} is mainly determined by (proportional to) the charge of the micelles. (The charge from the amphiphiles is a constitutive property of the micelles, and is therefore the suitable input parameter in the calculations. The reduction of charge due to electrostatic attraction of counterions - the ion binding - was calculated from the P-B equation giving results in good agreement with experimental

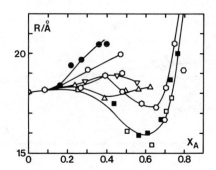

Figure 3. Change of SDS micelle size with mole fraction alcohol
in the micelle.
■ and □ : n-butanol at 0.13 and 0.35 M SDS;
◖ and ○ : n-pentanol at 0.13 and 0.035 M SDS;
△ : n-hexanol
▽ : n-heptanol;
○ : n-octanol;
● : n-octanoic acid, all at 0.035 M SDS.

findings.[20]) E_{el} depends less strongly on other parameters such
as micelle and saltconcentration, and since these varied only
slightly in the reported measurements, it can be understood that
E_{el} was roughly proportional to the constitutive micelle charge.
    It then follows from eqn (1)

$$\gamma \approx 2E_{el}/A = N_{SDS}/A = \sigma/e$$

    The constitutive charge density of the micelle, evaluated at
the level of the hydrophobic radius, should thus be proportional
to the interfacial tension γ. Several authors[20,21] have given a
value $\gamma_s \approx 0.0187$ J/m² for hydrocarbon-water. Jönsson found for
lamellar liquid crystals [20a] that γ decreases on addition of
alcohol according to

$$\gamma = \gamma_S X_S + \gamma_A X_A = \gamma_S - (\gamma_S - \gamma_A)X_A \qquad (2)$$

where the indices S and A stand for surfactant and alcohol, respectively, and the value of $\gamma_A = 0.004$ J/m². The ratio slope/inter-
cept of the line in Figure 4 is 0.77, which is very close to the

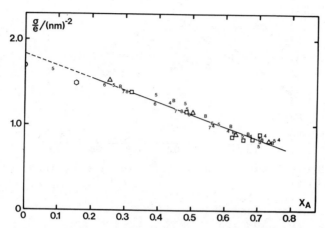

Figure 4. Charge density, or density of ionic headgroups, for SDS-alcohol micelles at varying mole fraction alcohol in the micelle. Numbers identify n-alcohols with the corresponding number of carbon atoms.
○: n-decanol;
B: benzyl alcohol;
□: t-butyl-alcohol;
△: n-butyl alcohol at 0.035 M SDS.
The SDS concentration in 4, 5, B, and □ was 0.13 M, otherwise 0.035 M.

value from eq. (2), 0.78.

A prediction from this discussion is that the surface charge density should remain unchanged on addition of hydrocarbons, as will be discussed next.

HYDROCARBON ADDITIVES

On solubilization of hydrocarbons both in SDS-micelles and in SDS-alcohol micelles, the surface charge density remained constant within errors. A very slight trend- decrease - in surface charge density could be observed for an aromatic hydrocarbon toluene, indicating a slight surface activity. In Figure 5, the hydrophobic radii for SDS additive micelles are compared for some different additives. The hydrocarbons induce a growth that is similar to what should be expected if the surface charge density had remained exactly constant. n-Octanol, which has about the same

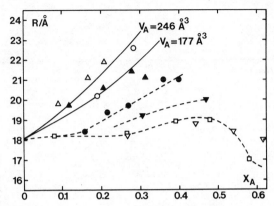

Figure 5. Hydrophobic radius of SDS micelles against mole-fraction
X_A of an additive.
  0: n-hexane ($V_A = 218$ A^3);
  △: n-heptane ($V_A = 246$ $Å^3$);
  ▲: toluene ($V_A = 177$ $Å^3$);
  ●: n-octanoic acid;
  ▼: n-octanol;
  ▽: n-pentanol;
  □: benzyl alcohol.

hydrophobic volume as n-heptane, induces a much smaller increase
in size. n-Octanoic acid is in between, illustrating the importance
of the headgroup. Benzylalcohol and n-pentanol are chemically very
different but have the same headgroup and about the same hydrophobic volume. They give closely the same size-composition behaviour.

THE MONOMER LENGTH AS A GEOMETRICAL CONSTRAINT

These results have some bearing on the question about the
size and shape of the small globular micelles. According to many
theoretical predictions[21-22], a spherical micelle should be limited
in size due to the necessity that one CH_3 fills up the center of
the micelle, and that all ionic headgroups are in contact with
water at the surface. The radius of the hydrophobic core of a
C_{12}-surfactant micelle should then be limited to 16.7 Å, which
corresponds to some 55 monomers. Larger micelles would have to be
nonspherical. The spherical shape is probably favourable electrostatically, and also entropically (since at a given surface area
per head group the spherial shape gives the smallest micelle).

The larger micelles would be in some sort of constrained state therefore. If this geometric constraint was important, a release of the constraint should result in an observable response. One way to release the constraint would be to add nonpolar compounds that could fill up the center of the micelle and allow a further spherical growth. In the size-composition diagrams for SDS-hydrocarbon micelles, however, there are no signs on such a sudden initial release of a constraint. Also, SDS-alcohol micelles and SDS micelles in the presence of 0.1 M NaCl show a smooth growth on addition of hydrocarbon.

The micelle growth on addition of salt is also smooth, Figure 1, up to about 0.4 M of NaCl. If the geometrical constraint really were important, the growth would have been expected to set in more suddenly and earlier.

Another way to release the geometrical constraint is by allowing some molecules of a longer surfactant homolog to enter the micelles. Size-composition diagrams for sodium tetradecyl- and dodecyl sulfate mixed micelles are shown in Figure 6. Again, there is nothing that suggests a transition from a constrained to a less constrained state.

There may be several reasons why the length, b, of the alkyl chain of the monomer is a less absolute constraint on the size of the sphere than what is commonly claimed. Hartley[23] gave some reasons for this; others were discussed recently:[17]

-- The micelle may have a very rough surface, and it is possible that there always is a depression somewhere, so that the surface-center distance there is less than b, although the mean radius is substantially larger[24].

-- A (transient) hole in the center of the micelle, with radius of 1-2 Å, is not prohibitively expensive in terms of free-energy cost.

-- A monomer head may occasionally become partly buried in the micelle. This is a realistic possibility for the alkylsulfates in particular, where the uncharged ester oxygen could be involved.[9]

MULTIVALENT COUNTERIONS

The interactions of multivalent counterions with anionic micelles is of considerable interest. The effect of various counterions on the size of SDS micelles show a great variability (Figure 7), which seems to be related mainly to specific, non-electrostatic interactions. This contrasts sharply with the quite uniform effect the bivalent counterions have on the cmc of $M(DS)_2$.[25]

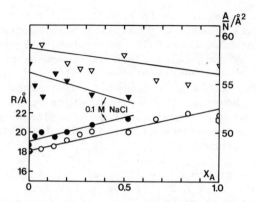

Figure 6. Hydrophobic radius, R(circles), and surface area per headgroup, A/N (triangles), for SDS - SC$_{14}$S mixed micelles as a function
of mole fraction SC$_{14}$S in the micelle, X$_A$. Filled symbols: .01 M NaCl,
open symbols: n$_o$ salt. The total concentration of surfactant was 0.060 M.

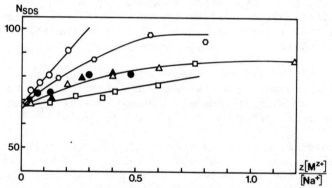

Figure 7. SDS micelle aggregation number as a function concentration
of added z-valent cation M^{z+} at 20°C. o: CaCl$_2$; ◖: MgCl$_2$; ■: Pb(NO$_3$)$_2$;
▲: Zn(ClO$_4$)$_2$; △: CdCl$_2$; □: TbCl$_3$; ●: Y(NO$_3$)$_3$. The SDS concentration
was in all cases close to 0.05 M.

The results in Figure 7 at high concentrations of added salt could
be in appreciable error due to complex formation, e.g. CdCl$^+$, CdCl$_2$
etc., and to liberation of the fluorescent probe, Ru(bpy)$_3^{2+}$ into
the water.

HYDROPHOBIC COUNTERIONS

Two types were tested: cationic surfactants of the alkyl-
ammonium- or alkyltrimethylammonium- type, and symmetric tetraalkylammonium salts and a tetraphenylphosphonium salt. In Figure 8,
the surface density of SDS headgroups (N$_{SDS}$/A$_{mic}$) is plotted vs.
composition. It is clearly seen that the cationic surfactants

leave the density of anionic headgroups constant. In other words, the addition of a cationic surfactant as counterion is about equivalent to separate additions of a small counterion and a hydrocarbon with the same hydrophobic volume as the cationic surfactant.

The bulky cations of the tetraalkylammonium and tetraphenylphosphonium salts do decrease the SDS headgroup density, however, as if they were pushing the headgroups apart. One possible way to account for this quantitatively is to assume that the hydrophobic counterions and the SDS headgroups contribute independently to the total micelle surface area.

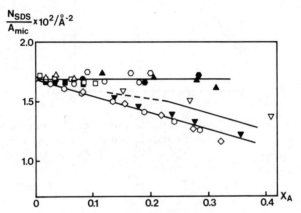

Figure 8. Surface density of SDS headgroups vs. mole fraction hydrophobic counterion at 20°C.

O: C_8AHCl	▼: $(C_4)_4$NNO$_3$
●: C_{10}AHCl	◇: $\emptyset_4$PCl
△: C_{14}TACl	◐: $(C_5)_4$NBr
□: C_{16} TABr	◯: n-heptane
■: C_{12}TACl	●: n-hexane
▽: $(C_2)_4$NCl	▲: toluene

This assumption results in the equation

$$\frac{A}{N_{SDS}} = A_{SDS} + A_A \frac{N_{SDS}}{N_A} \qquad (3)$$

where A is the total micelle area, A_{SDS} and A_A are area per SDS monomer and additive, respectively. Plots of this type are shown in Figure 9. Those for the tetraamyl- and tetrabutyl-ammonium salts are reasonably linear, whereas the tetraphenylphosphonium plot is curved. The values of the 'effective' areas A_A obtained in this way were not directly related to the size of the ions, as judged from molecular models.[17b]

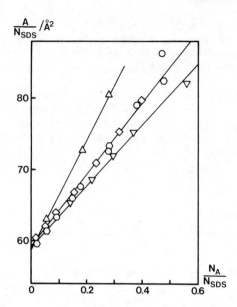

Figure 9. Area per SDS headgroup at the micelle surface with various amounts of added hydrophobic counterion or non-ionic surfactant. $\triangle$: E_8C_{12}; $\circ$: $\emptyset_4P^+$: $\bigcirc$: $(C_5)_4N^+$; $\diamondsuit$: E_5C_{12}; $\triangledown$: $(C_4)_4N^+$.

NON-IONIC SURFACTANT ADDITIVES

Due to the restriction of the method to anionic micelles, measurements were performed only on micelles with SDS as the main component using polyethylene dodecylethers, E_xC_{12}, x = 5,8, as additives. Plots according to Equation (3) were linear as shown in Figure 9. The effective areas change from 78 $\mathring{A}^2$ for E_8C_{12} to about 50 $\mathring{A}^2$ for E_5C_{12}, or as much as could be expected from the change of the headgroup size. Measurements on SDS-E_xC_{12} mixed micelles over the whole composition range, and at different temperatures would be of considerable interest.

ACKNOWLEDGEMENTS

J.-E.Löfroth has been of great help by checking the method for several SDS-additive compositions by time-resolved measurements. Many helpful discussions with the physical chemists in Lund, in particular with Bengt Jönsson, are greatfully acknowledged. Financial support has been provided by the Swedish Natural Science Research Council.

REFERENCES

*   Present address: The Institute of Physical Chemistry, Box 532
       S-751 21 Uppsala, Sweden.

1.   P. Stilbs and M. E. Moseley, Chem. Scripta, 15, 176 (1980).
2.   P. Stilbs, J. Colloid Interface Sci., 87, 385 (1982).
3.   P. Stilbs, J. Colloid Interface Sci., 80, 608 (1981).
4.   N. J. Turro and A. Yekta, J. Am. Chem. Soc., 100, 5951 (1978).
5.   M. Almgren and J.-E. Löfroth, J. Colloid Interface Sci., 81,
       486 (1981).
6.   M. Almgren and J.-E. Löfroth, J. Chem. Phys., 76, 2734 (1982).
7.   S. Atik, M. Nam, and L. Singer, Chem. Phys. Lett., 67, 75
       (1979).
8.   P. Lianos and R. Zana, J. Colloid Interface Sci., 84, 100 (1981).
9.   P. Lianos and R. Zana, J. Phys. Chem., 84, 3399 (1980).
10.   P. P. Infelta, Chem. Phys. Lett., 61, 88 (1979).
11.   N. A. Mazer, G. B. Benedek and M. C. Carey, J. Phys. Chem.,
       80, 1075 (1976); P. J. Missel, N. A. Mazer, G. B. Benedek,
       C. Y. Young and M. C. Carey, J. Phys. Chem., 84, 1044 (1980).
12.   M. Almgren and S. Swarup, J. Phys. Chem., 86, 4212 (1982).
13.   H. F. Huisman, Proc. Kon. Ned. Akad. Wetensch., B67, 367 (1964).
14.   D. A. Doughty, J. Phys. Chem., 83, 2621 (1979).
15.   J. P. Kratohvil, J. Colloid Interface Sci., 75, 271 (1980).
16.   M. van der Auweraer, C. Dederen, C. Palmans-Windels and F. C.
       DeSchryver, J. Am. Chem. Soc., 104, 1800 (1982).
17a.   M. Almgren and S. Swarup, J. Colloid Interface Sci., (in press).
17b.   M. Almgren and S. Swarup, J. Phys. Chem., (in press).
18a.   H. N. Singh and S. Swarup, Bull. Chem. Soc. Japan, 51, 1534
       (1978).
18b.   H. N. Singh, S. Swarup and S. M. Saleem, J. Colloid Interface
       Sci., 68, 128 (1979).
19.   P. Lianos and R. Zana, J. Phys Chem., 86, 1019 (1982).
20a.   B. Jönsson, Ph.D. Thesis, Lund, 1981.
20b.   B. Jönsson and H. Wennerström, J. Colloid Interface Sci., 80,
       482 (1981).
20c.   G. Gunnarsson, B. Jönsson and H. Wennerström, J. Phys. Chem.,
       84, 3114 (1980).
21.   C. Tanford, "The Hydrophobic Effect", Wiley, New York, 1973.
22a.   H. V. Tartar, J. Phys. Chem., 59, 1195 (1955).
22b.   J. N. Israelachvili, D. J. Mitchell and B. W. Ninham, J. Chem.
       Soc. Faraday Trans. 2, 72, 1525 (1976).
23.   G. S. Hartley,in "Micellization, Solubilization, and Micro-
       emulsions", K. L. Mittal, Editor, Vol. 1, p. 23, Plenum Press,
       New York, 1977.
24.   E. A. G. Aniansson, J. Phys. Chem., 82, 2805 (1978).
25.   P. Mukerjee and K. J. Mysels, "Critical Micelle Concentration
       of Aqueous Surfactant Systems", National Bureau of Standards,
       Washington, D. C., 1971.

FLUORESCENCE QUENCHING AGGREGATION NUMBERS IN A NON-IONIC MICELLE SOLUTION

J.-E. Löfroth* and M. Almgren**

Departments of Physical Chemistry
* Chalmers University of Technology and University of
  Göteborg, S-412 96 Göteborg, Sweden
**University of Uppsala, P.O.B. 532, S-751 21 Uppsala
  Sweden

Fluorescence methods to characterize amphiphile aggregation have attained much interest during the last few years. The main photophysical process has been the quenching of a fluorescent molecule residing in the aggregates. Both static (steady state illumination) and time resolved techniques have been employed.

We have estimated mean aggregation numbers with both static and time resolved methods (single photon counting) in a number of systems. Good agreement with results by others was normally found in systems with narrow size distributions. In other cases - big and polydisperse micelles, highly concentrated solutions, induced micelle growth - the results show severe disagreement with results from e.g. quasi elastic light scattering. Moreover, when values of the rate constants for interaction of probes and quenchers with the aggregates in these cases were estimated, the values could often be "unrealistic" in one way or another, despite their good inner consistency. This led us to further develop the theories describing the quenching situation, trying to take into account the effects of polydispersity, and to understand how a mismatch between model and reality will show up in the estimated values of various parameters.

INTRODUCTION

The reason for the strong light scattering observed close to the cloud point in aqueous solutions of non-ionic surfactants has been questioned recently. Does a real growth of the micelles occur or is it only an effect of intermicellar interactions close to the critical point? Corti and Degiorgio[1] favor the latter explanation, basing their arguments on results from classical and quasi elastic light scattering experiments. Nilsson et.al.[2] strongly support the former view, based on results from diffusion measurements. Since the fluorescence quenching method[3,4] would yield results that are dependent of intermicellar interactions in a quite different way than are results from thermodynamic measurements, we have undertaken a study of how fluorescence quenching aggregation numbers vary with temperature for a solution of hexa-ethyleneglycol mono n-dodecyl ether in water.

The interpretation of the results requires of necessity a theory that takes into account micelle polydispersity. Such a theory was recently developed,[5] and its main features will be reiterated here.

THEORY

The method based on the quenching of fluorescence to determine micelle sizes, as proposed by Turro and Yekta in 1978,[3] relies on a number of basic assumptions:
    i):   the residence time of the fluorescent and quenching molecules introduced in the micelles is much longer than the unquenched lifetime of the fluorescent probe;
    ii):   the channel for deactivation of the excited probe by quenching is much larger than other deactivation channels; the observed fluorescence is thus suggested to emanate from probes in micelles without quencher;
    iii): if $<x>_N$ is the average number of probes or quenchers in the micelles, the probability to find x of these molecules in a micelle is given by the Poisson distribution.[6-9]

These assumptions imply that the fluorescence intensity from a solution under low light intensity excitation is given by:

$$F = F_0 \exp(-[Q]/[M]) = F_0 \exp(-<x>_N)$$ (1)

where F, F_0 = fluorescence quantum yields in the presence and absence of quenchers, Q, in the micelle, M, respectively. Thus if $\ln(F_0/F)$ is plotted against [Q], the micelle concentration might be calculated from the slope of the straight line, and the aggregation number, s, is determined as

$$s = \frac{[\text{detergent}] - [\text{free monomer}]}{[M]} .$$

However, as was pointed out by Infelta,[10] assumption ii) is *decisive* for interpretation of the data with Equation (1). If the quenching in the micelles is not total but less efficient, as would be the case with increasing size of the micelles, the magnitude of the quenching rate constant, k_q, relative to the reciprocal of the unquenched lifetime, τ_0, of the probes must be considered. k_q was introduced to describe the quenching in a micelle with one quencher and xk_q as the quenching by x quenchers. A kinetic treatment of the reactiom scheme I leads to Equation (2) for the fluorescence decay after δ pulse excitation:

Reaction scheme I:
MP* $\to$ MP (+hν)   fluorescence and radiationless deactivation
MP*Q_X $\to$ MPQ$_X$ quenching in a micelle with x quenchers

MP* denotes a micelle with an excited probe and MP*Q_X a micelle with one excited probe and x quenchers. The rate constants for the two processes are $k_0 (=1/\tau_0)$ and xk_q respectively.

$$F(t) = F(0) \exp\{-A_2 t - A_3(1-\exp(-A_4 t))\} \tag{2}$$

where $A_2 = k_0$, $A_3 = <x>_N$ and $A_4 = k_q$.

Equation (2) is a special case of a more general situation in which assumption i) is replaced by a condition where the quenchers are free to move between the micelles and the surrounding media in the solubilization equilibrium during the excited lifetime:

$$MQ_X \underset{k_+}{\overset{xk_-}{\rightleftharpoons}} MQ_{X-1} + Q$$

Equation (2) is still valid but A_2, A_3, and A_4 are defined as

$$A_2 = k_0 + \frac{k_q k_-}{k_q + k_-} <x>_N; \quad <x>_N = \frac{k_+}{k_-} [Q]_{aq} = \frac{k_+}{k_-} \frac{[Q]}{1 + \frac{k_+}{k_-}[M]}$$

$$A_3 = \frac{(k_q)^2}{(k_q + k_-)^2} <x>_N$$

$$A_4 = k_q + k_-$$

These equations were suggested by Infelta et.al.[11] and also derived by Tachiya[12] and several other authors.[13,14]

Assumption iii) about Poissonian distribution means that the ratio k_-/k_+ is independent of x. Since k_-, k_q, and k_+ all would be expected to depend on the micelle size, the theory assumes that the micelles are almost monodisperse. For many systems studied and reported the validity of the narrow size distribution must be questioned, (see reference 5 and references therein). In one case we estimated aggregation numbers and rate constants for the solubilization equilibrium which were not in accord with results from studies with other methods.[15] We suggested[4] that k_q was inversely proportional to the micelle size and that the estimated aggregation numbers were close to the number average. In a later study[5] we showed that the number average is obtained only under certain conditions.

In that study two extreme types of distributions of quenchers on the polydisperse micelles were considered. Unfortunately, these were named the static and the dynamic case, respectively. In the first case the prototype is a system which contains micelles of different sizes, but where each micelle remains at the same size. A least biased set of assumptions in that case is the following:

I-a: The solubilizates show a Poissonian distribution over each subset of micelles at a given size, s.
I-b: The average occupancy number for micelles of size s is proportional to s:

$$\langle x \rangle_s = \eta s \tag{3}$$

where

η = micelle bound quenchers/micelle bound surfactants

$$= \frac{[Q]}{[\text{detergent}] - [\text{free monomer}]} \tag{4}$$

In the other case the polydisperse micelles are assumed to (II-a) change size quickly by exchange of monomers, whereas a quencher is assumed to stay very long in a given micelle. If we further assume that (II-b) the monomer exchange kinetics is totally unperturbed by the presence of the quenchers, we find that the average occupancy number must be independent of s:

$$\langle x \rangle = \eta \langle s \rangle \tag{5}$$

and the Poisson distribution must apply to the whole system as well as to each subset of micelles.

Since the two sets of assumptions lead to different equilibrium properties, the difference between the two cases is not only a question of kinetics (as the names seemed to suggest). Assumption II-b has thermodynamic implications which we do not understand well at present. It is not possible to predict therefore, which case that will best apply to a given system.

The presence of polydispersity, described by the distribution $A(s)$, also influences the quenching constant.[4,5]. Experimental and theoretical work[16-19] gives some support to the idea that

$$k_q = \kappa_q \, x/s \tag{6}$$

where κ_q is a second order rate constant as if the micelle was a small part of a macroscopic phase.

The total fluorescence intensity, $F(t)$, is obtained by summing up the contributions from every subset of micelles over the actual distribution. In case I we obtain:

$$F(t) = F(0) \, e^{-k_0 t} \, \frac{\sum_s s A(s) \, \exp\{-\eta s (1 - e^{-\kappa_q t/s})\}}{\sum_s s A(s)} \tag{7}$$

and in case II:

$$F(t) = F(0) \, e^{-k_0 t} \, e^{-\eta \langle s \rangle} \, \frac{\sum_s s A(s) \, \exp\{\eta \langle s \rangle \, e^{-\kappa_q t/\langle s \rangle}\}}{\sum_s s A(s)} \tag{8}$$

EXPERIMENTAL

Materials and Sample Preparation

Pyrene, P, (Aldrich 99%) was twice recrystallized from ethanol. N,N - dibutylaniline, DBA, (ICN,K&K) and hexa-ethyleneglycol mono n-dodecylether, $C_{12}E_6$, (Nikko), were used as supplied.

Stock solutions of $C_{12}E_6$ in distilled water, (HEGDE), of P in HEGDE, and of DBA in HEGDE were prepared; the pyrene solution was prepared by stirring a mixture of P in HEGDE for several days. Appropriate volumes of these stock solutions were then mixed to give the compositions in Table I:

Table I: Composition of samples.

sample number	0	1	2	3	4	5
[DBA] $\times 10^4$/M	0	1.33	2.50	5.00	7.50	9.99

$[P] = 9.9 \times 10^{-6}$/M, $[C_{12}E_6] = 0.045$ M

Fluorescence Measurements

Steady state emission spectra were recorded with an Aminco
SPF 500 corrected spectra spectrofluorometer. As the pyrene concentration was kept constant in all solutions, as assessed from
absorption spectra recorded on a Cary 219 (Varian) spectrophoto-
meter, the relative heights of the peaks in the emission spectra
for an unquenched and quenched solution could be used to determine
relative quantum yields. The fluorescence decay of pyrene was followed at 397±10 nm (Balzers R-UV interference filter) with the
single photon counting technique; the instrument has been described
in detail elsewhere.[20] For studies in the long wavelength region
a cut-off filter (Jena) with 50% transmission at 530 nm was used.
When the excitation profile and instrument response function was
recorded, the sample was replaced by a MgO – scatterer and no filter was used on the emission side. Excitation was always selected
by a monochromator (Jobin-Yvon H-10 UV) at 337±8 nm. The lamp was
a gated flashlamp (PRA 510B) filled with $H_2(g)$ at 0.7 atm pressure
and run at 5 kV, 30 kHz. With these settings and on the timescales
used, 3.103 and 1.385 ns channel^{-1}, the observed fluorescence could
be regarded as proportional to the true fluorescence decay after
background subtraction.[21] The data were analyzed by a fitting procedure based on a modified Levenberg – Marquardt algorithm.[22] Both
the reduced χ^2 test and a plot of the weighted residuals were used
as criteria for good fits. RUNS tests[23] were also made to accept
or discard the fits when the χ^2_ν values were higher than normally.
The measurements were made on air-equilibrated solutions at seven
temperatures: 17.4, 24.7, 34.4, 44.1, 46.2, 48.7, and 50.6 ^{0}C
(all at ±0.05 ^{0}C). Since the lifetime of pyrene in the excited
state is long enough, it experiences an average distribution of
oxygen.[24] Thus we have not considered the presence of oxygen as a
problem.

RESULTS AND DISCUSSION

Results are first presented for computer generated data.
These results form the basis for interpretation of the data from
the pyrene fluorescence quenching by dibutylaniline in HEGDE.

Computer Generated Data

Simulated data from single photon counting experiments were
generated.[5] Two types of size distributions with different widths,
σ, was assumed for the quenching in case I. In the first case we
regarded a Gaussian distribution, i.e.

$$A(s) = A(<s>_N) \exp\{-(s-<s>_N)^2/(2\sigma^2)\} \qquad (9)$$

in the other case we assumed an exponential distribution:

$$A(s) = A(s_0) \exp\{-(s-s_0)/\sigma\} \qquad s \geq s_0 \qquad\qquad (10)$$
$$A(s) = 0 \qquad\qquad\qquad\qquad\qquad\quad s \leq s_0$$

Values for the parameters used are given in Tables II and III respectively and the decays thus generated for the exponential case are shown in Figure 1. To mimic results from single photon counting experiments, Poissonian noise was added to the curves.[5]

Table II. Input and output parameters in simulated decay experiments. Gaussian distribution (Equations (7), (9), and (4)).

Common input: $k_0=2.0 \times 10^6 s^{-1}$, $\kappa_q=1.5 \times 10^9 s^{-1}$, $<s>_N=60$, $\sigma=60$

Curve number	G0	G1	G2	G3	G4	G5	G6 ($\sigma=25$)
Input η	0.000	0.003	0.006	0.010	0.015	0.020	0.030
Output A_3	.....	0.299	0.584	0.940	1.350	1.723	1.871
χ_ν^2	0.96	1.08	1.33	1.55	1.56	1.40	1.00

(From G0 A_2 was estimated to $2.0 \times 10^6 s^{-1} \pm 0.01 \times 10^6 s^{-1}$)

Table III. Parameters in simulated decay experiments. Exponential distribution (Equations (7), (10), and (4)).*

Common input: $k_0=2.0 \times 10^6 s^{-1}$, $\kappa_q=1.5 \times 10^9 s^{-1}$, $s_0=60$, $\sigma=400$

Curve number	E1	E2	E3	E4	E5	E6	E7	E8
Input η	0.001	0.002	0.004	0.006	0.008	0.010	0.020	0.030
Output A_3	0.639	1.151	1.943	2.551	3.049	3.475	5.069	6.309
χ_ν^2	1.13	1.13	1.01	1.69	3.01	4.57	9.94	10.4

*Curve number E0 is the same as G0.

Table IV. Aggregation numbers estimated from simulated decay experiments.

Curve number	G1	G2	G3	G4	G5	G6
$<s>_Q$	99.7	97.3	94.0	90.0	86.2	62.4

Curve number	E1	E2	E3	E4	E5	E6	E7	E8
$<s>_Q$	639.0	575.5	485.8	425.2	381.1	347.5	253.5	210.3

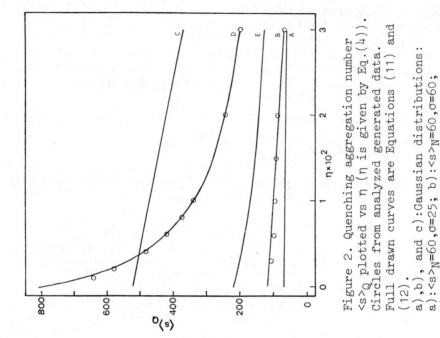

Figure 2. Quenching aggregation number $\langle s \rangle_Q$ plotted vs η (η is given by Eq.(4)). Circles from analyzed generated data. Full drawn curves are Equations (11) and (12).

a),b), and c):Gaussian distributions:
a): $\langle s \rangle_N = 60, \sigma = 25$; b): $\langle s \rangle_N = 60, \sigma = 60$;
c): $\langle s \rangle_N = 500, \sigma = 100$
d) and e): Exponential distributions:
d): $s_0 = 60, \sigma = 400$; e): $s_0 = 60, \sigma = 100$.

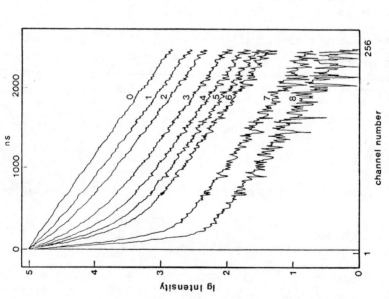

Figure 1. Computer generated decay curves. See Table III for data.

Inspection of the decays suggested that the quenching was described by Equation (2) and therefore the analysis was done by fitting this equation to the data to estimate A_2, A_3, and A_4. A_3 values are given in Tables II and III. From A_3, a quenching aggregation number, $<s>_Q$, is obtained as

$$<s>_Q = \frac{[\text{detergent}] - [\text{free monomer}]}{[Q]} A_3 = A_3/\eta$$

The estimated aggregation numbers are given in Table IV and are plotted against η in Figure 2 (circles at full drawn curves b) and d)).

Although it seemed appropriate from the long time behaviour of the decays to apply Equation (2), the fits of this model to the data are not acceptable. The reduced χ^2 values are too large to be accepted except at the lowest concentrations of quenchers. This was also verified by RUNS tests. Nevertheless, the estimated $<s>_Q$ values as a function of η perfectly follows the predictions for the different types of distributions, justifying the use of this analysis method. For a Gaussian distribution in case I, it was shown[5] that

$$<s>_Q = <s>_W - (\sigma^2/2)\eta \qquad (11)$$

and for the exponential situation

$$<s>_Q = s_0 + \frac{2}{\eta} \ln(1+\eta\sigma) - \frac{1}{\eta} \ln(1+ \frac{\eta s_0 \sigma}{s_0+\sigma}) \qquad (12)$$

In Figure 2 these two equations are illustrated for some cases (see figure legends). Thus if we apply Equation (2) to data in which polydispersity effects are present, and case I applies, a plot like Figure 2 would reveal these effects.

The ideas outlined in this Section were introduced to explain some peculiar results obtained earlier on systems where polydispersity most certainly was present or induced. A more critical test of the theory, its limitations and when the fluorescence quenching method is applicable, would be given by an investigation on the system $P/DBA/C_{12}E_6$ in water.

Pyrene Fluorescence in HEGDE

The probe-quencher pair used in this study was first tested on a 0.070 M solution of sodium dodecylsulphate, SDS, in water at 25 ^{0}C. A value around 60 for the aggregation number of SDS of this concentration is expected,[4] and from time resolved measurements we obtained 63±6. The quenching is described by a fast initial phase

in the pyrene fluorescence decay, followed by the steady state behaviour of unquenched fluorescence decay. k_q was estimated to be
$4.33 \times 10^7 \pm 0.56 \times 10^7 s^{-1}$, which gives $k_q \tau_0 = 7.62$. This indicates[10]
that steady state measurements on the quenching in SDS would give
acceptable aggregation numbers, if care is taken to avoid absorption of exciting light by the aniline. If this happens the quenched
fluorescence intensities measured would be too small and the estimated aggregation numbers too high, the quenching apparently being
too effective.

$C_{12}E_6$ in water is a non-ionic surfactant-water system, which
in the low concentration range shows a cloud point around 50 ^{0}C.
The cloud point for the present concentration (2% w/w) used is
47.8 ^{0}C.[25] Figure 3 presents the behaviour of the pyrene fluorescence decay measured at 398 nm (3a) and at wavelengths >500 nm (3b)
at two different temperatures. Corrected emission spectra are shown
in Figure 4. The relative heights of the peaks have not been corrected for the differences in absorption at 337 nm, where DBA ab-
sorbs some of the radiation. Therefore the trend in the intersection
between unquenched and quenched spectra around 490 nm is attributed
to the varying DBA concentration. The formation of an excited complex between pyrene and dibutylaniline is suggested, and is also
supported by the fact that the fluorescence in the long wavelength
region grows in slowly (Figure 3b). The concentration of pyrene
was low enough to make contributions from pyrene excimer fluorescence unimportant and moreover, if excimer fluorescence was present
it would also have been quenched by the DBA. Above the cloud point,
where there are two phases, the emission from the complex compared
to the pure pyrene decay reveals that the formation is similar to
the kinetic situation for exciplex formation in homogenous solutions.
(For these temperatures the fluorescence was measured from solutions
obtained by shaking the cells, so the photomultiplier viewed in fact
a dispersion of the two phases.) From a two component exponential
trial function fit to data at 50.6 ^{0}C for solution number 5, the
following kinetic behaviour for the two emissions were obtained:

$$F_{397nm}(t) = 0.30 \, e^{-t/19.9} + 0.70 \, e^{-t/6.6} \quad \text{(t in ns)}$$

$$F_{530nm}(t) = 0.52 \, e^{-t/17.1} - 0.48 \, e^{-t/5.9}$$

Considering the fact that no deconvolution was done to take into
account the finite width of the instrument response function, the
qualitative agreement is quite satisfactory between estimated decay
parameters and what is expected from an analysis of exciplex formation in a homogenous solvent.[26] From estimated preexponential
factors and decay constants, the complex lifetime is calculated to
be 14 ± 4 ns, and the complex formation constant to be 1.4×10^{11}
$\pm 0.3 \times 10^{11} M^{-1} s^{-1}$. Also, the complex dissociation leading to re-
generation of DBA and excited pyrene seems completely negligible.

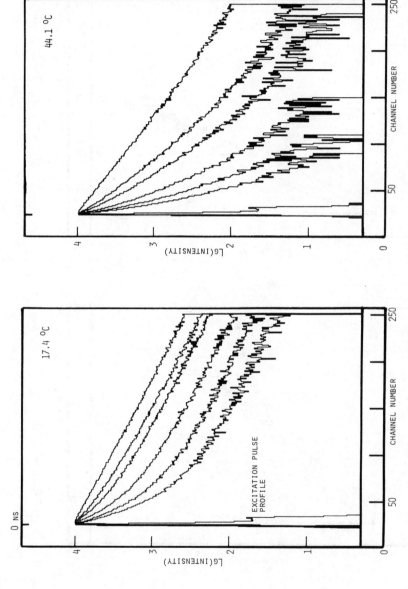

Figure 3a. Pyrene fluorescence decay at 397 nm at different temperatures. Each figure presents the decays of solutions number 0 (top) to 5 (bottom) of Table I. $\Delta t = 3.103$ ns channel^{-1}.

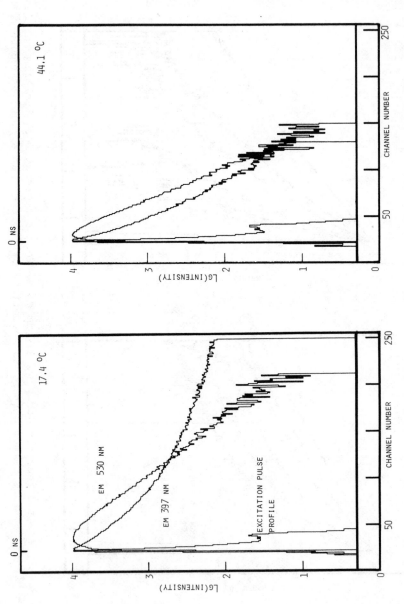

Figure 3b. Fluorescence decay at 397 nm and 530 nm of solution number 5 of Table I at different temperatures. $\Delta t = 1.385$ ns channel^{-1}

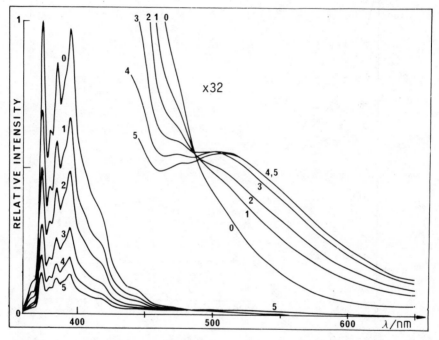

Figure 4. Corrected emission spectra. The studied system was pyrene in HEGDE at different concentrations of dibutylaniline (Table I). The temperature was 25 ^{0}C.

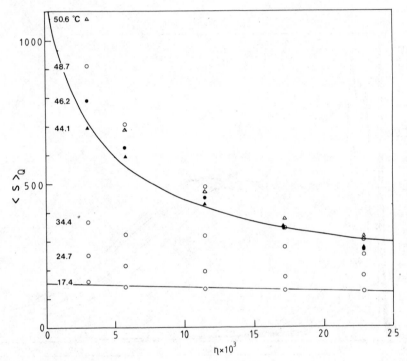

Figure 5. Estimated quenching aggregation number, $<s>_Q$, for $C_{12}E_6$ in water at different temperatures. η is given by Equation (4). Full drawn lines for temperatures 17.4 and 44.1 ^{0}C are Equations (11) and (12) respectively, with best fit parameters as given in Table V.

The formation constant was calculated by using the total concentration value of DBA in the solution. However, both pyrene and DBA seem to be confined in the micelles. For the 2% w/w solutions used, we may then calculate the *local* concentration of DBA to be higher than the total by a factor of about 50. This reduces the formation rate constant to about 3×10^9 $M^{-1}s^{-1}$, a reasonable value if the reaction is diffusion controlled in the micelles.

The features of the excited complex, whether it is an exciplex between singlet excited pyrene and DBA or not, and its kinetic analysis will be discussed further in a more detailed report on this system.*

The decays of the pyrene fluorescence (Figure 3a) were analyzed for each temperature and fitted with Equation (2). This seemed legitimate as the long time behaviour of the signal at 17.4 °C was the same for all concentrations of DBA, indicating that assumption i) is valid. The estimated values of $<s>_Q$ for each decay and temperature are presented in Figure 5. $<s>_Q$-values for one temperature were thus estimated by holding A_2 fixed and equal to k_0 (which were estimated for solution number 0) in the fitting procedure to obtain A_3 ($=\eta<s>_Q$) and A_4 ($=k_q$) of Equation (2). From the χ^2_ν and RUNS tests criteria it was found that this procedure did not give acceptable fits for temperatures higher than 17.4 and 24.7 °C. However, the same problems were met in the computer generated data analysis. Qualitatively there is remarkable agreement between Figure 5 and Figure 2 obtained by the computer simulated experiments. Best straight line fits, Equation (11), to the experimental data for 17.4, 24.7, and 34.4 °C give aggregation numbers and widths which would correspond to assumption of a Gaussian micelle size distribution, Table V. The trend in $<s>_W$, σ, and $<s>_N$ (estimated with $<s>_W = <s>_N + \sigma^2/<s>_N$) clearly shows an increase both in size, from $<s>_W = 156$ to 372, and in the degree of polydispersity, from $\sigma = 55$ to 102. Data at higher temperatures can not acceptably be fitted with straight lines. Instead we assumed an exponential size distribution and analyzed the $<s>_Q, \eta$ – data by Equation (12). The results are also given in Table V. For comparison, results from a Gaussian fit are also included. Keeping in mind the small number of data pairs available, the trend and discrepancies in the s_0 values obtained are of no significance, the s_0 values being very sensitive to small errors in the data. The most important findings are the increasing degree of polydispersity and the large limiting values obtained for $<s>_Q$ when $\eta \to 0$. As discussed in reference 5 $\lim_{\eta \to 0} <s>_Q$ should equal $<s>_W$.

---

*Later checks have shown that this emission may emanate from an impurity, sensitized by energy transfer on collisions with pyrene. The impurity must be about as good a quencher as DBA. The results and conclusions are only marginally affected.

The analysis so far reveals a significant growth in average size of the micelles and that the polydispersity is very much dependent on the temperature. The growth in volume was also reflected in the emission kinetics of the excited complex near and above the cloud point.

CONCLUSIONS

The fluorescence quenching method in combination with single photon counting is convenient and fast to determine aggregation numbers of assemblies and gives data for which the statistics are known to a high degree of confidence. However, the theory for interpretation of the data has several shortcomings. In this paper we have shown results obtained with a modified theory which takes into account the effects of polydispersity in aggregate size. We have found that increasing the temperature in solutions of hexaethyleneglycol mono n-dodecylether in water induces a growth of the aggregation number in parallel with an increase in polydispersity. The absolute values of the aggregation numbers are uncertain but the results are qualitatively in accord with computer generated data, in which known polydispersities were present.

It is not possible to decide from the present data whether the observed growth is due to a "true" growth of the aggregates or to a secondary aggregation of smaller micelles. However, if the latter is the case, the secondary aggregation must be very intimate, so that the solubilized species may pass between the different micelles without much delay. More detailed studies are in progress.

Table V. $<s>_Q, \eta$ - data from quenching of pyrene fluorescence by dibutylaniline in $C_{12}E_6$- water solutions analyzed by Equations (11) and (12).

Temperature (^{0}C)	Gaussian distribution (Equation (11))			Exponential distribution (Equation (12))		
	$<s>_W$	$<s>_N$	σ	s_0	σ	$<s>_{Q,\eta \to 0}$
17.4	156	133	55	....		
24.7	245	213	83	....		
34.4	372	341	102	....		
44.1	(719	656	204)	121	601	1200
46.2	(801	732	225)	82	767	1600
48.7	(908	837	244)	77	991	2000
50.6	(995	921	260)	24	1278	2600

ACKNOWLEDGEMENT

This work has been financially supported by grants from the Swedish Natural Science Research Council.

REFERENCES

1. M.Corti and V.Degiorgio, J.Phys.Chem.,85, 1442 (1981).
2. P.-G.Nilsson, H.Wennerström, and B.Lindman, (in press).
3. N.J.Turro and A.Yekta, J.Am.Chem.Soc.,100, 5951 (1978).
4. M.Almgren and J.-E.Löfroth, J.Colloid Interface Sci., 81, 486 (1981).
5. M.Almgren and J.-E.Löfroth, J.Chem.Phys., 76, 2734 (1982).
6. M.Almgren, Photochem.Photobiol., 15, 297 (1972).
7. M.Almgren, F.Grieser, and J.K.Thomas, J.Am.Chem.Soc., 101, 279 (1979).
8. D.J.Miller, U.K.A.Klein, and M.Hauser, Ber.Bunsenges.Phys. Chem., 84, 1135 (1980).
9. U.Khuanga, B.K.Selinger, and R.McDonald, Aust.J.Chem., 29, 1 (1976).
10. P.P.Infelta, Chem.Phys.Lett., 61, 88 (1979).
11. P.P.Infelta, M.Grätzel, and J.K.Thomas, J.Chem.Phys., 78, 190 (1974).
12. M.Tachiya, Chem.Phys.Lett., 33, 289 (1975).
13. M.D.Hatlee and J.J.Kozak, J.Chem.Phys., 72, 4358 (1980).
14. J.C.Dederen, M.Van der Auweraer, and F.C.de Schryver, Chem. Phys.Lett., 68, 451 (1979).
15. J.-E.Löfroth and M.Almgren, J.Phys.Chem., 86, 1636 (1982).
16. U.Gösele, U.K.A.Klein, and M.Hauser, Chem.Phys.Lett., 68, 291 (1979).
17. M.D.Hatlee, J.J.Kozak, G.Rothenberger, P.P.Infelta, and M. Grätzel, J.Phys.Chem., 84, 1508 (1980).
18. M.Van der Auweraer, J.C.Dederen, E.Geladi, and F.C.de Schryver, J.Chem.Phys., 74, 1140 (1981).
19. H.Sano and M.Tachiya, J.Chem.Phys., 75, 2870 (1981).
20. J.-E.Löfroth, S.Bergström, and D.Biddle, (submitted).
21. A.E.W.Knight and B.K.Selinger, Aust.J.Chem., 26, 1 (1973).
22. International Mathematical & Statistical Libraries, Inc., Houston, TX, routine ZXSSQ.
23. N.R.Draper and H.Smith, "Applied Regression Analysis", pp 95-99, Wiley, New York, 1966.
24. N.J.Turro, M.Aikawa, and A.Yekta, Chem.Phys.Lett., 64, 473 (1979).
25. R.R.Balmbra, J.S.Clunie, J.M.Corkhill, and J.F.Goodman, Trans.Faraday Soc., 58, 1661 (1962).
26. J.B.Birks, "Photophysics of Aromatic Molecules", p. 304, Wiley, London, 1970.

FLUORESCENCE QUENCHING EQUILIBRIA STUDIES

IN IONIC MICELLES IN AQUEOUS MEDIA

K. S. Birdi, M. Meyle and E. Stenby

Fysisk-Kemisk Institut,
The Technical University of Denmark
Building 206, DK 2800 Lyngby, Denmark

The fluorescence quenching equilibria of different molecules in anionic and cationic micelles in aqueous media have been reported in the current literature. In one particular study involving SDS (Sodium Dodecyl Sulfate) – $Ru(bipy)_3^{+2}$(tris(2,2'-bipyridyl)-ruthenium (II))-(9-methylanthracene), it has been asserted that the increase in aggregation number of SDS micelles on addition of NaCl could be estimated from fluorescence quenching studies. A reinvestiagtion was therefore carried out under varying experimental conditions. This study reportes that aggregation numbers cannot be estimated from spectral measurements (based on our experimental and thermodynamic analyses). Another fluorescence quenching system: SDS-Acridine orange-Methylene blue was also studied. This system also clearly showed that the fluorescence data are not dependent on aggregation number of micelles. The fluorescence quenching of anthracene and perylene in SDS was studied in micellar solutions on addition of Cu^{2+} and Ni^{2+}. The binding constants of Cu^{2+} and Ni^{2+} were calculated which agreed with the literature data. In cetyl trimethyl ammonium bromide (CTAB) micelles, the fluorescence of anthracene was enhanced on the addition of KCl, while no effect was observed if perylene was used. In cetyltrimethyl ammonium chloride (CTAC) the fluorescence of anthracene was quenched on addition of KBr, while no effect was observed in the case of perylene. The latter data clearly show that the probe anthracene is able to interact with the counter-ions in both SDS or CTAB or CTAC systems. On the other hand, the probe perylene behaves differently in CTAB and CTAC systems, as compared

645

to SDS-perylene system. This observation leads us to conclude that fluorescence probes are not present in micelles but as distinct (detergent-probe) complexes, or as
'mixed micelles'. The structure of these mixed micelles
is found to be different from that of the pure detergent
micelles, as one would also expect from thermodynamic
considerations.

INTRODUCTION

In the current literature the applications of fluorescence
techniques to micellar and micro-emulsion systems have been extensively reported.[1] The ionic micelles are considered as suitable systems ("compartments") which exhibit dual behavior: (1) these are
able to solubilize organic water-insoluble compounds, and (2) these
are able to bind very strongly counter-ions due to the presence of
large number of charges at the micelle surface, and charged micelles
are known to behave as macroions. These properties have been suggested to be very useful characteristics for solar energy storage.[2a,2b]
However, in some fluorescence quenching studies as reported in literature, the equilibria described have been found to be unsatisfac-
tory. The particular studies we have reinvestigated are:

I.   Anionic Micelles:
I.A.   The system studied was:[2]

Micelles=sodium dodecyl sulfate, SDS
Solvent=water with or without added NaCl
Fluorescent probe=Ru(bipy)$_3^{+2}$. (Donor=D)
Quencher (Q)=9-methyl anthracene.

The equilibria assumed are given as follows:[2]

$$(M)+(D) \rightleftharpoons (M-D), \quad K_{MD} = \frac{(M-D)}{(M)(D)} \quad (1)$$

$$(M-D)+(Q) \ (M-D-Q), \quad K_{MDQ} = \frac{(M-D-Q)}{(M-D)(Q)} \quad (2)$$

where (M) is the micelle concentration ($\frac{SDS_{total}-c.m.c}{N_f}$), N_f is the
aggregation number and c.m.c. is the critical micelle concentration.
(M-D) is a micelle with donor, D. (M-D-Q) is a micelle with donor
and quencher, K_{MD} and K_{MDQ} are the respective equilibrium constants.
It was assumed (without any direct proof)[2] that all (D) and (Q) are
solubilized or "attached" to micelles. The fluorescence measured is
contributed only by (M-D) species, whereas (M-D-Q) is completely
quenched (without proof). The distribution of probe, D, and Q, was
assumed to be present in each micelle of SDS, as well as it was of

the Poisson type.[1] Based on these assumptions, a simple relation was derived which enables one to estimate the value of the aggregation number, N_f:

$$F=F_o \cdot \exp\{-(Q)/(M)\} \qquad (3)$$

$$\ln\left(\frac{F_o}{F}\right) = \frac{(Q)N_f}{(SDS_{total}-c.m.c.)} \qquad (4)$$

where (Q) and (M) are total concentrations of quencher, and micellar SDS $(=(SDS_{total}-c.m.c.)/N_f)$, respectively. F_o and F are fluorescence with and without quencher, respectively.

The estimated N_f values for different concentrations of added NaCl in SDS systems at 25 °C and 60 °C were reported.[2] However, only the N_f data at 25 °C were compared with the literature data for N_w (N_w is the weight average aggregation number) as determined from light-scattering. This comparison was not satisfactory, since the light scattering N_w data used were later found to have been unfortunately[3] carried out on impure SDS (66 % $C_{12}H_{25}SO_4Na$) sample. This prompted us to re-investigate the validity of Equation (4) at 40 °C since systematic investigations of N_n (N_n is the number average aggregation number of micelles, as determined from membrane osmometry) have been reported[3] by us for these systems. From the considerations as derived from the thermodynamics of solubilization in micelles of organic water-insoluble probes, we will show the invalidity of relation given in Equation (4), since experimental data clearly show that the magnitude of N_n ($\approx N_w$) cannot be estimated from any spectral measurements.

I.B.   SDS-Acridine orange-Methylene blue

The charge transfer complex formed between SDS and Acridine orange gives enhanced fluorescence above the c.m.c. However, addition of Methylene blue (which is also known to form a charge transfer complex with SDS) gives rise to quenching. These data are shown to fit the relation in Equation (4).

II.   Cationic Micelles

The fluorescent quenching of probes of Anthracene and Perylene by counter-ions was investigated both in SDS and in cationic (CTAB or CTAC) micelles. These systems are of much interest, since it has been reported that the fluorescence of Anthracene is quenched by counter-ions both in SDS (i.e. Cu^{2+}, Ni^{2+})[4,5] and in CTAB (i.e. Cl^-).[6,7] On the other hand, fluorescence of Perylene is unaffected by counter-ion addition in cationic micelles (CTAB or CTAC).

EXPERIMENTAL

Sodium dodecyl sulfate (SDS) (Serva, Germany)(purity was found to be >99 % after the hydrolysis and analysis of the alcohol, by the same procedure as in "Anal. Biochem. 74, 620 (1976), K. S. Birdi); cetyl trimethyl ammonium bromide (CTAB) (Fluka, Germany) were used as supplied. Cetyl trimethyl ammonium chloride (CTAC) was prepared from CTAB by adding KCl to aqueous solution of CTAB. The precipitated CTAC was re-crystallized from water three times. The c.m.c. of CTAC was determined to be 0.3 g/liter, which is in agreement with the literature value.[8] All other chemicals used were of analytical purity grade.

The fluorescence intensities were measured using (i) Perkin-Elmer MPF-3 or (ii) Zeiss spectrofluorometer. In all these studies the fluorescence of a standard (the surfactant/probe solution with no quencher) was measured prior to measuring the intensity of a quenched sample. The temperature was kept constant by using thermostated cell holders.

The probes, naphthalene or anthracene or perylene, were added to aqueous solutions from their solutions in ethanol. This method was found to give the same fluorescence as where a known amount of probe was directly added. These solutions were sonicated for 30 minutes prior to measurements.

I.   RESULTS AND DISCUSSION

I.A. Anionic Micelles (SDS)

System: SDS-Donor(Ru bipy)$_3^{+2}$ Quencher(9-methyl) anthracene).

According to the donor-quencher model as described by Turro and Yekta[2] Equation (4), we should expect the calculated values of N_f to be the same as found from osmotic pressure, N_n, or light-scattering, N_w, (or any other colligative solution data). In Figure 1 we present the values of N_f from fluorescence quenching measurements and N_n from membrane osmometry data at 40 °C as a function of added NaCl. It is clear that while N_n increases by almost ten fold in the NaCl concentration range 0-0.8 M, the value of N_f from Equation (4) increases only by a factor of about two. A similar disagreement between N_f as estimated from Equation (4) and light-scattering data at 25 °C has been reported recently.[9]

In the following we will discuss the equilibria which show why one should not expect to be able to measure N_n or N_w of micelles from any spectral measurements.

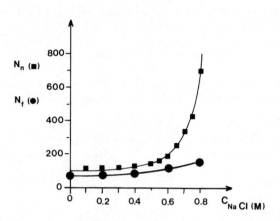

Figure 1. Dependence of N_f from Equation (4) and N_n (from osmometry data)[3] on added C_{NaCl}, at 40 °C.

[N_f data were carried out in the system containing: SDS=0.07 M, Ru(bipy)$_3^{+2}$ = 72 μM,   9-methyl anthracene = 0.82 M]

Aggregation Equilibria and Free Energy

Let us consider a system of SDS in aqueous media, where at concentrations >> c.m.c.:[10,11]

$$\rho_1 + \rho_1 \rightleftarrows \rho_2 \ , \quad K_2 = \frac{\rho_2}{(\rho_1)^2} \quad (5)$$

$$\rho_1 + \rho_1 + \rho_1 \rightleftarrows \rho_3 \ , \quad K_3 = \frac{\rho_3}{(\rho_1)^3} \quad (6)$$

$$\rho_1 + \rho_1 + \ldots \rho_1 \rightleftarrows \rho_i \ , \quad K_i = \frac{\rho_i}{(\rho_1)^i} \quad (7)$$

$$\Delta G_i^{\ominus} = -RT \ln K_i \quad (8)$$

where ρ_1 is the concentration of monomer species and ρ_i is the concentration of an aggregate made up of i monomers, the K_i and $\Delta G_i^{\ominus}$ are equilibrium constant and the standard free energy of aggregation, respectively. These equilibria may be simplified if we assume that $K_2=K_3=\ldots=K_i$, and that $\rho_1=$c.m.c., and if we further assume that the micelles are monodisperse, so that only aggregates, ρ_i, are present (this is supported by the fact that $N_w/N_n\sim 1$).

Thus, for pure micelles (i.e. in the absence of D or Q) the equilibria between monomers and micelle are given by K_i and $\Delta G_i^{\ominus}$.

Now, if one adds a probe (or some other similar component), the latter will give rise to new aggregation equilibria which now involves mixed micelle formation, i.e. detergent + probe:

if detergent: probe>>1

$$\rho_1+\rho_1+\text{probe}+\ldots\rho_j \rightleftharpoons [\rho_j-\text{probe}]+\rho_K \qquad (9)$$

and if detergent: probe>1:

$$\rho_1+\rho_1+\text{probe}\ldots+\rho_j \rightleftharpoons [\rho_i-\text{probe}] \qquad (10)$$

$$K_{j-\text{probe}}=\frac{[\rho_j-\text{probe}][\rho_K]}{(\rho_1)^K\,(\text{probe})}\ ,\ (\text{detergent: probe>>1}) \qquad (11)$$

$$=\frac{[\rho_i-\text{probe}]}{(\rho_1)^i\,(\text{probe})}\ ,\ (\text{detergent: probe>1}) \qquad (12)$$

$$\Delta G_{j-\text{probe}}^{\ominus}=-RT\ln K_{j-\text{probe}} \qquad (13)$$

It is clear that the addition of a probe to equilibria in Equations (5) to (7) would lead to a shift in all the equilibria. In other words, if the addition probes give rise to new aggregation equilibria, then c.m.c. and N would be different, as found experimentally also. This thus leads to the fact that $\Delta G_i^0 \neq \Delta G_{j-\text{probe}}^0$.

Further, the equilibria which are generally used in fluorescence, Equations (1) and (2), give equilibrium constants, K_{MD} and K_{MDQ}, which in turn give the standard free energy, $\Delta G_{MD}^{\ominus}=-RT\ln K_{MD}$ or $\Delta G_{MDQ}^{\ominus}=-RT\ln K_{MDQ}$. In all fluorescence analyses, in these equilibria, the results are analyzed by assuming that micelles, (M), behave as ideal solute. This is in contradiction to the colligative analyses, since all data show that micelle solutions are non-ideal (i.e. second virial coefficient, B_2, is non-zero, both from membrane osmometry and light-scattering).[3]

Solubilization in Micellar Solutions

The solubilization of water-insoluble organic compounds has been extensively investigated. However, it was only recently that the solubilization free energy was described in quantitative terms.[12] If a probe (such as naphthalene, anthracene, etc.) is present in excess and after the equilibrium is attained:[12]

$$\mu^{s}_{probe} = \mu^{w}_{probe} = \mu^{m}_{probe} \quad (14)$$

where μ^{s}_{probe}, μ^{w}_{probe} and μ^{m}_{probe} are the chemical potentials of the probe in solid (or liquid), water and micellar phases, respectively. This model assumes a pseudo phase model for the micellar systems.

The standard free energy of transfer for the probe from water phase to the micellar phase is given as:[12]

$$\Delta G^{\theta}_{w \to m} = -RT\ln K_{m-probe} \quad (15)$$

$$= -RT\ln \left(\frac{C^{m}_{probe}}{C^{w}_{probe}} \right) \quad (16)$$

where C^{m}_{probe} and C^{w}_{probe} are the concentrations of the probe in micellar and water phase, respectively (we can use concentrations instead of activities, assuming ideality). Our studies have clearly indicated that the magnitudes of quantity (C^{m}_{probe}) have no relation to the size of the micelle. Further, the magnitude of C^{m}_{probe} (= moles probe/mole detergent in micellar form) is independent of any change in aggregation number. Since, exactly the same thermodynamic analysis is applicable to fluorescence probes, then the micellar phase-model should be valid. The latter observation, therefore, clearly explains the absence of any dependence on aggregation number, as also found from recent experimental results, Figure 1.

It is also of interest to mention that if the magnitude of C^{m}_{probe} had a direct relation to the aggregation number, then one could have determined the value of N_n. This was suggested by some early investigators, but was clearly shown from our analyses[12] that N_n could have no relation to C^{m}_{probe}, as regards the thermodynamic equilibria given in Equation (14). In other words, there exists no evidence for the existence of such equilibria as: one probe per micelle.

Based on the above description of the solubilization of a probe in the micellar phase, it is clear that all solubilization equilibria are independent of aggregation number of pure micelles.

This thus indicates that aggregation numbers of pure micelles cannot be estimated from measurements carried out on mixed micellar systems, i.e. (M)+(D)(+(Q)).

Table I. The magnitude of C^m_{probe}, i.e. Moles Probe: Moles SDS Ratio in SDS Micellar Solutions.[12c]

Probe	C^m_{probe} (mole probe: moles SDS)
Anthracene	1:780
Naphthalene	1: 14
Azobenzene	1: 14
Phenathrene	1: 47

I.B.   The System Studied was:

SDS-Acridine orange (Donor)-Methylene blue (Acceptor, A)

A somewhat similar system, SDS-Acridine orange-10-dodecyl bromide (Donor)-Methylene blue (acceptor) has been studied by Usui and Goton.[13] The energy transfer was found to take place with a high efficiency from the photoexcited singlet state of acridine orange-10-dodecyl bromide in micelles to methylene blue (A). This study reports data where Donor, D, used was Acridine orange.

The fluorescence quenching of Acridine orange by Methylene blue was studied at various SDS and NaCl concentrations, at different temperatures.

In Figures 2-a,b we present some typical fluorescence data. It is seen that fluorescence quenching increases abruptly near the c.m.c. However, at concentrations above the c.m.c. the effect of added Methylene blue (MB) is rather small. This was also observed for Acridine orange-10-dodecyl bromide.[13] In the latter study,[13] the c.m.c. of SDS was found to be lower than in the pure SDS micelles. In our system, c.m.c. was found to remain unaffected. These data were analyzed by using the following Stern-Volmer relation (analogous to Equation (4)):

$$\ln\left(\frac{F_o}{F}\right) = \frac{a(C_{MB})}{(C_{SDS}-c.m.c.)} \quad (17)$$

where SDS in the micellar phase (in monomeric units) is $(C_{SDS}-c.m.c.)$ (C_{MB}) is the concentration of Methylene blue added, and a is a constant ($\equiv N_f$ in Equation (4)). The plots of $\ln(F_o/F)$ versus (C_{MB}) are linear for varying $(C_{SDS}-c.m.c.)$, Equation (17), see figure 3 and Table II.

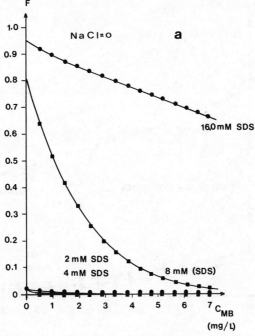

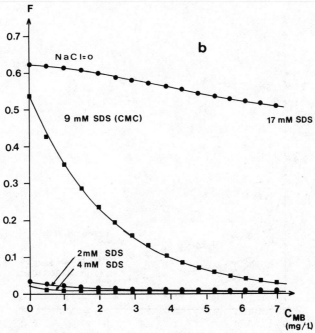

Figure 2. Variation of fluorescence (F) on addition of quencher, C_{MB}, to SDS solutions, (a) 25 °C, (b) 40 °C (with different NaCl concentration). ($C_{Acridine \ orange}$ = 3µM).

(continued)

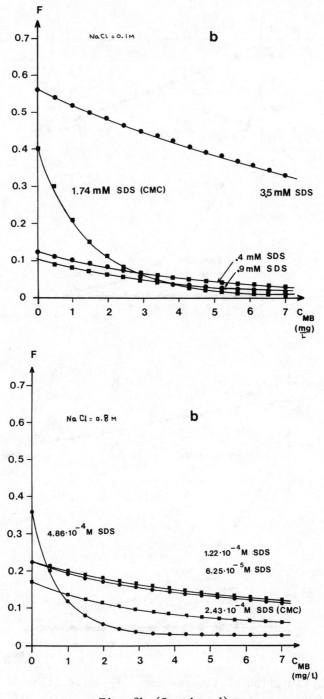

Fig. 2b (Continued)

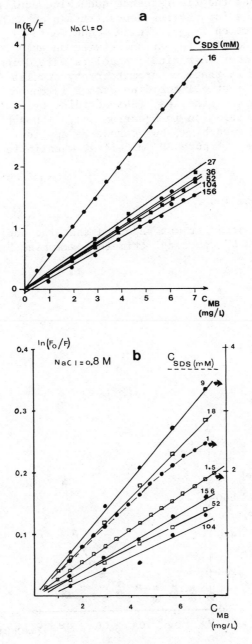

Figure 3. Plots of $\ln(F_0/F)$ versus C_{MB} (quencher) for varying concentration of SDS. (a) 25 °C (NaCl=0), (b) 40 °C (NaCl=0.8 M).

These data show that the relation in Equation (4) (or Equation (17)) describes the fluorescence quenching equilibria. The magnitude of constant a for varying concentration of NaCl was found to be (weakly) dependent on C_{NaCl}, Figure 4 and Table II. In these systems charge-transfer complex formation between an anionic detergent (SDS) and a cationic donor (Acridine orange) is well documented.[13,14] Further, our gel filtration chromatography studies (unpublished) clearly indicate that all Acridine orange is bound to (SDS). This indicates that mixed micelles, (SDS-Acridine orange), are formed. In other words, the relation in Equation (4), although valid as regards the fluorescence quenching, but cannot be applied to estimate the aggregation number of pure SDS micelles, when in fact mixed micelles are present.

II. Cationic Micelles

In a recent study it was reported that the addition of counter-ions, such as F^-, to cetyl trimethyl ammonium bromide, (CTAB),

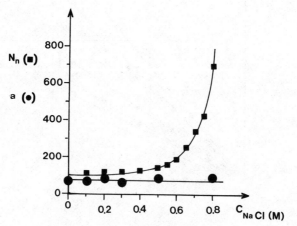

Figure 4. Variation of constant a in Equation (17) and N_n in SDS Systems, with added C_{NaCl}, at 40 °C.

Table II. Values of constant a in Equation (17) for SDS-Acridine orange-Methylen blue systems.

C_{NaCl} (M)	25 $^{\circ}$C		40 $^{\circ}$C				
	0	0.2	0	0.2	0.4	0.6	0.8
a	45	68	43	60	83	100	129
Correlation coefficient (r^2)	0.99		0.97				

gave rise to an increase in the fluorescence of some probes (naphthalene, anthracene) while no change was observed in the case of perylene. This study was carried out further by using CTAB and CTAC (cetyl trimethyl ammonium chloride). The fluorescence change of probe, anthracene, on addition of (Br$^-$) or (Cl$^-$) to CTAC or CTAB is shown in Figure 5. As expected, the fluorescence decreases as stronger quencher, (Br$^-$), replaces a weaker quencher, (Cl$^-$), in the case of (CTAC). The reverse is observed for the CTAB system. These data show that fluorescence quenching is useful as regards the replacement of different counter-ions at the micellar surface. These equilibria in CTAB and CTAC micelles have been studied. A more thorough analysis will be given in a later report.

It is of interest to mention that the fluorescence of perylene is neither affected by (Cl$^-$) in CTAB nor by (Br$^-$) in CTAC micellar systems, Figure 5. In other words, this observation can only be explained by accepting the model as proposed here, i.e. the probe and detergent form mixed micelles. The structure of mixed micelle of (perylene-CTAB) or (perylene-CTAC) is different than the complexes with other probes, so that the fluorescence quenching by counter-ion is absent. The fluorescence life-time, τ, (in CTAB) data suggest no evidence for any relation between τ and the fluorescence quenching[6] ($\tau_{naphthalene}$=53 ns; $\tau_{anthracene}$=5 ns; $\tau_{perylene}$=6.4 ns).

We have further investigated the comparative fluorescence quenching of different probes, naphthalene, anthracene and perylene, in (i) cationic and (ii) anionic micelles by the anion and cation counter-ions, respectively.

The fluorescence quenching of different probes in SDS micelles by different counter-ions is given in Figure 6. It is seen that in these systems, perylene is quenched in the same manner as the other probes. The binding constants for Cu^{2+} and Ni^{2+} were found to be the same as reported in the literature ($K_{Cu2+} \approx K_{Ni2+} \sim 9000 \ M^{-1}$). These data are in contrast to the observation in Figure 5, where it was shown that perylene did not interact with counter-ions in CTAB or CTAC micelles. If we accept the above described micelle solubilization model[2,4,5,6,7] both in SDS and CTAB or CTAC, then the probe, perylene, is solubilized in the hydrophobic interior of the micelles.

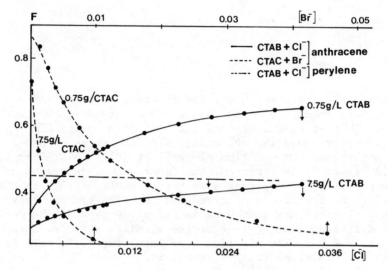

Figure 5. Variation of fluorescence, F, (i) on addition of KCl to CTAB, and (ii) on addition of KBr to CTAC. Fluorescence probes: Anthracene and perylene.

Hence, it should be situated in a similar manner, both in SDS or CTAB or CTAC micelles. However, if we have differences in quenching for any particular probe, then this clearly supports our postulate that solubilization being a mixed micelle formation, leads to different complex formation between any probe and a detergent. This mixed micelle is different in structure than the pure micelle and thus cannot provide information about the aggregation number of pure micelles. In fact, studies show that as the concentration of a solubilizate (probe) increases in relation to detergent, both the c.m.c. and aggregation number change. This means that the standard free energy of system changes from that of a pure micelle (Equation (8)) to that of a mixed micelle (Equation (13)).

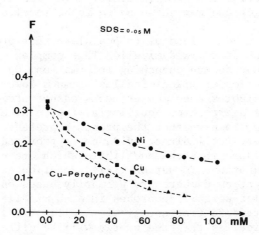

Figure 6. Fluorescence change, F, of anthracene probe on addition of Cu^{2+} or Ni^{2+} to SDS solutions. (25 °C).

CONCLUSION

The various fluorescence studies which have been tried to estimate micellar aggregation numbers are shown to be suspect[2,9,15,16]. In all the fluorescence quenching models[2,4,5,6,7,9,13,14,15,16] it is assumed that the aggregation equilibria of pure micelles remain unaffected on the addition of a probe (such as naphthalene, anthracene, 9-methyl anthracene, etc.). This assumption is in contradiction to the fact that since addition (or solubilization) of a probe gives rise to mixed micelle formation, then the standard free energy of aggregate formation would be different than that in the case of pure micelles. The latter observation is supported by the fact that the addition of benzene or hexane leads to a drastic change in the c.m.c. and aggregation number.

Further, the fluorescence measurements are carried out in mixed micelles and as such provide only information about these systems, i.e. mixed micelles. If aggregation numbers could possibly be estimated from relation in Equation (4), then it would also suggest that solubilization of water-insoluble organic compounds in micelles could provide the same information (as was already suggested a few decades ago, and was shown by us to be invalid).[12]

The above studies lead us to postulate that probe-detergent complex is formed as a mixed micelle. This complex is the one that provides the fluorescence quenching information. On the other hand, if the concentration of mixed micelles is much lower than the pure micelles, then fluorescence measurements cannot provide any information about the aggregation equilibria of pure micelles. Further, it remains to be shown whether the probe is present in micelles or in mixed micelles. Until direct evidence is provided, our postulate as given above, describes the equilibria which are consistent with thermodynamic equations (Equation (16)). The fluorescence data for measuring aggregation numbers are primarily based on the assumption that pure micelles solubilize probes in each micelle, without any effect on the standard free energy of micellization. We have shown that in mixed micelles the free energy will be different from that in the pure micelles.

These arguments allow us to conclude that micellar aggregation numbers cannot be estimated from fluorescence measurements, where the presence of probes could lead to change in the standard free energy of micelle formation.

ACKNOWLEDGEMENTS

The Danish Council for Scientific and Industrial Research (STVF) is thanked for the partial support of this project.

REFERENCES

1.   J. A. Fendler and E. J. Fendler, "Catalysis in Micellar and Macromolecular Systems", Academic Press, New York, 1975.
1a. V. N. J. Turro, M. Grätzel and A. M. Braun, Angew. Chem., 92, 712 (1980).
2.   N. J. Turro and A. Yekta, J. Am. Chem. Soc. 100, 5951 (1978).
2a. M. Calvin, Science 184, 375 (1974).
2b. J. S. Connolly, editor, "Photochemical Conversion and Storage of Solar Energy", Academic Press, New York, 1981.
3.   K. S. Birdi, S. U. Dalsager and S. Backlund, J. C. S. Faraday I 76, 2035 (1980).

4. H. W. Ziemiecki, R. Holland and W. R. Cherry, Chem. Phys. Letters 73, 145 (1980).

5. H. W. Ziemiecki and W. R. Cherry, J. Am. Chem. Soc. 103, 4479 (1981).

6. H. D. Burrows, S. J. Formosinho and M. F. J. F. Paiva, J. Photochem. 12, 285 (1980).

7. H. D. Burrows, S. J. Formosinho and M. F. J. R. Paiva, E. J. Rasburn, J. C. S. Faraday II 76, 685 (1980).

8. P. Mukerjee and K. J. Mysels, "Critical Micelle Concentration of Surfactant Systems", Nat. Stand. Ref. Data, NBS-36, Washington, D. C., 1971.

9. M. Almagren and J. E. Löfroth, J. Colloid Interface Sci. 81, 486 (1981).

10. A. Ben-Naim, "Hydrophobic Interactions", Plenum Press, New York, 1980.

11. K. S. Birdi, presented at the Indian Science Congress, 69th Session, Mysore, 1982.

12a. K. S. Birdi, Colloid Polymer Sci. 254, 1059 (1976);

12b. K. S. Birdi, in "Micellization, Solubilization and Microemulsion", K. L. Mittal, Editor, Vol. I, p. 151, Plenum Press, New York, 1977;

12c. K. S. Birdi and J. Steinhardt, Biochim. Biophys. Acta 534, 219 (1976).

13. Y. Usui and A. Goton, Photochem. and Photobiol. 29, 165 (1979)

14. H. Sato, M. Kawasaki and K. Kasatani, J. Photochem. 17, 243 (1981).

15. P. Lianos and R. Zana, J. Phys. Chem. 84, 3339 (1980).

16. P. K. F. Koglin, D. J. Miller, J. Steinwandel and M. Hansen, J. Phys. Chem. 85, 2363 (1981).

FLUORESCENCE QUENCHING IN MICELLAR SYSTEMS

F.C. De Schryver[*], Y. Croonen, E. Geladé, M. Van der
Auweraer, J.C. Dederen, E. Roelants and N. Boens
Department of Chemistry
K.U.Leuven
Celestijnenlaan 200 F, B-3030 Heverlee, Belgium

The use of the excited state properties and
particularly of the fluorescence of an added probe in
non homogeneous systems such as micelles, anionic as
well as cationic, and inverse micelles has expanded tre-
mendously over the past few years. Using the single photon
counting technique and steady state fluorescence it is
possible to determine a number of important parameters
such as the aggregation number, the rate constants of
exit and entrance of compounds and the quenching rate
constants in these systems.

INTRODUCTION

Can the study of the luminescence of the excited state of a probe
added to or built in a micellar system provide information on
structural or dynamic aspects of these aggregates ?
Before answering this question one should state that excited state
luminescence can be studied using either continuous or pulsed excitation. The former mode usually leads to spectral information providing intensities of emission of the probe as a function of the
wavelength or wavenumber while the latter permits the observation
of the time dependence of the intensity of emission either at one
wavelength or, in time resolved emission spectroscopy, of the total
emission spectrum.
In the present discussion emphasis will be placed on fluorescence,
although the possible use of phosphorescence has been pointed out.[1]
Within the limitations typical of external probing[2] one can try to
relate the spectral changes (intensity or wavelength) or the time

dependent characteristics (decay parameters) to properties of the
micellar system such as polarity[3], viscosity[4], partitionic or size.
In this paper emphasis will be placed on the use of fluorescence
quenching, a technique well established in homogeneous medium, for
obtaining information on micellar systems.
On the basis of the following assumptions
    a) an equilibrium between the micelle and the added probe and
quencher,
    b) a Poisson distribution, first proposed by Almgren[5], of
probes and quencher at low concentrations,
    c) no of minimal influence of the added probe or quencher in
the concentration range used on the micellar properties,
    d) residence of the probe in the micelle within the excited
state lifetime,
    e) description of the quenching as a monomolecular process,
the following kinetic scheme for fluorescence quenching can be proposed :

Scheme I

Process	Rate (n and j = 0,1,2,...)
1) $P_n^* \rightarrow P_n$	$k_f[P_n^*]$
2) $P_n^* \rightarrow P_n$	$n\,k_{qm}[P_n^*]$
3) $P_n^* + Q_w \rightarrow P_{n+1}^*$	$k_m^+[P_n^*][Q_w]$
4) $P_n^* \rightarrow P_{n-1}^* + Q_w$	$n\,k_m^-[P_n^*]$

P_n is a micelle with n quenchers and a probe

k_f rate constant of fluorescence decay in a micellar solution
    without quencher

k_{qm} rate constant for intramicellar quenching when only one
    quencher is present in the micelle

k_m^- rate constant for a quencher to leave the micelle

k_m^+ rate constant for a quencher to enter the micelle

This scheme leads to a time dependence of the intensity of the
following form[6]

$$I_t = A_1 \exp[-A_2 t - A_3(1-\exp(-A_4 t))] \qquad \text{Equation (1)}$$

The meaning of the parameters experimentally accessible from single
photon counting experiments has been described more explicitly.[7]
The form of the decay curves depends on the relative value of the
different rate constants. It is important to note that the time dependence is complex (not exponential) if $k_{qm} > k_m^-$ but simplifies
to an exponential decay if $k_{qm} \ll k_m^-$. If k_m^- is very small, that is
if the quencher will not exit from the micelle in the lifetime of

the excited state of the probe, the time dependent intensity at long
times will be independent of the quencher concentration and will be
equal to the inverse of the probe lifetime in absence of quencher.
These situations are represented in Figure 1.
Furthermore the following expression can be derived for the ratio
of the emission intensity in absence (I_0) and in the presence of
variable quencher concentrations $(I)^8$

$$\frac{I_0}{I} = \left(\frac{A_2}{k_f}\right)\left\{1 - A_3 e^{-A_3} \int_0^1 x^{A_2/A_4} e^{A_3 x}\, dx\right\}^{-1} \qquad \text{Equation (2)}$$

the parameters A are related to Equation (1) and k_f equals one
over the lifetime of the probe in absence of quencher. This
relation can be checked with quenching data obtained from
fluorimetry (continuous excitation).
The validity of Scheme I and its usefulness will be looked into for three types of micelles : an anionic one (SDS), a cationic
one (CTAC) and an inverse micellar system (DAP). Furthermore quenching can lead to excited state complex formation. The properties
of this complex will provide information on the polarity in the
the neighbourhood of waterpools in inverse DAP micelles.

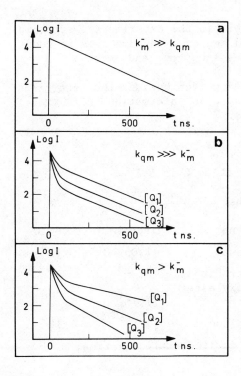

Fig.1

Influence of the value of the
rate constant k_{qm} relative to
k_m^- on the form of the decay
curves

a) $k_m^- \gg k_{qm}$: one-exponential
   decay
b) $k_{qm} \ggg k_m^-$: time dependent
   intensity at long times independent of quencher concentration
c) $k_{qm} > k_m^-$: time dependence
   is complex

RESULTS AND DISCUSSION

a) Sodium Dodecylsulfate (SDS) : Micelle Forming Anionic Surfactant

In view of the existing knowledge of SDS[9] the validity of the above
mentioned scheme was checked using a number of probes and neutral
quenchers (Table I) by analysing the fluorescence decay as measured
by single photon counting. For the quenchers mentioned in Table I
$k_{qm} > k_m^-$ except for 1,3,5-benzenetricarboxylic acid trihexylester
where $k_{qm} >>> k_m^-$.

Table I. N_{agg} of SDS in the Concentration Range Between 1 and 5
10^{-2} M at 24°C.

Probe	Quencher	N_{agg}
1-methylpyrene	m-dicyanobenzene	66
	1,3,5-benzenetricarboxylic acid trihexylester	62
pyrene	m-dicyanobenzene	73
diethylindolo indole	αCN-naphthalene	66
dibutylindolo carbazole	α-cyanonaphthalene	66

From these data it is clear that, within the experimental error involved in this method, the observed aggregation number is in good
agreement with other methods[10] using low probe and quencher
concentrations.
The analysis according to Scheme I provides besides the determination of the mean aggregation number, also information on the rate
constants k_m^-, k_m^-, 7a and b, and 18a.
In Table II pertinent data related to neutral quenchers are assembled.

Table II. k_m^+, k_m^- at 24°C in SDS

Probe	Quencher	$k_m^-(s^{-1} \, lmol^{-1})$	$k_m^- (s^{-1})$
pyrene	m-dicyanobenzene	1.1×10^{10}	7.6×10^6
1-methylpyrene	m-dicyanobenzene	1.05×10^{10}	5×10^6
diethylindoloindole	1,3,5-benzenetricarboxylic acid trihexylester	-[a]	-[a]
diethylindoloindole	m-dicyanobenzene	1×10^{10}	5×10^6
diethylindoloindole	α-cyanonaphthalene	-[b]	5×10^5

(a) $k_m^- <<< k_{qm}$, no exit within the excited state lifetime;
(b) could not be determined

These results clearly show that the exit rates are probe independent
but decrease as the size of the aromatic part or hydrophobic part
of the quencher increases. This is of course related to the
partition between an organic and an aqueous phase. The values obtained are in agreement with minimum values for aromatic molecules
obtained by phosphorescence measurements.[11] Extrapolating these
data to the probes used one can safely say that $k_m^- <<< 1/\tau^\circ$ where
τ° represents the lifetime of the singlet excited state of the probe in absence of quencher. The rate constant of entrance for a
neutral probe can be seen to be diffusion controlled.
Another parameter which can be determined is the quenching rate constant k_{qm}. The physical meaning of the quenching process in a micelle has been described by a theoretical model.[12] The value of
the quenching constant can be related to the quenching in homogeneous medium. [12,18a] If quenching occurs by electron transfer the value of the rate constant can be related to the free energy changes
upon quenching.[7a] A difference in k_{qm} comparing a probe and two
quenchers, which in homogeneous medium are shown to be diffusion
controlled, reflects differences in average solubilization site.
This explains why k_{qm} for α-cyanonaphthalene in micellar medium is
smaller than k_{qm} for p-cyanotoluene although the former is for the same
probe certainly a better quencher. Using the system 1-methylpyrene-meta-
dicyanobenzene one can further investigate the influence of ionic
strength, temperature and micelle concentration on the different
parameters.[13] This is exemplified for the mean aggregation number
in Table III. It is important to note that k_m^- and k_m^+ are for a
neutral quencher independent of the ionic strength and the micelle
concentration.

Table III.   Influence of Ionic Strength, SDS Concentration, Temperature and Counterion on the Mean Aggregation Number of SDS.

a) Ionic Strength (a)	N_{agg}
0 M NaCl	66
0.6 M NaCl	182
b) SDS Concentration (b)	
0.07 M SDS	63
0.73 M SDS	91
c) Temperature (c)	
t = 20.4°C	66
t = 51.5°C	58
d) Counterion	
0.01 M CaCl$_2$ at 51°C	90

(a) see also References 10c, 14, 15, 17; (b) see also Reference 16;
(c) see also Reference 14.

As stated in the Introduction the validity of the Scheme (I) can be checked against fluorescence quenching measurements using continuous irradiation. From the comparison of the calculated I_0/I value, based on single photon counting data, with the experimentally obtained Stern-Volmer plot one observes that in the quenching of 1-methylpyrene with metadicyanobenzene a deviation occurs for quencher concentrations exceeding 7.8×10^{-4}M. This can be related to static quenching.[13]

So far, only neutral quenchers have been discussed. Arenes such as pyrene or 1-methylpyrene can also be quenched using metal ions such as Cu^{2+}, Tl^+ or Ag^+.[18] If the attention is focussed on Cu^{2+} one observes that at low quencher concentration the mean aggregation number is uninfluenced ($N_{agg} = 62$). Contrary to data obtained with the neutral quencher it is found that the value of k_m^- depends on the micelle concentration (Table IVb). We described this effect to be due to a hopping process[18a] between two micelles. Almgren[18b] indicated that the dependence of k_m^- on the micelle concentration could be related to the influence of the electrostatic field of the surrounding micelles. That the ionic strength influences the value of k_m^- for ionic quenchers can be seen in Table IVa.

Table IV. Influence of Micelle Concentration and Ionic Strength upon k_m^- of Cu^{2+} in SDS at 24°C.

a. Ionic Strength

[NaCl]	$k_m^- \ (s^{-1})$
0 M	1.2×10^5
0.3 M	1.3×10^7

b. Micelle Concentration

1.8×10^{-4} M	2.4×10^5
3.1×10^{-3} M	3.6×10^6
4.68×10^{-3} M	9.4×10^6

These data indicate that the electrostatic potential strongly influences the value of the exit rate for ionic quenchers.

b) Aggregates of Hexadecyltrimethylammoniumchloride in Water (CTAC)

Much less information is available on micelles based on hexadecyltrimethylammonium halogenides. Values of the mean aggregation for
CTAB reported vary from 61 to 96.[19] In the type of experiments
used here to determine the different parameters describing the
micellar system the bromide will also quench. Therefore
CTAC was chosen as detergent. Little information on the mean
aggregation number of CTAC is published.[20] To determine the mean
aggregation number the fluorescence quenching of 1-methylpyrene in
CTAC micelles (detergent concentration between 3 10^{-2} and 8 10^{-2}M)
was studied using three kinds of quenchers. The first quencher,
I^-, an efficient ionic quencher, has a $k_{qm} > k_{\bar{m}}$ and the rate constant for exit, $k_{\bar{m}}$, is dependent on the micelle concentration
(ionic strength). The second quencher, metadicyanobenzene, has a
$k_{qm} > k_{\bar{m}}$ but no dependence of $k_{\bar{m}}$ on the micelle concentration is
observed. The third quencher, N-methyl-N-decylaniline, is an effi-
cient non-ionic but strongly lipophilic quencher and will during the
lifetime of the excited state remain in the micelle. The kinetics
of quenching is hence determined only by the Poisson distribution
of probe and quencher. This means that in Equation 1

   A_2 = 1/lifetime of the probe in absence of quenchers

   A_3 = <n> , the mean quencher occupancy

   $A_4 = k_{qm}$

The mean aggregation number obtained for these three quenchers
equals 115 $\pm$ 5 at 24°C. This value is higher than the one reported[20] extrapolated at the CMC.

It is furthermore noteworthy that $k_{\bar{m}}$ for metadicyanobenzene in CTAC
is,within experimental error,identical to the value in SDS.

c) Aggregates of Dodecylammoniumpropionate (DAP) in Apolar Media :
                Inverse Micelles in Apolar Media

Excited state properties of probes and in particular fluorescence
quenching can also be used in the study of inverse micellar systems.[21,22] Atik and Thomas[22] and we[23] were recently able to analyze the fluorescence intensity as a function of time by a four
parameter fit (Equation 1).
Using a probe which is bound to the micelle surface by a polar head
group such as 2-(1-naphthyl)acetic acid and iodide or nitrate ions,
dissolved in the waterpool, itwas possible to determine the mean
aggregation number for DAP at low water concentration[23]   (Table
V).
Using a series of probes of the general structure ω -(1-naphthyl)
$(CH_2)_n$-COOH and triethylamine as quencher exciplex emission could
be observed[24] and related to the dielectric constant of the environment where complex formation occurred (Table VI).

Table V. Mean Aggregation Number of DAP at 24°C using 2-(1-Naphthyl) Acetic Acid as Probe.

$\dfrac{[H_2O]}{[DAP]}$	Quencher	N_{agg}
1.375	I^-	30
	NO_3^-	28
2.75	I^-	40
	NO_3^-	37
4.125	I^-	67

Table VI. Exciplex Emission in an Inverse Micellar System.

Probe	$\lambda_{exciplex}$(nm) maximum	ε
1-Me naphthalene	408	2.0058
12(1-naphthyl)dodecanoic acid	413	2.2
6(1-naphthyl)hexanoic acid	423	2.65
4(1-naphthyl)butyric acid	427	2.9
2(1-naphthyl)acetic acid	not observed	> 5

$[H_2O]/[DAP]$ = 1.375; [DAP] = 0.12 M; cyclohexane
[Triethylamine] = 2.1×10^{-2} M.

CONCLUSION

Luminescence properties of properly chosen probes quenched by ionic or non-ionic quenchers lead to information on the mean aggregation number, the exit and entrance rate of the quencher and the rate constant for fluorescence quenching. It is a method of choice to obtain detailed information on parameters that influence them.

ACKNOWLEDGEMENTS

The authors are indebted to the FKFO and the University Research Fund for financial support to the laboratory. Fellowships by the NFWO (to MvdA, EG and NB) and by the IWONL (YC, JCD) are gratefully acknowledged.

REFERENCES

1. T. Okubo, N.J. Turro, J.Phys.Chem., 85, 4034 (1981)
2. N.J. Turro, M. Grätzel and A.M. Braun, Agnew. Chem., 19, 675 (1980)
3. a) K. Kalyanasundaram, J.K. Thomas, J.Am.Chem.Soc., 99, 2039 (1977); b) R. Nakajima, Bull.Chem.Soc.Japan, 50, 2473 (1977)
4. a) M. Grätzel, J.K. Thomas, J.Am.Chem.Soc., 95, 6885 (1973); b) H.J. Pownall, L.C. Smith, J.Am.Chem.Soc., 95, 3136 (1973); c) R.C. Dorrance, T.F. Hunter, Faraday Trans.I, 68, 1312 (1972) d) S.S. Atik, L.A. Singer, J.Am.Chem.Soc., 101, 6759 (1979)
5. M. Almgren, Photochem. and Photobiol., 15, 297 (1972)
6. a) N.J. Turro and A. Yekta, J.Am.Chem.Soc., 100, 595 (1978); b) K. Kalyanasundaram, M. Grätzel and J.K. Thomas, J.Am.Chem. Soc., 97, 3915 (1975); c) S.S. Atik, M. Nam and L. Singer, Chem.Phys.Lett., 67, 75 (1979); d) M. Grätzel, K. Kalyanasundaram and J.K. Thomas, J.Am.Chem.Soc., 96, 7877 (1974)
7. a) M. Van der Auweraer, J.C. Dederen and F.C. De Schryver, J. Am.Chem.Soc., 104, 1800 (1982); b) J.C. Dederen, M. Van der Auweraer, F.C. De Schryver, Chem.Phys.Lett., 68, 451 (1979); c) A.M. Tachiya, Chem.Phys.Lett., 33, 289 (1975); d) M. Tachiya, J.Chem.Phys., 76(1), (1982); e) P.P. Infelta, M. Grätzel and J.K. Thomas, J.Phys.Chem., 78, 190 (1974)
8. a) P.P. Infelta, Chem.Phys.Lett., 61, 88 (1979); b) F. Grieser, Chem.Phys.Lett., 83(1), (1981)
9. B. Lindman and H. Wennerström, "Topics in Current Chemistry", N.87, Springer-Verlag, Heidelberg, pp 1-85 (1980).
10. a) K.J. Mysels and L.H. Princen, J.Phys.Chem., 63, 1699 (1959); b) E.A.G. Aniansson, S.N. Wall and M. Almgren, J.Phys.Chem., 80, 905 (1976); c) P. Lianos and R. Zana, J.Phys.Chem., 84, 3339 (1980); d) D.J. Miller, U.K.A. Klein and M. Hauser, Ber. Bunsenges.Phys.Chem., 84, 1135 (1980); e) P. Lianos, J. Lang, C. Strazielle and R. Zana, J.Phys.Chem., 86, 1019 (1982); f) K. Shinoda, "Colloïdal Surfactants" Academic Press (1963); g) M. Aikawa, A. Yekta and J. Turro, Chem.Phys.Lett., 68, 285 (1979)
11. M. Almgren, F. Grieser and J.K. Thomas, J.Am.Chem.Soc., 101, 279 (1979)
12. a) M. Van der Auweraer, J.C. Dederen, E. Geladé and F.C. De Schryver, J.Chem.Phys., 74, 1140 (1981); b) H. Sano and M. Tachiya, J.Chem.Phys., 75(6), (1981); c) M.D. Hatlee, J.J. Kozak, G. Rothenberger, P.P. Infelta and M. Grätzel, J.Phys. Chem., 84, 1508 (1980); d) V. Gösele, U.K.A. Klein and M. Hauser, Chem.Phys.Lett., 68, 291 (1979); e) M. Tachiya, Chem.Phys. Lett., 69, 605 (1980)
13. Y. Croonen, E. Geladé, M. Van der Zegel, M. Van der Auweraer, H. Vandendriessche, F.C. De Schryver and M. Almgren, submitted
14. a) S. Hayashi and S. Ikeda, J.Phys.Chem., 84, 749 (1980); b) N.A. Mazer, G.B. Benedek and M.C. Carey, J.Phys.Chem., 80, 1075

672   F. C. DE SCHRYVER ET AL.

(1976); c) P.J.Missel, N.A.Mazer, G.B. Benedek, C.Y.Young and
M.C.Carey, J.Phys.Chem. 84, 1044 (1980)

15. a) D.J.Miller, Ber.Bunsenges.Phys.Chem., 85, 337 (1981); b) P.K.F.
Koglin, D.J.Miller, J.Steinwandel and M. Hauser, J.Phys.Chem.
85, 2363 (1981)

16. a) J.B. Hayter and J. Penfold, J.Chem.Soc.Faraday Trans. I, 77,
1851 (1981); b) P.Lianos, J.Lang, C. Strazielle and R.Zana,
J.Phys.Chem., 86, 1019 (1982)

17. M. Almgren and J.E. Lofroth, J.Colloid.Interface Sci., 85, 928
(1981)

18. a) J.C. Dederen, M. Van der Auweraer and F.C.De Schryver, J.Phys.
Chem., 85 (1981); b) M. Almgren, G. Gunnarson and P.Linse, Chem.
Phys.Lett., 85, 451 (1982) and references cited herein; c) H.W.
Ziemiecki, R. Holland and W.R. Cherry, Chem.Phys.Lett., 73, 145
(1980); d) H.W. Ziemiecki, W.R. Cherry, J.Am.Chem.Soc., 103, 4479,
(1981); e) F. Grieser and R. Tausch-Treml, J.Am.Chem.Soc., 102,
7259 (1980); f) M.A.J. Rodgers and M.F. da Silva e Wheeler, Chem.
Phys.Lett., 53, 165 (1978); M.A.J. Rodgers and M.F. da Silva e
Wheeler, Chem.Phys.Lett., 43, 587 (1976)

19. a) B.Selinger and A. Watkins, Chem.Phys.Lett., 56, 99 (1978);
b) M. Aikawa, A. Yekta and N.J. Turro, Chem.Phys.Lett., 68, 285
(1979); c) E.J. Fendler and J.F. Fendler, in "Advances in Physical
Chemistry", Ed. by Gold, Vol.8, 271-397 (1970); d) R.C. Dorrance
and T.F. Hunter, J.C.S. Faraday Trans. I, 70, 1572 (1974); e) N.J.
Turro and Y. Tanimoto, Photochem. Photobiol., 34, 157 (1981)

20. a) F. Reiss-Husson and V. Luzatti, J.Phys.Chem., 68, 3504 (1964)
b) H. Singh and W.L. Hinze, Anal.Lett., 15, 221 (1982); c) P.
Mukerjee and K.J. Mysels, NSRDS-NBS-36, Washington D.C. (1971);
d) S.S. Atik, M. Nam and L.A. Singer, Chem.Phys.Lett., 67, 75
(1979); e) Reference 19(e)

21. a) J.H. Fendler, Acc.Chem.Res., 9, 153 (1976); b) M. Wong and
J.K. Thomas, in "Micellization, Solubilization and Microemulsions",
K.L. Mittal, Editor, vol.2, p.647, Plenum Press, N.Y. (1977); c)
P.D.I. Fletcher and B.H. Robinson, Ber.Bunsenges.Phys.Chem., 85,
863 (1981); d) E. Keh and B. Valeur, J.Colloid Interface Sci., 79,
465 (1981); e) D.J. Miller, U.K.A. Klein and M. Hauser, J.C.S.
Faraday Trans. I, 73, 1654 (1977); f) M. Wong, J.K. Thomas and
M.Grätzel, J.Am.Chem.Soc., 98, 2391 (1976); g) H.F. Eicke, J.C.W.
Shepherd and A. Steinemann, J.Colloid Interface Sci., 56, 168 (1976);
h) G.D. Correll, R.N. Cheser, III, F. Nome and J.H. Fendler, J.Am.
Chem.Soc., 100, 1254 (1978); i) M.A.J. Rodgers and J.C. Becker, J.
Phys.Chem., 84, 2762 (1980); j) T.F. Hunter and A.I. Younis, Fara-
day Trans. 1, 75, 550 (1979); k) R.Mc Neil and J.K. Thomas, J.
Colloid Interface Sci., 83, 57 (1981)

22. a) S.S. Atik and J.K. Thomas, Chem.Phys.Lett., 79, 351 (1981);
b) S.S. Atik and J.K. Thomas, J.Am.Chem.Soc., 103, 3543 (1981);
c) S.S. Atik and J.K. Thomas, J.Am.Chem.Soc., 103, 4367 (1981);
d) S.S. Atik and J.K. Thomas, J.Phys.Chem., 85, 3921 (1981)

23. E. Geladé and F.C. De Schryver, J.Photochem., 18, 223 (1982)

24. E. Geladé, N. Boens and F.C. De Schryver, J.Am.Chem.Soc.,
104, 6288 (1982)

FT-IR STUDIES OF AQUEOUS SURFACTANTS:

THE TEMPERATURE INDUCED MICELLE FORMATION

H.H. Mantsch, V.B. Kartha and D.G. Cameron

Division of Chemistry
National Research Council of Canada
Ottawa, Ontario, K1A 0R6, Canada

Fourier transform infrared spectroscopy is presented as a technique with great potential for the study of aqueous surfactants, especially when investigating structural changes brought about by a change in temperature or concentration. Micelle formation in aqueous surfactants can be induced by either an increase in concentration through the critical micelle concentration range, or by an increase in temperature through the critical micellization temperature. Due to the concentration range involved, (i.e. 0.1 to 0.3 M surfactant), infrared spectroscopy is particularly suitable for the study of the temperature-induced micellization, and we discuss herein details of the structural changes observed at the critical micellization temperature of a number of anionic surfactants. The thermotropic phase behavior of these aqueous surfactants was monitored via structural changes specific to the hydrophobic acyl chains and those of the hydrophilic carboxylate and sulfate head group moieties. The temperature-induced changes in infrared spectral parameters such as frequency and bandwidth characterize the critical micellization temperature of these surfactants as a phase transition from a conformationally ordered, solid-like phase to an isotropic micellar phase. The concept of a critical micellization temperature is discussed in relation to that of the Krafft point and the critical micelle concentration.

INTRODUCTION

Despite a very old history, as far as spectroscopic
techniques are concerned, conventional dispersive infrared
spectroscopy has been used very little for the study of aqueous
surfactants. The prime reason is the high infrared absorption
of water, or heavy water, which limit the energy throughput to
the extent that dispersive instruments are only of limited use.
However, the advent of Fourier transform infrared (FT-IR)
spectroscopy with a very high energy throughput and the multi-
plexing advantage, coupled with the use of fast, liquid nitrogen
cooled detectors, have made it now possible to explore areas
previously dismissed as being incapable of study by infrared
spectroscopy. One of these areas is that of aqueous
surfactants.

We have recently reported on the application of FT-IR
spectroscopy to the study of the concentration range of the
critical micelle concentration (CMC) of a series of short chain
sodium n-alkanoates, ranging from propionate to decanoate.[1,2]
These studies have provided valuable information concerning the
self-association of such systems and the structural changes
observed at the CMC. In the present report we discuss the
application of FT-IR spectroscopy to the study of long chain
n-alkanoates and n-alkyl sulfates.

METHODOLOGY

Materials and sample preparation. The sodium n-alkanoates
were commercial samples (purity ≥99) obtained from Nu Chek,
Elysian, MN. The lithium, ammonium, sodium, potassium, rubidium
and cesium hexadecyl sulfates were prepared in our laboratory
from n-hexadecanol, concentrated sulfuric acid and the
corresponding alkali carbonates. Sodium dodecyl sulfate was
obtained from Calbiochem-Behring, La Jolla, CA. Sodium 2,2-
dideuterio hexadecyl sulfate was a gift from Dr. L.C. Leitch of
the Division of Chemistry, NRC, Ottawa.

Aqueous micellar solutions were prepared at concentrations
of 0.1 or 0.2 M by heating a mixture of the solid surfactant
with the required amount of water (or D_2O) until a clear
solution resulted, a drop of which was deposited onto a heated
cell window. Since all studies involved aqueous solutions it
was necessary to use water insoluble cell windows. We generally
employ calcium or barium fluoride, occasionally also cadmium
telluride or zinc selenide cell windows and use 2.5 to 25 μm
thick Mylar spacers. These thin cells are difficult to prepare
by the conventional amalgam technique, therefore we utilize a
demountable Harrick cell, shown in Figure 1, which has a number

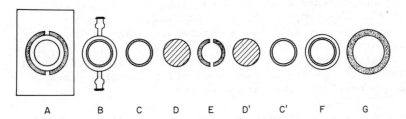

Figure 1. Description of the modular Harrick infrared cell
consisting of stainless steel cell holder (A), stainless steel
cup with female Luer fittings (B), rubber O-rings (C and C'),
25×2 mm CaF_2 or BaF_2 cell windows (D and D'), 2.5 to 25 μm thick
Mylar spacers (E), stainless steel retaining ring (F), and
knurled screw collar (G); the cell is assembled from left to
right.

of advantages: (i) it is round, being designed to match the
circular optics used in FT instruments rather than the linear-
slit optics of grating instruments; (ii) it is readily assembled
and demounted which reduces the danger of damage to the windows;
(iii) the spacers are inexpensive Mylar film, available in a
wide range of thicknesses; (iv) the cells are sealed by two
rubber O-rings rather than by the spacers, which provide an
extremely good seal and permit maintaining the sample for an
extended period of time at temperatures up to 80°C.

Spectra acquisition and data handling. In general our
experiments involve collecting the infrared spectra of the same
system at various temperatures below, at and above the
corresponding phase transition and monitoring changes in band
parameters as a function of temperature. Reproducible results
can be obtained only when the temperature is increased, as the
aqueous surfactants show a hysteresis in time when decreasing
the temperature. To overcome this problem we cool our samples
for a considerable time prior to the measurements and proceed
always from low to high temperatures. The cell is placed in a
thermostated mount with three degrees of freedom, and an
alignment laser is used to center the beam on the sample.
Spectra are usually recorded at up to 40 different temperatures,
the range and temperature increments being selected according to
the particular sample under study. In our experimental setup

this process is completely under the control of the spectrometer computer, which records a spectrum, increments the temperature, waits for temperature equilibration, then records another spectrum[3,4]. The temperature is monitored by a copper-constantan thermocouple located against the edge of the cell windows. The thermocouple output is routed to a printer via a Newport digital pyrometer and the printer is triggered by the spectrometer at the end of each spectral collection, thus providing a complete temperature record.

Infrared spectra reported herein were recorded on three different Fourier transform infrared spectrometers (Digilab FTS-11, Digilab FTS-15 and BOMEM DAC-02), all equipped with high sensitivity mercury cadmium telluride detectors. Typically, five hundred scans were coadded, zero filled once, triangularly apodized and Fourier transformed to yield a resolution of either 1 or 2 cm^{-1}. The infrared spectral data were processed by computer programs developed in our laboratories,[5] which allow us to measure temperature-induced changes in frequency or bandwidth of 0.1 cm^{-1}.

RESULTS AND DISCUSSION

The alkali salts of long chain carboxylates and n-alkyl sulfates are practically insoluble in water. However, when heated above a certain temperature, the critical micellization temperature (CMT), an optically clear micellar solution is formed which upon cooling solidifies into an opaque curd. While this behaviour is typical of aqueous ionic surfactants,[6,7] the structural changes and the nature of the phase transition at the CMT are not well characterized.

It is worthwhile mentioning here that the CMT is different from the Krafft point, the latter being the point where the CMC curve, which is only slightly affected by temperature, intersects the solubility curve, which is strongly temperature-dependent. Thus, the Krafft point is always observed at a lower temperature than the CMT, and can be considered as the critical micellization temperature observed at the critical micelle concentration.

At very low concentrations, up to a few weight percent surfactant, the CMT generally shows a steep concentration-dependence and increases with temperature; however, over the range 5 to 15 weight percent surfactant the critical micellization temperature is fairly constant for a given system. It is this concentration range (between 0.1 and 0.3 M) that we have chosen to study by FT-IR spectroscopy.

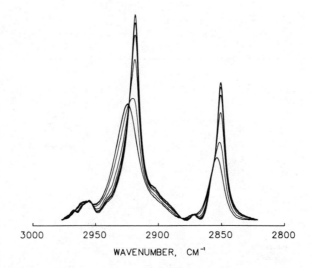

WAVENUMBER, CM⁻¹

Figure 2. Temperature-induced variation of the carbon-hydrogen
stretching bands in the infrared spectra of 0.2 M sodium
hexadecyl sulfate; shown are eight individual spectra recorded
between 38 and 45°C and plotted in steps of 1°C; peak heights
decrease with increasing temperature, the first and last two
spectra are practically superimposed.

Temperature-dependence of the acyl chain modes. In the case
of amphiphilic surfactants it is possible to monitor separately
infrared bands characteristic of the hydrophobic region of the
acyl chains and bands characteristic of the hydrophilic region
of the head group. Figure 2 shows a series of eight individual
infrared spectra of sodium hexadecyl sulfate in the region of
the C-H stretching modes at temperatures between 38 and 45°C.
All bands are due to acyl chain C-H bonds. The strong bands at
2920 and 2850 cm⁻¹ are the antisymmetric and symmetric CH_2
stretching modes, respectively; the weaker bands at 2955 and
2873 cm⁻¹ are the asymmetric CH_3 stretching modes of the
terminal methyl group. All of these bands exhibit temperature-
dependent variations in frequency and width which can be related
to structural changes at the molecular level.

Figure 3 illustrates the temperature-dependence of the
frequency (closed symbols) and the width (open symbols) of the
2850 cm⁻¹ band in sodium hexadecyl sulfate (SHS), and that in
sodium dodecyl sulfate (SDS). The frequency of this vibrational
mode is constant up to a certain temperature, shows a steep
increase at a certain temperature (which is different for SHS
and SDS), after which it again remains constant. Such frequency

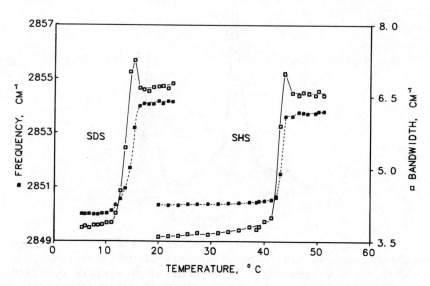

Figure 3. Temperature-dependence of the frequency (closed symbols) and full-width at three quarter height (open symbols) of the symmetric CH_2 stretching bands of 0.2 M sodium dodecyl sulfate (SDS) and of 0.2 M sodium hexadecyl sulfate (SHS).

versus temperature profiles can be conveniently used to determine the critical micellization temperature or the corresponding phase transition temperature of the system under study. According to the midpoints of the sigmoidal frequency versus temperature curves in Figure 3, the CMT's of aqueous 0.2 M SHS and SDS are centered at 44 and 15°C, respectively. These temperatures are in excellent agreement with the critical micellization temperature values obtained by the temperature-dependence of the optical density at 580 nm.[8]

The absolute frequency values at temperatures below that of the corresponding CMT are characteristic of conformationally highly ordered acyl chains, as found in solid hydrocarbons, whereas the values at temperatures above that of the CMT are characteristic of conformationally disordered acyl chains with a high content of gauche conformers, such as found in liquid hydrocarbons.

The bandwidth of this CH_2 stretching mode increases drastically throughout the temperature range of the phase transition but is constant at temperatures below and above the CMT. The large increase in bandwidth is typical for the melting of solid hydrocarbons and for the acyl chain melting phase transition of aqueous phospholipids. However, from Figure 3 it can be seen that the changes in frequency and width of the CH_2 stretching bands are not concerted. The maximum rate of change of the bandwidth is observed at a lower temperature than that of the frequency; furthermore, a decrease in bandwidth is coupled with an increase in frequency at the high temperature end of the phase transition. This phenomenon can be explained through the overlap of bands from two species with very different band-widths, and the resultant unusual band contour reflects only the summation of the two band contours. We have found similar bandwidth profiles in studies of the monomer aggregation of short chain n-alkanoates at the critical micelle concentration,[1,2] and in studies of the temperature-induced micellization of aqueous soaps.[9,10] This observation is indicative of a gradual change from the solid surfactant with narrow bands to the isotropic micellar surfactant with much broader CH_2 stretching bands. In fact, the width in temperature of the changes in frequency and bandwidth are related to the degree of cooperativity, a narrow temperature range being indicative of a high cooperativity of the corresponding phase change.

In addition to the conformationally-sensitive CH_2 stretching modes we have also studied the CH_2 bending modes around 1470 cm^{-1}, which are sensitive to the interchain interactions determined by the packing pattern of the acyl chains at temperatures below that of the CMT. This CH_2 bending mode results from the out-of-phase scissoring vibration of all-trans CH_2 segments and provides information specific to the packing of the acyl chains in solid-like phases.[11]

Figure 4 shows this band in SHS (A), in SHS-d_2 (B) and in rubidium hexadecyl sulfate (C). The infrared spectra of the micellar phase (broken lines) are identical for all hexadecyl sulfates with the CH_2 scissoring band at 1467.5 cm^{-1}, which is characteristic for rotationally disordered acyl chains. The infrared spectra in this region of the solid surfactants, however, are quite different (solid lines). A comparison of A and B in Figure 4 shows that the weaker band at 1480 cm^{-1} is due to the scissoring of the α-methylene group next to the sulfate head group; this band is also present in the rubidium salt in Figure 4C.

To our knowledge there are no diffraction studies available on the crystal structure of solid sodium hexadecyl sulfate; nevertheless, the presence of a single strong CH_2 scissoring

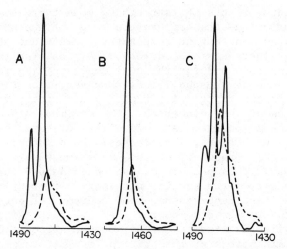

Figure 4. Infrared spectra in the CH_2 scissoring region (1490
to 1430 cm^{-1}, from left to right) of sodium hexadecyl sulfate
(A), of sodium α-d_2 hexadecyl sulfate (B), and of rubidium
hexadecyl sulfate (C) at temperatures below the CMT (solid
lines) and at temperatures above the CMT (broken lines).

band in sodium hexadecyl sulfate at 1473 cm^{-1} allows us to
suggests that the acyl chain packing in solid SHS is triclinic.

In the infrared spectra of the rubidium hexadecyl sulfate,
the CH_2 scissoring band is split at all temperatures below the
CMT, as shown in Figure 4C; the two components are at 1472 and
1467 cm^{-1}. Such crystal field (or factor group) splitting has
only been observed when the polymethylene chains are in a
monoclinic or orthorhombic unit cell, which have a common
orthorhombic subcell;[11] this allows us to suggest that the solid
rubidium salt of hexadecyl sulfate has the acyl chains packed in
an orthorhombic subcell. The infrared spectra in the CH_2
scissoring region of cesium hexadecyl sulfate at temperatures
below the CMT are similar to those of the sodium salt,
suggesting that the acyl chain packing also is triclinic,
whereas the spectra of the lithium, potassium and ammonium
hexadecyl sulfates at temperatures below that of the
corresponding CMT are similar to that of the rubidium salt,
suggesting that these salts have the acyl chains packed in an
orthorhombic subcell. These predictions, however, have to await
confirmation by an X-ray analysis of the solid alkali hexadecyl
sulfates.

Figure 5 demonstrates further that, similar to the temperature profiles obtained from the CH_2 stretching modes, the frequency versus temperature curves obtained from the CH_2 scissoring mode of sodium hexadecyl sulfate can also be used to determine the corresponding critical micellization temperature.

Temperature dependence of the head group modes. The temperature-induced micelle formation of ionic surfactants can also be monitored via changes observed in the vibrational modes of the hydrophilic head group moiety. As these bands originate from polar bonds such as sulfur-oxygen or carbon-oxygen bonds in SO_3 or CO_2 head groups, the corresponding infrared bands are usually quite strong.

Figure 6 shows the thermotropic behavior of sodium hexadecyl sulfate in the region of the SO_3 stretching bands. At temperatures below that of the CMT the asymmetric SO_3 stretching mode gives two strong bands around 1270 and 1207 cm^{-1}, Figure 6A, while the symmetric SO_3 stretching mode gives the strong band at 1070 cm^{-1}, Figure 6B. These bands change only slightly with temperature; however, at the CMT the narrow asymmetric SO_3 stretching bands collapse into two broad bands at 1245 and 1217 cm^{-1}; the symmetric SO_3 stretching band broadens and shifts to 1061 cm^{-1} at temperatures above that of the CMT.

Both the asymmetric and symmetric SO_3 stretching bands show clear isosbestic points, indicating that the equilibrium between

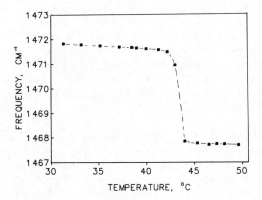

Figure 5. Temperature dependence of the frequency of the CH_2 scissoring band of sodium hexadecyl sulfate.

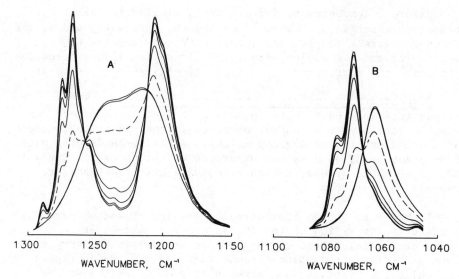

Figure 6. Temperature induced changes in the infrared spectra of 0.2 M sodium hexadecyl sulfate in the region of the asymmetric (A) and symmetric SO_3 stretching modes (B); shown are eight individual spectra taken in 1°C intervals between 38 and 45°C; the broken line corresponds to a spectrum recorded at 43°C.

the two structures is continuously changing throughout the temperature range of the critical micellization temperature. In fact, the dashed line in Figure 6, at 43°C, corresponds approximately to the midpoint of this change.

The infrared bands in the region of the asymmetric SO_3 stretching modes show considerable fine structure in the solid surfactants and the frequencies and bandshapes differ considerably among the individual alkali hexadecyl sulfates. This behaviour is determined to a large extent by the different electrostatic interactions and the counterion binding in these solid surfactants, and will be discussed elsewhere.[12] The infrared spectra of the micellar phase are identical regardless of the temperature or the nature of the counterions in these alkali hexadecyl sulfates.

There is only a single symmetric SO_3 stretching band in each of these alkali salts and its exact frequency in the solid surfactants can be correlated with the ionic crystal radii of the corresponding counterion.[12] The shoulder band at 1078 cm^{-1} on the high frequency side of the symmetric SO_3 stretching band in Figure 6B disappears upon deuteration of the α-methylene group in sodium hexadecyl sulfate, and is assigned to a

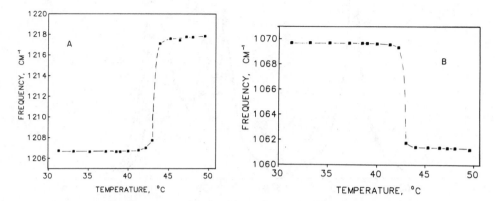

Figure 7.   Temperature dependence of the frequency of the asymmetric (A) and symmetric SO_3 stretching bands (B) of sodium hexadecyl sulfate.

carbon-carbon stretching mode. All micellar type spectra are identical in the region of the symmetric SO_3 stretching bands, regardless of the nature of the counterion.

The temperature-induced change in structure can also be followed by monitoring the temperature dependence of the frequency of the asymmetric or that of the symmetric SO_3 stretching bands as illustrated in Figure 7. The low frequency component of the asymmetric SO_3 stretching mode increases abruptly at the CMT from 1207 to 1217 cm^{-1}, Figure 7A, while the frequency of the symmetric SO_3 stretching mode decreases at the CMT from 1070 to 1061 cm^{-1}, Figure 7B. From the data in Figures 6 and 7 it can be seen that the thermal response of the hydrophilic head group parallels that of the hydrophobic acyl chains.

Figure 8 illustrates the vibrational modes of another head group, those originating from the carboxylate moiety of sodium oleate.[10] At temperatures below that of the CMT of 0.38 sodium oleate (i.e. 28°C), the antisymmetric CO_2 stretching mode is observed as a very strong infrared band at 1559 cm^{-1}. The high intensity of this band indicates a high charge distribution across the vibrating OCO bonds. The thermotropic behavior of

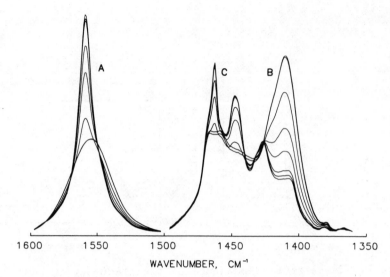

Figure 8. Temperature induced changes in the infrared spectra
of 0.38 M sodium oleate in the region of the CO_2 stretching
modes. A, antisymmetric CO_2 stretching band; B, symmetric CO_2
stretching bands; C, CH_2 scissoring bands. Shown are eight
individual spectra recorded in 2°C intervals between 20 and
34°C; the intensity of the antisymmetric CO_2 stretching bands in
part A is scaled down by a factor of two compared to that in
part B and C.

this vibrational mode is evident in Figure 8A. As the
temperature is raised through the range of the phase transition,
the band decreases in intensity, broadens and is comprised of
two components until above the CMT. In the micellar phase it
consists of a single, considerably broadened band shifted to
1554 cm^{-1}.

At temperatures below that of the CMT, the symmetric CO_2
stretching mode is observed as a medium strong band at 1426
cm^{-1}. Furthermore, a second symmetric CO_2 stretching mode is
observed as a weaker band at 1405 cm^{-1}, due to the fraction of
surfactant in solution (the solubility of sodium oleate at room
temperature is ~10 percent). This band increases dramatically
in intensity on transition to the micellar phase, Figure 8B.

Interestingly, the thermotropic behavior of the two
carboxylate modes is quite different. There is a major gain in

intensity of the symmetric CO_2 stretching mode on transition to the micellar phase, accompanied by an unusually large (17 cm^{-1}) shift in frequency, while the antisymmetric CO_2 stretching mode undergoes a decrease in intensity and only a moderate shift in frequency (5 cm^{-1}) upon micellization. These observations may be rationalized in the following way: From the crystal structure analysis of anhydrous carboxylate soaps with monovalent cations,[13,14] it is known that the cation and oxygen atoms form double ionic layers and that the cations of one half of a double layer are located on the bisecting line of the OCO angle of the other half of the double layer, i.e. in the direction of the symmetry coordinate vector of the symmetric CO_2 stretching mode. Upon micellization the two layers must be separated from each other, leading to the loss of an interaction partner, the sodium ion, for the symmetric CO_2 stretching mode. The loss of interaction, in turn, causes the large shift in frequency, while the ionization, or rather deneutralization, can account for the large increase in intensity of the symmetric CO_2 stretching mode.

The antisymmetric CO_2 stretching mode, on the other hand, is much less affected by the temperature-induced micellization of the solid surfactant, since the position of the cation in the crystal, located between neighbouring CO_2 goups of the same half layer, is not that different in the micelles.

As mentioned earlier, we have also studied the formation of micelles from monomers in the case of short chain n-alkanoates (C_6 to C_{10}),[1,2] in which case the antisymmetric CO_2 stretching band was found to shift much more than the symmetric CO_2 stretching band; furthermore, the shift at the CMC was to higher frequencies, whereas the shift at the CMT is to lower frequencies. These facts taken together all suggest that in the micelles the monovalent sodium ion is preferentially located between neighbouring CO_2 groups, which is in agreement with results from a sodium NMR study.[7] The situation is different in the case of the alkali sulfates due to the threefold symmetry of the SO_3 group as compared to the twofold symmetry of the CO_2 group. The totally symmetric SO_3 stretching mode in sodium hexadecyl sulfate also shows a considerable shift (9 cm^{-1}) to lower frequencies upon micellization; however, there is no major change in intensity.

Kinetics of micelle formation. One way to obtain information on the kinetics of the temperature-induced micelle formation is to conduct experiments which determine the rate of change of the system under study as a function of temperature. Variations resulting from a change in an external parameter such as temperature, can be monitored with the Fourier transform technique by calculating the difference between infrared spectra

recorded before and after the change has occurred. The
resultant infrared difference spectra deviate from zero
absorbance only in the regions where spectral changes have
occurred and only in the temperature range over which the system
has changed. We have thus monitored the rate of change of
various infrared spectral parameters via a number of individual
difference spectra.

Figure 9 shows four series of infrared difference spectra
obtained from the regions of the CH_2 stretching (A), the CH_2
scissoring (B), and the asymmetric (C) and symmetric (D) SO_3
stretching modes of sodium hexadecyl sulfate. In each case one
observes first a gradual increase, followed by an abrupt
decrease in the rate of change of these vibrational modes. The
gradual formation of

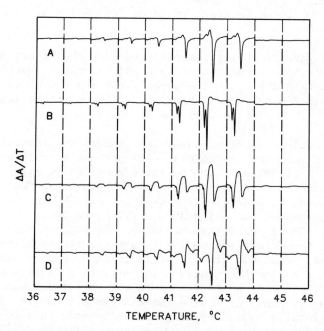

Figure 9. Infrared difference spectra in the 2900-2800 cm^{-1}
region of the CH_2 stretching bands (A), the 1500-1400 cm^{-1}
region of the CH_2 scissoring bands (B), the 1320-1220 cm^{-1}
region of the asymmetric SO_3 stretching bands (C) and the 1120-
1020 cm^{-1} region of the symmetric SO_3 stretching bands (D) of
SHS; all spectra are normalized with respect to temperature, the
temperatures used in the subtraction are indicated on the bottom
axis; lower temperature spectra were always subtracted from
higher temperature spectra.

micelles shows that throughout a certain temperature range
micelles coexist with the solid surfactant and that the phase
transition at the CMT does not follow first order kinetics.

The critical micellization temperature. The concept of a
"critical micellization temperature"[15] is a more recent
concept than that of the critical micelle concentration and
holds a number of interesting features worthy of exploration.
It is not surprising to find certain differences between the
formation of micelles from monomers via an increase in
concentration, and the micelle formation from a solid surfactant
via an increase in temperature. In order to understand these
differences one has to consider the phase boundaries in each
case. The micellar phase boundary at the CMC can be derived
simply from the corresponding monomer-micelle equilibrium,
whereas the micellar phase boundary at the CMT has to be
obtained on the basis of coupling the monomer-micelle
equilibrium to the monomer-solid surfactant equilibrium.

The solubility of the solid surfactant in its monomeric form
is a weakly increasing function of temperature, i.e. until the
monomer concentration reaches the CMC. Once the formation of
micelles has commenced, the concentration of monomers in
solution starts to decrease; this results in the release of
more monomers from solid into solution to maintain the solid-
monomer equilibrium. Thus, the solid surfactant continues to
release monomers in order to maintain the solid-monomer
equilibrium; this equilibrium, however, can never be reached
because it is coupled to the monomer-micelle equilibrium which
leads to the sequestration of monomers within micelles. The
release of monomers thus continues until the entire solid
surfactant is transformed into micelles. Only at the Krafft
point, which corresponds to the intersection of the solubility
curve with that of the critical micelle concentration, are
solid-monomer and monomer-micelle equilibria in equilibrium.

There is a distinct lack of experimental information on the
CMT phase boundaries and the factors which affect this phase
transition. The concentration range involved in such studies
i.e. 5 to 15 weight percent solid surfactant (or 0.1 to 0.3 M)
makes infrared and particularly FT-IR spectroscopy an extremely
useful tool, whereas FT-IR is less suitable for the study in the
concentration range of the CMC of long chain surfactants (around
10^{-3} M).

Both the CMC and the CMT of ionic surfactants are strongly
dependent on the length of the alkyl chain. The CMC
decreases,[6,16] while the CMT increases with increasing chain
length. For instance, the CMT of 0.2 M SDS is 15°C, while that

of 0.2 M SHS is 44°C. Surfactants with unsaturated alkyl chains
have considerably lower CMT's. For example, the CMT of 0.3 M
sodium oleate is observed at 28°C, while that of 0.38 M sodium
laurate, with a much shorter alkyl chain, is at 34°C. The CMT's
of 0.1 M sodium myristate, sodium palmitate and sodium stearate
are 53, 65 and 74°C, respectively.[8,17] These temperatures
parallel the trend observed for the Krafft point phenomenon, the
respective Krafft points are 20°C for sodium laurate, 34°C for
sodium myristate, 45°C for sodium palmitate and 54°C for sodium
stearate,[18] whereas the Krafft point of sodium oleate is below
20°C.[17] The CMC decreases by about one order of magnitude when
the alkyl chains increase by two carbon atoms.[19]

The critical micellization temperature also depends on the
structure of the hydrophilic head group and the nature of the
counterion. For the same acyl chain and counterion, the
sulfates have lower CMT's than the corresponding carboxylates.
For example, the CMT's of sodium dodecyl and hexadecyl sulfate
are 15 and 44°C, respectively, while those of sodium dodecyl and
hexadecyl carboxylates are 34 and 65°C, respectively.

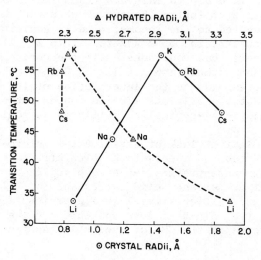

Figure 10. Transition temperature of 0.1 M lithium, sodium,
potassium, rubidium and cesium hexadecyl sulfate, plotted versus
the crystal and hydrated radii of the corresponding alkali
ions.

The effect of a change in the counterion is different for carboxylates and sulfates. In the carboxylates the potassium salts have lower CMT's than the sodium salts, whereas the situation is reversed in the case of the corresponding sulfates. The micellization temperature of the sodium salt of ascorbyl palmitate is also lower than that of the corresponding potassium salt.[9]

Finally, we have also undertaken an investigation, by FT-IR spectroscopy, of the dependence of the critical micellization temperature of hexadecyl sulfates on the nature of the counterion. Preliminary results on the critical micellization temperature for the most common alkali counterions are shown in Figure 10. The cation binding in the solid hexadecyl sulfates has been shown to exhibit a systematic dependence on the ionic crystal radii, as demonstrated by the linear decrease in the frequency of the symmetric SO_3 stretching mode from lithium to cesium.[12] The critical micellization temperature, however, is not a simple function of the crystal or hydrated radii of these alkali salts, as shown in Figure 10. The transition temperature in this series of alkali salts depends on the critical balance between the energy required to disrupt the crystal structure and the energy gained by micelle formation, the latter being a complex function of the hydration energy, the radius of the hydrated cations, the size and shape of the micelles etc. A detailed discussion of these factors will be reported elsewhere.[8]

REFERENCES

1. J. Umemura, D. G. Cameron and H. H. Mantsch, J. Phys. Chem., 84, 2272 (1980).
2. J. Umemura, H. H. Mantsch and D. G. Cameron, J. Colloid Interface Sci., 83, 558 (1981).
3. D. G. Cameron, H. L. Casal and H. H. Mantsch, J. Biochem. Biophys. Methods 1, 21 (1979).
4. D. G. Cameron and G. M. Charette, Appl. Spectrosc., 35, 224 (1981).
5. D. G. Cameron, J. K. Kauppinen, D. J. Moffatt and H. H. Mantsch, Appl. Spectrosc., 36, 245 (1982).
6. B. Lindman and H. Wennerström, "Micelles. Amphiphilic Aggregation in Aqueous Solution", in Topics in Current Chemistry, F. L. Boschke, Editor, Vol. 87, p. 1, Springer-Verlag, New York, 1980.
7. H. Wennerström and B. Lindman, Physics Reports 52, 1 (1979).
8. H. H. Mantsch, V. B. Kartha and D. G. Cameron, (1982), unpublished data.

9.   H. Sapper, D. G. Cameron and H. H. Mantsch, Can. J. Chem.,
     $\underline{59}$, 2543 (1981).
10. D. G. Cameron, J. Umemura, P. T. T. Wong and H. H. Mantsch,
     Colloids and Surfaces $\underline{4}$, 131 (1982).
11. R. G. Snyder, J. Chem. Phys. $\underline{71}$, 3229 (1979).
12. V. B. Kartha, L. C. Leitch and H. H. Mantsch, (1983), Can.
     J. Chem., submitted for publication.
13. V. Vand, T. R. Lomer and A. Lang, Acta Cryst., $\underline{2}$, 214
     (1949).
14. J. H. Dumbleton and T. R. Lomer, Acta Cryst., $\underline{19}$, 301
     (1969).
15. N. A. Mazer, G. B. Benedek and C. M. Carey, J. Chem. Phys.,
     $\underline{80}$, 1075 (1976).
16. K. Shinoda, Pure Appl. Chem., $\underline{52}$, 1195 (1980).
17. J. W. McBain and W. C. Sierichs, J. Amer. Oil Chem. Soc.,
     $\underline{25}$, 221 (1948).
18. K. Shinoda, Y. Minegeshi and H. Arai, J. Phys. Chem., $\underline{80}$,
     1987 (1976).
19. C. Tanford, "The Hydrophobic Effect: Formation of Micelles
     and Biological Membranes", 2nd ed., p. 66, Wiley-
     Interscience, New York, 1980.

ABOUT THE CONTRIBUTORS

Here are included biodata of only those authors who have
contributed to this volume. Biodata of contributors to Volumes
2 and 3 are included in those volumes.

Lawrence P. Aggerbeck is Chargé de Recherche, Centre de
Génétique Moleculaire, CNRS, Gif sur Yvette, France. He received
his Ph.D. in Biochemistry and M.D. degree in 1974 from the University of Chicago Pritzker School of Medicine and postgraduate
medical training at Evanston Hospital, Northwestern University.
His research has been mainly in the area of serum lipoprotein
structure.

Thomas Ahlnas is a graduate student in the Department of
Physical Chemistry I, University of Lund, Sweden.

Mats Almgren is presently Professor in the Department of
Physical Chemistry, University of Uppsala, Sweden. He started as
a photochemist at Chalmers University of Technology, Goteborg,
Sweden and later worked in the kinetics of micellization at the
University of Goteborg. During 1977-1978 he was Visiting Pro-
fessor in the Department of Chemistry and Radiation Laboratory,
University of Notre Dame, and in 1980 was a Docent in Physical
Chemistry at the University of Lund.

Gösta Arvidson has been Assistant Professor in the Department of Physiological Chemistry, Biomedical Center in Uppsala,
Sweden since 1975. Received M.D. degree from Lund University and
has published mainly in the field of phospholipids metabolism and
membrane structure.

Piero Baglioni is with the Institute of Physical Chemistry,
University of Florence, Italy. He graduated in Chemistry in 1977.
His main field of interest is colloid and interface chemistry
with particular attention to the properties of liquid surfaces,
and currently he is interested in the application of ESR spectroscopy to surface and colloid science.

691

D. Balasubramanian is Deputy Director of the Centre for Cellular and Molecular Biology, Hyderabad, India. He received his Ph.D. degree in 1965 from Columbia University. He is a biophysical chemist with interests in biomolecular spectroscopy, protein and peptide conformations, lipid and surfactant assemblies, and in the recent technique of photoacoustic spectroscopy. He was awarded the Indian Chemical Society Yeddanapalli Medal for outstanding physical chemist (1977), and the Bhatnagar Prize for chemical sciences (1981). He is a Fellow of the Indian Academy of Sciences and is on the editorial board of Current Science and of the Journal of Biosciences.

B. Bedwell is with the Department of Chemical Engineering, University of Michigan, Ann Arbor, MI.

Saddek Belaid is a Research Associate in the Department of Physical Chemistry of the Nuclear Research Center of Saclay (France). He obtained his Docteur-Ingénieur degree in 1980. He is currently concerned with the molecular dynamics of organophosphorus coordination compounds in different phases.

W. J. Benton is with the Department of Chemical Engineering, Rice University, Houston, TX.

K. S. Birdi is Assistant Professor ("lektor" in Danish) at the Technical University of Denmark, Lyngby, Denmark. His research interests are monolayers, biological macromolecules, and micelles (solubilization).

F. D. Blum is currently with the Department of Chemistry, Drexel University, Philadelphia. He received his Ph.D. degree in Chemistry from the University of Minnesota, and was a member of the University of Minnesota program on fundamentals of enhancing petroleum recovery.

N. Boens is currently a postdoctoral fellow of the Belgian National Science Foundation at the Catholic University of Leuven, Heverlee, Belgium. His interest at present is photoelectrochemistry and photochemistry in membranes.

João B. S. Bonilha is Assistant Professor at the University of Sao Paulo at Riberao Preto, Brazil. He received his Ph.D. degree in Chemistry and did one year postdoctoral work with Prof. D. Whitten. He has ten publications in organic and photochemistry and is currently interested in photochemistry in organized systems.

Jean Brondeau is Assistant at the Faculte des Sciences of the University of Nancy, France. He received his D.Sc. in 1981. His principal research interests are NMR relaxation from both theoretical and experimental points of view, and instrumentation.

Bernard Brun is Professor of Physical Chemistry at the University of Sciences and Techniques, Montpellier, France. He is also Director of the Associated Laboratory "Physicochimie des Systemes Polyphases". He received his Doctorat Sciences in 1967 from the University of Montpellier. His research interests are micellar and microemulsion systems, monolayers and bilayers, and adsorption on surfaces.

C. A. Bunton is Professor of Chemistry at the University of California, Santa Barbara. His B.Sc. and Ph.D. degreess are from the University college, London. He has had visiting appointments at universities in Europe and North and South America. His research interests include micellar catalysis and inhibition.

B. Cabane obtained his Ph.D. at the University of Paris Sud (Orsay) in 1971. Part of the thesis was with Prof. Friedel on the structure of liquid tellurium; the other part was with Prof. de Gennes on the NMR of thermotropic liquid crystals. His main field of interest is the study of the interactions between polymers and surfactants. He works as a "chargé de recherche" for CNRS.

David Cameron is an Associate Research Officer, Molecular Spectroscopy Section, National Research Council, Ottawa, Canada.

Daniel Canet is Professor of Physcial Chemistry at the Faculte des Sciences of the University of Nancy, France. He received his D.Sc. in 1973. Presently he is the leader of a research group whose essential activities concern NMR relaxation, instrumentation and methodology, application of NMR spectroscopy to molecular dynamics, surfactant and biological systems.

Claude Chachaty is currently in charge of the high resolution NMR and ESR laboratory of the Department of Physical Chemistry at the Nuclear Research Center of Saclay (France) which he joined in 1962. He obtained his Docteur-es-Science degree in 1968. He works mainly in the field of paramagnetic complexes, polymers and liquid crystalline phases.

J. Carvolin is Maitre de Recherche au CNRS at the Laboratoire de Physique des Solides, Universite Paris Sud, Orsay, France. Holds Docteur es Sciences degree, and is Maitre de Conference a l'Ecole Polytechnique.

Mario Corti is affiliated with the Quantum Optics Section of CISE (Segrate-Milano, Italy). He is also Professor at the Istituto di Fisica dell'Universita de Milano. In the last seven years, he has been working on quasielastic light scattering from pure liquids and micellar solutions.

Y. Croonen is currently working toward doctoral degree at the Catholic University of Leuven, Heverlee, Belgium.

H. T. Davis is Professor of Chemical Engineering and Materials Science, University of Minnesota, Minneapolis. He was a member of the University of Minnesota program on fundamentals of enhancing petroleum recovery.

J. C. Dederen is currently affiliated with an industry, and was with K.U., Leuven, Belgium.

Vittorio Degiorgio is Professor of Quantum Electronics at the University of Pavia, Italy. He has published in the area of quantum optics and statistical physics. His current research interests are in the applications of laser-light scattering techniques to micellar and macromolecular solutions, and in phase transitions and instabilities in open systems.

Emma de Lacey is currently studying for a Ph.D. in the Department of Applied Mathematics, Australian National University, Canberra. She received a B.Sc. (Hons.) in Physics in 1978 at the University of Canterbury, New Zealand.

F. C. DeSchryver has been Professor at Leuven University, Belgium since 1973. He is at present Chairman of the European Photochemistry Association, and of the IUPAC Commission on Photochemistry. The research interest of his group is centered around the excited state behavior and ranges from multichromophoric systems to supramolecular aggregates.

Ken Dill is currently Assistant Professor of Pharmaceutical Chemistry and Pharmacy, University of California, San Francisco. Before his current position, he was Assistant Professor of Chemistry at the University of Florida. He received his Ph.D. degree from the Biology Department, University of California, San Diego, and did postdoctoral research in the Chemistry Department, Stanford University.

R. B. Dorshow is presently working on Ph.D. research in the Physics Department at UC Santa Barbara. He received B.S. in 1977 from the University of Minnesota. He is especially interested in micelles and microemulsions by dynamic and classical light scattering and has authored 7 publications.

Maurice Drifford is in charge of the Light Scattering Laboratory of the Department de Physico-chemie, CEA Saclay, France. He received his D.Sc. in 1967 from the University of Paris. His research interests are electrostatic interactions in aqueous solutions (polyelectrolytes, micelles).

Robert Duplessix is a Maitre Assistant at Universite Louis Pasteur in Strasbourg; he also does research at Centre de Recherches sur les Macromolecules in Strasbourg. His Ph.D. thesis was on the study of polymer networks through small angle neutron scattering and inelastic scattering of light. His current interests, mainly centered on neutron scattering, include the structure of colloidal particles and the study of polymer chain configurations through the use of partially labeled macromolecules.

Per-Olaf Eriksson is a graduate student in Physical Chemistry at the University of Umea, Umea, Sweden.

Enzo Ferroni has been Professor of Physical Chemistry, Institute of Physical Chemistry, University of Florence, Italy since 1959. He was Dean of the Faculty of Sciences from 1969 to 1975 and Rector from 1976 to 1979 at this same university. He graduated in Chemistry in 1944. His main field of interest is physical chemistry of solid and liquid surfaces and colloidal systems.

Krister Fontell has been at the Chemical Center, University of Lund, Sweden since 1973 and is currently Associate Professor (Docent) sponsored by Swedish Natural Science Council. Prior to coming to Lund, he was at the Laboratory for Surface Chemistry in Stockholm (1963-1972). He received his Ph.D. in Physical Chemistry from Åbo Akademi, Åbo, Finland. He has about 100 publications in the field of colloid and surface chemistry especially micellar solutions and amphiphilic liquid crystalline systems.

Elias I. Franses has since January 1982 been a Union-76 Assistant Professor in the School of Chemical Engineering, Purdue University, which he joined in 1979 after receiving his Ph.D. from the University of Minnesota. In 1981 he received an Outstanding Junior Faculty Award from Arco Oil and Gas Company. He has published in the areas of surfactant phase behavior, ultralow interfacial tension for enhanced oil recovery, liquid crystals, micellization, vesicles, and optical and electrical properties of aqueous solutions and dispersions.

Peter Fromherz has had Chair for Biophysics in the Physics Faculty of the University of Ulm, Ulm, W. Germany since 1981. He carried out his graduate study in Physical Chemistry at the University of Marburg and received his Ph.D. degree in 1969 for his thesis research in the laboratory of H. Kuhn. He was Assistant at the Max-Planck-Institute for Biophysical Chemistry at Göttingen from 1970-1981. His interests include instrumentation for Langmuir-Blodgett films, lipid-protein interaction, dye-probes for membrane potentials, natural membrane processes, and micelles and vesicles.

K. N. Ganesh is a bio-organic chemist, and obtained his Ph.D. from Delhi University (1977) and from Cambridge University, England in 1980. He has published in the areas of the structures of shellac resins, and constituents, and on quinone-capped porphyrins. His special interests are NMR spectroscopy of biological and organic molecules, and synthesis and studies of biopolymers.

E. Gelade is currently working toward doctoral degree at the Catholic University of Leuven, Heverlee, Belgium.

David Gruen is presently a Queen Elizabeth II Fellow in the Department of Applied Mathematics, Australian National University, Canberra. He received his Ph.D. degree in the Physiological Laboratory, University of Cambridge, England developing a theoretical model of the uptake of n-alkanes into lipid bilayer membranes. He is the author of 15 papers.

Erdogan Gulari is Associate Professor of Chemical Engineering at the University of Michigan. He received his Ph.D. in Chemical Engineering from California Institute of Technology in 1973. He is the author of 22 papers in the area of liquid solutions. He is currently engaged in research in heterogeneous catalysis, surfactant solutions, non-ideal simple solutions and solution dynamics of polymers.

L. Guldbrand is a Research Student at the University of Lund working in the field of phase equilibria and electrostatic interactions in lamellar systems. Received B.Sc. degree from the University of Stockholm.

Thaddée Gulik-Krzywicki is Maitre de Recherche in the CNRS, Gif sur Yvette, France. He received his Ph.D. in Physical Chemistry in 1964 from the University of Warsaw, Poland. Since 1965 his main interest has been the structural study of biological membranes and of their main components (lipids and proteins) and has published over 100 papers.

Helene Gutman is a graduate student in the Department of Physical Chemistry, Lund University, Lund, Sweden. Received M.Sc. in Biochemistry from UCLA and has published in the field of biological and synthetic model membranes.

Genevieve Haouche is currently working for her Doctorat of Specialty in the Associated Laboratory "Physicochemie des Systemes Polyphases", University of Sciences and Techniques, Montpellier, France.

Marc Hayoun is a graduate student at the University of Paris.

Y. Hendrikx is Chargee de Recherche au CNRS at the Laboratoire de Physique des Solides, Universite Paris Sud, Orsay, France. Holds Docteur es Sciences degree.

Ulf Henriksson is Research Associate in the Department of Physical Chemistry, The Royal Institute of Technology, Stockholm, Sweden, where Doctor of Technology was received in 1975. Has published on the applications of NMR spectroscopy in amphiphilic systems.

Heinz Hoffmann is Professor of Physical Chemistry at the University of Bayreuth, West Germany. He obtained his Ph.D. degree in 1962 from the University of Karlsruhe, West Germany. He spent several years at the Case Western Reserve University in Cleveland, Ohio as a Postdoctoral Fellow and later as a Research Associate. In 1969 he returned to the University of Erlangen-Nurnberg and accepted his present position in 1975.

Annick Jehanno is a graduate student at the Ecole Centrale, Chatenay Malabry, France.

Lennart B.-Å. Johansson is a Research Assistant at the Swedish Natural Science Research Council. He received his Doctor of Technology in 1981 from the Lund Institute of Technology. He has published mainly in the field of polarized light spectroscopy with applications to amphiphilic systems and model membranes.

Bengt Jonsson is a Research Associate in Physical Chemistry, University of Lund, Sweden. His research interests are in the theoretical chemistry of surfactant systems, especially electrostatic theory.

Bo Jonsson is Assistant Professor in Physical Chemistry, University of Lund, Sweden. His research interests are in the field of computer simulations of liquids and solutions.

Manfred Kahlweit is Director at the Max-Planck-Institute for biophysical Chemistry in Goettingen, West Germany. He obtained his Ph.D. in 1954 from the University of Goettingen and has worked for five years as research assistant with the late Carl Wagner. His current research interests include kinetics of nucleation and aging of precipitates, micellization and microstructure of ternary and quaternary systems. He has 80 publications to his credit.

E. W. Kaler is currently with the Department of Chemical Engineering, University of Washington, Seattle. He received his Ph.D. degree in Chemical Engineering and Materials Science from the University of Minnesota, Minneapolis. He was a member of the University of Minnesota program on fundamentals of enhancing petroleum recovery.

Nicole Kamenka is Chargée de Recherches at the Associated Laboratory "Physicochimie des Systemes Polyphases", University of Sciences and Techniques, Montpellier, France. Received Doctorat Sciences degree in 1970 from Ecole de Chimie de Montpellier. Research interests are protein, micellar and microemulsion solutions.

Robert F. Kamrath is currently working toward his Ph.D. degree in Chemical Engineering at Purdue University. His research interests include microstructure of liquid crystals and micelles, and has published on the thermodynamics of micellization in mixed surfactant systems.

V. B. Kartha is a Senior Research Officer at the Bhabha Atomic Research Centre, Bombay, India. He was a Visiting Scientist at the National Research Council, Ottawa, Canada.

Ali Ahmad Khan has be a Research Fellow in the Chemical Center, University of Lund, Sweden since December 1978. Prior to his current position he was a Senior Lecturer in Chemistry in Cape Coast University, Ghana (1967-1978). He received his Ph.D. degree in 1967 from the University of Manchester, England. His biography has been cited in Dictionary of African Biography (1971); and Ghana Who's Who, 1972. He is the author of about 25 research papers.

P. K. Kilpatrick is currently with the Department of Chemical Engineering, North Carolina State University, Raleigh. He received his Ph.D. degree in Chemical Engineering and Materials Science, University of Minnesota, Minneapolis. He was a member of the University of Minnesota program on fundamentals of enhancing petroleum recovery.

Tomas Klason is a graduate student in the Department of Physical Chemistry, the Royal Institute of Technology, Stockholm, Sweden. He has published on the applications of NMR spectroscopy in amphiphilic systems.

Boleslaw Kozankiewicz is a Research Associate in the Molecular Photophysics Group, Institute of Physics, Polish Academy of Sciences, Warsaw. He received his Ph.D. degree in Physics from the University of Warsaw in 1978, and during 1979-1980 he was a Postdoctoral Fellow at the Max-Planck-Institute for Biophysical Chemistry in Gottingen. His research interests include photophysics of molecular crystals and electron donor-acceptor complexes.

Wolfgang Kühnle is a Research Associate at the Max-Planck-Institute for biophysical Chemistry in Gottingen, West Germany.

He received his Ph.D. degree in Radiation Chemistry in 1964 from the Technical University of Hannover. His present research includes the synthesis of novel organic compounds for spectroscopic studies and preparative organic photochemistry.

P. Laggner is currently a Staff Scientist at the European Molecular Biology Laboratory, Hamburg Outstation at DESY where he has been since 1981. Also since 1978 he has been the Vice-Director of the Institut für Röntgenfeinstrukturforschung, Austrian Academy of Sciences, Research Centre, Graz, Austria and will be its Head in 1983. He studied chemistry and physics at the University of Graz and received his Ph.D. degree in 1971. He has been the recipient of the following awards: Sandoz Research Award, 1976; Research Award of Styria, 1977; and Rudolf Wegscheider Award of Austrian Academy of Sciences, 1979.

John C. Lang is a Research Physical Chemist at Miami Valley Laboratories of the Procter and Gamble Company, Cincinnati, OH. He received his Ph.D. degree in 1972 from Cornell University and did postdoctoral research at both the University of Reading, England (1972-1973) and Cornell University (1974-1975) before joining Procter and Gamble. His interests are in the areas of critical phenomena, magnetic resonance, dynamic light scattering, thermodynamics of phase equilibria and phase transitions, surfactant physical chemistry and interfacial phenomena, and has published in these areas.

K. Larsson is Professor of Food Technology, University of Lund, Sweden and has published about 100 papers on lipid structure.

Erich Lessner is currently working as Research Assistant at the Max-Planck-Institute for biophysical Chemistry in Goettingen, West Germany. He obtained his Ph.D. in 1981 from the University of Goettingen and is the author of three publications.

Göran Lindblom is Professor of Physical Chemistry and Head of the Division of Physical Chemistry at the University of Umea, Umea, Sweden. He received his Doctor of Technology in 1974 from the Lund Institute of Technology. His publications deal with applications of NMR and polarized light spectroscopy to the study of structure, dynamics and phase behavior in amphiphilic systems, biological membranes and model membranes.

Bjorn Lindman is Professor of Physical Chemistry, University of Lund, Sweden. His research interests are in the fields of micelles, microemulsions, and other surfactant systems, ion binding to polyelectrolytes and biomacromolecules and applications of NMR spectroscopy to physico-chemical problems.

P. Linse is with the Division of Physical Chemistry I, Chemical Center, Lund, Sweden.

Jan-Erik Lofroth is University Lecturer at the University of Goteborg and Chalmers University of Technology, Goteborg, Sweden. Received Ph.D. degree in 1982 and has published in the area of surfactants and micelles.

Lee Magid is Associate Professor of Chemistry at the University of Tennessee. She obtained her Ph.D. at the University of Tennessee in 1973. Her research interests include the application of small-angle scattering techniques, NMR and related physicochemical measurements to the elucidation of aggregate structure and aggregate-aggregate interactions in micellar and microemulsion systems.

Henry H. Mantsch is Senior Research Officer and Head of the Molecular Spectroscopy Section, National Research Council, Ottawa, Canada. He is also honorary Professor of Chemistry at Carleton University, Ottawa. He received his Ph.D. degree in Physical Organic Chemistry in 1964 from the University of Cluj, Romania, and Dr.Sc. degree in Biophysical Chemistry in 1975 from the University of Munchen. He is a Fellow of the Canadian Institute of Chemistry, and Royal Society of Canada, and in 1980 he was the recipient of the Humboldt Award for the promotion of international scientific cooperation. He has over 150 research publications to his credit.

Jean-Pierre Marchal is Maitre-Assistant at the Faculte des Sciences of the University of Nancy, France. He obtained his D.Sc. degree in 1979. He was initially interested in applications of multinuclear NMR to the study of peptides, but presently he is engaged in surfactant studies and initiating a project dealing with solid state NMR.

Giacomo Martini is with the Institute of Physical Chemistry, University of Florence, Italy where he has been Assistant Professor since 1971. He graduated in Chemistry in 1967. His main field of interest is ESR spectroscopy with particular attention to relaxation mechanisms of paramagnetic probes, catalytic systems, adsorption of liquids on solid supports.

Kazuhiro Matsushita is a Senior Staff Scientist, NMR Application Laboratory, JEOL, Tokyo, Japan. He has published more than 25 papers dealing with the application of sophisticated NMR techniques to basic research, natural products, surfactants in solution, and biological systems.

Fredric M. Menger is Professor of Chemistry at Emory University, Atlanta where he has been since 1965. He obtained his Ph.D. degree with Prof. David Lemal at Wisconsin and carried out

postdoctoral work with Prof. M. Bender at Northwestern. He has
published in many areas of bioorganic chemistry including intramolecular catalysis, charge-transfer complexation, steric effects
in endo cavities, synthetically useful reactions at crystal and
polymer surfaces, new probe compounds for biological systems, and
synthesis of hexapus molecules. In 1979 he authored an Accounts
of Chemical Research article developing a theory of micellar
structure.

M. Meyle is an M.Sc. student at Fysisk-Kemisk Institut,
Technical University of Denmark, Lyngby, Denmark.

Clarence A. Miller is Professor of Chemical Engineering at
Rice University, Houston, TX. After receiving his Ph.D. degree
from the University of Minnesota, he spent twelve years on the
faculty at Carnegie-Mellon University before assuming his present
position in 1981. His research interests are in interfacial
phenomena, with particular interest in interfacial stability,
microemulsions, and enhanced oil recovery.

W. G. Miller is Professor of Chemistry, University of
Minnesota, Minneapolis. He was a member of the Univeristy of
Minnesota program on fundamentals of enhancing petroleum
recovery.

P. Mitra is doing his doctoral research in the area of
micelles, reverse micelles and liposomes. He has published on
these topics and on the hydrophobic effect.

Kashmiri Lal Mittal is presently employed at the IBM Corporation in Hopewell Junction, NY. He received his M.Sc. (First
Class First) in 1966 from Indian Institute of Technology, New
Delhi, and Ph.D. in Colloid Chemistry in 1970 from the University
of Southern California. In the last ten years, he has organized
and chaired a number of very successful international symposia
and in addition to this three-volume set, he has edited 14 more
books as follows: Adsorption at Interfaces, and Colloidal Dispersions and Micellar Behavior (1975); Micellization, Solubilization, and Microemulsions, Volumes 1 & 2 (1977); Adhesion Measurement of Thin Films, Thick Films and Bulk Coatings (1978); Surface
Contamination: Genesis, Detection, and Control, Volumes 1 & 2
(1979); Solution Chemistry of Surfactants, Volumes 1 & 2 (1979);
Solutions Behavior of Surfactants - Theoretical and Applied
Aspects, Volumes 1 & 2 (1982); Physicochemical Aspects of Polymer
Surfaces, Volumes 1 & 2 (1983); and Adhesion Aspects of Polymeric
Coatings (1983). In addition to these volumes he has published
about 50 papers in the areas of surface and colloid chemistry,
adhesion, polymers, etc. He has given many invited talks on the
multifarious facets of surface science, particularly adhesion, on
the invitation of various societies and organizations in many
countries all over the world, and is always a sought-after speak-

er. He is a Fellow of the American Institute of Chemists and
Indian Chemical Society, is listed in American Men and Women of
Science, Who's Who in the East, Men of Achievement and other
reference works. He is or has been a member of the Editorial
Boards of a number of scientific and technical journals. Presently he is Vice President of the India Chemists and Chemical Engineers Club.

Marjanca Nemec is a researcher at the J. Stefan Institute,
Ljubljana, Yugoslavia, working in the field of biophysics of
membranes by means of EPR.

D. F. Nicoli has been Assistant Professor in the Department
of Physics, University of California, Santa Barbara since 1976.
He received his Ph.D. in Solid State Physics with Prof. M.
Tinkham in 1973 from Harvard University followed by postdoctoral
work (1973-1976) at MIT with Prof. G. Benedek. His specialties
are micelles and microemulsions by light scattering, fluores-
cence-based immunoassay development, scientific instrumentation,
and polymer gels. He is the author of more than 40 papers.

K. J. Nietering is currently with the Ford Motor Company,
Detroit. He received his M.S. degree in Chemistry from the
University of Minnesota, Minneapolis. He was a member of the
University of Minnesota program on fundamentals of enhancing
petroleum recovery.

P. G. Nilsson is a research student in the field of nonionic surfactants. Received M.Sc. from the University of Lund.

Lars Ödberg is research director at the Institute for Surface Chemistry, Stockholm, Sweden. He obtained his Ph.D. from the
Royal Institute of Technology, Stockholm in 1970 doing research
in the NMR field. During the period 1971-1978 he was lecturer at
the University of Linköping. He has published about 35 papers in
the field of NMR, light scattering and surface chemistry.

Hirofumi Okabayashi is Professor of Chemistry at Nagoya
Institute of Technology, Nagoya, Japan. He received his Ph.D.
degree in Chemistry in 1968 from Osaka University. His research
interests have centered on the conformational changes in
surfactant molecules and biochemical substances and he has
published about 40 papers.

M. Francesca Ottaviani is with the Institute of Physical
Chemistry, University of Florence, Italy. Graduated in Chemistry
in 1974 and the main field of interest is ESR spectroscopy with
particular attention to adsorbed species onto solid surfaces,
dispersed systems and relaxation mechanisms.

Heikki Palonen is Research Scientist at the Rauma Repola Paper Mills, Rauma, Finland. He received his M.Sc. degree from Helsinki University and is a Ph.D. student at the Institute for Surface Chemistry, Stockholm, Sweden.

J. E. Puig is currently with the Instituto Mexicano del Petroleo, Mexico City. He was a member of the University of Minnesota program on fundamentals of enhancing petroleum recovery.

Marie-Claude Puyal is a postgraduate student in the Associated Laboratory "Physicochimie des Systemes Polyphases", University of Sciences and Techniques, Montpellier, France. She received Doctor of Specialty in 1980 from the University of Montpellier.

M. Rawiso is Attache de Recherche au CNRS at ILL, Grenoble, France.

Heinz Rehage is Diplom-Chemiker and Assistent at the chair of Physical Chemistry at the University of Bayreuth, West Germany. He received his Ph.D. degree from the University of Bayreuth in 1982.

Brigitte Robin-Lherbier is a graduate student at the Laboratoire de Chimie Theorique, University of Nancy, France where she received a D.E.A. degree in 1981.

E. Roelants is currently working toward doctoral degree at the Catholic University of Leuven, Belgium.

Kirk Schanze is a graduate student at the University of North Carolina, Chapel Hill, NC. He received B.S. in 1979 from Florida State University and has coauthored four papers.

Milan Schara is Head of the Biophysical Laboratory of the J. Stefan Institute, Ljubljana, Yugoslavia. His fields of research include biophysics of membranes, and chemistry and physics of condensed matter.

Walter Schorr is Diplom-Physiker and Assistent at the chair of Physical Chemistry at the University of Bayreuth, West Germany. He received his Ph.D. degree in 1982 from the University of Bayreuth.

L. E. Scriven is Professor of Chemical Engineering and Materials Science, University of Minnesota, Minneapolis. He was a member of the University of Minnesota program on fundamentals of enhancing petroleum recovery.

Olle Sodermann is a Postdoctoral Fellow in the Department of Physical Chemistry I, University of Lund, Sweden.

E. Stenby is an M.Sc. student at the Fysisk-Kemisk Institut, Technical University of Denmark, Lyngby, Denmark.

Per Stenius is Director of the Institute for Surface Chemistry, Stockholm, Sweden and Associate Professor of Colloid Chemistry at the Royal Institute of Technology, Stockholm. He received his Ph.D. from the University Åbo Akademi (Turku), Finland in 1973. He is a member of the Royal Academy of Engineering Sciences in Sweden. He serves on the editorial boards of Journal of Colloid and Interface Science, and Colloids and Surfaces and is an associate member of the IUPAC commission on Colloid Science. He has published about 90 papers on association colloids, polymer latexes, colloidal stability and surface characterization.

Goran Ström is Research Scientist and Section Manager at the Institute for Surface Chemistry, Stockholm, Sweden. He received his B.Sc. degree from the University of Stockholm in 1973 and has published about 8 papers in the field of association colloids and colloid chemistry.

Shanti Swarup is currently a Postdoctoral Fellow in the Department of Physical Chemistry, University of Uppsala, Sweden. Before moving to Uppsala, he was at the University of Lund. He received his Ph.D. from Aligarh Muslim University, India. His current research interests are to utilize photophysical and photochemical techniques for structural determination of micelles and microemulsions and has published 16 papers.

Herbert Thurn is Diplom-Physiker and Assistent at the chair of Physical Chemistry at the University of Bayreuth, West Germany.

Roberto Triolo is Associate Professor of Physical Chemistry at the University of Palermo, Italy. He received his degree in Chemistry in 1964 from the University of Palermo and in 1971 was awarded the "Libera Docenza" in Electrochemistry. He has published in many areas including electrolyte and conductance, wide angle x-ray and neutron scattering from ionic fluids, statistical mechanics of ionic fluids, and small angle neutron scattering and light scattering of micellar solutions.

M. Van der Auweraer is currently a Postdoctoral Fellow of the Belgian National Science Foundation at the Catholic University of Leuven, Heverlee, Belgium. His interest at present is photoelectrochemistry and photochemistry in membranes.

Harald Walderhaug is a graduate student in the Department of
Physical Chemistry I, University of Lund, Sweden.

H. Wennerstrom has been Professor of Physical Chemistry at
the University of Stockholm since 1980. He received his Ph.D.
from the University of Lund in 1974. His research fields are NMR
relaxation theory, electrostatic interations in liquids, and
thermodynamic properties of surfactant-water systems.

David G. Whitten is M.A. Smith Professor of Chemistry,
University of North Carolina, Chapel Hill which he joined in
1966. He received his Ph.D. degree in Organic Chemistry in 1963
from Johns Hopkins University and has since been the recipient of
many awards and fellowships including Sloan Foundation Fellowship
(1970); John Van Geuns Fellowship (1973); Special U.S. Scientist
Award of the Alexander von Humboldt Foundation (1975); Distinguished Visiting Lecturer, University of Texas (1978); Invited
Visiting Professor, E.P.F.L., Lausanne, Switzerland (1980); and
Japan Society for the Promotion of Science Fellowship (1982). He
has been an invited speaker at many national/international meet-
ings and research organizations, and is the chairman for the Xth
IUPAC Symposium on Photochemistry (1984), and vice-chairman for
the Gordon Research Conference on Micelles and Macromolecular
Catalysis (1983). His research interests are mechanisms of photochemical reactions in solution, chemistry at surfaces and in
organized media, synthetic applications of solid state photochemistry, and biophysics and photobiology, and has published
more than 130 papers.

Joseph R. Winkle is a graduate student at the University of
North Carolina, Chapel Hill. He received his B.S. in 1978 from
Davidson College.

Klaas Zachariasse is a Research Associate at the Max-
Planck-Institute for biophysical chemistry in Gottingen, West
Germany. He received his Ph.D. degree in Physical Chemistry in
1971 from the Free University in Amsterdam. His research interests include kinetics of inter- and intramolecular excimer formation and investigations of micelles and biological membranes with
probe molecules.

Thomas Zemb is in the Department de Physico-Chimie, Saclay,
France and he is working on his doctoral thesis at the University
of Paris-Sud. His research interests are NMR and diffusion techniques applied to micellar systems.

INDEX